Gene Expression Systems

Gene Expression Systems

Using Nature for the Art of Expression

Edited by

Joseph M. Fernandez and James P. Hoeffler
Invitrogen Corporation
Carlsbad, California

ACADEMIC PRESS
San Diego London Boston New York Sydney Tokyo Toronto

This book is printed on acid-free paper. ∞

Academic Press
a division of Harcourt Brace & Company
525 B Street, Suite 1900, San Diego, California 92101-4495, USA
http://www.apnet.com

Academic Press
24-28 Oval Road, London NW1 7DX, UK
http://www.hbuk.co.uk/ap/

Library of Congress Catalog Card Number: 98-85621

International Standard Book Number: 0-12-253840-4

Transferred to digital printing 2006

98 99 00 01 02 03 MM 9 8 7 6 5 4 3 2 1

CONTENTS

CONTRIBUTORS

Numbers in brackets indicate the pages on which the authors' contributions begin.

Rula Abbud [367], Department of Pharmacology, Case Western Reserve University, Cleveland, Ohio 44106

Roger N. Beachy [429], Division of Plant Biology, The Scripps Research Institute, La Jolla, California 92037

Robert P. Bennett [259], Invitrogen Corporation, Carlsbad, California 92008

Marc Better [95], XOMA Corporation, Santa Monica, California 90404

Jürgen Brosius [45], Institute for Experimental Pathology, Center for Molecular Biology of Inflammation (ZMBE), University of Münster, D-48149 Münster, Germany

E. S. Cole [399], Genzyme Corporation, Framingham, Massachusetts 01701

James M. Cregg [157], Department of Biochemistry and Molecular Biology, Oregon Graduate Institute of Science and Technology, Portland, Oregon 97291

John M. Curling [399], John Curling Consulting AB, S-75329 Uppsala, Sweden

Russell Durbin [9], Department of Hematology and Oncology, Wexner Institute for Pediatric Research, Children's Hospital, Columbus, Ohio 43205

Y. Echelard [399], Genzyme Transgenics Corporation, Framingham, Massachusetts 01701

Joseph M. Fernandez [1], Invitrogen Corporation, Carlsbad, California 92008

Eugenio Ferrari [65], Genencor International, Inc., Palo Alto, California 94304

Michael Galleno [331], Invitrogen Corporation, Carlsbad, California 92008

S. Groet [399], Genzyme Transgenics Corporation, Framingham, Massachusetts 01701

M. Harvey [399], Genzyme Transgenics Corporation, Framingham, Massachusetts 01701

James P. Hoeffler [1], Invitrogen Corporation, Carlsbad, California 92008

Robert B. Kirkpatrick [289], Department of Gene Expression Sciences, SmithKline Beecham Pharmaceuticals, King of Prussia, Pennsylvania 19406

H. M. Meade [399], Genzyme Transgenics Corporation, Framingham, Massachusetts 01701

Brian Miller [65], Genencor International, Inc., Palo Alto, California 94304

Glen R. Nemerow [111], Department of Immunology, The Scripps Research Institute, La Jolla, California 92037

John H. Nilson [367], Department of Pharmacology, Case Western Reserve University, Cleveland, Ohio 44106

Christopher K. Raymond [193], Department of Protein Expression Technology, ZymoGenetics, Inc., Seattle, Washington 98102

Marijane Russell [235], Invitrogen Corporation, Carlsbad, California 92008

Allan Shatzman [289], Gene Expression Sciences Department, SmithKline Beecham Pharmaceuticals, King of Prussia, Pennsylvania 19406

August J. Sick [331], Invitrogen Corporation, Carlsbad, California 92008

T. E. Smith [399], Genzyme Transgenics Corporation, Framingham, Massachusetts 01701

Mark F. Stinski [211], Department of Microbiology, University of Iowa, Iowa City, Iowa 52242

Andreas Voloudakis [429], Department of Cell Biology, The Scripps Research Institute, La Jolla, California 92037

Daniel J. Von Seggern [111], Department of Immunology, The Scripps Research Institute, La Jolla, California 92037

Yanhai Yin [429], Department of Medicine and School of Medicine, University of California, San Diego, La Jolla, California 92037

M. W. Young [399], Genzyme Transgenics Corporation, Framingham, Massachusetts 01701

C. A. Ziomek [399], Genzyme Transgenics Corporation, Framingham, Massachusetts 01701

Introduction

SO MANY POSSIBILITIES: HOW TO CHOOSE A SYSTEM TO ACHIEVE YOUR SPECIFIC GOAL

Joseph M. Fernandez and James P. Hoeffler
Invitrogen Corporation, Carlsbad, California 92008

The Art of Expression

Our understanding of gene expression has increased greatly since the mid-1980s, resulting in new developments in protein expression system technology. A key element in creating efficient and economic expression systems has been the construction of vectors that include, along with the gene of interest, the appropriate promoter and other regulatory sequences. Recombinant DNA techniques have enabled unique pairings of promoters and structural genes in a wide variety of vectors for expression of desired recombinant proteins. These new gene combinations are currently being utilized in numerous prokaryotic and eukaryotic organisms to produce recombinant products of both academic and industrial importance. Notably, all the elements that constitute an expression system—structural genes, control sequences, markers, and inducers—are present in nature; scientists have only had to mix and match them from a variety of organisms to create diversity and flexibility in the regulation of gene expression, much as a painter mixes colors on a palette to obtain the desired shades. The result is a wide range of effective expression systems available to researchers to achieve their specific objectives: from high-level expression of recombinant products for large-scale production, to subtle expression for studying protein function in the cell, from prokaryotes to transgenic animals. Selecting the most suitable system depends on a series of parameters, such as time, resources, and intended use. Often, after weighing advantages and disadvantages, there is not always a clear winner, but there may be some areas of overlap among alternatives. The choice is indeed so wide and complex that selecting the right system has become an art in itself, further emphasizing the fact that Nature is a palette for the art of expression.

Purpose of This Book

The aim of this book is to provide the latest information on state-of-the-art protein expression systems and to help the reader select one that will best suit individual goals and resources. Leading scientists in the field review the most popular prokaryotic and eukaryotic expression systems: from bacteria (Section I) to yeast (Section II) and to insects (Section III), mammalian cell cultures (Section II), and transgenic animals and plants (Section IV). Advantages and disadvantages of each system are surveyed and summarized for easy reference. Each chapter illustrates how a system works and what proteins are most likely to work well, addressing potential problems and suggesting solutions. Although some information is offered on methods for generating recombinants, detailed protocols are beyond the scope of this book and, for these, most authors provide references. In light of the increasingly important role of protein expression, most chapters also contribute insight into future developments. Finally, in an effort to provide further guidance to the reader, prokaryotic and eukaryotic systems are compared for commonly desirable characteristics in a convenient, easy-to-read chart.

Selecting a Suitable Expression System: Considerations

When selecting a protein expression system, a number of considerations must be made, including the intended use, time frame, availability of resources, and the characteristics of the recombinant product. These considerations will affect the choice of expression system and type of promoter to be used. As for the intended use, this may be the isolation of the recombinant protein or the study of protein function in a cell or organism. If the objective is large-scale production of the gene product, then a yeast such as *Pichia pastoris* (discussed in Chapter 6) or a transgenic animal system (see Chapter 14) may be suitable options. Both systems allow for high-level production of exogenous eukaryotic proteins, but differ in a number of characteristics related to additional considerations. If time and cost are of concern, the yeast offers the advantages of rapid growth in inexpensive medium and easy handling of microbes, which translates into economical high-level production of the gene product. A transgenic system, conversely, involves higher costs and time-consuming activities associated with transgenic animal development, dairying, and testing; this system, however, is able to produce correctly processed protein and would therefore be a better choice when the expressed protein requires posttranslational modifications not possible in the yeast.

When the user's objective is to study the function of a recombinant protein in a cell or organism, the ability to regulate expression may be important, leading to the choice of a system that allows for maximum optimization. There are numerous stable and transient expression systems, for regulated or constitutive production, that can be used for functional studies. The choice will depend on factors such as the need for posttranslational processing to obtain a biologically active product, toxicity of the recombinant product on the host, the amount of protein needed, and the time frame of the study. Viruses, for instance, make powerful tools for the expression of heterologous gene products in higher eukaryotes because of their high transfection efficiencies and the high level of recombinant protein expression. The Sindbis virus (reviewed in Chapter 10) is an example of an effective vector system particularly suitable for transient expression, as infection with this virus results in eventual death of the host cells. In contrast, retroviruses integrate into the host genome and therefore represent an attractive system if stable protein expression is desired, allowing for optimization of a cell line.

One important consideration is the effect of the recombinant protein on the host. If the protein is toxic to or if it inhibits growth of the host cells, a transient system or a system that uses an inducible promoter may be desirable. In Chapter 4, the reader will find information on the arabinose expression system, which allows for the tightly regulated expression of recombinant prokaryotic and eukaryotic proteins in bacteria, and Chapter 9 discusses inducible mammalian expression systems that enable regulated inducible expression, including prokaryotic control elements in eukaryotic cells, as well as the novel insect system based on regulatory elements from *Drosophila.*

Finally, to help the reader in the identification of a suitable choice, we have provided a quick comparison chart (Table 1) in

Table 1
Comparison of Expression Systems

	Expression system			
Desired characteristics	Bacteria	Yeast	Insect	Mammalian cell culture
Cell growth	Rapid	Rapid	Slow	Slow
Complexity of growth medium	Minimum	Minimum	Complex	Complex
Cost of growth medium	Low	Low	High	High
Expression level	High	Low to high	Low to high	Low to moderate
Extracellular expression	Secretion to periplasm	Secretion to medium	Secretion to medium	Secretion to medium
Posttranslational modifications				
Protein folding	Refolding usually required	Refolding may be required	Proper folding	Proper folding
N-linked glycosylation	None	High mannose	Simple, no sialic acid	Complex
O-linked glycosylation	No	Yes	Yes	Yes
Phosphorylation	No	Yes	Yes	Yes
Acetylation	No	Yes	Yes	Yes
Acylation	No	Yes	Yes	Yes
γ-Carboxylation	No	No	No	Yes

which the expression systems featured in this book are compared for commonly desired characteristics. Each system category (bacteria, yeast, insect, mammalian cell cultures) is reviewed, respectively, in Section I, II, III and IV of this book. Whatever the challenge, selecting the right system is an important key to success.

Genomics and the Future of Protein Expression Systems

Developments in DNA array technology have made it possible to define gene expression profiles as never before, enabling the simultaneous analysis of thousands of genes, large-scale gene discovery, and mapping of genomic DNA clones. DNA arrays are used to measure the expression levels of prokaryotic and eukaryotic genes and can quickly elucidate the correlation between gene expression and biochemical pathways. In addition, they have provided information on the expression patterns of many previously unknown genes. DNA chip technology is advancing rapidly, with applications in diagnostics (mutation detection), gene discovery, gene expression, and mapping, as well as pharmaceutical development. Because of the abundance of data generated by this technology, protein expression systems will be increasingly more important, as researchers study the structure and function of gene products identified by DNA arrays. However, with the plethora of expression systems available, one thing to bear in mind is that there is always an element of unpredictability in the behavior of any given protein in a system. No matter how advanced the state of technology and how carefully all parameters are evaluated when selecting a system, no system is totally predictable and ultimately; the only sure way to find out if a system will work is to try and experiment with it. As Mark Ratner pointed out in Bio/Technology 1989, "Expression systems are protein specific. You must be able to play around with each one, insert your gene of choice, tweak it, and then see what you've got."

Section I

PROKARYOTIC EXPRESSION SYSTEMS

1

GENE EXPRESSION SYSTEMS BASED ON BACTERIOPHAGE T7 RNA POLYMERASE

Russell Durbin
Department of Hematology and Oncology
Wexner Institute for Pediatric Research
Children's Hospital, Columbus, Ohio 43205

Gene Expression Systems: Using Nature for the Art of Expression

Introduction

In recent years expression systems using the bacteriophage T7 DNA-dependent RNA polymerase (RNAP) have been used with increasing popularity to produce a large number of proteins in bacterial hosts. This chapter reviews the development of these systems and the biochemistry of transcription by T7 and related RNAPs, with particular attention to aspects relevant to artificial gene expression systems. The advantages of T7-based gene expression systems derive from the independence of the RNAP from the biology of the host cell. Largely due to the fact that the target gene does not share the host RNAP with thousands of host genes, extraordinarily high yields of recombinant proteins have been achieved with these systems. That same independence from the host provides the opportunity—indeed, often may impose a requirement—for the phage RNAP activity to be regulated by human design (see Table 1). New information available on the structure and function of the T7 fami-

Table 1
Advantages and Disadvantages of T7 Expression Systems

Advantages
Higher levels of expression than other systems
Potentially toxic target plasmids can be constructed and maintained in the absence of T7 RNAP
Expression can be controlled both at the level of T7 RNAP synthesis and target transcription—allowing for very tight control
Disadvantages
Dependence on an auxiliary gene (T7 RNAP) for expression
Extra measures may be required to repress T7 RNAP transcription or activity in order to control toxicity of target plasmids
Insolubility of recombinant proteins associated with very high expression levels

ly of RNAPs may influence future developments in artificial gene expression.

Studier *et al.* (1990) reviewed the state of the art of T7 RNAP-based expression systems, with an emphasis on technical and practical aspects. Since then, the popularity of these systems has driven the commercial development of several vectors specifically designed to address challenges peculiar to particular kinds of applications. For practical information on the choice of systems or for troubleshooting when problems arise, the reader should consult the manuals provided by commercial suppliers.

T7 RNAP in Its Natural Habitat

Bacteriophage T7 was discovered independently in the 1940s by Demerec and Fano (1945) and by Delbruck (1946). It is classified as a member of the "T7-like phages" (often referred to as genus *podovirus*) within the Podoviridae family of bacteriophages, together with coliphage T3, *Salmonella* phage SP6, and *Klebsiella* phage K11. Members of this genus have double-stranded linear DNA genomes, tailed polyhedral capsids, and an exclusively lytic replication cycle. They share characteristic features of virion morphology, genome organization, and replication strategy (Fig. 1). An evolutionary relationship can be inferred from amino acid sequence homologies among the phage RNAPs (Fig. 2).

Studier's (1983) review of the structure of the T7 genome and its replication cycle still stands as a good starting point for the reader interested in the biological context of this enzyme. The following is a broad overview abstracted therefrom, relying on references therein. In the course of infection, the "left" end of the genome enters the host first and class I promoters (i.e., promoters similar to those of the host) there are immediately recognized by the host RNA transcription machinery. The last (right-most) host-specific promoter directs the transcription of gene 1,[1] which encodes the phage RNAP. Two phage-encoded proteins, the products of genes

[1]The first 19 T7 genes identified were assigned integral numbers, from left to right on the genome. Subsequently, identified genes were given decimal numbers, thus preserving consistency between nomenclature and genomic structure. All genes are transcribed in the same direction: left to right.

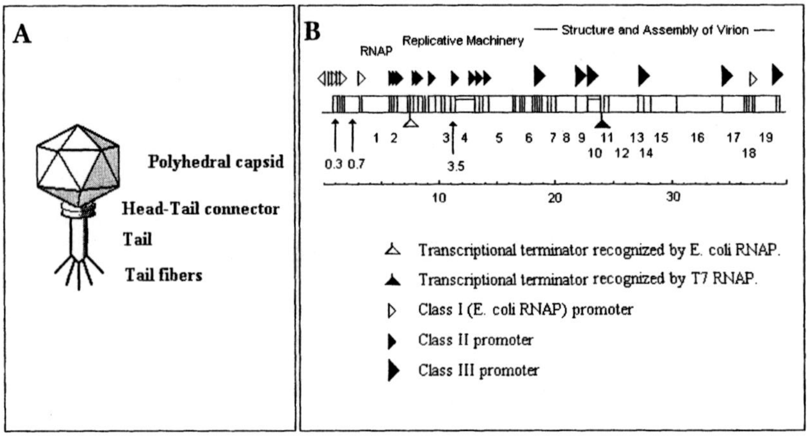

Figure 1 Podovirus virion and genome structure. (A) Schematic drawing of a podovirus virion structure. (B) Genetic structure of bacteriophage T7 (Dunn *et al.,* 1983; Studier, 1983). Boxes indicate open reading frames. Numbers below the bar indicate T7 genes. For clarity, only integrally numbered genes are labeled, except for genes 0.3, 0.7, and 3.5 (T7 lysozyme), which are discussed in the text. The scale indicates size in kilobase pairs.

0.7 and 2, inactivate host RNAP, resulting in the shutdown of all host, as well as class I phage, transcription, including that of the phage RNAP itself. All phage promoters to the right of gene 1 are recognized by the phage RNAP and not by the host RNAP. Promoters driving phage genes are divided into two groups: class II and class III (Fig. 1). Class III promoters share a completely conserved 23 bp sequence (Fig. 3). Class II promoters share a consensus sequence with class III promoters, but differ in 2 to 7 bp. Class III promoters drive the most abundantly expressed genes (e.g., those for the capsid proteins), whereas class II promoters drive less abundantly expressed genes. In addition to the promoters driving the expression of phage genes, there are two promoters involved in DNA replication: φOL resembles a class II promoter and φOR has the consensus sequence of a class III promoter.

The product of T7 gene 3.5 is a lysozyme. Independently of its enzyme activity, this protein binds to and inhibits the phage RNAP (Moffatt and Studier, 1987; Kumar and Patel, 1997; Zhang and Studier, 1997). Because accumulation of lysozyme occurs late in the phage infection, its inhibition of RNAP may serve to prevent the unnecessary loss of nucleoside triphosphates in superfluous transcription.

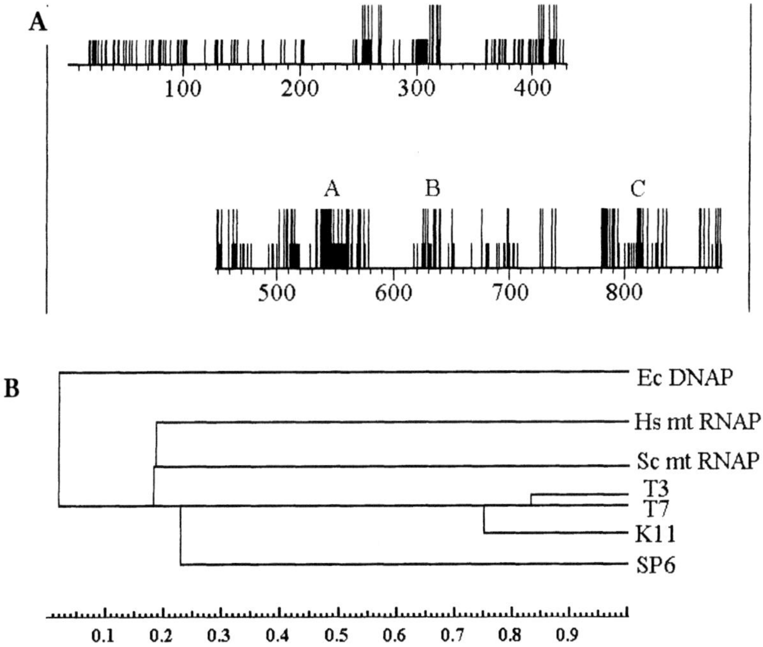

Figure 2 RNAP family ties. (A) Alignment of phage and mitochondrial RNA polymerases. Sequences were aligned using the MACAW 2.0.4 computer algorithm (Schuler *et al.*, 1991; available at ftp://ncbi.nlm.nih.gov/pub/.). Short vertical bars indicate residues conserved among four bacteriophage RNAPs (T7, T3, K11 and SP6). Tall vertical bars indicate residues conserved in the four phage RNAPs and five mitochondrial RNAPs (*Saccharomyces cerevisiae, Neurospora crassa, Chenopodium album, Arabidopsis thaliana,* and *Homo sapiens*). (B) Sequence similarities among polymerases. The right-most node connecting any two enzymes corresponds to a sequence similarity index between them. Nodes connecting multiple enzymes represent the average of the indices of all the pairs they comprise. The index was calculated as follows: Alignment scores for the blocks indicated in (a) were summed for each pair of enzymes, including self-alignments. A matrix of the summed scores was then normalized by dividing each element by the smaller of the two corresponding self-alignment scores. (Sequence identity corresponds to an index of 1.0.) *E. coli* DNA polymerase I is included to exemplify sequences with no, or extremely distant, evolutionary relatedness to the RNAPs.

In addition to their obvious role in transcription, podoviral RNAPs are involved in the injection of the phage genome into the host cell, DNA replication, encapsidation, and possibly other aspects of phage replication (Zhang and Studier, 1995). It is of considerable interest that an ancestral gene for these polymerases apparently played a major, although still mysterious, role in the history of life on earth: amino acid sequence homology with mitochondrial

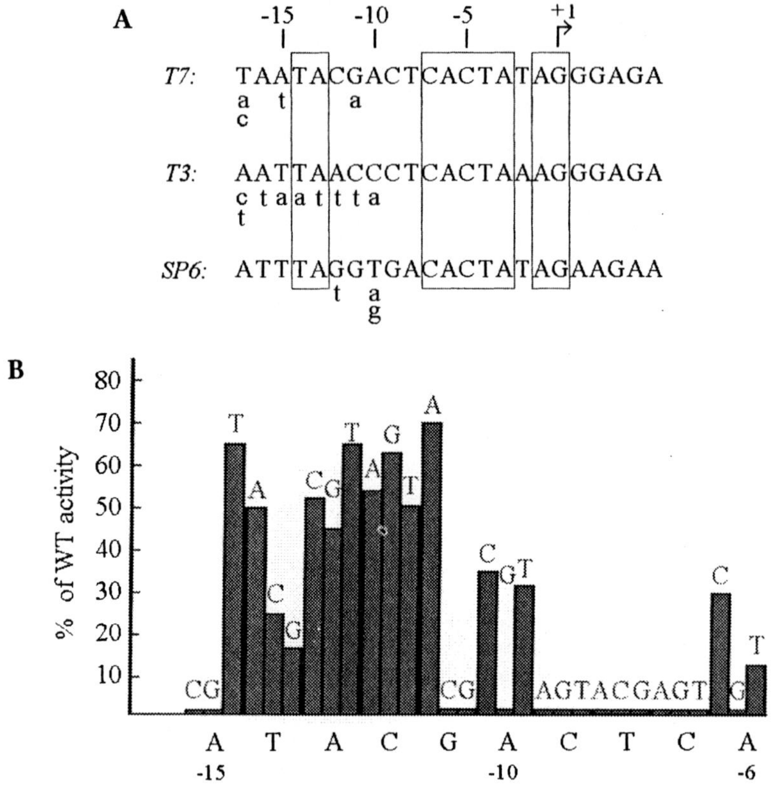

Figure 3 Promoter Specificity. (A) Consensus promoter sequences of three podoviruses. (Only the "top," or nontemplate sequence is shown.) Lower case characters below the consensus upper case characters indicate alternative nucleotides found by "*in vitro* evolution" by Breaker *et al.*, 1994. (B) Base hierarchy. The effect on transcription efficiency of all possible substitutions from base pair –15 to –6. Transcription efficiencies are expressed as a percentage, relative to transcription of the wild type promoter (Diaz *et al.*, 1993).

RNAPs indicates that the two gene families had a common origin (Fig. 2; Cermakian *et al.*, 1996, 1997).

Biochemistry of T7 RNAP

Most studies of podoviral RNAP function have focused on the prototype, T7 RNAP. In broad outline, the biochemistry of T3 and SP6 RNAPs is consistent with T7 RNAP. Of the more recently discovered phage K11 RNAP, fewer details are available. Based on its sequence homology with the coliphages T3 and T7, however, it is expected to resemble them more closely than the Salmonella phage, SP6 (Fig. 2; Dietz *et al.*, 1990).

T7 RNAP is a single polypeptide chain of 883 amino acids (M_r ~ 100,000 Da). N-terminal and C-terminal domains (about 20 and 80% of the molecule, respectively) are connected by a protease-sensitive region. The determination of the crystal structure of T7 RNAP at 3.3 Å resolution (Sousa *et al.*, 1993), together with hundreds of biochemical and genetic research studies, provides a more thorough picture of structure–function relationships for this relatively simple polymerase than is currently available for any of the more complicated multisubunit cellular RNAPs (Fig. 4). Indeed, T7 RNAP serves as a model for the function, not only of cellular RNAPs, but also of DNA polymerases (e.g., HIV reverse transcriptase and *Escherichia coli* DNA polymerase I) with which it shares functional, as well as three-dimensional structural, homologies (McAllister, 1993; McAllister and Raskin, 1993; Sousa, 1996). The cycle of events involved in transcription is depicted schematically in Fig. 5.

Promoter Binding

The nature of the promoter recognition by T7 RNAP has been the subject of numerous biochemical and genetic studies. This interaction has been divided into two phases: (1) specific binding of the RNAP to the promoter, a process dominated by thermodynamics, and (2) initiation events, largely governed by kinetic considerations. The binding phase is sensitive to changes mainly in the left portion of the promoter (base pairs –15 to –6).

Promoter competition experiments have been performed to determine the contribution to binding specificity of each promoter base pair from –15 to –6. From a large number of such experiments a "base hierarchy" has emerged that summarizes the dependence of RNAP specificity on each base pair within the promoter (Fig. 3; Chapman and Burgess, 1987; Diaz *et al.*, 1993). A T7 promoter in which a single base pair, –11, was changed from the canonical T7 C[2] to the T3 G was no longer recognized by T7 RNAP, but was recognized by T3 RNAP.

[2]Base pairs in the promoter are conventionally identified by naming the nucleotide in the "top," i.e., *nontemplate*, strand and are numbered relative to the transcriptional start site.

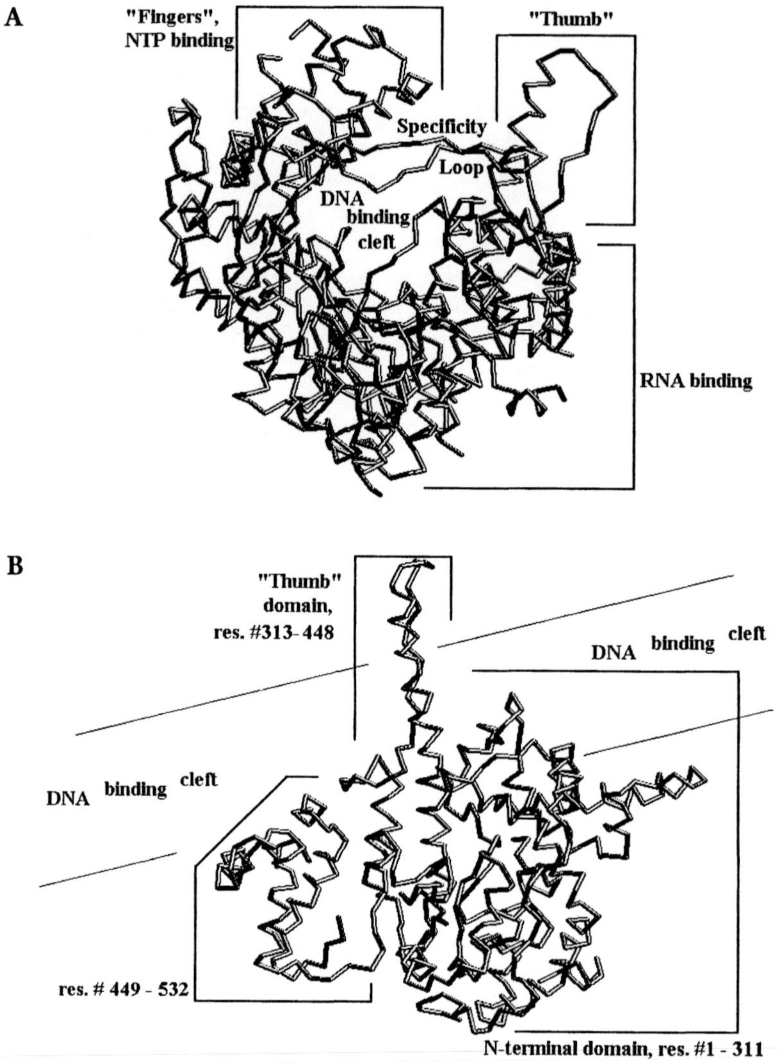

Figure 4 Three dimensional structure of T7 RNAP. The T7 RNAP crystal structure, as determined by Sousa *et al.* (1993) is represented as a backbone trace generated by the RASMOL computer program (Sayle and Milner-White, 1995; coordinates for T7 RNAP are available at http://www.ncbi.nlm.nih.gov/cgi-bin/Structure/mmdbsrv). Domains within the structure are often identified by their resemblance to a cupped right hand. The specificity loop corresponds to the little finger, and the base of the DNA binding cleft to the palm. (A) The complete enzyme structure as seen from behind. (Template DNA is envisioned as perpendicular to the plane of the page, and transcription proceeds away from the reader.) (B) The right wall of the enzyme in panel A, consisting of three N-terminal domains: residues #1 to #532. (C) The right wall of the enzyme, as in (B), "exploded" to show domains. Residues conserved in all podoviral and mitochondrial enzymes are indicated as space-filled alpha carbons. Residue #148, which is implicated in nascent RNA exit (He *et al.*, 1997a), is also indicated. (D) The left wall of the enzyme: residues #533–883. (E) The left wall of the enzyme, with conserved residues indicated by space-filled alpha carbons, and with subdomains exploded, as in (C).

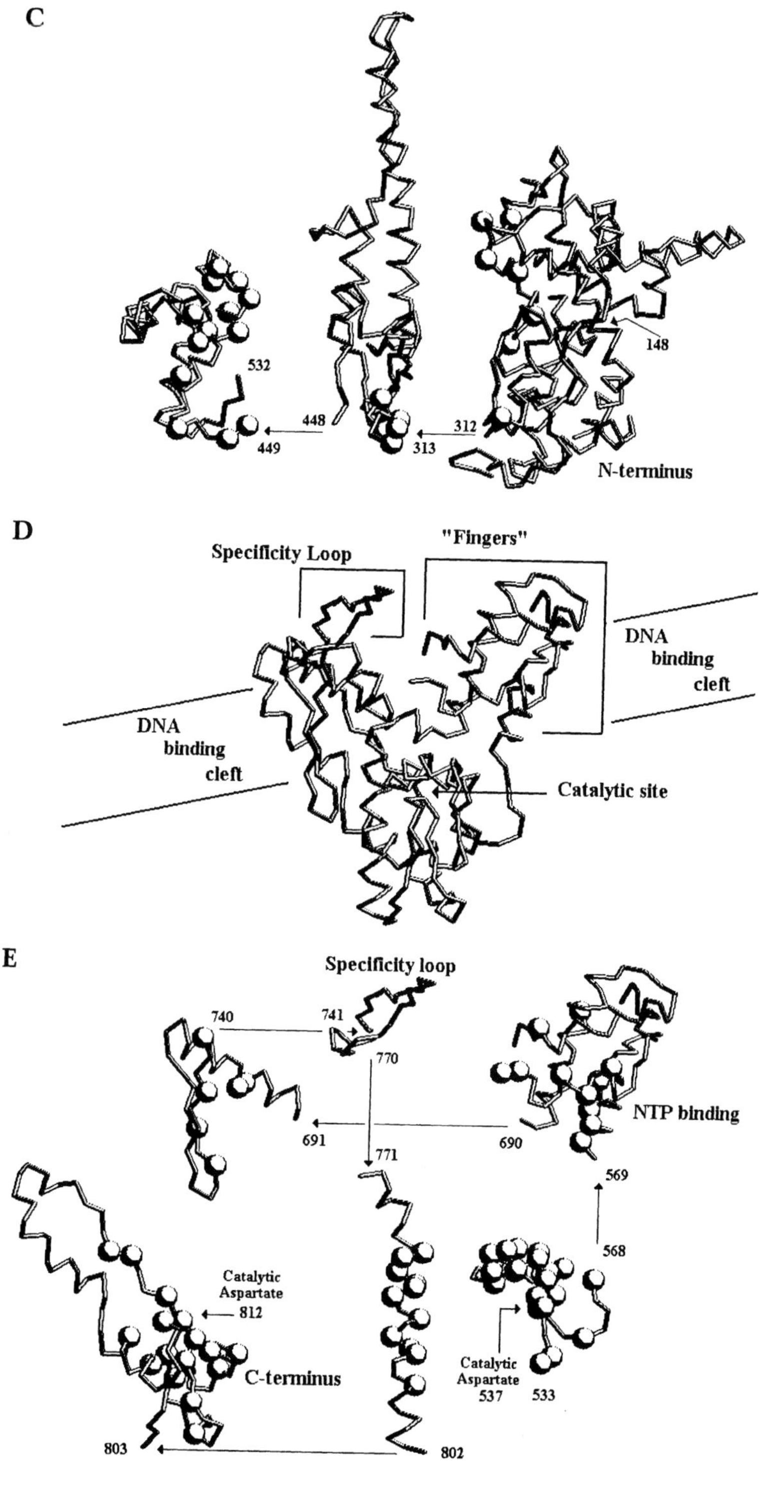
C
148
532
448
449
312
313
N-terminus
D
"Fingers"
Specificity Loop
DNA binding cleft
DNA binding cleft
Catalytic site
E
Specificity loop
740
741
770
NTP binding
691
690
771
569
568
Catalytic Aspartate 812
C-terminus
Catalytic Aspartate 537
533
803
802

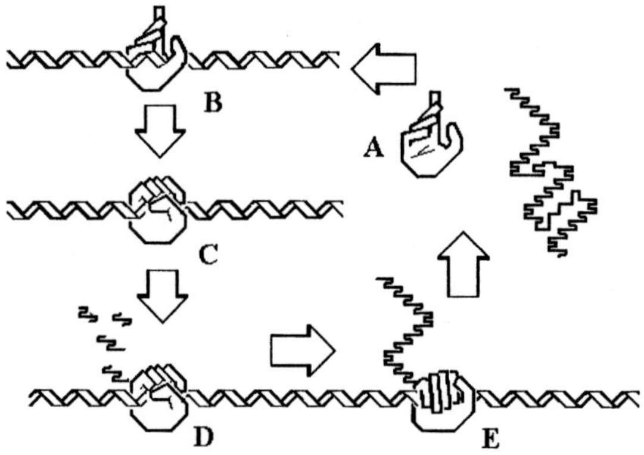

Figure 5 Schematic diagram of transcription cycle. T7 RNAP is symbolized by an open right hand. (A) The enzyme recognizes a promoter sequence on template DNA, and (B) undergoes a conformational rearrangement in the course of tight binding, envisioned as the enzyme gripping the template. While the enzyme is still associated with the promoter, it enters a phase of "abortive cycling" (C) in which short nascent RNAs are repeatedly started and released as they reach a length of up to a dozen nucleotides. The enzyme is eventually either released, or (D) enters the processive phase, in which the enzyme elongates the transcript by some 200 nucleotides per second until it reaches a termination signal (E), resulting in the release of the nascent RNA, the breakup of the RNAP:DNA complex, and availability of enzyme for another round of transcription.

Another approach to the study of promoter structure–function relations was taken by Breaker *et al.* (1994), who devised a system in which randomized promoter sequences were allowed to evolve *in vitro* toward optimal efficiency with T7, T3, or SP6 RNAP. Perhaps not surprisingly, the natural promoter for each RNAP dominated the pool of sequences selected by this procedure. Variations at some of the positions, however, were tolerated (Fig. 3).

T7 promoter DNA is probably recognized by T7 RNAP from one face of the DNA helix, making specific contacts in the major groove to both strands in the upstream portion of the promoter (Muller *et al.*, 1989; Li *et al.*, 1996; Rong *et al.*, 1998). Unwinding of the DNA in the initiation region appears to be essentially simultaneous with binding in the upstream part of the promoter (Ujvari and Martin, 1996). In the initiation region, the RNAP apparently interacts only with the template strand and seems to bind more tightly to targets in which the initiation region is unwound (Maslak and Martin, 1993; Diaz *et al.*, 1996).

It should be noted that the high degree of specificity of each podoviral RNAP for its cognate promoter is defined under conditions of promoter excess. Both *in vitro* and *in vivo,* the presence of a large excess of polymerase over promoter sites can result in promoter saturation. The dissociation constant for the RNAP–promoter complex is estimated to be about 10^{-8} *M,* whereas the concentration of RNAP in a bacterial cell overexpressing podoviral RNAP may be as high as 10^{-5}–10^{-4} *M*. Under these conditions, transcription by T7 RNAP from a suboptimal promoter (e.g., a T3 or SP6 promoter) could equal or even exceed transcription from a T7 promoter. (R. K. Durbin, unpublished observations; Dobrikova *et al.*, 1996).

Initiation

The efficiency of transcriptional initiation is reflected in the magnitude of the rate constant (k_{cat}). This parameter is sensitive to variations in the part of the promoter that includes the transcriptional start site, –5 to +6.

Because the consensus T7 promoter overlaps the transcribed region, the six 5′-most nucleotides of all class III transcripts are identical. Artificial promoters have been constructed in which these positions have been altered, and some rules have been inferred. (1) Efficiency of transcription is particularly dependent on the identity of the nucleotide occupying the +1 position: (G>A≫T,C). The second position can also be critical. Thus, while initiation of transcripts starting 5′-AG . . . Is reasonably efficient (perhaps 10% of optimal), transcription from an otherwise similar promoter starting with AU was undetectable (Milligan *et al.*, 1987; R. K. Durbin, unpublished). For these reasons, the design of the junction between promoter and transcript sequences in T7 expression vectors requires special care. The dependence of the efficiency of initiation on the 5′ terminal sequence of the transcript may be less pronounced for SP6 than for T7 RNAP (Jobling *et al.*, 1988; Cunningham *et al.*, 1991). The sequence downstream of the second transcribed nucleotide seems to have little influence on transcription efficiency (Milligan *et al.*, 1987), although it may have a subtle role in this regard (He *et al.*, 1997a; Breaker *et al.*, 1994).

Like other RNAPs, T7 RNAP is poorly processive until its nascent transcript reaches about 12 nucleotides in length (Martin *et al.*, 1988). After binding the promoter, each enzyme molecule typically makes several abortive transcripts: oligoribonucleotides corresponding to the 5′ sequence of the transcript that are released from the enzyme:template complex (Diaz *et al.*, 1996; Jia and Patel, 1997). Escape from "abortive cycling" into the elongation phase of transcription may be the overall rate-limiting step in the transcription of at least some templates and, again, is influenced by the nature of the 5′ sequence (Jia and Patel, 1997; Lopez *et al.*, 1997).

Elongation

Once the enzyme has escaped abortive cycling, elongation of the transcript by T7 RNAP is extraordinarily fast: about 200–260 nucleotides/sec (Bonner *et al.*, 1994), (cf. 40 nucleotides/sec for *E. coli* RNAP; Gotta *et al.*, 1991). T7 RNAP has proven to be less sensitive to impedance by regions of RNA secondary structure, possibly in part because the kinetics of elongation are more competitive with the kinetics of secondary structure formation than is the case with slower enzymes. Although this effect may be generally advantageous, Makarova *et al.* (1995) observed that stability of mRNA was *inversely* correlated with elongation rate, presumably because the rate of RNA formation can affect its secondary structure. Once past the initiation phase, transcription by T7 RNAP is also apparently less sensitive to obstruction by protein–DNA interactions than is transcription by *E. coli* RNAP (Deuschle *et al.*, 1986; Giordano *et al.*, 1989).

Termination

Two types of sequence have been identified that trigger transcriptional termination by podoviral RNAPs. The first is typified by TΦ, a sequence that follows the Φ10 gene approximately 60% from the left end of the genome. This type of sequence, similar to ρ-independent terminators of *E. coli* RNAP, encodes an RNA sequence with a predicted stem–loop structure followed immediately by a run of U's (Fig. 6). It should be noted, however, that no sequence has been found that terminates T7 transcription with 100% efficiency. Efficiency of termination of TΦ ranges from about 20 to 90% and is dependent on distant as well as local sequence context (Macdonald *et al.*, 1994).

The second type of terminator was discovered fortuitously in the course of experiments in which T7 RNAP was used to tran-

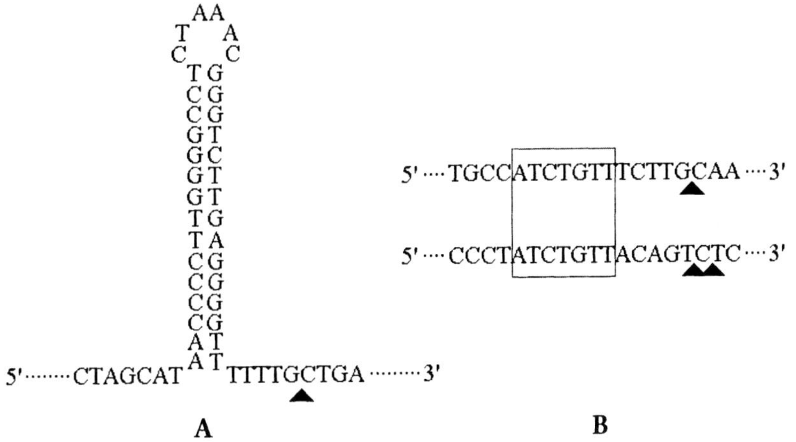

Figure 6 Structure of sequences associated with T7 transcriptional termination. (A) The phage terminator, TΦ, represented as an RNA hairpin structure. (B) Type II terminators, represented as the DNA sequence of the nontemplate strand. Top: PTH terminator. Bottom: T7 concatamer junction (Lyakhov *et al.*, 1997). Triangles indicate position of 3′-most transcribed nucleotides.

scribe human preproparathyroid hormone (PTH) cDNA (Mead *et al.*, 1986). The transcripts unexpectedly terminated at a specific site within the coding region. A heptameric consensus sequence within the PTH terminator has been found in other sequences associated with T7 transcription termination (Lyakhov *et al.*, 1997; Fig. 6). In troubleshooting poorly performing expression constructs, it may be useful to scan the sequence of the transcript, if known, for regions bearing similarity to the PTH family of terminators and/or to examine the products of *in vitro* transcription for shorter than expected transcripts. Although such unanticipated termination signals are rare, mutant enzymes that fail to recognize them may prove useful for minimizing this nuisance (Lyakhov *et al.*, 1997).

RNA Binding

The physical relationship between enzyme and nascent RNA is probably critical in the maintenance of processivity and in the appropriate response to termination signals. T7 RNAP, in the absence of template DNA, binds free RNA, and this binding activity is thought to involve the same site as does the binding of nascent RNA (Diaz *et al.*, 1996). Certain mutations within the N-terminal domain, or proteolytic "nicking" of the enzyme at amino acid position 178, affect RNA binding and involve changes in processivity and termination (He *et al.*, 1997a; Muller *et al.*, 1988).

Structure–Function Relationships

The correlation of biochemical phenotypes with molecular alterations in many mutant T7 RNAPs provides at least a tentative functional map of the molecule.

The N terminus of the enzyme seems to be among the most malleable regions of the T7 RNAP. It has been possible to insert additional sequences at or near the N terminus of the enzyme without detectably affecting any aspect of polymerase function (Dunn *et al.*, 1988; Zhou *et al.*, 1990; Gross *et al.*, 1992). Among members of the podoviral polymerase family, the N terminus is relatively poorly conserved (Fig. 2). The N-terminal 20% of T7 RNAP seems to be involved in RNA binding, processivity, and response to termination signals (Muller *et al.*, 1988; He *et al.*, 1997a; Lyakhov *et al.*, 1997). These observations point to a region of the enzyme structure from which the nascent RNA probably emerges (Fig. 4).

In contrast to the N terminus, missense mutations, truncations, and fusions at the extreme C terminus impair enzyme function dramatically (Mookhtiar *et al.*, 1991; Gross *et al.*, 1992; Gardner *et al.*, 1997). The C terminus is relatively well conserved across diverse members of the podoviral RNA family and is buried in the interior of the T7 RNAP crystal structure.

A combination of genetic and biochemical analyses has identified two specific amino acid: promoter base pair contacts. The asparagine at position 748 (N748) of T7 RNAP apparently contacts the major groove of the base pair at promoter positions –11 and –10 (Raskin *et al.*, 1992). The glutamine at position 758 (Q758) contacts the base pair at promoter position –8 (Rong *et al.*, 1998). By site-directed mutagenesis of either of these two sites, the promoter specificity of the enzyme can be altered. In the crystal structure, amino acid residues 742–773 form a loop that extends across a large cleft, which is assumed to cradle the template DNA (Sousa *et al.*, 1993; Fig. 4). The loop has been dubbed the "specificity loop" in honor of these contacts. The existence of several other specificity-determining contacts can be inferred from structure–function studies of promoters containing base pair substitutions and nucleotide analogs (Jorgensen *et al.*, 1991; Diaz *et al.*, 1993; Lee and Kang, 1993; Li *et al.*, 1996). The amino acids involved in these contacts, however, have not bee identified. Other promoter-recognizing amino acids are expected to be located elsewhere on the specificity loop and/or in the DNA-binding cleft.

The catalytically active site of the enzyme has been associated with aspartic acid residues No. 537 and No. 812 (Osumi-Davis *et al.*, 1992; Woody *et al.*, 1996). In the crystal structure, these two residues lie in close proximity to one another in the DNA-binding cleft, near the extreme C terminus (Fig. 4).

Mutations in residues No. 639 and No. 641 have been associated with altered discrimination between ribo- and deoxyribonucleotides (Kostyuk *et al.*, 1995; Sousa and Padilla, 1995). Presumably, these residues are among those that properly position the incoming nucleotide relative to the template DNA strand and the 3′ end of the growing RNA. Consistent with this observation, Knoll *et al.* (1992) observed that T7 RNAP photoaffinity labeled with nucleotide analogs was labeled at three distinct peptides corresponding to amino acid positions Nos. 314–362, 550–666, and 751–861.

Podoviral RNAPs as Tools in Biotechnology

Several features of the podoviral RNAPs combine to make them especially attractive for use in biotechnology. First, they have a high specificity for their cognate promoters and virtually none for any unrelated sequence identified thus far. Second, and conversely, host cell RNAPs completely fail to recognize the phage promoters. For this reason, even highly toxic genes prefaced by a phage promoter can be cloned in hosts that lack a podoviral RNAP gene. Third, T7 and related RNAPs are single subunit proteins capable of carrying out all phases of transcription. Thus a single portable gene confers on its host the ability to transcribe virtually any fragment of DNA bearing the corresponding phage promoter. Finally, because fully functional T7 and related RNAPs have themselves been overexpressed and purified in quantity, there is a wealth of information on their structure and function (see earlier). A number of mutant T7 RNAPs have been developed that may find practical applications in biotechnology.

Prior to 1984, biochemical studies of T7 and related bacteriophage RNAPs depended largely on purification of the enzyme from infected bacteria. The enzyme is produced in a natural infection only in catalytic quantities, about 0.05% of the soluble protein of an infected culture (Butler and Chamberlin, 1982). Moreover, expression of phage RNAP in a bacterial host involves other, often undesired, aspects of a phage infection, such as death and lysis. After the cloning and overexpression of gene 1, it became possible to produce large amounts of phage RNAP for purification (two to three orders of magnitude more than from an equivalent number of phage-infected cells) and to express it independently of other phage genes.

The first widespread use of the phage RNAPs in biotechnology was for *in vitro* transcription of DNA containing phage promoter. This made possible the large-scale production of a number of interesting RNAs and the routine production of "riboprobes" (Melton *et al.*, 1984). SP6 RNAP was the first phage RNAP to be widely used for this purpose. Even before the first podoviral RNAP was cloned, SP6 RNAP produced from phage-infected bacteria was available commercially. The cloning of T7 (Davanloo *et al.*, 1984), T3 (Morris *et al.*, 1986), and SP6 (Kotani *et al.*, 1987) RNAPs greatly increased the practicality of *in vitro* transcription, and several vectors have been developed specifically for this purpose. Some of these vectors [e.g., Bluescript (Stratagene) and pGem (Promega)] have two different phage promoters in opposite orientation so that either strand of any DNA fragment cloned between them can be transcribed *in vitro.* Podoviral RNAPs have been used to transcribe entire infectious RNA viral genomes from engineered DNA plasmid templates (van der Werf *et al.*, 1986; Rice *et al.*, 1987), making it possible to use recombinant DNA technology in RNA virus research.

Advances in our understanding of specificity determinants (Raskin *et al.*, 1993) and other structural determinants of polymerase function (Patra *et al.*, 1992; He *et al.*, 1997a; Lyakhov *et al.*, 1997), together with technological advances in RNAP production and purification (He *et al.*, 1997b), have greatly simplified and accelerated the production of "designer RNAPs": enzymes specifically engineered to recognize novel promoter sequences or to have other properties useful for *in vitro* transcription.

T7-Based Gene Expression Systems

McAllister *et al.* (1981) demonstrated that plasmids containing class III T7 promoters were transcribed on infection of the host cell with

T7 phage. From this result, it was evident that the polymerase generated by phage infection could act in *trans* on "bystander" DNA that contains the cognate promoter. After the cloning of T7 gene 1, systems for artificial gene expression were developed by Studier and colleagues, who described a series of "pET" vectors (plasmids for Expression by T7 RNAP) (Studier and Moffatt, 1986; Rosenberg *et al.*, 1987; Studier *et al.*, 1990). These plasmids served as the foundation for the development of subsequent vectors, now available commercially (Novagen, WI), that incorporate a variety of features useful in biotechnology. Most of the reports in the literature describing T7-based expression in *E. coli* rely on this family of vectors. Separately, Tabor and Richardson (1985) developed a similar vector, pT7-1, whose descendants, especially pT7-7, have also been used in a number of studies calling for foreign gene expression in *E. coli.*

Since the first descriptions of T7-based gene expression systems, their use has figured in over a thousand research reports in the biomedical literature. What follows is not an exhaustive catalog, therefore, but rather a sampling chosen to illustrate typical experiences with typical problems, as well as to highlight a few developments, which, although not yet incorporated into widely used T7 expression systems, may yet be of general interest.

RNAP Delivery

In expression systems that depend on the bacterial RNAP for transcription, the only extraneous genetic element that must be introduced into the host is the expression target itself. T7 RNAP-dependent systems, however, must make provision for the expression of the RNAP. There are a number of strategies by which this has been accomplished.

Perhaps the simplest method for delivering T7 RNAP would be to infect *E. coli,* carrying target genes under T7 promoter control, with T7 phage. Unfortunately, though, the viral agenda includes items ill suited to high-level foreign gene expression, not the least of which is cell death within minutes of infection. Another approach has been to incorporate T7 gene 1 into defective nonlytic phages, λ CE6 and m13HEP (Studier *et al.*, 1990; Chen *et al.*, 1996). This approach avoids the often vexatious problem of leaky expression due to less than absolute repression of T7 gene 1. Optimizing conditions for phage infection can be tricky, however, and this strategy has not been widely used.

Currently, the most common strategy for T7 RNAP delivery in *E. coli* makes use of BL21(DE3), a strain in which the gene for T7 gene 1 is present in the genome as an excision-deficient λ lysogen. The parent strain, BL21, was originally chosen for expression of T7 RNAP because it is defective in *omp*T, a gene encoding an outer membrane protease that cleaves a site in the RNAP (Grodberg and Dunn, 1988). BL21 (like other B strains of *E. coli*) is also deficient in *lon*1, a cytoplasmic protease that prevents the accumulation of many foreign proteins in *E. coli.* The T7 RNAP gene in BL21(DE3) is under the control of the lacUV5 promoter. This promoter confers on the system inducibility by galactosides [most commonly isopropyl thiogalactoside (IPTG)] and insensitivity to catabolite repression, which is important in providing maximal expression in rich culture media. One disadvantage to the standard, IPTG-inducible, BL21(DE3) system is that the very low expression of T7 RNAP that occurs even in the absence of inducer can be deleterious in the presence of extremely toxic target genes (see later).

Lebedeva *et al.* (1994) designed a thermoinducible T7-based system. Bhandari and Gowrishankar (1997) described a system in which the expression of T7 RNAP is regulated by an osmoresponsive promoter and can be induced by elevating the salt concentration in the medium.

Another device that has been used to provide T7 RNAP has been referred to as an "autogene," in which the T7 RNAP gene is placed under the control of its own promoter (Dubendorff and Studier, 1991b). Thus the presence of a trace of T7 RNAP triggers an autocatalytic expansion of its own expression. A target gene, also prefaced by the T7 promoter, in a cell experiencing this explosion of polymerase activity is also saturated transcriptionally.

Design of Target Plasmids for Expression in *E. coli*

The original pET plasmids illustrate the minimal requirements of a T7 expression vector (Fig. 7). Promoters that have been used for these systems have typically been the consensus, i.e., class III, promoter at least through the first transcribed base pair. The position of the promoter in the plasmid was chosen carefully in order to minimize transcription from cryptic upstream host promoters. Promoters in other positions and orientations resulted in higher background expression, i.e., expression in the absence of T7 RNAP (Brown and Campbell, 1993; Giordano *et al.*, 1989).

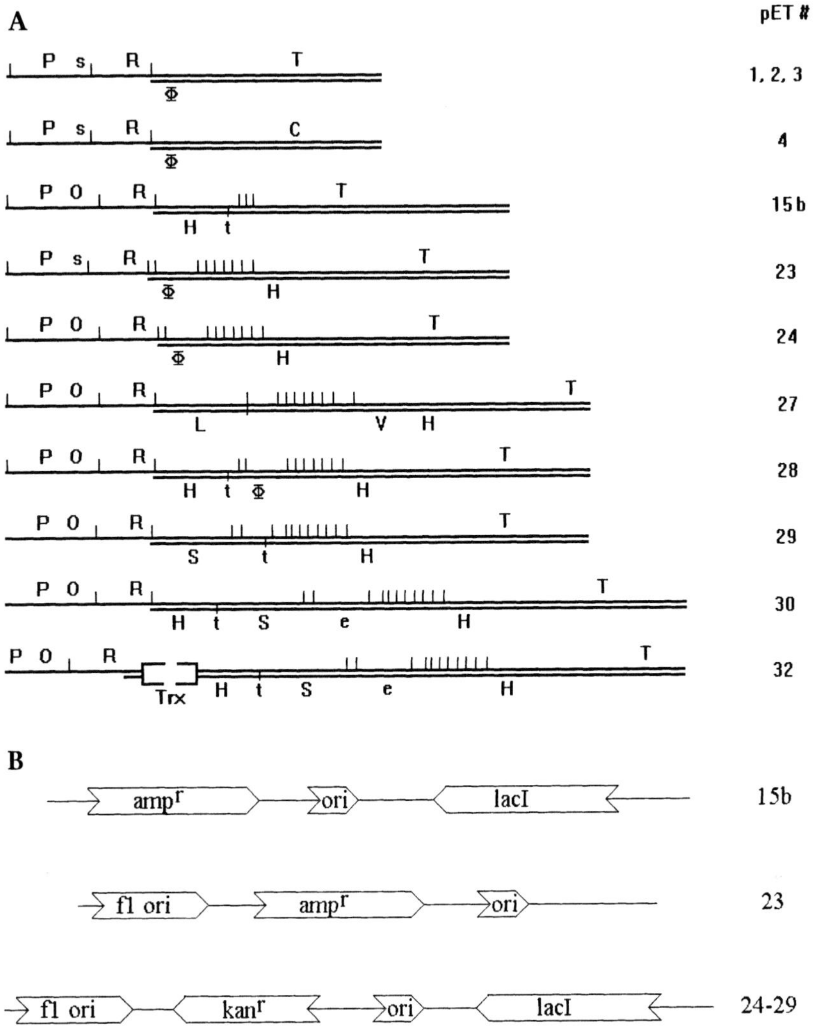

Figure 7 Structures of pET bacterial expression vectors. (A) Elements immediately surrounding the cloning site are indicated. Symbols above the horizontal bars indicate signals in the DNA or RNA. C, Rnase III cleavage site; P, T7 promoter; O, lac operator; R, ribosomal-binding site; T, transcriptional terminator. Upward vertical ticks indicate unique restriction sites. The double horizontal bar indicates the translated portion of the vector. Symbols below the double bars indicate signals within the peptide. H, oligohistidine tag; Φ, T7 Φ10 N-terminal tag; V, herpesvirus epitope tag; S, S peptide tag; L, pelB periplasmic export leader; Trx, thioredoxin fusion partner. Downward vertical ticks with lowercase letters indicate proteolytic cleavage sites: t, thrombin; e, enterokinase; x, export leader peptidase. (B) Plasmid replication and antibiotic resistance elements of the pET vectors. Amp, ampicillin resistance; Kan, kanamycin resistance; Ori, ColE1 origin of replication, from pBR322. Transcriptional terminators recognized by *E. coli* RNAP are indicated, Te. Data from Rosenberg *et al.* (1987) and from Novagen WI (www.novagen.com).

Expression vectors are divided into two types: transcription vectors and translation vectors. Transcription vectors lack translational start signals. Expression of protein products from these vectors depends on the presence of the appropriate start signals in the cloned DNA inserts. In translation vectors, a translational initiation site (a ribosome-binding site juxtaposed with an ATG codon) is engineered downstream of the promoter and just upstream of a cloning site. This is necessary for the expression of 5′-truncated bacterial genes, eukaryotic cDNAs, or any other DNA inserts that lack the canonical bacterial translational start signals. In the original pET vectors, these sequences were derived from the T7 phage gene 10, which encodes the major capsid protein, Φ10, and is the most abundantly expressed gene of the bacteriophage. Proteins programmed by these vectors are actually chimeric proteins containing the N-terminal 11 amino acid of Φ10 fused to the polypeptide encoded by the inserted DNA. The gene 10 sequences that are present in these vectors apparently act as a "translational enhancer," increasing the efficiency of translation by extending the complementarity of the ribosomal-binding site with the bacterial 16S rRNA (Olins and Rangwala, 1989). The Φ10 peptide at the N terminus of recombinant proteins can be useful as an "epitope tag." Sequences encoding a number of other peptide tags have been engineered into pET vectors (Fig. 7). These sequences are useful for purification and detection by affinity with a variety of proteins and ligands. Specific protease cleavage sites have also been introduced so that, after purification, the peptide tag can be cleaved, leaving—ideally—the authentic terminus of the engineered protein.

One of the most frequently encountered problems in expression systems is the unpredictable toxicity to the host cell of foreign proteins. In the absence of very tight regulation, this toxicity often leads to difficulty in constructing and maintaining the expression vector. Podoviral RNAP-based systems offer an advantage in these cases, as the recombinant plasmid can be constructed and maintained in the absence of cognate RNAP. Cells used for expression of such constructs can remain viable as long as the gene for the phage RNAP is repressed thoroughly. Even in tightly regulated inducible systems, however, a toxic gene product can lead to the demise of the host cell so shortly after induction that only very low yields of recombinant protein can accumulate. In other cases, the recombinant protein is unstable in the bacterial host. The BL21 strain of *E. coli* was chosen in part for its lack of proteases, OmpT and lon1, that might be expected to prevent the accumulation of recombinant protein. The use of export signals may be helpful both in cases where the recombinant protein is unstable in the intracellular environment and in cases where intracellular accumulation of the recombinant protein destabilizes the host cell. By appending an export signal to the N terminus of the recombinant protein, it is possible to direct the protein to the periplasmic space of the host cell.

In many of the expression vectors, a T7 phage-specific transcriptional terminator is placed downstream of the cloning site (Fig. 7). This feature is not strictly necessary because, in bacteria, the long tandemly repeated RNA resulting from circular transcription of a plasmid can bind ribosomes and initiate translation internally. Transcriptional terminators may be desirable, however, if the transcription of downstream sequences is to be minimized. The efficiency of termination can be increased by tandem repeats of the terminator sequence. With the recent definition of the minimal class II terminator sequence, it should be possible to string together multiple copies to achieve arbitrarily low levels of read-through transcription.

Typical Yields

Yields of recombinant proteins differ, of course, depending on the protein itself as well as the system used to express it. Reported yields of purified recombinant proteins from T7 promoters in *E. coli* have typically been in the range of tens or even hundreds of milligrams per liter of culture, constituting up to 60% of the total bacterial protein after induction (Table 2).

Inclusion Bodies

Sequestration of recombinant protein in insoluble inclusion bodies seems to be a function not only of properties inherent in some proteins, but also of the rate of accumulation. As a result, the extraordinary protein yields of T7-based systems may often be achieved at the expense of solubility (e.g., Sun *et al.*, 1997). In many cases, the insolubility of the protein has been advantageous, in that a substantial purification can be achieved simply by collecting the insoluble material from bacterial cell lysates. It has often been possible to resolubilize, renature, and further purify proteins from these in-

Table 2
Representative Yields of Recombinant Proteins in *E. coli*

Protein	Yield[a]	Method[b]	Reference
HSV protease	8–10	Sol	Apeler *et al.*, 1997
Yeast DNA polymerase delta subunit	25*	Ren	Brown *et al.*, 1993
Aristocholene synthase	20	Sol	Cane *et al.*, 1993
T7 DNA ligase	60–70	Sol	Doherty *et al.*, 1996
RuvC resolvase	3	Sol	Dunderdale *et al.*, 1994
Plastocyanin	38	Sec	Ejdeback *et al.*, 1997
Apolipoprotein domain	>100	Sec	Fisher *et al.*, 1997
Ca^{2+} binding protein in human prostate cancer	150	Sol	Gribenko *et al.*, 1997
Human globin	7.5	Ren	Hernan *et al.*, 1992
Bacterial acyl carrier protein	120	Sol	Hill *et al.*, 1995
Bacterial Leader peptidase	1	Sol	Kim *et al.*, 1995
Protein methyltransferase	4	Sol	MacLaren *et al.*, 1995
Protein tyrosine phosphatase	2.5–3.5	Sol	Pei *et al.*, 1993
Acrosomal antigen SP-10	20	Ren	Reddi *et al.*, 1994
Alpha-sarcin	40–50	Ren	Rathore *et al.*, 1997
Alpha-sarcin	2–3	Sec	Rathore *et al.*, 1997
Ribonucleotic toxin	45	Ren	Rathore *et al.*, 1996
Phosphoenolpyruvate mutase	5	Sol	Seidel *et al.*, 1992
Phospholipase from B cereus	30–40	Ren	Tan *et al.*, 1997
Ribosomal protein	3–7	Sol	Vysotskaya *et al.*, 1994
Beta troponin T	40–60	Ren	Wu *et al.*, 1995
Clathrin ATPase	5–20	Sol	Wang *et al.*, 1993
Plasminogen activator	87	Ren	Zhou *et al.*, 1997

[a]Yields are in mg purified protein per liter culture, and reflect not only the efficiency of expression, but also efficiency of purification. Estimates of recombinant protein as a percentage of total cell protein are in the range of 3–60%. *Reported in mg/gram cells, converted to mg/*L* assuming 5 grams cells/*L* culture.

[b]Ren = renaturation of protein recovered as insoluble inclusion bodies. Sol = purified from soluble fraction. Sec = secreted into periplasmic space and/or medium.

clusion bodies (Brown *et al.*, 1997). In other cases, proteins in inclusion bodies have resisted all attempts to renature them.

Hydrophobic Proteins

A common observation is that expression of hydrophobic foreign proteins is toxic to the bacterial host or, at best, results in intractable, insoluble inclusion bodies. One solution to this problem suitable to some applications is to express selectively the hydrophilic domains of the protein (Sisk *et al.*, 1992). In some cases it has been possible to accumulate fairly high levels of hydrophobic proteins in bacterial membranes (San Francisco *et al.*, 1989; Hamood and Iglewski, 1990). The mechanism(s) by which recombinant hydrophobic proteins exert toxic effects on the host bacteria is not clear. Miroux and Walker (1996), however, were able to select mutant strains of *E. coli* BL(23) that survived overexpression of a membrane protein. The resulting strain was able to tolerate the overexpression of other membrane proteins that were toxic to the parental strain.

Fusion Partners

The first "translation vectors" described by Rosenberg *et al.* (1987) programmed the synthesis of fusion proteins, the N-terminus of which was derived from the T7 capsid protein Φ10. This "T7 tag" has been used for immunodetection. More recently, pET plasmids have been developed (and are commercially available; Novagen, WI) which incorporate other "fusion partners" designed to facilitate detection and purification, as well as specific protease cleavage sites, making possible the generation of recombinant proteins with native or near-native N-termini by *in vitro* proteolysis (Fig. 7).

Too Much of a Good Thing?

One of the most attractive features of T7-based gene expression systems is the extraordinary yield of recombinant protein that can be produced. This productivity, however, may not be helpful if, as is often the case, the recombinant product is toxic to the cell.

Incomplete Repression of Phage RNAP In principle, placing the RNAP under inducible control prevents the expression of T7 promoter-dependent genes in the absence of induction. In practice, however, the basal, uninduced expression of T7 RNAP can be sufficient to produce substantial levels of gene product. For instance, BL21(DE3), even in the absence of lac inducer, supports the replication of T7 phage deficient in gene 1 (Zhang and Studier, 1995). The practical consequence of this leakiness is the instability of plasmids bearing T7 promoters in cells carrying T7 gene 1, even under repression. To address this problem, a number of strategies have been devised to further tighten the control of phage promoter-driven target genes.

Giordano *et al.* (1989) fused a lac operator to a phage promoter driving a target gene in cells in which the RNAP was under the control of the lacUV5 promoter. Thus, the few molecules of phage polymerase synthesized in the absence of galactoside inducer are effectively blocked at the target gene. The same strategy has been incorporated into the design of most of the commercially available pET expression vectors (Dubendorff and Studier, 1991a; Fig. 7).

The toxicity of plasmids encoding ill-tolerated products is exacerbated by the fact that they are usually, as is the case with the pET series, relatively high copy number plasmids derived from pBR322. Dersch *et al.* (1994) described a set of plasmids similar to the pET series, but based on the low copy number plasmid pSC101. Recombinant genes that proved toxic in higher copy number plasmids were tolerated by *E. coli* in the context of the lower copy number plasmids.

Another strategy takes advantage of T7 phage biology: the ability of the phage lysozyme to inhibit the phage RNAP. BL21(DE3) was modified by transformation with a plasmid that directs the constitutive expression of T7 lysozyme (Studier, 1991). This plasmid is compatible with plasmids (such as the pET series) bearing the colE1 ori and confers chloramphenicol resistance, allowing for subsequent transformation by ampicillin or kanamycin resistance-conferring plasmids. In the absence of induction, an excess of lysozyme saturates the trace concentrations of T7 RNAP resulting from incomplete repression at the lacUV5 promoter, suppressing any T7 transcriptional activity. On induction, however, the concentration of T7 RNAP exceeds that of lysozyme and the suppression is overcome.

Toxicity on Induction Even when the chronic problems of plasmid instability and toxicity in the absence of induction have been addressed, toxicity *on* induction can still limit the yield of an expression system. Indeed, in such cases the more efficiently the recombinant protein is produced, the more quickly the cell dies after induction. One approach to this problem has been to use export peptide signals (see earlier). Using such a system, Fisher *et al.* (1997) were able to purify over 100 mg/liter of a coli-toxic recombinant protein directly from the medium after clearing it of cells. (Once exported to the periplasmic space, the protein evidently passed through the outer membrane into the medium.) In such a system, the bacteria, on induction, devote virtually all of their resources to the synthesis of the recombinant protein, yet experience no intracellular accumulation and therefore minimal toxicity.

Another approach again takes advantage of T7 phage biology. Brown and Campbell (1993) note that the product of T7 gene 4 accumulates to only very low levels in cells in which it is expressed artificially, i.e., from a plasmid rather than from the phage genome. The same product accumulates to much higher levels in the course of a T7 infection, even though transcriptional and translational competition from other phage genes might be expected to have the opposite effect. They hypothesized that a protein-level "restriction system," like DNA restriction systems, might defend the host against invaders. Just as T7 and T3 genes 0.3 encode products that defeat the host EcoK and EcoB DNA restriction systems (gene 0.3), these authors further hypothesized that one or more of the early genes of T7 might have an analogous function relative to protein restriction. After attempts to express a yeast DNA polymerase gene failed, they achieved high-level expression when T7 RNAP was delivered by infection with a defective T7 phage (T7SC). This phage lacks genes 3 and 6, thus—unlike wild-type phage—cannot complete the lytic cycle and does not degrade nonphage DNA.

Posttranscriptional Rate-Limiting Steps Given the very high concentrations of T7 RNAP in, for example, induced BL21(DE3), and the fact that the only T7 promoter typically present is that driving the recombinant gene, transcription is seldom, if ever, the rate-limiting step in the synthesis of recombinant protein in a T7 expression system. In fact, as discussed earlier, excess RNAP could have the unwanted effect of reducing promoter specificity. In order to limit toxicity and reduce the formation of insoluble inclusion bodies, it may be desirable to *make* transcription the rate-limiting step. This might be accomplished by using a promoter weaker than lacUV5 to drive expression of the RNAP, using a variant T7 RNAP with lower activity, or using a variant T7 promoter. Many such variant enzymes and T7 promoters have been described.

Codon usage may differ considerably between the expression host and the organism from which a recombinant sequence is derived. Although hundreds of foreign proteins have been expressed at high levels in *E. coli*, heavy reliance on low-abundance tRNAs may limit translation in some cases. One radical solution to this problem is to systematically mutate the offending codons to synonymous codons more typical of *E. coli* (Hernan *et al.*, 1992; Apeler *et al.*, 1997; Ejdebeck *et al.*, 1997). A special case of mismatched codon usage arises from a bias toward CGN arginine codons in *E. coli* and toward AGA/G codons in eukaryotic organisms. In *E. coli*, the tRNA complementary to AGA/G codons is in low abundance and plays a critical role in DNA replication (it is encoded by *E. coli* gene dnaY). Seidel *et al.* (1992) engineered a supplementary dnaY gene into *E. coli* in an effort to accommodate the translation of sequences rich in AGA/G codons. The general utility of this approach remains to be tested.

Non-*E. coli* Bacterial T7 Expression Systems

Bacteriophage T7 is a coliphage, and the overwhelming majority of recombinant proteins expressed using T7 RNAP have been expressed in *E. coli*. It may be useful to use other hosts to express non-*E. coli* recombinant proteins, either to study their behavior in their natural contexts or to avoid the instability or incompatibility of the foreign proteins or their genes in *E. coli*. Inducible gene expression systems using T7 RNAP have been developed in *Bacillus subtilis* (Conrad *et al.*, 1996) and in *Pseudomonas* species (Brunschwig and Darzins, 1992; Herrero *et al.*, 1993).

Eukaryotic Gene Expression Systems Based on Podoviral RNAPs

Vectors designed to express recombinant genes under T7 promoter control in *E. coli* should, in principle, direct transcription of the same genes in eukaryotic cells that express T7 RNAP. The fact that prokaryotic and eukaryotic translation signals differ, however, prevents the straightforward use of vectors such as the pET series for gene expression in eukaryotic cells. In particular, the translational machinery of eukaryotic cells generally requires mRNA to be "capped," i.e., linked with guanosine, 5′ to 5′, through a triphosphate bridge to the 5′-most nucleotide of the transcript. Further, the guanosine cap must be methylated, at least at the N7 position. The enzyme activities that normally accomplish these tasks are restricted to the nucleus and usually act only on the products of host RNA polymerase II. Finally, the eukaryotic translational system generally requires the initiation codon be in the proper sequence context (Kozak, 1978).

The first, and still most commonly used, solution to this problem involved recombinant vaccinia virus (VV) to express the phage RNAP gene (Fuerst *et al.*, 1986). Mammalian cells can be coinfected with recombinant VVs, one of which has been modified to include a target gene coupled to a phage promoter and the other the corresponding phage gene 1 under VV promoter control. The enzymes that VV encodes for the cytoplasmic capping and methylating its own mRNA apparently act in *trans* on the phage transcripts, allowing for abundant expression of recombinant proteins. Recombinant VV expressing a nuclear-targeted T3 RNAP has also been used to express a nuclear gene under phage promoter control (Rodriguez *et al.*, 1990). The lac operator repressor system has also proven to be functional in eukaryotic podoviral RNAP systems (Deuschle *et al.*, 1989; Rodriguez *et al.*, 1990).

Britton *et al.* (1996) reported the development of another T7 RNAP-expressing poxvirus, fowlpox virus (FPV). This system has the advantage that the viral replication cycle in mammalian cells is blocked after the early phase. Thus the recombinant FPV provides the T7 RNAP and the accessory enzymes required for translational activation of FPV and T7 transcripts, but not the later viral products responsible for cytopathic effect. Other viral systems for delivery of T7 RNAP and target genes without the cytopathic effect of VV have been based on recombinant adenovirus (Tomanin *et al.*, 1997) and baculovirus (Yap *et al.*, 1997).

Another approach to phage RNAP-based expression in animal cells has taken advantage of yet another viral strategy for circumventing the normal requirement that all functional mRNA arise from host nuclear RNAP II activity. Typical of the picornavirus family, the genomic sense RNA of encephalomyocarditis virus (EMCV) of mice contains a structural element of some 600 nucleotides at its 5′ end that directs cap-independent translation of the sequence immediately downstream. This "IRES" (internal ribosome entry site) has been used to render the transcripts of phage RNAP translatable in animal cells. Prefacing target sequences with the EMCV IRES has been found not only to enhance the translational efficiency of T7 transcripts in recombinant VV-infected cells (Elroy-Stein *et al.*, 1989), but to render podoviral RNAP transcripts translatable in the absence of any other accessory enzymes (Elroy-Stein and Moss, 1990; Zhou *et al.*, 1990). In principle, this should make possible the expression of foreign genes while minimally perturbing the physiology of the host mammalian cell. The EMCV IRES has also been recruited to craft a T7 RNAP autogene expression system widely applicable to eukaryotic cells (Deng and Wolff, 1994; Chen *et al.*, 1995).

Finally, a couple of T7-based, prokaryotic–eukaryotic "shuttle vector" designs have been described. These plasmids are designed to direct high-level expression in either *E. coli* or in animal cells. One design places a highly active cytomegalovirus (CMV) promoter upstream of a T7 promoter, which, in turn, is placed upstream of the target sequence (Schneider *et al.*, 1997). Thus the gene can be transcribed in a prokaryotic environment by T7 RNAP and in a mammalian nuclear environment by host RNAP II. Another design combines the EMCV IRES and the prokaryotic ribosome-binding site, directing the synthesis of the same translatable transcript, either in *E. coli* or in the cytoplasm of an animal cell, provided the presence of T7 RNAP (He *et al.*, 1995).

Conclusions

The high productivity of systems using T7 RNAP is probably the feature that most often motivates its choice in studies calling for recombinant gene expression. However, the flexibility of regulatory controls available because T7 RNAP is nonessential to the host may be at least as important. T7-based systems currently in use take advantage of transcriptional control of the RNAP (e.g., lac inducibility) and posttranslational control (lysozyme). Future developments may be expected to expand on these regulatory options and to take advantage of the growing understanding of the relationship between structure and function of podoviral RNAPs.

References

Apeler, H., Gottschalk, U., Guntermann, D., Hansen, J., Massen, J., Schmidt, E., Schneider, K. H., Schneidereit, M., and Rubsamen-Waigmann, H. (1997). Expression of natural and synthetic genes encoding herpes simplex virus 1 protease in *Escherichia coli* and purification of the protein. *Eur. J. Biochem.* **247,** 890–895.

Bhandari, P., and Gowrishankar, J. (1997). An *Escherichia coli* host strain useful for efficient overproduction of cloned gene products with NaCl as the inducer. *J. Bacteriol.* **79,** 4403–4406.

Bonner, G., Lafer, E. M., and Sousa, R. (1994). The thumb subdomain of T7 RNA polymerase functions to stabilize the ternary complex during processive transcription. *J. Biol. Chem.* **269,** 25129–25136.

Breaker, R. R., Banerji, A., and Joyce, G. F. (1994). Continuous in vitro evolution of bacteriophage RNA polymerase promoters. *Biochemistry* **33,** 11980–11986.

Britton, P., Green, P., Kottier, S., Mawditt, K. L., Penzes, Z., Cavanagh, D., and Skinner, M. A. (1996). Expression of bacteriophage T7 RNA polymerase in avian and mammalian cells by a recombinant fowlpox virus. *J. Gen. Virol.* **77,** 963–967.

Brown, L. R., Deng, J., Noll, D. M., Mori, N., and Clarke, N. D. (1997). Construction and overexpression of a synthetic gene for human DNA methylguanine methyltransferase: Renaturation and rapid purification of the protein. *Protein Express. Purif.* **9,** 337–345.

Brown, W. C., and Campbell, J. L. (1993). A new cloning vector and expression strategy for genes encoding proteins toxic to *Escherichia coli. Gene* **127,** 99–103.

Brown, W. C., Duncan, J. A., and Campbell, J. L. (1993). Purification and characterization of the *Saccharomyces cerevisiae* DNA polymerase delta overproduced in *Escherichia coli. J. Biol. Chem.* **268,** 982–990.

Brunschwig, E., and Darzins, A. (1992). A two-component T7 system for the overexpression of genes in *Pseudomonas aeruginosa. Gene* **111,** 35–41.

Butler, E. T., and Chamberlin, M. J. (1982). Bacteriophage SP6-specific RNA polymerase. I. Isolation and characterization of the enzyme. *J. Biol. Chem.* **257,** 5772–5778.

Cane, D. E., Wu, Z., Proctor, R. H., and Hohn, T. M. (1993). Overexpression in *Escherichia coli* of soluble aristolochene synthase from *Penicillium roqueforti. Arch. Biochem. Biophys.* **304,** 415–419.

Cermakian, N., Ikeda, T. M., Cedergren, R., and Gray, M. W. (1996). Sequences homologous to yeast mitochondrial and bacteriophage T3 and T7 RNA polymerases are widespread throughout the eukaryotic lineage. *Nucleic Acids Res.* **24,** 648–654.

Cermakian, N., Ikeda, T. M., Miramontes, P., Lang, B. F., Gray, M. W., and Cedergren, R. (1997). On the evolution of the single-subunit RNA polymerases. *J. Mol. Evol.* **45,** 671–681.

Chapman, K. A., and Burgess, R. R. (1987). Construction of bacteriophage T7 late promoters with point mutations and characterization by in vitro transcription properties. *Nucleic Acids Res.* **15,** 5413–5432.

Chen, C., Huang, H., Yang, X., Xia, Q., Li, B., and Wang, Y. (1996). T7-promoter-based *Escherichia coli* expression system induced with bacteriophage M13HEP. *Chin. J. Biotechnol.* **12,** 207–213.

Chen, X., Li, Y., Xiong, K., Xie, Y., Aizicovici, S., Snodgrass, R., Wagner, T. E., and

Platika, D. (1995). A novel nonviral cytoplasmic gene expression system and its implications in cancer gene therapy. *Cancer Gene Ther.* **2,** 281–289.

Conrad, B., Savchenko, R. S., Breves, R., and Hofemeister, J. (1996). A T7 promoter-specific, inducible protein expression system for *Bacillus subtilis. Mol. Gen. Genet.* **250,** 230–236.

Cunningham, P. R., Weitzmann, C. J., and Ofengand, J. (1991). SP6 RNA polymerase stutters when initiating from an AAA . . . sequence. *Nucleic Acids Res.* **19,** 4669–4673.

Davanloo, P., Rosenberg, A. H., Dunn, J. J., and Studier, F. W. (1984). Cloning and expression of the gene for bacteriophage T7 RNA polymerase. *Proc. Natl. Acad. Sci. U.S.A.* **81,** 2035–2039.

Delbruck, M. (1946). Bacterial viruses or bacteriophages. *Biol. Rev. Cambridge Philos. Soc.* **21,** 30–40.

Demerec, M., and Fano, U. (1945). Bacteriophage-resistant mutants in *Escherichia coli. Genetics* **30,** 119–136.

Deng, H., and Wolff, J. A. (1994). Self-amplifying expression from the T7 promoter in 3T3 mouse fibroblasts. *Gene* **143,** 245–249.

Dersch, P., Fsihi, H., and Bremer, E. (1994). Low-copy number T7 vectors for selective gene expression and efficient protein overproduction in *Escherichia coli. FEMS Microbiol. Lett.* **123,** 19–26.

Deuschle, U., Gentz, R., and Bujard, H. (1986). lac Repressor blocks transcribing RNA polymerase and terminates transcription. *Proc. Natl. Acad. Sci. U.S.A.* **83,** 4134–4137.

Deuschle, U., Pepperkok, R., Wang, F. B., Giordano, T. J., McAllister, W. T., Ansorge, W., and Bujard, H. (1989). Regulated expression of foreign genes in mammalian cells under the control of coliphage T3 RNA polymerase and lac repressor. *Proc. Natl. Acad. Sci. U.S.A.* **86,** 5400–5404.

Diaz, G. A., Raskin, C. A., and McAllister, W. T. (1993). Hierarchy of base-pair preference in the binding domain of the bacteriophage T7 promoter *J. Mol. Biol.* **229,** 805–811.

Diaz, G. A., Rong, M., McAllister, W. T., and Durbin, R. K. (1996). The stability of abortively cycling T7 RNA polymerase complexes depends upon template conformation. *Biochemistry* **35,** 10837–10843.

Dietz, A., Weisser, H. J., Kossel, H., and Hausmann, R. (1990). The gene for Klebsiella bacteriophage K11 RNA polymerase: Sequence and comparison with the homologous genes of phages T7, T3, and SP6. *Mol. Gen. Genet.* **221,** 283–286.

Dobrikova, E. Y., Pletnev, A. G., Karamyshev, V. N., and Morozova, O. V. (1996). T7 DNA-dependent RNA polymerase can transcribe RNA from tick-borne encephalitis virus (TBEV) cDNA with SP6 promoter. *FEBS Lett.* **382,** 327–329.

Doherty, A. J., Ashford, S. R., Subramanya, H. S., and Wigley, D. B. (1996). Bacteriophage T7 DNA ligase. Overexpression, purification, crystallization, and characterization. *J. Biol. Chem.* **271,** 11083–11089.

Dubendorff, J. W., and Studier, F. W. (1991a). Controlling basal expression in an inducible T7 expression system by blocking the target T7 promoter with lac repressor. *J. Mol. Biol.* **219,** 45–59.

Dubendorff, J. W., and Studier, F. W. (1991b). Creation of a T7 autogene. Cloning and expression of the gene for bacteriophage T7 RNA polymerase under control of its cognate promoter. *J. Mol. Biol.* **219,** 61–68.

Dunderdale, H. J., Sharples, G. J., Lloyd, R. G., and West, S. C. (1994). Cloning, overexpression, purification, and characterization of the *Escherichia coli* RuvC Holliday junction resolvase. *J. Biol. Chem.* **269,** 5187–5194.

Dunn, J. J., and Studier, F. W. (1983). Complete nucleotide sequence of bacteriophage T7 DNA and the locations of T7 genetic elements. *J. Mol. Biol.* **166,** 477–535.

Dunn, J. J., Krippl, B., Bernstein, K. E., Westphal, H., and Studier, F. W. (1988). Targeting bacteriophage T7 RNA polymerase to the mammalian cell nucleus. *Gene* **68,** 259–266.

Ejdeback, M., Young, S., Samuelsson, A., and Karlsson, B. G. (1997). Effects of codon usage and vector-host combinations on the expression of spinach plastocyanin in Escherichia coli. *Protein Express. Purif.* **11,** 17–25.

Elroy-Stein, O., and Moss, B. (1990). Cytoplasmic expression system based on constitutive synthesis of bacteriophage T7 RNA polymerase in mammalian cells. *Proc. Natl. Acad. Sci. U.S.A.* **87,** 6743–6747.

Elroy-Stein, O., Fuerst, T. R., and Moss, B. (1989). Cap-independent translation of mRNA conferred by encephalomyocarditis virus 5′ sequence improves the performance of the vaccinia virus/bacteriophage T7 hybrid expression system. *Proc. Natl. Acad. Sci. U.S.A.* **86,** 6126–6130.

Fisher, C. A., Wang, J., Francis, G. A., Sykes, B. D., Kay, C. M., and Ryan, R. O. (1997). Bacterial overexpression, isotope enrichment and NMR analysis of the N-terminal domain of human apolipoprotein E. *Biochem. Cell Biol.* **75,** 45–53.

Fuerst, T. R., Niles, E. G., Studier, F. W., and Moss, B. (1986). Eukaryotic transient-expression system based on recombinant vaccinia virus that synthesizes bacteriophage T7 RNA polymerase. *Proc. Natl. Acad. Sci. U.S.A.* **83,** 8122–8126.

Gardner, L. P., Mookhtiar, K. A., and Coleman, J. E. (1997). Initiation, elongation, and processivity of carboxyl-terminal mutants of T7 RNA polymerase. *Biochemistry* **36,** 2908–2918.

Giordano, T. J., Deuschle, U., Bujard, H., and McAllister, W. T. (1989). Regulation of coliphage T3 and T7 RNA polymerases by the lac repressor-operator system. *Gene* **84,** 209–219.

Gotta, S. L., Miller, O. L., Jr., and French, S. L. (1991). rRNA transcription rate in *Escherichia coli. J. Bacteriol.* **173,** 6647–6649.

Gribenko, A., Lopez, M. M., Richardson, J. M., III, and Makhatadze, G. I. (1998). Cloning, overexpression, purification, and spectroscopic characterization of human S100P. *Protein Sci.* **7,** 211–215.

Grodberg, J., and Dunn, J. J. (1988). ompT encodes the *Escherichia coli* outer membrane protease that cleaves T7 RNA polymerase during purification. *J. Bacteriol.* **170,** 1245–1253.

Gross, L., Chen, W. J., and McAllister, W. T. (1992). Characterization of bacteriophage T7 RNA polymerase by linker insertion mutagenesis. *J. Mol. Biol.* **228,** 488–505.

Hamood, A. N., and Iglewski, B. H. (1990). Expression of the *Pseudomonas aeruginosa* toxA positive regulatory gene (regA) in *Escherichia coli. J. Bacteriol.* **172,** 589–594.

He, B., McAllister, W. T., and Durbin, R. K. (1995). Phage RNA polymerase vectors that allow efficient gene expression in both prokaryotic and eukaryotic cells. *Gene* **164,** 75–79.

He, B., Rong, M., Durbin, R. K., and McAllister, W. T. (1997a). A mutant T7 RNA polymerase that is defective in RNA binding and blocked in the early stages of transcription. *J. Mol. Biol.* **265,** 275–288.

He, B., Rong, M., Lyakhov, D., Gartenstein, H., Diaz, G., Castagna, R., McAllister, W. T., and Durbin, R. K. (1997b). Rapid mutagenesis and purification of phage RNA polymerases. *Protein Express. Purif.* **9,** 142–151.

Hernan, R. A., Hui, H. L., Andracki, M. E., Noble, R. W., Sligar, S. G., Walder, J. A., and Walder, R. Y. (1992). Human hemoglobin expression in *Escherichia coli*: Importance of optimal codon usage. *Biochemistry* **31,** 8619–8628.

Herrero, M., de Lorenzo, V., Ensley, B., and Timmis, K. N. (1993). A T7 RNA polymerase-based system for the construction of Pseudomonas strains with phenotypes dependent on TOL-meta pathway effectors. *Gene* **134,** 103–106.

Hill, R. B., MacKenzie, K. R., Flanagan, J. M., Cronan, J. E., Jr., and Prestegard, J. H. (1995). Overexpression, purification, and characterization of *Escherichia coli* acyl carrier protein and two mutant proteins. *Protein Express. Purif.* **6,** 394–400.

Jia, Y., and Patel, S. S. (1997). Kinetic mechanism of transcription initiation by bacteriophage T7 RNA polymerase. *Biochemistry* **36,** 4223–4232.

Jobling, S. A., Cuthbert, C. M., Rogers, S. G., Fraley, R. T., and Gehrke, L. (1988). In vitro transcription and translational efficienty of chimeric SP6 messenger RNAs devoid of 5′ vector nucleotides. *Nucleic Acids Res.* **16,** 4483–4498.

Jorgensen, E. D., Durbin, R. K., Risman, S. S., and McAllister, W. T. (1991). Specific contacts between the bacteriophage T3, T7, and SP6 polymerases and their promoters. *J. Biol. Chem.* **266,** 645–651.

Kim, Y. T., Muramatsu, T., and Takahashi, K. (1995). Leader peptidase from *Escherichia coli*: Overexpression, characterization, and inactivation by modification of tryptophan residues 300 and 310 with N-bromosuccinimide. *J. Biochem.* **117,** 535–544.

Knoll, D. A., Woody, R. W., and Woody, A. Y. (1992). Mapping of the active site of T7 RNA polymerase with 8-azidoATP. *Biochim. Biophys. Acta* **1121**(3), 252–260.

Kostyuk, D. A., Dragan, S. M., Lyakhov, D. L., Rechinsky, V. O., Tunitskaya, V. L., Chernov, B. K., and Kochetkov, S. N. (1995). Mutants of T7 RNA polymerase that are able to synthesize both RNA and DNA. *FEBS Lett.* **369,** 165–168.

Kotani, H., Ishizaki, Y., Hiraoka, N., and Obayashi, A. (1987). Nucleotide sequence and expression of the cloned gene of bacteriophage SP6 RNA polymerase. *Nucleic Acids Res.* **15,** 2653–2664.

Kozak, M. (1978). How do eukaryotic ribosomes select initiation regions in messenger RNA? *Cell (Cambridge, Mass.)* **15,** 1109–1123.

Kumar, A., and Patel, S. S. (1997). Inhibition of T7 RNA polymerase: Transcription initiation and transition from initiation to elongation are inhibited by T7 lysozyme via a ternary complex with RNA polymerase and promoter DNA. *Biochemistry* **36,** 13954–13962.

Lebedeva, M. I., Rogozhkina, E. V., Tsyba, N. A., and Mashko, S. V. (1994). A new T7 RNA polymerase-driven expression system induced via thermoamplification of a recombinant plasmid carrying a T7 promoter-*Escherichia coli* lac operator. *Gene* **142,** 61–66.

Lee, S. S., and Kang, C. (1993). Two base pairs at –9 and –8 distinguish between the bacteriophage T7 and SP6 promoters. *J. Biol. Chem.* **268,** 19299–19304.

Li, T., Ho, H. H., Maslak, M., Schick, C., and Martin, C. T. (1996). Major groove recognition elements in the middle of the T7 RNA polymerase promoter. *Biochemistry* **35,** 3722–3727.

Lopez, P. J., Guillerez, J. H., Sousa, R., and Dreyfus, M. (1997). The low processivity of T7 RNA polymerase over the initially transcribed sequence can limit productive initiation in vivo. *J. Mol. Biol.* **269,** 41–51.

Lyakhov, D. L., He, B., Zhang, X., Studier, F. W., Dunn, J. J., and McAllister, W. T. (1997). Mutant bacteriophage T7 RNA polymerases with altered termination properties. *J. Mol. Biol.* **269,** 28–40.

Macdonald, L. E., Durbin, R. K., Dunn, J. J., and McAllister, W. T. (1994). Characterization of two types of termination signal for bacteriophage T7 RNA polymerase. *J. Mol. Biol.* **238,** 145–158.

MacLaren, D. C., and Clarke, S. (1995). Expression and purification of a human recombinant methyltransferase that repairs damaged proteins. *Protein Express. Purif.* **6,** 99–108.

Makarova, O. V., Makarov, E. M., Sousa, R., and Dreyfus, M. (1995). Transcribing of Escherichia coli genes with mutant T7 RNA polymerases: Stability of lacZ mRNA inversely correlates with polymerase speed. *Proc. Natl. Acad. Sci. U.S.A.* **92,** 12250–12254.

Martin, C. T., and Coleman, J. E. (1987). Kinetic analysis of T7 RNA polymerase-promoter interactions with small synthetic promoters. *Biochemistry* **26,** 2690–2696.

Martin, C. T., Muller, D. K., and Coleman, J. E. (1988). Processivity in early states of transcription by T7 RNA polymerase. *Biochemistry* **27,** 3966–3974.

Maslak, M., and Martin, C. T. (1993). Kinetic analysis of T7 RNA polymerase transcription initiation from promoters containing single-stranded regions. *Biochemistry* **32,** 4281–4285.

McAllister, W. T. (1993). Structure and function of the bacteriophage T7 RNA polymerase (or, the virtues of simplicity). *Cell Mol. Biol. Res.* **39,** 385–391.

McAllister, W. T., and Raskin, C. A. (1993). The phage RNA polymerases are related to DNA polymerases and reverse transcriptases. *Mol. Microbiol.* **10,** 1–6.

McAllister, W. T., Morris, C., Rosenberg, A. H., and Studier, F. W. (1981). Utilization of bacteriophage T7 late promoters in recombinant plasmids during infection. *J. Mol. Biol.* **153,** 527–544.

Mead, D. A., Skorupa, E. S., and Kemper, B. (1986). Single-stranded DNA "blue" promoter plasmids: A versatile tandem promotersystem for cloning and protein engineering. *Protein Eng.* **1,** 67–74.

Melton, D. A., Krieg, P. A., Rebagliati, M. R., Maniatis, T., Zinn, K., and Green, M. R. (1984). Efficient in vitro synthesis of biologically active RNA and RNA hybridization probes from plasmids containing a bacteriophage SP6 promoter. *Nucleic Acids Res.* **12,** 7035–7056.

Milligan, J. F., Groebe, D. R., Witherell, G. W., and Uhlenbeck, O. C. (1987). Oligoribonucleotide synthesis using T7 RNA polymerase and synthetic DNA templates. *Nucleic Acids Res.* **15,** 8783–8798.

Miroux, B., and Walker, J. E. (1996). Over-production of proteins in *Escherichia coli*: Mutant hosts that allow synthesis of some membrane proteins and globular proteins at high levels. *J. Mol. Biol.* **260,** 289–298.

Moffatt, B. A., and Studier, F. W. (1987). T7 lysozyme inhibits transcription by T7 RNA polymerase. *Cell (Cambridge, Mass.)* **49,** 221–227.

Mookhtiar, K. A., Peluso, P. S., Muller, D. K., Dunn, J. J., and Coleman, J. E. (1991). Processivity of T7 RNA polymerase requires the C-terminal Phe882-Ala883-COO or "foot." *Biochemistry* **30,** 6305–6313.

Morris, C. E., Klement, J. F., and McAllister, W. T. (1986). Cloning and expression of the bacteriophage T3 RNA polymerase gene. *Gene* **41,** 193–200.

Muller, D. K., Martin, C. T., and Coleman, J. E. (1988). Processivity of proteolytically modified forms of T7 RNA polymerase. *Biochemistry* **27,** 5763–5771.

Muller, D. K., Martin, C. T., and Coleman, J. E. (1989). T7 RNA polymerase interacts with its promoter from one side of the DNA helix. *Biochemistry* **28,** 3306–3313.

Olins, P. O., and Rangwala, S. H. (1989). A novel sequence element derived from bacteriophage T7 mRNA acts as an enhancer of translation of the lacZ gene in *Escherichia coli. J. Biol. Chem.* **264,** 16973–16976.

Osumi-Davis, P. A., de Aguilera, M. C., Woody, R. W., and Woody, A. Y. (1992). Asp537, Asp812 are essential and Lys631, His811 are catalytically significant in bacteriophage T7 RNA polymerase activity. *J. Mol. Biol.* **226,** 37–45.

Patra, D., Lafer, E. M., and Sousa, R. (1992). Isolation and characterization of mutant bacteriophage T7 RNA polymerases. *J. Mol. Biol.* **224,** 307–318.

Pei, D., Neel, B. G., and Walsh, C. T. (1993). Overexpression, purification, and characterization of SHPTP1, a Src homology 2-containing protein-tyrosine-phosphatase. *Proc. Natl. Acad. Sci. U.S.A.* **90,** 1092–1096.

Raskin, C. A., Diaz, G., Joho, K., and McAllister, W. T. (1992). Substitution of a single bacteriophage T3 residue in bacteriophage T7 RNA polymerase at position 748 results in a switch in promoter specificity. *J. Mol. Biol.* **228,** 506–515.

Raskin, C. A., Diaz, G. A., and McAllister, W. T. (1993). T7 RNA polymerase mutants with altered promoter specificities. *Proc. Natl. Acad. Sci. U.S.A.* **90,** 3147–3151.

Rathore, D., Nayak, S. K., and Batra, J. K. (1996). Expression of ribonucleolytic toxin restrictocin in *Escherichia coli*: Purification and characterization. *FEBS Lett.* **392,** 259–262.

Rathore, D., Nayak, S. K., and Batra, J. K. (1997). Overproduction of fungal ribotoxin alpha-sarcin in *Escherichia coli*: Generation of an active immunotoxin. *Gene* **190,** 31–35.

Reddi, P. P., Castillo, J. R., Klotz, K., Flickinger, C. J., and Herr, J. C. (1994). Production in *Escherichia coli,* purification and immunogenicity of acrosomal protein SP-10, a candidate contraceptive vaccine. *Gene* **147,** 189–195.

Rice, C. M., Levis, R., Strauss, J. H., and Huang, H. V. (1987). Production of infectious RNA transcripts from Sindbis virus cDNA clones: Mapping of lethal mutations, rescue of a temperature-sensitive marker, and in vitro mutagenesis to generate defined mutants. *J. Virol.* **61,** 3809–3819.

Rodriguez, D., Zhou, Y. W., Rodriguez, J. R., Durbin, R. K., Jimenez, V., McAllister, W. T., and Esteban, M. (1990). Regulated expression of nuclear genes by T3 RNA polymerase and lac repressor, using recombinant vaccinia virus vectors. *J. Virol.* **64,** 4851–4857.

Rong, M., He, B., McAllister, W. T., and Durbin, R. K. (1998). Promoter specificity determinants of T7 RNA polymerase. *Proc. Natl. Acad. Sci. U.S.A.* **95,** 515–519.

Rosenberg, A. H., Lade, B. N., Chui, D. S., Lin, S. W., Dunn, J. J., and Studier, F. W. (1987). Vectors for selective expression of cloned DNAs by T7 RNA polymerase. *Gene* **56,** 125–135.

San Francisco, M. J., Tisa, L. S., and Rosen, B. P. (1989). Identification of the membrane component of the anion pump encoded by the arsenical resistance operon of R-factor R773. *Mol. Microbiol.* **3,** 15–21.

Sayle, R. A., and Milner-White, E. J. (1995). RASMOL: Biomolecular graphics for all. *Trends Biochem. Sci.* **20,** 374.

Schneider, S., Georgiev, O., Buchert, M., Adams, M. T., Moelling, K., and Hovens, C.

M. (1997). An epitope tagged mammalian/prokaryotic expression vector with positive selection of cloned inserts. *Gene* **197,** 337–341.

Schuler, G. D., Altschul, S. F., and Lipman, D. J. (1991). A workbench for multiple alignment construction and analysis. *Proteins: Struct., Funct., Genet.* **9,** 180–190.

Seidel, H. M., Pompliano, D. L., and Knowles, J. R. (1992). Phosphonate biosynthesis: Molecular cloning of the gene for phosphoenolpyruvate mutase from *Tetrahymena pyriformis* and overexpression of the gene product in *Escherichia coli. Biochemistry* **31,** 2598–2608.

Sisk, W. P., Bradley, J. D., Kingsley, D., and Patterson, T. A. (1992). Deletion of hydrophobic domains of viral glycoproteins increases the level of their production in *Escherichia coli. Gene* **112,** 157–162.

Sousa, R. (1996). Structural and mechanistic relationships between nucleic acid polymerases. *Trends Biochem. Sci.* **21,** 186–190.

Sousa, R., and Padilla, R. (1995). A mutant T7 RNA polymerase as a DNA polymerase. *EMBO J.* **14,** 4609–4621.

Sousa, R., Chung, Y. J., Rose, J. P., and Wang, B. C. (1993). Crystal structure of bacteriophage T7 RNA polymerase at 3.3 angstrom resolution. *Nature (London)* **364,** 593–599.

Studier, F. W. (1983). Organization and expression of bacteriophage T7 DNA. *Cold Spring Harbor Symp. Quant. Biol.* **47**(pt 2), 999–1007.

Studier, F. W. (1991). Use of bacteriophage T7 lysozyme to improve an inducible T7 expression system. *J. Mol. Biol.* **219,** 37–44.

Studier, F. W., and Moffatt, B. A. (1986). Use of bacteriophage T7 RNA polymerase to direct selective high-level expression of cloned genes. *J. Mol. Biol.* **189,** 113–130.

Studier, F. W., Rosenberg, A. H., Dunn, J. J., and Dubendorff, J. W. (1990). Use of T7 RNA polymerase to direct expression of cloned genes. *In* "Methods of Enzymology" (D. V. Goeddel, ed.), Vol. 185, pp. 60–89. Academic Press, San Diego, CA.

Sun, Y., Jiang, Y., Zhang, P., Zhang, S. J., Zhou, Y., Li, B. Q., Toomey, N. L., and Lee, M. Y. (1997). Expression and characterization of the small subunit of human DNA polymerase delta. *J. Biol. Chem.* **272,** 13013–13018.

Tabor, S., and Richardson, C. C. (1985). A bacteriophage T7 RNA polymerase/promoter system for controlled exclusive expression of specific genes. *Proc. Natl. Acad. Sci. U.S.A.* **82,** 1074–1078.

Tan, C. A., Hehir, M. J., and Roberts, M. F. (1997). Cloning, overexpression, refolding, and purification of the nonspecific phospholipase C from *Bacillus cereus. Protein Expr. Purif.* **10,** 365–372.

Tomanin, R., Bett, A. J., Picci, L., Scarpa, M., and Graham, F. L. (1997). Development and characterization of a binary gene expression system based on bacteriophage T7 components in adenovirus vectors. *Gene* **193,** 129–140.

Ujvari, A., and Martin, C. T. (1996). Thermodynamic and kinetic measurements of promoter binding by T7 RNA polymerase. *Biochemistry* **35,** 14574–14582.

van der Werf, S., Bradley, J., Wimmer, E., Studier, F. W., and Dunn, J. J. (1986). Synthesis of infectious poliovirus RNA by purified T7 RNA polymerase. *Proc. Natl. Acad. Sci. U.S.A.* **83,** 2330–2334.

Vysotskaya, V., Tischenko, S., Garber, M., Kern, D., Mougel, M., Ehresmann, C., and Ehresmann, B. (1994). The ribosomal protein S8 from *Thermus thermophilus* VK1. Sequencing of the gene, overexpression of the protein in *Escherichia coli* and interaction with rRNA. *Eur. J. Biochem.* **223,** 437–445.

Wang, C., and Lee, M. R. (1993). High-level expression of soluble rat hsc70 in *Escherichia coli*: Purification and characterization of the cloned enzyme. *Biochem. J.* **294**(Pt. 1), 69–77.

Woody, A. Y., Eaton, S. S., Osumi-Davis, P. A., and Woody, R. W. (1996). Asp537 and Asp812 in bacteriophage T7 RNA polymerase as metal ion-binding sites studied by EPR, flow-dialysis, and transcription. *Biochemistry* **35,** 144–152.

Wu, Q. L., Jha, P. K., Du, Y., Leavis, P. C., and Sarkar, S. (1995). Overproduction and rapid purification of human fast skeletal beta troponin T using *Escherichia coli* expression vectors: Functional differences between the alpha and beta isoforms. *Gene* **155,** 225–230.

Yap, C. C., Ishii, K., Aoki, Y., Aizaki, H., Tani, H., Shimizu, H., Ueno, Y., Miyamura, T., and Matsuura, Y. (1997). A hybrid baculovirus-T7 RNA polymerase system for recovery of an infectious virus from cDNA. *Virology* **231,** 192–200.

Zhang, X., and Studier, F. W. (1995). Isolation of transcriptionally active mutants of T7 RNA polymerase that do not support phage growth. *J. Mol. Biol.* **250,** 156–168.

Zhang, X., and Studier, F. W. (1997). Mechanism of inhibition of bacteriophage T7 RNA polymerase by T7 lysozyme. *J. Mol. Biol.* **269,** 10–27.

Zhou, Y., Giordano, T. J., Durbin, R. K., and McAllister, W. T. (1990). Synthesis of functional mRNA in mammalian cells by bacteriophage T3 RNA polymerase. *Mol. Cell. Biol.* **10,** 4529–4537.

Zhou, A., Jiang, X., Dou, F., Zhu, D., and Xu, X. (1997). Renaturation, purification, and characterization of human plasminogen activator inhibitor type 2 (PAI-2) accumulated at high level in *Escherichia coli. J. Biochem.* **121,** 930–934.

2

EXPRESSION VECTORS EMPLOYING THE *trc* PROMOTER*

Jürgen Brosius

Institute for Experimental Pathology, Center for Molecular Biology of Inflammation (ZMBE), University of Münster, D-48149 Münster, Germany

*This chapter focuses on the *trc* promoter, and examples given almost exclusively cover *trc* expression vectors. The citations given here may therefore not necessarily include the first examples of a given vector or process improvement.

Introduction

The classification of expression vectors according to transcription promoters is somewhat arbitrary and has historical reasons originating from a time when a strong promoter was considered to be the most important asset of an expression vector. It later became clear that elements important for efficient initiation of translation were equally important and that other sequences around the transcriptional and translational start sites and sequences even further downstream can play an important role in determining the levels of overexpression in *Escherichia coli.*

Not infrequently, drastic differences in the expression of the same gene in one expression vector versus another could be observed. This was often falsely attributed to the contribution of the promoters employed. Most likely, the fortuitous juxtaposition of sequences from the vector with sequences from the gene to be expressed was the sole or main parameter determining the success of high-level expression (Shatzman and Rosenberg, 1987; Brosius, 1988). For example, sequences downstream from the promoter, including elements important for the initiation of translation, cannot be judged as a constant contributor to a guaranteed high level of expression, but may function with different efficiency depending on the gene sequences inserted downstream by forming different secondary structures.

Creation of the *trc* promoter stemmed from the desire to further improve the strength of a promoter. Although the hybrid *trp/lac* fusion or *tac* promoter, first constructed by de Boer *et al.* (1982, 1983), featured the consensus -35 and -10 sequences, their spacing of 16 bp was 1 bp short of the consensus distance. Insertion of 1 or 2 bp led the *trc* and *tic* promoters, respectively. As expected, the *tic* promoter with a spacing of 18 bp between the two consensus sequences exhibited only about 65% of the activity of the *tac* promoter *in vivo* (Brosius *et al.*, 1985). *In vitro* the *trc* promoter (17-bp spacing) was, as expected, the most efficient promoter in open complex formation (Mulligan *et al.*, 1985). *In vivo*, however, using two independent reporter genes (including 4.5S RNA), the *trc* promoter was shown to be somewhat less powerful (~90%) than the original *tac* promoter (Brosius *et al.*, 1985). It is now known that the strongest promoters in *E. coli* do not necessarily adhere to the consensus sequence (Deuschle *et al.*, 1986; Kammerer *et al.*, 1986). Unlike the binding constant that improves when the promoter elements approach the consensus sequences and distance, the rates of isomerization and promoter clearance are less dependent on a consensus sequence.

Apart from the promoters belonging to the *trp*/*lac* fusion family, a number of other suitable promoter systems have been employed in expression vectors, such as the *trp* promoter and the thermoinducible λP_L promoter (reviewed by Brosius, 1988), as well as promoters from T-type bacteriophages (see, e.g., Deuschle *et al.*, 1986; Kammerer *et al.*, 1986; Bujard *et al.*, 1987; Giordano *et al.*, 1989; Studier *et al.*, 1990). For an excellent review on the relative merits of different *E. coli* expression systems, the reader is referred to Makrides (1996).

The drawback of *tac* and *trc* promoters, as with many promoters, is their incomplete repression in the uninduced state. This is often unproblematic unless the foreign gene product to be expressed is toxic to the host cell, leads to plasmid instability, decreases the cell growth rate, or leads to death (Makrides, 1996). The relative merits of different promoter systems are described briefly by Lutz and Bujard (1997) and comprehensively by Makrides (1996). For systems achieving an almost complete repression of promoter activity yet excellent levels of protein expression in the induced state, the reader should refer to Lutz and Bujard (1997) and to the chapter in this book on the *ara*BAD system by Guzman.

The reason why vectors featuring the *trc* promoter became very popular expression vehicles compared to vectors harboring the *tac* promoter probably lay in an additional feature that was incorporated into the first described *trc* promoter vector: Although both *tac* and *trc* promoter vectors employ the *lacZ* ribosome-binding site (RBS), the *trc* promoter vector, pKK233-2 (Amann and Brosius, 1985), features a unique *Nco*I restriction site (C/CATGG) downstream from the RBS such that the embedded ATG start codon is in a near optimal distance (8 bp) to the ATG. The vector could be prepared by digestion at the *Nco*I site followed by a fill-in reaction with the Klenow fragment of T4 DNA polymerase. Any open reading frame can now be fused to the exposed ATG start codon. Coincidentally, the translational start in eukaryotic genes is often surrounded by sequences that contain an *Nco*I site (Kozak, 1987).

A further reason for the popularity of the *trc* promoter was that the addition of the inducer isopropyl-β-D-thiogalactopyranoside (IPTG) on a laboratory scale is a very convenient way to induce the promoter. In contrast, thermoinduction often requires a second incubator or shaking water bath. An additional discussion of the relative merits of thermoinduction is discussed later. Up until the mid-1980s the use of a *trc* promoter containing expression vectors was limited to particular *E. coli* strains (I^q or I^{sq} strains) that overproduce the *lac* repressor if regulation was to be achieved. Since the introduction of the *lac* repressor gene (*lacI*) onto vectors containing the *trc* promoter for controlling transcription such as vectors of the pTrc99 series (Amann *et al.*, 1988) or the pSE series (see Table 1), virtually any *E. coli* strain (e.g., protease-deficient strains) can serve as a host for regulated expression.

Attempts to improve the repression of the *trc* or *tac* promoter have failed thus far (J. Comiskey, U. Niewerth, M. Nettermann, and J. Brosius, unpublished observations). Replacement of the natural *lac* operator downstream from the -10 region with a symmetrical or ideal operator (Simons *et al.*, 1984) failed because (a) repression was still incomplete and (b) expression levels at full induction were significantly below in comparison to those of the original promoters. Placement of the ideal operator between the -10 and the -35 consensus sequences (Lanzer and Bujard, 1988) of the *tac* promoter had a similar unwanted effect. By increasing the level of *lac* repressor in the host cell, e.g., by altering the *lac*I promoter into the stronger *lac*I^q promoter, Andrews *et al.* (1996) reported significantly better repression from the *trc* promoter. However, the fully induced level was compromised by a factor of five. Other approaches to regulate the activity of the *trc* promoter better include the use of plasmid replicons (such as in pACYC184) that reduce the copy number, thus increasing the ratio of repressor to operator (Bingham, 1991).

In many instances, regulation of an expression system by a temperature increase may not be desired because of the pleiotropic effects, such as a marked increase of proteolytic activity during heat shock response (Goff *et al.*, 1984; Goff and Goldberg, 1987). Furthermore, it has been established in many cases that folding of the desired protein to its native form is optimized at lower growth temperatures (Schein, 1989; Cabilly, 1989; Totsuka and Fukazawa, 1993; Sim *et al.*, 1996).

For situations where the aforementioned drawbacks are not relevant and temperature shift as the mode of regulation is favored, temperature-sensitive *lac* repressors (Bukrinsky *et al.*, 1988) have been used in expression vectors (Adari *et al.*, 1995; Andrews *et al.*, 1996; Hasan and Szybalski, 1995; Yabuta *et al.*, 1995). The repressor allele described by Bukrinsky *et al.* (1988) lost its ability to be

Table 1
Compilation of Selected Plasmids Featuring the *trc* Promoter[a]

Plasmid	Relevant features	Size (kb)	Commercial supplier	Reference
pKK233-2	Expression vector; *trc* promotor can be regulated with IPTG; *rrn*B transcription terminators; *Nco*I cloning site contains translational ATG start codon in 8 bp distance to RBS; two additional cloning sites	4.6	Pharmacia Clontech	Amann and Brosius (1985)
pSE220	As pKK233-2; short polylinker replaced with SL2 superlinker	4.9		Brosius (1989, 1992)
pSE280	As pSE220; remnants of *tet* gene removed	3.9	Invitrogen	Brosius (1989, 1992)
pKK388-1	As pKK233-2; *rrn*B anti-terminator region distal to *trc* promotor/*lac* operator; *tet* gene restored to confer resistance; optimized RBS; pUC19 polylinker distal to *Nco*I cloning site	5.1	Clontech	Brosius (1988)
pTrc99A, B, C	As pKK233-2; nonessential pBR322 vector sequences including remnants of *tet* gene removed; carries *lacI* repressor gene for vector use in wider strain range; pUC19 polylinker	4.2	Pharmacia	Amann *et al.* (1988)
pSE380	As pTrc99A; *lacI* repressor gene; pUC19 polylinker replaced with SL2 superlinker	4.5	Invitrogen	Brosius (1989, 1992)
pSE420	As pSE380; *lacI* repressor gene; *trc* promoter/*lac* operator followed by *rrn*B antiterminator; T7 gene 10 translational enhancer; A/T-rich regions flanking optimized RBS; mini cistron upstream from CCATGG (= start codon in *Nco*I clining site) for efficient translational restart; superlinker SL2	4.6	Invitrogen	Brosius (1992)

(*continues*)

Table 1
Continued

Plasmid	Relevant features	Size (kb)	Commercial supplier	Reference
pTrcHisA, B, C	As pSE420; following the ATG start codon within *Nco*I site is a His_6 tag for protein purification; enterokinase cleavage site; polylinker (*Bam*HI, *Xho*I, *Sac*I, *Bgl*II, *Pst*I, *Kpn*I, *Eco*RI, *Bst*BI, *Hin*dIII); three versions A, B, C available to ensure in-frame translation of insert; *Sac*I site absent in pTrcHisC version	4.4	Invitrogen	—
pTrcHis2A, B, C	As pSE420; ATG embedded in *Nco*I site followed by three versions of a polylinker (*Bam*HI, *Xho*I, *Sac*I, *Bgl*II, *Pst*I, *Kpn*I, *Eco*RI, *Bst*BI, *Hin*dIII, *Sma*I) to ensure in-frame translation of insert; in versions B and C the *Sma*I site is replaced by *Xba*I or *Sna*BI, respectively), *myc* epitope for protein detection; His_6 tag at 3′ end of polypeptide	4.4	Invitrogen	—
pThioHisA, B, C	As pSE420; no mini cistron; ATG start followed by modified thioredoxin protein containing a histidin patch; the His-patch-thioredoxin fusion partner is followed by an enterokinase cleavage site, followed by three versions of a polylinker (*Kpn*I, *Nsi*I, *Sac*I, *Xho*I, *Bgl*II, *Bst*BI, *Eco*RI, *Sac*II, *Not*I, *Stu*I, *Xba*I, *Sal*I, *Pst*I); in versions B and C the *Nsi*I site is replaced by *Nco*I or *Sph*I, respectively. The *rrn*B terminators have been replaced by the *aspA* transcription terminator.	4.4	Invitrogen	—

[a]All vectors confer resistance to ampicillin. *rrn*B T_1 and T_2 transcriptional terminators are located distal to the polylinkers.

chemically modulated with IPTG. Another *lac* repressor mutant described by Ward *et al.* (1995) is both thermosensitive and chemically inducible.

Vectors for *trc*-Driven Expression

Vectors employing the *trc* promoter, such as pKK233-2, pKK388-1, pTrc99A, pSE380, and pSE420, probably have been used, as judged from their citations, for more than a thousand expression studies. For some of these examples, see Table 2. Many attempts to further improve the performance of vectors containing *trc* promoters did not target the efficiency of transcription, but other stages during expression of a foreign protein in *E. coli*. By increasing, for example, the preferred *E. coli* codons from 43 to 85% in a synthetic gene encoding human interleukin 2, Williams *et al.* (1988) could increase its expression more than 16-fold when compared to the native cDNA. Furthermore, the localization of the recombinant protein within different host cell compartments may affect its level of accumulation and tertiary structure. Ferraz *et al.* (1991) demonstrated that variants of human β-tropomyosin expressed in pKK233-2 showed differences in expression levels, suggesting that mRNA or protein structure was responsible for these effects.

If full repression of the promoter is not necessary with a given foreign gene, several groups have resorted to another measure to further boost expression levels of the recombinant protein: An increase of the plasmid copy number by using origins of replications from pUC-plasmid-derived vectors has been shown to be very effective. Ishijima *et al.* (1991) have increased the level of expression of rat phosphoribosylpyrophosphate synthetase subunits I and II by at least 2.5-fold. Beernink and Tolan (1992) have increased the level of rabbit muscle fructose-1,6-biphosphate aldolase by a factor of six to about 40% of total *E. coli* protein, and Huh *et al.* (1996) have enhanced the expression of human lipocortin (annexin-1) to 20% of total *E. coli* protein. Iwamoto *et al.* (1996) have obtained 10 mg/liter of purified dust mite allergen methionyl-Der f2 by similarly increasing the copy number of the expression vector.

The track record of the *trc* promoter in expression cases, the ease of induction under laboratory conditions, and the modular arrange-

Table 2
Examples of Efficient Protein Expression with Vectors Employing the *trc* Promoter

Protein expression	Protein/source organism	Reference
28 mg/liter (800 mg/liter in a fed-batch culture system)	Human T-cell receptor Vb5.3	Andrews *et al.* (1996)
6 mg/liter	Spinach acyl carrier protein I	Beremand *et al.* (1987)
10 mg/liter or 7.7% total cell protein	Diphtheria toxin/MSH fusion protein	Bishai *et al.* (1987)
[a]	Bovine rotavirus protein NCVP2	Bremont *et al.* (1987)
[a]	Human thioltransferase	Chrestensen *et al.* (1995)
[a] (soluble)	Rat protein disulfide isomerase	De Sutter *et al.* (1994)
20% soluble fraction	*Escherischia coli* threonine deaminase	Eisenstein (1991)
[a] (periplasmic expression)	Pallidipin (*Triatoma pallidipennis*) with leader peptide of alkaline phosphatase or the periplasmic form of cyclophilin	Haendler *et al.* (1995)
[a]	*Bacillus stearothermophilus* purine nucleoside phosphorylase I	Hamamoto *et al.* (1997)
1.5% total	Chicken prolactin	Hanks *et al.* (1989)
20% total	*Aspergillus nidulans* QUTE gene	Hawkins and Smith (1991)
[a]	Human apolipoprotein CII	Holtfreter and Stoffel (1988)
[a]	Horse elastase inhibitor	Kordula *et al.* (1993)
20 mg/liter soluble	Human factor XIII A-chain	Lai *et al.* (1994)
50% total or 55 mg/liter (pure)	*E. coli* uridine-diphosphate-*N*-acetylmuramate:L-alanine ligase	Liger *et al.* (1995)
4% soluble	*E. coli* deoR repressor	Mortensen *et al.* (1989)
10–15% total	Rat γ-D-crystallin	Ooki *et al.* (1994)
5.7% soluble	Bovine cardiac fatty acid-binding protein	Oudenampsen *et al.* (1990)
Several milligrams per liter (pure)	Human calmodulin-like protein	Rhyner *et al.* (1992)

Table 2
Continued

Protein expression	Protein/source organism	Reference
[a]	Bovine tracheobronchial phenol sulfotransferase	Schauss *et al.* (1995)
35–40%; total 14–21% soluble	Human interferon-γ variants	Slodowski *et al.* (1991)
[a]	Soybean glycinin precursor	Utsumi *et al.* (1988)
30% total	*A. nidulans* 3-dehydroquinate synthase	van den Hombergh *et al.* (1992)
5 mg/45 g cells	Soybean reductase	Welle and Schröder (1992)
[a]	*Staphylococcus epidermis* penicillin-binding protein 2′	Westendorp and Reynolds (1993)
23–70%; 200–800 mg/liter pure	*Salmonella typhimurium* tryptophan synthase	Yang *et al.* (1996)

[a]Designation of large quantities, high level, significant amount, abundant, preparative scale, or strong expression.

ment of the vectors employed have been the bases for modifications of the vectors. Features such as tags and fusion partners for easy purification, signal sequences for transport, and improved folding in cell compartments other than the cytoplasm have been introduced. A short description of examples of the *trc* promoter expression vector family follows (see also Table 1). All of the following vectors are based on the vector pBR322 and hence have a similar copy number in the host cell. Except for one example (pThioHis), the *rrn*B transcription terminators T_1 and T_2 (Brosius *et al.*, 1981; Brosius, 1984) are always located distal to the polylinker cloning sites. The benefit of transcriptional termination downstream from genes to be overexpressed had first been realized by Stüber and Bujard (1982). Terminators prevent read-through from the *trc* promoter into other plasmid genes or regulatory regions, and the secondary structures at the 3′ ends of mRNAs may enhance their stability. Because there were no convenient restriction sites available for its removal at the time, the *rrn*B 5S rRNA gene is still present proximal to the terminators (Brosius *et al.*, 1981). The presence of this small RNA gene does not seem to affect protein expression levels in a negative way.

pKK233-2

pKK233-2 is ampicillin resistant and employs the *lacZ* RBS. At a distance of 8 bp the ATG start codon is embedded in an *Nco*I restriction site (C/CATGG). A short polylinker containing unique *Pst*I and *Hin*dIII sites follows. When the *Nco*I site is filled in it exposes the sequence CCATG. Any blunt-ended gene or gene fragment can be fused to this blunt end. If in frame, the vector may enable direct expression of the desired protein. Initially the polypeptide will be translated as a Met derivative. In some cases, the N-terminal amino acid is removed from all or a fraction of the recombinant protein by the host cell (Shen *et al.*, 1993). In many eukaryotic genes the ATG start codon lies within an *Nco*I site.

pSE220

pSE220 is identical to pKK233-2, except that the short *Nco*I, *Pst*I, and *Hin*dIII polylinker has been replaced with the SL2 superlinker (Brosius, 1989, 1992) containing, among other sites, all 64 uninterrupted palindromic hexameric restriction sites. This leaves many choices to insert a foreign gene with a blunt-ended N terminus and a sticky-ended restriction enzyme recognition site downstream from the open reading frame.

pSE280

pSE280 is identical to pSE220, except that remnants of the truncated tetracycline resistance gene have been removed between the *Eco*RI and *Ava*I restriction sites. In the process, these sites were deactivated by a fill-in reaction. This manipulation reduced the size of the vector by about 1 kb.

pKK388-1

pKK388-1 is identical to pKK233-2, except that the *rrn*B antiterminator region follows the promoter operator (Brosius, 1988.) This element may prevent the RNA polymerase from pausing or terminating in areas of the inserted gene that feature prominent secondary structures. In addition, the truncated tetracycline resistance gene is restored and is transcribed by an operator-deficient *lac*UV5 promoter. Furthermore, the *lacZ* RBS has been exchanged with a synthetic optimized RBS (Ringquist *et al.*, 1992) where eight instead of four nucleotides can potentially interact with the 3′ end of 16S rRNA during the initiation of translation. Finally, the pUC19 polylinker (Yanisch-Perron *et al.*, 1985) featuring *Eco*RI, *Sac*I, *Kpn*I, *Sma*I, *Bam*HI, *Xba*I, *Sal*I, *Pst*I, *Sph*I, and *Hin*dIII sites has been placed downstream from the *Nco*I site.

pTrc99A,B,C

pTrc99A,B,C has been constructed by Amann *et al.* (1988) and its main difference in comparison to pKK233-2 is the replacement of the truncated tetracycline resistance gene by the *lacI* gene encoding the *lac* repressor. Whereas the use of all the aforementioned plasmids was restricted to *E. coli* strains that overexpress the *lac* repressor (*lac*I^{q} or *lac*I^{sq}), if repression of the promoter was necessary, the pTrc99 series allowed for the first time the use of any *E. coli* strain as a host with desired features, such as the lack of certain proteases to protect the recombinant protein from degradation (e.g., Gottesman, 1990). Distal to the *Nco*I site containing the ATG start codon, the polylinker from pUC19 (Yanisch-Perron *et al.*, 1985) has been inserted. The distance between the *Nco*I site and the first site of the pUC19 polylinker exists in three variants, pTrc99A,B,C, differing by 1 bp, thus allowing to continue an open reading frame between the ATG start codon and the inserted DNA fragment in any of the restriction sites of the pUC19 linker. It should be noted that the *lacI* gene on the pTrc99 series of plasmids was assumed to be the *lacI*^{q} allele with an up mutation in the -10 region of the *lacI* promoter, thus producing more transcripts. Sequence analysis, however, showed that the pTrc series contained the wild-type *lacI* gene (Adari *et al.*, 1995).

pSE380

pSE380 is identical to pTrc99A (Amann *et al.*, 1988) except that the pUC19 polylinker has been replaced by the SL2 superlinker (Brosius, 1989).

pSE420

With the exception of pKK388-1, all of the aforementioned vectors feature the sometimes suboptimal *lacZ* RBS for the initiation of translation. In pSE420, several features that potentially enhance the initiation of translation were introduced (Brosius, 1992). First, the RBS of pKK388-1 was introduced. It is flanked on both sides by A/T-rich sequences to reduce the formation of secondary structures with gene sequences downstream. Such structures may reduce the efficiency of translation initiation by occluding the RBS and/or the ATG start codon. Second, a mini cistron was introduced upstream (Schoner *et al.*, 1986). This feature places the initial translational start site at a greater distance from sequences and/or structures in the foreign gene that may negatively influence the translational start (see earlier discussion). In such cases, translation of the desired gene product can be improved when "jump started" at an upstream mini cistron followed by a translational restart when compared to a direct translational start at the desired open reading frame. Third, the translational enhancer from bacteriophage T7 gene 10 (Olins *et al.*, 1988) has been placed upstream from the mini cistron. Also, the *rrn*B antiterminator is located between the *trc* promoter/*lac* operator and the translational enhancer. The presence of the *lacI* gene allows regulation of the *trc* promoter in any *E. coli* strain.

The following vectors are derived from pSE420 and have features built into the open reading frame (next to the multiple cloning sites) that are useful for detection and single-step purification of the desired protein.

pTrcHisA,B,C Following the start codon, six codons for the amino acid histidine have been inserted. This N-terminal tag allows one-step purification of the polypeptides on a metal-chelating resin. For subsequent removal of the His_6 tag, an enterokinase cleavage site has been encoded prior to a polylinker that will accommodate the DNA fragment to be expressed. This polylinker is available in all three reading frames yielding pTrcHisA, B, and C.

pTrcHis2A,B,C This series of vectors accommodates tags, including His_6 at the 3′ end of the expressed polypeptide chain. The polylinker for insertion follows the *Nco*I site. If this site can be employed (e.g., via filling in the *Nco*I site blunt end ligation of the insert), the N terminus of the desired polypeptide can be expressed without additional sequences, with the possible exception of an additional methionine residue. The multiple cloning site is followed by codons for the decapeptide Glu–Gln–Lys–Leu–Ile–Ser–Glu–Glu–Asp–Leu, which is an epitope for an anti-*myc* antibody. This antibody can be used for immunocytochemical detection of the fusion protein. The *myc* epitope is followed by the His_6 tag. In order to ensure that the DNA fragment with the desired sequence leads in frame into the aforementioned C-terminal tags, the polylinker is available in three variants yielding pTrcHisA, B, and C.

pThioHisA,B,C The vectors pThioHisA,B,C are a variant of pTrcHisA,B,C. The desired polypeptide is expressed as a fusion protein featuring a modified thioredoxin moiety. This domain keeps the fusion protein soluble, eliminates the need for solubilization and refolding procedures (LaVallie *et al.*, 1993), and allows purification of the fusion protein on metal-chelating resins. The extra sequence can be removed from the desired polypeptide via an enterokinase cleaveage site: Asp–Asp–Asp–Asp–Lys. The codons for this site are followed by three variants of a multiple cloning site yielding pThioHisA, B, and C, ensuring in-frame translation of the desired polypeptide with the His-patch-thioredoxin fusion partner. In the vectors of the pThioHis series the *rrn*B transcriptional terminators are replaced by the *asp*A terminator.

Additional Hosts for *trc*-Driven Expression

There are also a number of examples where the *trc* promoter was shown to be functional in other gram-negative and gram-positive bacteria. Expression in *Vibrio cholerae* of the Shiga-like toxin I B subunit at 22 mg/liter has been achieved by Acheson *et al.* (1993). *Anabaena* PCC 7937 plastocyanin or *E. coli* β-glucuronidase was expressed efficiently in the cyanobacterium *Synechococcus* PCC7942 under the *trc* promoter (Geerts *et al.*, 1994, 1995). Furthermore, the *lasA* protease was expressed in *Pseudomonas aeroginosa* under the *trc* promoter. Rat glutathione *S*-transferase was expressed in *Salmonella typhimurum* NM5004 or TA1535 (Oda *et al.*, 1996; Thier *et al.*, 1995); β-galactosidase was also expressed in *S. typhimurum* (Ervin *et al.*, 1993). Furthermore, the *Bordella pertussis* P.69 antigen was expressed in *S. typhimurum* (Strugnell *et al.*, 1992; Velterop *et al.*, 1995). Finally, using a promoter-probe vector (pPL603), the *trc* promoter was also found to be active in *Bacillus subtilis* (Osburne and Craig, 1986).

Outlook for Further Improvements of Expression Vectors Employing the *trc* Promoter

As mentioned earlier, incomplete repression of the *trc* promoter is probably its main disadvantage in situations that initially demand a complete shut off of recombinant protein expression. (For a summary of advantages and disadvantages of the *trc* promoter system, the reader is referred to Table 3.) Such a need arises, for example, when the expressed foreign protein is toxic to the host strain, even at low concentrations. The rate of complex formation between the RNA polymerase and the *trc* promoter is very high, which may be one of the reasons why the *trc* promoter may be difficult to regulate with any of the tested repressor combinations (Lutz and Bujard, 1997). Promoters with a lower rate of complex formation, such as the bacteriophage λP_L and the *lac* promoter, have been shown to be better suited for achieving a high regulatory range (Lutz and Bujard, 1997). A number of other measures that custom improve (depending on the desired application) *trc* expression vectors were shown by Andrews

Table 3
Advantages and Disadvantages of the *trc* Promoter in Overexpression of Proteins in *E. coli*

Advantages	Disadvantages
Strong promoter	Promoter cannot be completely repressed; system is not recommended for expression of proteins that are toxic to the host
Well-defined and extensively tested system	For large-scale preparations, IPTG is expensive; for clinical uses, traces of IPTG may be toxic
Easy inducibility with IPTG	High temperature will induce heat shock, leading to degradation (effects can be ameliorated by using protease-deficient strains)
Using temperature-sensitive *lac* repressors, induction can be achieved by temperature increase	
Promoter often functions in hosts other than *E. coli*	

et al. (1996). (1) Replacement of the β-lactamase antibiotic resistance gene with the tetracycline resistance gene (as in the pK series described by Andrews *et al.*, 1996), as traces of the potentially allergenic penicillin derivatives should be avoided during bacterial growth of strains overexpressing therapeutic proteins. (2) The *lacI* gene carrying the wild-type promoter can be exchanged with alleles producing greater amounts of the *lac* repressor. This measure, as shown in pKBiq-Vβ5.3, improves regulation of the *trc* promoter but reduces the levels of recombinant protein after induction by a factor of five (Andrews *et al.*, 1996). (3) The *lacI* allele located on the expression vector can be exchanged with a temperature-sensitive allele, which eliminates the need to use IPTG, a substance that is toxic and expensive. The use of protease-deficient strains (Goff and Goldberg, 1987; Gottesman, 1990) with this thermoinducible system minimizes the proteolysis of recombinant proteins due to the heat shock response as demonstrated with vectors pKTB-Vβ5.3 or pKTBi-Vβ5.3 (Andrews *et al.*, 1996). (4) Signal sequences such as the *E. coli* heat-stable enterotoxin II (STII) signal sequence for protein secretion (as in pK-Vβ5.3, pKT-Vβ5.3, pKB-Vβ5.3, or pKTB-Vβ5.3; Andrews *et al.*, 1996) are a further option in improving the yields of a correctly folded protein in expression systems. As with all systems, examination of a variety of host strains may be beneficial. (5) The future will bring further sophistication of fusion partners for convenient and economic one-step purification schemes. In addition, we may see improvements in the efficiency of cleavage between the fusion partner and the desired polypeptide. All these existing and anticipated improvements could be integrated into one of the aforementioned *trc* expression vectors (see Table 2) to offer the researcher a wide variety of vector variants for the most economic and efficient expression of a desired recombinant protein.

Acknowledgments

I thank Dr. S. Makrides for useful comments on an earlier version and Dr. B. Flott-Rahmel for help preparing the manuscript.

References

Acheson, D. W., Calderwood, S. B., Boyko, S. A., Lincicome, L. L., Kane, A. V., Donohue-Rolfe, A., and Keusch, G. T. (1993). Comparison of Shiga-like toxin I

B-subunit expression and localization in *Escherichia coli* and *Vibrio cholerae* by using *trc* or iron-regulated promoter systems. *Infect. Immun.* **61**, 1098–1104.

Adari, H., Andrews, B., Ford, P. J., Hannig, G., Brosius, J., and Makrides, S. C. (1995). Expression of the human T-cell receptor V β 5.3 in *Escherichia coli* by thermal induction of the *trc* promoter: Nucleotide sequence of the *lacIts* gene. DNA. *Cell. Biol.* **14**, 945–950.

Amann, E., and Brosius, J. (1985). "ATG vectors" for regulated high-level expression of cloned genes in *Escherichia coli. Gene* **40**, 183–190.

Amann, E., Ochs, B., and Abel, K. J. (1988). Tightly regulated *tac* promoter vectors useful for the expression of unfused and fused proteins in *Escherichia coli. Gene* **69**, 301–315.

Andrews, B., Adari, H., Hannig, G., Lahue, E., Gosselin, M., Martin, S., Ahmed, A., Ford, P. J., Hayman, E. G., and Makrides, S. C. (1996). A tightly regulated high level expression vector that utilises a thermosensitive *lac* repressor: Production of the human T cell receptor V β 5.3 in *Escherichia coli. Gene* **182**, 101–109.

Beernink, P. T., and Tolan, D. R. (1992). Construction of a high-copy 'ATG vector' for expression in *Escherichia coli. Protein Express. Purif.* **3**, 332–336.

Beremand, P. D., Hannapel, D. J., Guerra, D. J., Kuhn, D. N., and Ohlrogge, J. B. (1987). Synthesis, cloning, and expression in *Escherichia coli* of a spinach acyl carrier protein-I gene. *Arch. Biochem. Biophys.* **256**, 90–100.

Bingham, A. H. (1991). Low copy number vectors for expression of fused genes to β-galactosidase in *Escherichia coli. FEMS Microbiol. Lett.* **63**, 239–246.

Bishai, W. R., Rappuoli, R., and Murphy, J. R. (1987). High-level expression of a proteolytically sensitive diphtheria toxin fragment in *Escherichia coli. J. Bacteriol.* **169**, 5140–5151.

Bremont, M., Charpilienne, A., Chabanne, D., and Cohen, J. (1987). Nucleotide sequence and expression in *Escherichia coli* of the gene encoding the nonstructural protein NCVP2 of bovine rotavirus. *Virology* **161**, 138–144.

Brosius, J. (1984). Plasmid vectors for the selection of promoters. *Gene* **27**, 151–160.

Brosius, J. (1988). Expression vectors employing lambda-, *trp-*, *lac-*, and *lpp-* derived promoters. *Bio/Technology* **10**, 205–225.

Brosius, J. (1989). Superpolylinkers in cloning and expression vectors. *DNA* **8**, 759–777.

Brosius, J. (1992). Compilation of superlinker vectors. *In* "Methods in Enzymology" (R. Wu, ed.), Vol. 216, pp. 469–483. Academic Press, San Diego, CA.

Brosius, J., Ullrich, A., Raker, M. A., Gray, A., Dull, T. J., Gutell, R. R., and Noller, H. F. (1981). Construction and fine mapping of recombinant plasmids containing the *rrn*B ribosomal RNA operon of *E. coli. Plasmid* **6**, 112–118.

Brosius, J., Erfle, M., and Storella, J. (1985). Spacing of the -10 and -35 regions in the *tac* promoter. Effect on its *in vivo* activity. *J. Biol. Chem.* **260**, 3539–3541.

Bujard, H., Gentz, R., Lanzer, M., Stueber, D., Mueller, M., Ibrahimi, I., Haeuptle, M. T., and Dobberstein, B. (1987). A T5 promoter-based transcription-translation system for the analysis of proteins *in vitro* and *in vivo*. *In* "Methods Enzymology" (R. Wu, ed.), Vol. 155, 416–433. Academic Press, Orlando, FL.

Bukrinsky, M. I., Barsov, E. V., and Shilov, A. A. (1988). Multicopy expression vector based on temperature-regulated *lac* repressor: Expression of human immunodeficiency virus *env* gene in *Escherichia coli. Gene* **70**, 415–417.

Cabilly, S. (1989). Growth at sub-optimal temperatures allows the production of functional, antigen-binding Fab fragments in *Escherichia coli. Gene* **85**, 553–557.

Chrestensen, C. A., Eckman, C. B., Starke, D. W., and Mieyal, J. J. (1995). Cloning, expression and characterization of human thioltransferase (glutaredoxin) in *E. coli*. *FEBS Lett.* **374**, 25–28.

de Boer, H. A., Comstock, L. J., Yansura, D. G., and Heynecker, H. L. (1982). Construction of a tandem *trp-lac* promoter and a hybrid *trp-lac* promoter for efficient and controlled expression of the human growth hormone gene in *Escherischia coli*. pp. 462–481 (R. L. Rodriguez and M. J. Chamberlin, eds.). Praeger, New York.

de Boer, H. A., Comstock, L. J., and Vasser, M. (1983). The *tac* promoter: A functional hybrid derived from the *trp* and *lac* promoters. *Proc. Natl. Acad. Sci. U.S.A.* **80**, 21–25.

De Sutter, K., Hostens, K., Vandekerckhove, J., and Fiers, W. (1994). Production of enzymatically active rat protein disulfide isomerase in *Escherichia coli*. *Gene* **141**, 163–170.

Deuschle, U., Kammerer, W., Gentz, R., and Bujard, H. (1986). Promoters of Escherichia coli: A hierarchy of *in vivo* strength indicates alternate structures. *EMBO J.* **5**, 2987–2994.

Eisenstein, E. (1991). Cloning, expression, purification, and characterization of biosynthetic threonine deaminase from *Escherichia coli*. *J. Biol. Chem.* **266**, 5801–5807.

Ervin, S. E., Small, P. A., Jr., and Gulig, P. A. (1993). Use of incompatible plasmids to control expression of antigen by *Salmonella typhimurium* and analysis of immunogenicity in mice. *Microb. Pathog.* **15**, 93–101.

Ferraz, C., Sri Widada, J., and Liautard, J. P. (1991). Purification and characterization of recombinant tropomyosins. *J. Chromatogr.* **539**, 465–473.

Geerts, D., Schubert, H., de Vrieze, G., Borrias, M., Matthijs, H. C., and Weisbeek, P. J. (1994). Expression of *Anabaena* PCC 7937 plastocyanin in *Synechococcus* PCC 7942 enhances photosynthetic electron transfer and alters the electron distribution between photosystem I and cytochrome-c oxidase. *J. Biol. Chem.* **269**, 28068-28075.

Geerts, D., Bovy, A., de Vrieze, G., Borrias, M., and Weisbeek, P. (1995). Inducible expression of heterologous genes targeted to a chromosomal platform in the cyanobacterium *Synechococcus* sp. PCC 7942. *Microbiology* **141**, 831–841.

Giordano, T. J., Deuschle, U., Bujard, H., and McAllister, W. T. (1989). Regulation of coliphage T3 and T7 RNA polymerases by the *lac* repressor-operator system. *Gene* **84**, 209–219.

Goff, S. A., and Goldberg, A. L. (1987). An increased content of protease La, the *lon* gene product, increases protein degradation and blocks growth in *Escherichia coli*. *J. Biol. Chem.* **262**, 4508–4515.

Goff, S. A., Casson, L. P., and Goldberg, A. L. (1984). Heat shock regulatory gene *htp*R influences rates of protein degradation and expression of the *lon* gene in *Escherichia coli*. *Proc. Natl. Acad. Sci. U.S.A.* **81**, 6647–6651.

Gottesman, S. (1990). Minimizing proteolysis in *Escherichia coli:* Genetic solutions. *In* "Methods in Enzymology" (D. V. Goeddel, ed.), Vol. 185, pp. 119–129. Academic Press, San Diego, CA.

Haendler, B., Becker, A., Noeske-Jungblut, C., Kratzschmar, J., Donner, P., and Schleuning, W. D. (1995). Expression of active recombinant pallidipin, a novel platelet aggregation inhibitor, in the periplasm of *Escherichia coli*. *Biochem. J.* **307**, 465–470.

Hamamoto, T., Okuyama, K., Noguchi, T., and Midorikawa, Y. (1997). Molecular cloning and expression of purine nucleoside phosphorylase I gene from Bacillus stearothermophilus Th 6-2. *Biosci. Biotechnol. Biochem.* **61**, 272–275.

Hanks, M. C., Talbot, R. T., and Sang, H. M. (1989). Expression of biologically active recombinant-derived chicken prolactin in *Escherichia coli. J. Mol. Endocrinol.* **3**, 15–21.

Hasan, N., and Szybalski, W. (1995). Construction of *lacIts* and *lacIqts* expression plasmids and evaluation of the thermosensitive *lac* repressor. *Gene* **163**, 35–40.

Hawkins, A. R., and Smith, M. (1991). Domain structure and interaction within the pentafunctional AROM polypeptide. *Eur. J. Biochem.* **196**, 717–724.

Holtfreter, C., and Stoffel, W. (1988). Expression of normal and mutagenized apolipoprotein CII in procaryotic cells. Structure-function relationship. *Biol. Chem. Hoppe-Seyler* **369**, 1045–1054.

Huh, K. R., Cho, E. H., Lee, S. O., and Na, D. S. (1996). High level expression of human lipocortin (annexin) 1 in *Escherichia coli. Biotechnol. Lett.* **18**, 163–168.

Ishijima, S., Kita, K., Ahmad, I., Ishizuka, T., Taira, M., and Tatibana, M. (1991). Expression of rat phosphoribosylpyrophosphate synthetase subunits I and II in *Escherichia coli.* Isolation and characterization of the recombinant isoforms. *J. Biol. Chem.* **266**, 15693–15697.

Iwamoto, N., Nishiyama, C., Yasuhara, T., Saito, A., Yuuki, T., Okumura, Y., and Okudaira, H. (1996). Direct expression of Der f2, a major house dust mite allergen, in *Escherichia coli. Int. Arch. Allergy. Immunol.* **109**, 356–361.

Kammerer, W., Deuschle, U., Gentz, R., and Bujard, H. (1986). Functional dissection of *Escherichia coli* promoters: Information in the transcribed region is involved in late steps of the overall process. *EMBO J.* **5**, 2995–3000.

Kordula, T., Dubin, A., Schooltink, H., Koj, A., Heinrich, P. C., and Rose-John, S. (1993). Molecular cloning and expression of an intracellular serpin: An elastase inhibitor from horse leucocytes. *Biochem. J.* **293**, 187–193.

Kozak, M. (1987). An analysis of 5′-noncoding sequences from 699 vertebrate messenger RNAs. *Nucleic Acids Res.* **15**, 8125–8148.

Lai, T. S., Santiago, M. A., Achyuthan, K. E., and Greenberg, C. S. (1994). Purification and characterization of recombinant human coagulant factor XIII A-chains expressed in *E. Coli. Protein Express. Purif.* **5**, 125–132.

Lanzer, M., and Bujard, H. (1988). Promoters largely determine the efficiency of repressor action. *Proc. Natl. Acad. Sci. U.S.A.* **85**, 8973–8977.

LaVallie, E. R., DiBlasio, E. A., Kovacic, S., Grant, K. L., Schendel, P. F., and McCoy, J. M. (1993). A thioredoxin gene fusion expression system that circumvents inclusion body formation in the *E. coli* cytoplasm. *Bio/Technology* **11**, 187–193.

Liger, D., Masson, A., Blanot, D., van Heijenoort, J., and Parquet, C. (1995). Overproduction, purification and properties of the uridine-diphosphate-N-acetylmuramate:L-alanine ligase from *Escherichia coli. Eur. J. Biochem.* **230**, 80–87.

Lutz, R., and Bujard, H. (1997). Independent and tight regulation of transcriptional units in *Echerischia coli via* the LacR/O, the TetR/O and AraC/I_1-I_2 regulatory elements. *Nucleic Acids Res.* **25**, 1203–1210.

Makrides, S. C. (1996). Strategies for achieving high-level expression of genes in *Escherichia coli. Microbiol. Rev.* **60**, 512–538.

Mortensen, L., Dandanell, G., and Hammer, K. (1989). Purification and characterization of the *deo*R repressor of *Escherichia coli. EMBO J.* **8**, 325–331.

Mulligan, M. E., Brosius, J., and McClure, W. R. (1985). Characterization *in vitro* of the effect of spacer length on the activity of *Escherichia coli* RNA polymerase at the TAC promoter. *J. Biol. Chem.* **260**, 3529–3538.

Oda, Y., Yamazaki, H., Thier, R., Ketterer, B., Guengerich, F. P., and Shimada, T. (1996). A new *Salmonella typhimurium* NM5004 strain expressing rat glutathione S-transferase 5-5: Use in detection of genotoxicity of dihaloalkanes using an SOS/umu test system. *Carcinogenesis (London)* **17**, 297–302.

Olins, P. O., Devine, C. S., Rangwala, S. H., and Kavka, K. S. (1988). The T7 phage gene *10* leader RNA, a ribosome-binding site that dramatically enhances the expression of foreign genes in *Escherichia coli. Gene* **73**, 227–235.

Ooki, K., Amuro, N., Shimizu, Y., and Okazaki, T. (1994). High level expression of rat γ-D-crystallin in *Escherichia coli. Biochimie* **76**, 398–403.

Osburne, M. S., and Craig, R. J. (1986). Activity of two strong promoters cloned into *Bacillus subtilis. J. Gen. Microbiol.* **132**, 565–568.

Oudenampsen, E., Kupsch, E. M., Wissel, T., Spener, F., and Lezius, A. (1990). Expression of fatty acid-binding protein from bovine heart in *Escherichia coli. Mol. Cell. Biochem.* **98**, 75–79.

Rhyner, J. A., Koller, M., Durussel-Gerber, I., Cox, J. A., and Strehler, E. E. (1992). Characterization of the human calmodulin-like protein expressed in *Escherichia coli. Biochemistry* **31**, 12826–12832.

Ringquist, S., Shinedling, S., Barrick, D., Green, L., Binkley, J., Stormo, G. D., and Gold, L. (1992). Translation initiation in *Escherichia coli:* Sequences within the ribosome-binding site. *Mol. Microbiol.* **6**, 1219–1229.

Schauss, S. J., Henry, T., Palmatier, R., Halvorson, L., Dannenbring, R., and Beckmann, J. D. (1995). Characterization of bovine tracheobronchial phenol sulphotransferase cDNA and detection of mRNA regulation by cortisol. *Biochem. J.* **1**, 209–217.

Schein, C. H. (1989). Production of soluble recombinant proteins in bacteria. *Bio/Technology* **7**, 1141–1149.

Schoner, B. E., Belagaje, R. M., and Schoner, R. G. (1986). Translation of a synthetic two-cistron mRNA in *Escherichia coli. Proc. Natl. Acad. Sci. U.S.A.* **83**, 8506–8510.

Shatzman, A. R., and Rosenberg, M. (1987). Expression, identification, and characterization of recombinant gene products in *Escherichia coli. In* "Methods in Enzymology" (S. L. Berger and A. R. Kimmel, eds.), Vol. 152, pp. 661–673. Academic Press, Orlando, FL.

Shen, T.-J, Ho, N. T., Simplaceanu, V., Zou, M., Green, B. N., Tam, M. F. and Ho, C. (1993). Production of unmodified human adult hemoglobin in *Escherichia coli. Proc. Natl. Acad. Sci. U.S.A.* **90**, 8108–8112.

Sim, B. J., Tan, D. S. H., Liu, X. A., and Sim, T. S. (1996). Production of high levels of soluble recombinant *Streptomyces clavuligerus* isopenicillin n synthase in *Escherichia coli. J. Mol. Catal., B: Enzymatic* **2**, 71–83.

Simons, A., Tils, D., von-Wilcken-Bergmann, B., and Müller-Hill, B. (1984). Possible ideal *lac* operator: *Escherichia coli lac* operator-like sequences from eukaryotic genomes lack the central G: C pair. *Proc. Natl. Acad. Sci. U.S.A.* **81**, 1624–1628.

Slodowski, O., Böhm, J., Schöne, B., and Otto, B. (1991). Carboxy-terminal truncated rhuIFN-γ with a substitution of Gln133 or Ser132 to leucine leads to higher biological activity than in the wild type. *Eur. J. Biochem.* **202**, 1133–1140.

Strugnell, R., Dougan, G., Chatfield, S., Charles, I., Fairweather, N., Tite, J., Li, J. L., Beesley, J., and Roberts, M. (1992). Characterization of a *Salmonella typhimurium* aro vaccine strain expressing the P.69 antigen of *Bordetella pertussis. Infect. Immun.* **60**, 3994–4002.

Studier, F. W., Rosenberg, A. H., Dunn, J. J., and Dubendorf, J. W. (1990). Use of the

T7 RNA polymerase to direct expression of cloned genes. *In* "Methods in Enzymology" (D. V. Goeddel, ed.), Vol. 185, pp. 60–89. Academic Press, San Diego, CA.

Stüber, D., and Bujard, H. (1982). Transcription from efficient promoters can interfere with plasmid replication and diminish expression of plasmid specified genes. *EMBO J.* **1**, 1399–1404.

Thier, R., Müller, M., Taylor, J. B., Pemble, S. E., Ketterer, B., and Guengerich, F. P. (1995). Enhancement of bacterial mutagenicity of bifunctional alkylating agents by expression of mammalian glutathione S-transferase. *Chem. Res. Toxicol.* **8**, 465–472.

Totsuka, A., and Fukazawa, C. (1993). Expression and mutation of soybean β-amylase in *Escherichia coli. Eur. J. Biochem.* **214**, 787–94.

Utsumi, S., Kim, C. S., Sato, T., and Kito, M. (1988). Signal sequence of preproglycinin affects production of the expressed protein in *Escherichia coli. Gene* **71**, 349–358.

van den Hombergh, J. P., Moore, J. D., Charles, I. G., and Hawkins, A. R. (1992). Overproduction in *Escherichia coli* of the dehydroquinate synthase domain of the *Aspergillus nidulans* pentafunctional AROM protein. *Biochem. J.* **284**, 861–867.

Velterop, J. S., Dijkhuizen, M. A., van't Hof, R., and Postma, P. W. (1995). A versatile vector for controlled expression of genes in *Escherichia coli* and *Salmonella typhimurium. Gene* **153**, 63–65.

Ward, G. A., Stover, C. K., Moss, B., and Fuerst, T. R. (1995). Stringest chemical and thermal regulation of recombinant gene expression by vaccinia virus vectors in mammalian cells. *Proc. Natl. Acad. Sci. U.S.A.* **92**, 6773–6777.

Welle, R., and Schröder, J. (1992). Expression cloning in *Escherichia coli* and preparative isolation of the reductase coacting with chalcone synthase during the key step in the biosynthesis of soybean phytoalexins. *Arch. Biochem. Biophys.* **293**, 377–381.

Westendorp, M. O., and Reynolds, P. E. (1993). Construction and overexpression in *Escherichia coli* of genetically engineered derivatives of penicillin-binding protein 2′ of *Staphylococcus epidermidis. FEMS Microbiol. Lett.* **112**, 87–91.

Williams, D. P., Regier, D., Akiyoshi, D., Genbauffe, F., and Murphy, J. R. (1988). Design, synthesis and expression of a human interleukin-2 gene incorporating the codon usage bias found in highly expressed *Escherichia coli* genes. *Nucleic Acids Res.* **16**, 10453–10467.

Yabuta, M., Onai-Miura, S., and Ohsuye, K. (1995). Thermo-inducible expression of a recombinant fusion protein by *Escherichia coli lac* repressor mutants. *J. Biotechnol.* **39**, 67–73.

Yang, L., Ahmed, S. A., and Miles, E. W. (1996). PCR mutagenesis and overexpression of tryptophan synthase from *Salmonella typhimurium:* On the roles of β 2 subunit Lys-382. *Protein. Express. Purif.* **8**, 126–136.

Yanisch-Perron, C., Vieira, J., and Messing, J. (1985). Improved M13 phage cloning vectors and host strains: Nucleotide sequences of the M13mp18 and pUC19 vectors. *Gene* **33**, 103–119.

3

Bacillus EXPRESSION: A GRAM-POSITIVE MODEL

Eugenio Ferrari and Brian Miller
Genencor International, Inc.
Palo Alto, California 94304

Gene Expression Systems: Using Nature for the Art of Expression

Introduction

The presence of enzymes useful in large volume industrial applications has made *Bacillus* a very important manufacturing organism. Proteases used as additives in the laundry detergent formulation are alkaline proteases of *Bacillus* origin. In addition, most of the amylases used in the conversion of corn starch to high fructose corn syrup are derived from *Bacillus*. Overall, *Bacillus* enzymes represent approximately 60% of the more than $1 billion industrial enzyme market. This is due not only to the fact that *Bacillus* enzymes show good performance in the target application and to the relative ease at which *Bacillus* can be cultivated at very high density in large-scale fermentor using inexpensive carbon and nitrogen sources, but also because it secretes such enzymes into the culture medium. Secretion of these enzymes represent a very important aspect of the economics associated with these processes. The isolation of such enzymes from the supernatant allows reduced costs. In additon to being efficient, industrial enzymes have to be inexpensive. Typically the amylases used in the corn starch processing cost less than $200 dollars a kilogram of enzyme.

Over the years, *Bacillus licheniformis*, for its proteases and amylases, and *B. amyloliquefaciens* for its amylases, have been developed for enzyme production. Via a reiteration of mutagenesis and screening, it has been possible to obtain mutants capable of expressing and secreting several grams per liter of their natural enzymes in a fermentation process that lasts approximately 3 to 4 days.

Concomitant with the development of *B. licheniformis* and *B. amyloliquefaciens* as organisms for the production of industrial enzymes, *B. subtilis* genetics was being developed. *B. subtilis* has become the best characterized gram-positive microorganism and, after *Escherichia coli*, is the best characterized microbe. A number of tools are available that make *Bacillus* very easy to genetically manipulate. Moreover, the recent completion of the sequencing of its genome (Kunst *et al*., 1997) will certainly stimulate molecular genetics and physiology studies.

A number of other Bacilli, e.g., *B. brevis*, *B. megaterium*, and *B. alkalophilus*, have been proposed as alternate hosts for enzyme production.

This chapter gives an overview of the results generated in the area of gene expression in *Bacillus*. However, because of the abundance of data on this subject, this chapter will limit itself, in most cases, to briefly mention several of the aspects inherent with the molecular genetic work associated with gene expression. The interested reader is referred, throughout the text, to a number of scientific articles and reviews for a more in-depth reading.

Bacilli Used as Industrial Production Organisms

Some members of the *Bacillus* genus are potentially very attractive hosts for gene expression and secretion. In addition to their ability to easily grow at very high density using inexpensive carbon and nitrogen sources and being capable of secreting large amounts of enzymes, there is a great deal of information on their transcriptional and translational regulation and they appear to have no major codon bias. Furthermore, because of their lack of pathogenicity and long and safe use as a production organism, *Bacillus* enzyme preparations have been allowed a GRAS status (Generally Recognized As Safe) by the Food and Drug Administration of the United States.

Because of sensitivity issues concerning the pricing associated with industrial enzymes, producers are reluctant to reveal the productivity of the industrial strains at the fermentation plant. However, the price at which some of these enzymes are available on the market safely assumes that, in some instances, the final fermentation yields can be as high as 10–20 g/liter of enzymes. The variability in the yield may depend on the type of enzyme produced, the ease of secretion, the fermentation protocol followed, and the length of the fermentation time, which ranges from 50 to 100 hr.

Although the information on the yields at the fermentor level is scant, there is a wealth of information on the expression of gene products at the bench level. Table 1 shows some of the representative yields reported in the scientific literature, both for homologous as well as heterologous gene products. One must be very careful in interpreting these data. A number of issues, which include ease of secretion, presence of proteases in the medium, promoter used for the expression, and optical density of the culture, can dramatically affect the final yield and stability of a particular gene product.

Table 1
Examples of Expression of Heterologous and Homologous Proteins in Bacilli

	Origin	Yield	*Bacillus* sp.	References
Eukaryotic genes				
Epidermal growth factor	Human	3 g/liter	*B. brevis*	Udaka and Yamagata (1993a)
Growth hormone	Human	0.2 g/liter	*B. brevis*	Udaka and Yamagata (1993a)
Growth hormone	Human	0.2 g/liter	*B. subtilis*	Honjo *et al.* (1987)
Growth hormone	Tuna	0.24 g/liter	*B. brevis*	Udaka and Yamagata (1993a)
IL–3	Human	0.10 g/liter	*B. licheniformis*	van Leen *et al.* (1991)
Pepsinogen	Swine	0.01 g/liter	*B. brevis*	Takao *et al.* (1989)
Salivary α-amylase	Human	0.06 g/liter	*B. brevis*	Udaka and Yamagata (1993a)
Isomerase (disulfide)	*Humicola insolens*	0.3 g/liter	*B. brevis*	Udaka and Yamagata (1993a)
Taka-amylase	*Aspergillus orizae*	0.02 g/liter	*B. brevis*	Udaka and Yamagata (1993a)
Bacterial genes				
α-amylase	*B. licheniformis*	3.5 g/liter	*B. brevis*	Udaka and Yamagata (1993)
α-amylase	*B. licheniformis*	1.0 g/liter	*B. subtilis*	Sloma *et al.* (1988)
β-lactamase	*B. licheniformis*	140m g/liter	*B. subtilis*	Yoshimura *et al.* (1986)
Cholera toxin B subunit	*Vibrio cholerae*	1.4 g/liter	*B. brevis*	Udaka and Yamagata (1993a)
Pertussis toxin S1	*B. pertussis*	100m g/liter	*B. subtilis*	Saris *et al.* (1990)
Pertussis toxin S4	*B. pertussis*	0.5m g/liter	*B. subtilis*	Saris *et al.* (1990)
Protein A	*S. aureus*	1 g/liter	*B. subtilis*	Fahnestock and Fisher (1986)
Subtilisin	*B. amyloliquefaciens*	60 mg/liter	*B. subtilis*	Ferrari *et al.* (1993)

Bacillus subtilis

B. subtilis is the best-studied member of the genus *Bacillus*. Its genetics and physiology have been the subject of a large number of publications and, because of its close relatedness to other members of the genus, e.g., *B. licheniformis* and *B. amyloliquefaciens*, it can be considered as a paradigm for industrial bacilli. The completion of the sequence of its genome (Kunst *et al.*, 1997) will presumably expedite its genetic characterization. It should be possible to quickly identify genes that are turned on or off in particular growth conditions, which in turn will also allow researchers to build production strains, both for enzymes as well as for chemicals, with improved performance. *B. subtilis* strains capable of producing riboflavin (Perkins and Pero, 1993; Sauer *et al.*, 1996) and folic acid (Eichler *et al.*, 1997) at a level high enough to be competitive with chemical industrial processes have been described.

Transformation via competent cells was the first procedure developed (Anagnostopoulos and Spizizen 1961) to simplify genetic manipulation and mapping studies in *Bacillus*. It allows one to introduce any desired DNA fragment into this microbe, either chromosomal or of plasmid origin, and to easily manipulate the chromosome. Because of this technique it is possible to readily introduce mutations in practically any gene and study the effect of such mutations on the growth and productivity of the host. Other techniques available for cloning and/or genetic studies are protoplast transformation and phage-mediated transduction.

Bacillus licheniformis

B. licheniformis has a long history as a host for the production of industrial products (de Boer *et al.*, 1994). Although considered relatively closely related to *B. subtilis* (Priest, 1993), *B. licheniformis* is a true facultative anaerobe capable of a type of mixed acid fermentation (Shariati *et al.*, 1995). The level of DNA homology between *B. licheniformis* and *B. subtilis* has been estimated at 10–15% (Priest, 1993). The report of mixed acid–butanediol fermentation by *B. subtilis* (Nakano *et al.*, 1997) may reveal more about the similarities between *B. subtilis* and *B. licheniformis*.

Compared to *B. subtilis*, the amount of genetic map data for *B. licheniformis* is limited and a circular genetic map has not been demonstrated. However, enough information exists to allow general comparisons of *B. licheniformis* with other bacilli. This analysis reveals a rough similarity in the position of many loci to regions of the chromosome. Since the initial mapping work of the Thorne lab, little further work in this area has been reported (Kowalski and Thorne, 1979).

B. licheniformis possesses two indigenous enzymes, α-amylase and an alkaline protease (Carlsberg), which are valuable industrial products (Priest, 1989). Along with use as a host for its own enzymes, *B. licheniformis* has been used to produce heterologous proteins from other types of bacteria and eucaryotes (Simonen and Palva, 1993).

Although not utilized as extensively as a research organism like *B. subtilis, B. licheniformis* has continued to be studied in relation to its industrial potential.

Other Bacilli

A few other members of the *Bacillus* species have been used in the production of industrial enzymes (e.g., *B. amyloliquefaciens*) or have the potential of becoming industrial workhorses.

B. amyloliquefaciens has been used in several decades in the industrial production of α-amylase. Its similarity to *B. subtilis* is so high that it has been identified as a different species only as recently as 1967 (Welker and Campbell, 1967). *B. amyloliquefaciens* is very difficult to transform via competent cells and it has very active DNA restriction/modification systems. Because of these problems, its genetic characterization has lagged behind the other two closely related Bacilli and is not widely used for the production of recombinant DNA products.

B. megaterium is also an interesting member of the *Bacillus* species for gene expression. It is relatively well characterized (Vary, 1994), an efficient transformation system is available, and it offers the advantage over *B. subtilis* and *B. licheniformis* that replicative plasmids appear to be stably maintained. Furthermore, *B. megaterium* appears to lack alkaline proteases and generally the level of proteases expressed and secreted in the medium is low (Vary, 1994). The low level of proteases makes it a very attractive organism for the expression of heterologous proteins. *B. megaterium* has been used for some time in the production of certain industrial enzymes.

As a production organism, *B. brevis* (Udaka *et al.*, 1989; Udaka and Yamagata, 1993a) has been studied and developed primarily in

Japan. Literature reports claim that this *Bacillus* can efficiently secrete eukaryotic as well as prokaryotic proteins (Udaka and Yamagata, 1993a,b) (see Table 1). Similar to *B. megaterium*, *B brevis* also seems to be deprived of major extracellular proteolytic activity and therefore the product secreted in the media appears to be stable and not subject to degradation. Other than the natural low level of proteases, the export mechanism used for the production of recombinant proteins in this microbe is different than the one used by the other bacilli we are familiar with. *B. brevis* releases into the media primarily two proteins that are components of the outer wall protein (OWP) and the middle wall protein (MWP). These genes are transcribed as an operon, *cwp*, by multiple tandem promoters located upstream of the *mwp* gene and their gene products are synthesized during logarithmic growth. In the early stationary phase the protein layer starts to shed, releasing increased amounts of proteins in the medium. During the stationary phase the cells continue to synthesize and shed the cell wall proteins that accumulate in the medium as extracellular proteins at levels reaching up to 30 g/liter. The target gene product to be expressed can be fused, for example, to the *mwp* signal peptide under the control of the *cwp* promoter. This allows the target gene to be transcribed at the same high level as the wall proteins and to be translocated outside by a similar "shedding" mechanism. This represents a very interesting feature peculiar to *B. brevis* and is completely different, as far as we know, from the secretion mechanism in other Bacilli. This may also explain the relatively high level of eukaryotic proteins sometime obtained.

Like many other bacilli, *B. brevis* is a harmless microbe and is commonly found in soil, milk, and cheese, therefore, from the regulatory point of view, it should be considered safe.

Genetic Manipulation in Bacilli

Transformation via Competent Cells

During the transition from vegetative cells to sporulation, the great majority of the laboratory strains of *B. subtilis* can uptake DNA when part of the population reaches a metabolic state called competence (Dubnau, 1993). The fraction of the population becoming competent can be optimized using one of several simple and well-established procedures (Ferrari and Hoch, 1989; Cutting and Van der Horn, 1990). In optimal conditions, approximately 10–20% of the population reaches competence (Somma and Polsinelli, 1970). During this state, several high molecular weight chromosomal DNA fragments are bound very efficiently by competent cells and, after being clipped to an approximate size of 30 kb (Dubnau and Cirigliano, 1972), are transported inside the cell as single-stranded DNA. If the degree of homology is high, the incoming DNA can replace the homologous region of the host chromosome. Because or the relatively small size of DNA being taken up, transformation via competence is very useful for the fine mapping of chromosomal mutations. Furthermore, because several molecules are transported inside the cell, it is possible to introduce two unlinked chromosomal markers into the same strain, which can sometimes facilitate strain constructions.

Although the characteristics of competent cell transformation, i.e., the DNA taken up being small in size and single stranded, are extremely useful for mapping studies and chromosomal manipulation, they hamper the transformability of Bacilli with ligation mixtures. Because of the paring of the DNA at the binding stage, monomeric plasmids, as well as ligation mixtures, cannot be used to successfully transform *B. subtilis*-competent cells (Canosi *et al.*, 1978). Only plasmid multimers, either present in the preparation from most *E. coli* strains or generated *in vitro* via ligation (Mottes *et al.*, 1979) or via polymerase chain reaction (Shafikani *et al.*, 1997), can efficiently transform *Bacillus*. This is a major obstacle to the preparation of random plasmid libraries directly in this organism. One must first prepare the libraries in *E. coli* and then transform *Bacillus* with such libraries. One of the major drawbacks to this approach is the possible toxicity to *E. coli* often exhibited by genes from bacilli when present in a multicopy vector. In this case the problem can be overcome either via protoplast transformation or by taking advantage of plasmids maintained in *E. coli* at very low copy.

Similar to *B. subtilis*, laboratory strains of *B. licheniformis* can also be efficiently transformed via the competent cells procedures (Prestidge and Spizizen, 1969; McCuen and Thorne, 1971). The efficiency of transformation is comparable to the one obtained in *B. subtilis* and is presumably carried out by the cells in a similar manner. *B. licheniformis* is also poorly transformable by plasmid

DNA, and all the techniques and strategies described earlier can be applied to its manipulation.

Protoplast Transformation of *Bacillus*

Transformation of *Bacillus* via protoplasts (for a review, see Bron, 1990) is a vital methodology in situations were the Bacillus strains of interest cannot be transformed by natural competence methods. In brief, protoplast transformation involves producing a population of protoplasts by the enzymatic digestion of the *Bacillus* cell wall in the presence of an osmotic stabilizing medium. The DNA of interest is introduced into the cell in the presence of polyethylene glycol. The protoplasts are incubated in a osmotically stable medium until regeneration of the cell wall.

Although much work with the standard *B. subtilis* derivatives has been performed using natural competence, there are instances where protoplast transformation is needed. *B. subtilis* I168 derivatives bearing certain mutations and many non-I168 *B. subtilis* strains cannot be transformed by standard natural competence methods or have a much reduced frequency of transformation. In addition, it has been reported that the natural competence transformation of plasmids and recombinant plasmids into *B. subtilis* can be many orders of magnitude lower than for determinants located on the chromosome.

For *Bacillus* species other than *B. subtilis,* protoplast transformation may be the only available way to perform transformation. There are reports of natural competence among *B. licheniformis* strains (McCuen and Thorne, 1971, Prestidge and Spizizen, 1969). However, most published reports using other *Bacillus* species have relied on protoplast transformation methods. In some instances, attempts to use natural competence transformation have been reported with negative results (Jensen and Hulett, 1989). Because many non-*B. subtilis* species have important industrial uses, successful protoplast transformation may be the most important technique for strain construction. Protoplast transformation has been reported for such *Bacillus* species as *B. licheniformis* (Pragai *et al.*, 1994), *B. amyloliquefaciens* (Vehmaanpera, 1988), *B. stearothermophilus* (Zhang *et al.*, 1988), *B. anthracis* (Makino *et al.*, 1987), *B. thuringiensis* (Crawford *et al.*, 1987, Temeyer, 1987), and other *Bacillus* (Mann *et al.*, 1986).

Before using protoplast transformation, the type of DNA used must be considered. Unlike natural transformation, protoplast transformation has not been effective in the transfer of determinants carried on chromosomal DNA or on nonreplicating plasmids. In general, protoplast transformation requires replicating plasmids. One advantage of protoplasts is that plasmid monomers (such as ligations) and linear plasmids can be used whereas natural competence requires multimeric forms of plasmids to be successful (Chang and Cohen, 1979; Pragai *et al.*, 1994). It should be noted that there is a technique called protoplast fusion that has been used to induce genetic recombination between strains of *Bacillus* as shown using alkalophilic *Bacillus* (Aono *et al.*, 1994).

Bacteriophage-Mediated Transduction

Bacteriophage-mediated transduction has played a very important role in the development of the genetic map of both *B. subtilis* and *B. licheniformis*. Because of the relatively short length of the chromosomal DNA fragment integrated in the chromosome via competent cell transformation and the lack of an efficient conjugation, it would have been impossible to create a complete linkage map of the *B. subtilis* chromosome without the aid of the PBS1-mediated transduction (Lepesant-Kejzlarova *et al.*, 1975). PBS1 is a large pseudo temperate phage capable of packing up to 300 kb into its head (Hemphill and Whiteley, 1975). This represents roughly 7% of the *B. subtilis* chromosome, and because of this it has been possible to create the linkage of markers located very distantly in the chromosome. A kit of nine strains, prepared by Dedonder and collaborators, is still the most efficient genetic way to localize mutations not linked to any known markers via transformation.

One of the major limitations in the use of PBS1 is the fact that it can infect only motile bacteria. Therefore, in the case when the motility is impaired due to certain chromosomal mutations, this approach is not always applicable. Furthermore, the preparation of PBS1 lysates with high transducing efficiency is not always straightforward. The use of high phage titer lysates does not always result in high transduction efficiency. The protocol provided in the Appendix is, in the authors' opinion, very reliable in producing lysates with good transduction efficiencies.

B. licheniformis also has a phage very similar in size to PBS1.

Such a phage (SP15) has been used for genetic studies in *B. licheniformis* (Hemphill and Whiteley, 1975), as well as to study heterologous recombination in *B. subtilis.*

Plasmid Vectors

Replicating Plasmids

A number of plasmids are available for cloning experiments and genetic manipulation in *Bacillus* (Bron, 1990; Janniere *et al.*, 1993). In general, plasmids developed for *B. subtilis* also work for the other *Bacillus* species. The first generation of *Bacillus* plasmids were isolated from other gram-positive organisms and include plasmids such as pUB110, pE194, and pC194. These plasmids have been used for a variety of work in *Bacillus,* and there have been studies on the mode of replication of these plasmids. These plasmids have been reported to have various limitations because of their sigma replication modes, such as structural and segregational instability. A number of these gram-positive plasmids have been developed into useful shuttle vectors with polylinker cloning sites to allow cloning in *E. coli* and transfer to *Bacillus.* One finding from this work is that many *Bacillus* sequences are toxic in *E. coli* (Bron, 1990). It is thus sometimes preferable to bypass *E. coli* when cloning *Bacillus* genes.

Based on a plasmid origin of replication from a stable, low-copy number *B. subtilis* plasmid, the pHP13 vector system has been shown to offer improvements over the plasmids described earlier. During shotgun cloning experiments, it was easier to obtain large inserts with pHP13 in *B. subtilis* compared to the gram-positive plasmid systems (Haima *et al.*, 1987).

Other systems with increased stability have been constructed using the replication region of pAMβ1. This plasmid uses a θ mode of replication and can carry large (33 kb reported) DNA fragments stably (Janniere *et al.*, 1990) .

A number of antibiotic resistance genes have been used in *Bacillus,* including resistance to chloramphenicol, erythromycin, neomycin, spectinomycin, phleomycin, and tetracycline (Steinmetz and Richter, 1994; Chary *et al.*, 1997).

A beginning strategy for overexpression of *Bacillus* genes or the expression of heterologous genes would be the use of one of the replicating plasmids containing a *Bacillus* functional promoter and secretion sequences fused to the gene of interest. As an example, McGrath *et al.* (1997) used a pUB110 derivative with a *B. subtilis apr* promoter and a fusion signal sequence derived from both *B. subtilis* and *B. amyloliquefaciens* to drive the expression and secretion of a human chymase in milligram quantities.

Temperature-Sensitive Plasmids

Plasmids carrying a temperature-sensitive origin of replication for *Bacillus* have been isolated and used successfully to integrate the plasmid into the chromosome via a cloned insert DNA that bears homology to a region on the chromosome (Youngman, 1990). Unlike nonreplicating integrative plasmids, temperature-sensitive integrative plasmids can be easily introduced into *Bacillus* by protoplast transformation. After the introduction into the cell of a temperature-sensitive plasmid bearing a homologous cloned region of DNA and an antibiotic resistance gene, the culture temperature is raised above the permissive temperature, which will force the integration of the plasmid into the chromosome and allow the easy selection of such integrants in the presence of antibiotic. Temperature-sensitive plasmids constructed for *Bacillus* typically use the replication origin from pE194. Although this origin naturally ceases replication above 45°C, a useful mutant has been isolated that is sensitive at above 38°C (pE194Ts) (Bron, 1990). The uses of temperature-sensitive plasmids have included plasmid curing, controlling the expression of genes cloned on the plasmid, insertional mutagenesis, amplification of genes on the chromosome, and a transposon mutagenesis system that uses Tn917 (Bron, 1990, Youngman, 1990). Temperature-sensitive plasmids have also been used in *Bacillus* to create deletions of specific genes (Fleming *et al.*, 1995). pE194 can also integrate into the *Bacillus* chromosome in a random manner at the restrictive temperature and could introduce multiple copies of cloned genes into the chromosome (Bron, 1990). Vectors have also been developed for gene inactivation and replacement using other temperature-sensitive plasmid origins from gram-positive bacteria that should be applicable in Bacilli (Biswas *et al.*, 1993).

Integrative Vectors

The lack of an efficient transposon system in *Bacillus* and the availability of an efficient transformation mechanism have prompted the development of integrative plasmids as a tool for genetic analysis. Integrative plasmids are usually shuttle vectors that can replicate in *E. coli* and carry an antibiotic resistance that allows their selection in *B. subtilis* (Perego, 1993). On transformation and proper selection, these vectors integrate into the chromosome as long as they carry a fragment of DNA homologous to a region of DNA in the *Bacillus* chromosome. This allows to target precisely the integration into a specific gene and create gene deletion or interruption (Stahl and Ferrari 1984) and to easily carry out several other genetic manipulation, e.g., creation of partial diploidy (Ferrari and Hoch 1983), deletion of a large fragment of genome (Barilla' *et al.*, 1994), or induction of mini artificial chromosomes (Itaya and Tanaka, 1997). One tremendous advantage of this technique is that it allows one to integrate and stably maintain in the chromosome the construct used for the production of enzymes. Because of the relative instability of most free replicating plasmids in Bacilli, it would be extremely difficult to build efficient production strains carrying the genes to be expressed in such replicative vectors. The integration of such constructs in the chromosome assures their stability in the population.

Transcriptional Regulation

aprE Promoter

The major industrial product from Bacilli in terms of a dollar value is an alkaline protease, also known as subtilisin, which is used as an additive to laundry detergent. The annual production of such enzymes is estimated to have a total value in excess of $300 million a year. The best industrial producers of subtilisin, mostly derived from *B. licheniformis*, have been isolated via an iterative process of mutagenesis/screening over the span of several years. Because of the relative difficulty in carrying out genetic manipulation of industrial Bacilli, not much is known about the mutations responsible for making these strains so efficient at synthesizing and secreting subtilisin. Most of the work done in understanding the complex regulatory network that controls subtilisin expression has been carried out in *B. subtilis*.

The appearance of subtilisin activity in the medium is concomitant with the onset of sporulation (Ferrari *et al.*, 1993) . This is due to the repression, at the transcriptional level, of the products of at least two genes, *abrB* (Brehm *et al.*, 1973) and *scoC* (Dod and Balassa, 1978). *abrB,* which binds to the region of the promoter spanning from –60 to 15 (Strauch *et al.*, 1989), is synthesized constitutively throughout the vegetative growth and its synthesis is repressed by the appearance of Spo0A-P (Perego *et al.*, 1988; Strauch *et al.*, 1990). The *scoC* gene product, which also acts as a transcriptional repressor (Ferrari *et al.*, 1988; Henner *et al.*, 1988a), binds to four different sites, three upstream of the *aprE* promoter and a fourth one to the region between the –35 and the –10 (Perego and Hoch, 1988; Kallio *et al.*, 1991). A third level of repression seems to be exerted by SinI, which is believed to bind in the region around the –230 region (Gaur *et al.*, 1991). Other than repressors, the region upstream of the *aprE* promoter can be populated by a number of transcriptional activators. The most dramatic effect seems to be exerted by the product of the *degU* gene. *degU,* together with *degS,* forms an operon whose gene products belong to the family of two-component signaling systems (Henner *et al.*, 1988b; Kunst *et al.*, 1988). Mutations in either one of these two genes may affect the level of expression of a number of secreted degradative enzymes, as well as numerous other pleiotropic effect. Among these are diminished competent cell transformability and loss of motility (Kunst *et al.*, 1974).

Two other genes have been shown to act as transcriptional activators for the subtilisin expression, *degQ* and *degR* (Henner *et al.*, 1988a). Similar to *degU,* they appear to exert their activity around –150. It is important to note that despite efforts made in several laboratories, binding of any of these transcriptional activators to the upstream of the *aprE* promoter has never been proven.

Amylase Promoter

The amylase promoter represents another class of growth phase and nutrition-regulated promoters of *Bacillus*. The amylase promoter of *B. subtilis* is turned on at the end of exponential growth and its expression is repressed by glucose (de Vos *et al.*, 1997). The amylase promoter is useful in the industrial expression of secreted proteins during the long stationary production phase (Arbige *et al.*, 1993). A number of vectors using the amylase promoter and amylase secretion signals have been described (Behnke, 1992). Consideration must be given to the requirements for secretion from *Bacillus,* such as signal sequence structure and the proteolytic activities of the host *Bacillus* strain (Wong, 1995). Proteins produced from amylase promoters in *Bacillus* have included various types derived from humans and other mammals (Behnke, 1992, Nakazawa *et al.*, 1991; Novikov *et al.*, 1990).

Other Promoters

It is worthwhile to mention here two other promoters that have been studied extensively or have been used for gene expression work in *B. subtilis*.

One of them is the promoter of the *sacB* gene, possibly the best characterized and most extensively studied promoter in *B. subtilis* (Steinmetz, 1993). Like amylase and subtilisin, the product of the *sacB* gene, levansucrase, is an enzyme secreted by *B. subtilis*. However, unlike the other two enzymes, it does not have any known commercial application nor does it appear to be temporally regulated, e.g., transcribed after or at the onset of sporulation. Similar to the *aprE* promoter, transcription of the *sacB* promoter is strongly enhanced in DegU(Hy) mutants as well as in DegQ(Hy) mutants.

A second promoter to be mentioned here is the *spac* promoter, which is the result of a fusion between a promoter from the *B. subtilis* phage SP01 and the *lac* operator sequence (Yansura and Henner, 1984). When such a hybrid promoter is carried by the same plasmid carrying the *lac* repressor expressed by a constitutive promoter, the *spac* promoter is active only when induced by isopropyl-β-D-thiogalactopyranoside (IPTG). This type of inducible promoter is extremely useful when one wants to express a given gene at a desired time of the cell cycle for production purposes, especially if such gene product has a deleterious effect on growth.

Homologous vs Heterologous Gene Expression

Although there seems to be little or no problem expressing and secreting homologous genes or genes naturally secreted by other gram-positive microbes, a number of problems arise when one tries to achieve efficient heterologous gene expression and secretion (Benkhe, 1992; Simonen and Palva, 1993). Part of the problem is certainly associated with secretion, and the simplistic point of view that all that is needed is a signal sequence has been abandoned a long time ago. Despite numerous studies focused on secretion in both prokaryotes and eukaryotes, a rational approach to improving the secretion of heterologous proteins is not yet available. Furthermore, in *Bacillus* the study of the secretion mechanism lags behind that carried out in other organisms, such as *E. coli.* A consortium of European laboratories has started a more focused effort in this area. Several components of the secretion machinery have already been identified, whereas others have not been found and may be missing. Those found so far seem to have a good degree of homology with the ones described in other bacteria. One striking difference seems to be the presence of an unusually high number of signal peptidases (Tjalsma *et al.*, 1997). What the significance of this is and if all the signal peptidases have a specific functional role in secretion have not been elucidated thus far.

In summary, regarding the current level of knowledge and the complexity of the secretion process, there is no simple recipe that can be followed to obtain the efficient export of foreign proteins from Bacilli. Nor it is possible to predict the likelihood, for any given heterologous protein, that such proteins will be secreted efficiently.

Another problem that seems to interfere with the efficient export of heterologous gene products in Bacilli is the presence of a number of proteases that may have an effect on the stability of the newly synthesized product. Although this does not seem to be a widespread problem with most naturally secreted bacterial enzymes, it may certainly play a role when dealing with the low-level expression of mammalian gene products. A number of proteases have been cloned and deleted strains have been made (Pero and Sloma, 1993; Wang and Doi, 1992; Wong 1995). Because of the lack of information on their use for commercial production, the true potential of these multiply deleted strains is difficult to assess.

Conclusions

This chapter has tried to summarize the wealth of data generated over the past 40 years of *Bacillus* genetic studies. Because of the abundance of information available, we were forced to be concise. However, to be thorough, references to original papers as well as to some of the numerous reviews that deal in more detail with different aspects of the *Bacillus* genetics have been dispersed throughout the text. In addition, we have focused more on the areas of expertise that we are familiar with and in which we have had first-hand experience.

Because of the exquisite characterization of the system, *Bacillus* has the potential of becoming a very valuable organism for the expression of protein. The industrial strains have already been proven capable of very high yields and low cost because of the well-characterized fermentation as well as their ability to secrete proteins in the medium (Table 2). A number of tools are available that allow a relative ease of manipulation. *Bacillus,* however, is not yet a ready-to-use, off-the-shelf system for every expression need.

Table 2
Advantages and Disadvantages of *Bacillus* Expression

Advantages	Disadvantages
Efficient transformation via natural competence	Industrial strains not easy to transform
B. subtilis is genetically well characterized.	
Laboratory strains easy to manipulate	
B. subtilis genome fully sequenced	
Availability of numerous plasmids	Poor efficiency of transformation of ligation mixtures.
Capable of secreting proteins into the medium	Poor secretion of most heterologous proteins; degradation problems
Availability of several signal sequences	
Well-established fermentation protocol	Genetics and physiology of the fermentation not well understood
Can grow efficiently on inexpensive feedstock	
GRAS status of its products	

Most of the genetic techniques and approaches described in this chapter belong to what the authors would like to call the traditional approaches, the pregenomic approaches. Although these tools and approaches will still play an important role in the day-to-day strain manipulation, it is almost certain that they will lose importance in the genetic analysis of bacteria. While it is still difficult to evaluate the consequences that the availability of the complete blueprint of a microbe will have on its genetics, it is certain that it will change dramatically the speed and the way in which we will study physiology and genetics. Two papers are cited to demonstrate what can be achieved using DNA array technologies in terms of the amount of data generated and the completeness of analysis of any given genome (DeRisi *et al.*, 1997; de Saizieu *et al.*, 1998). Because of the impact on the genetics analysis, these new technologies will allow us to better understand the production capability of cells and dramatically affect the way we look at and solve scientific problems.

Appendix

A. *Bacillus subtilis* Protoplast Transformation

The protocol of Chang and Cohen (1979) has been the most widely used for *B. subtilis* protoplast transformation using covalently closed circular, linear, and ligated plasmid DNA. The method reported here is essentially as that described by Bron (1990).

1. The strain to be transformed can be started in two ways: on a plate for O/N growth or in a O/N shake flask. The plate is typically LA agar at 37°C. The shake flask is usually 10 ml of filter-sterilized PAB (Difco antibiotic medium No. 3) at 30°C.
2. In the morning of the experiment, inoculate 10 or 20 ml of filtered PAB into a 250-ml shake flask. Use 0.5 ml of a O/N shake flask culture or a loop of cells from a fresh O/N plate. Grow shaking at 37°C.
3. Grow the cells to OD_{550} 0.5–1.0. Transfer culture to a 50 ml Oak Ridge polycarbonate centrifuge tube. Pellet cells at 10,000 rpm for 10 min at room temperature.
4. Vacuum off medium, paying attention to the fact that the

pellet may not be compact. A little medium left behind is not a problem.

5. Resuspend pellet in 4.5 or 5 ml (see below) of SMMP and vortex.

6. Add lysozyme to a final concentration of 1 mg/ml. This can be done in one of two ways. One is to add the dry powder to the tube (5 ml). The other is to make up 5 mg lysozyme in 0.5 ml SMMP. Filter-sterilize this and add 0.5 ml to 4.5 ml of cells.

7. Incubate cells at 37°C with slow shaking. At 30 min and every 15 min after, check under a phase-contrast microscope to determine the percentage of protoplasts. When >99% protoplasts, pellet at 4000 RPM for 15 min.

8. Carefully vacuum off the media and gently resuspend the cells in 5 ml of SMMP. If the cells do not come off the wall of the tube, use a pipette to gently scrap and wash the cells off. Rock the tube to get all cells in solution.

9. For transformation, place 1.5 ml of PEG 40% into a new, clean, dry Oak Ridge tube. Place 10–40 μl (0.2–2 μg) of the plasmid to be transformed into a Eppendorf tube.

10. Add 0.5 ml of protoplasts to the plasmid tube and then draw it all up and pipette the mixture on top of the PEG. Mix for 2 min by gently rocking the tube. Extra protoplasts can be frozen as is in 0.5 ml aliquots at –70°C and used in the future.

11. Add 5 ml of SMMP and mix gently.

12. Spin down cells at full speed (4000 rpm) for 15 min at room temperature

13. Vacuum off medium on the side away from the pellet (the pellet is usually very diffuse and faint) and add 1 ml of SMMP.

14. Slow shake (100 rpm) the tube at 37°C for about 90 min. Check the tube several times and rotate it around to make sure all the cells come off the tube wall.

15. Plate to DM3 regeneration medium with antibiotic (use a plate with no drug and screen for a phenotype by replica plating). In general, use 0.2 ml/plate and spread with a plate spreader. Be gentle as the medium is a little soft. Dry plates for 15 min in a hood. Then invert and incubate the plates at 30 or 37°C depending on the plasmid transformed. Transformants begin to appear the end of the next day and are white in color. To test for real transformants, pick them to LA plus the antibiotic of interest. Sometimes there will be some background colonies that will not grow on LA plus antibiotic plates.

Media and Solutions The indicated origin of some components was found to be important. Milli-Q H_2O was used. Use detergent-free glassware throughout.

2× SMM (per liter): 342 g sucrose (1 *M*), 4.72 g sodium maleate (0.04 *M*), 8.12 g $MgCl_2 \cdot 6H_2O$ (0.04 *M*), pH to 6.5, divide into 50-ml portions, and autoclave for 15 min.

4× PAB: 7 g Difco antibiotic medium No. 3 per 100 ml. Filter sterilize

SMMP: Mix equal volumes of 2× SMM and 4× PAB.

PEG 40%: 10 g polyethyleneglycol 6000 (Serva or Fluka) in 25 ml 1× SMM. Autoclave for 15 min.

DM3 Protoplast Regeneration Plates

Agar: 4% BBL agar. Autoclave for 15 min.

Sodium succinate: 270 g/liter (1 *M*), pH to 7.3. Autoclave for 15 min.

Phosphate buffer: 3.5 g K_2HPO_4 and 1.5 g KH_2PO_4 per 100 ml. Autoclave for 15 min.

$MgCl_2$: 20.3 g $MgCl_2 \cdot 6H_20$ per 100 ml. Filter sterilize.

Casamino acids: 5% (w/v) solution. Filter sterilize.

Yeast extract (Difco): 10 g per 100 ml. Autoclave for 15 min.

Glucose: 20% (w/v) solution. Filter sterilize.

Bovine serum albumin (BSA): 2% (w/v) BSA. Filter sterilize.

Mix components at 60°C in 500-ml bottle: 250 ml sodium succinate, 50 ml casamino acids, 25 ml yeast extract, 50 ml phosphate buffer, 15 ml glucose, 10 ml $MgCl_2$, 2.5ml BSA, and 100 ml melted agar.

Add appropriate antibiotics: 5 μg/ml chloramphenicol, 5 μg/ml tetracycline, 1 μg/ml erythromycin, and 200 μg/ml kanamycin (higher concentrations needed for DM3 media)

LA agar contains per liter: 10 g Bacto-tryptone, 5 g Bacto-yeast extract, 0.5 g NaCl, and 17.5 g agar

B. *Bacillus licheniformis* Protoplast Transformation

The method of Pragai *et al.* (1994) was reported to be successful for a number of strains and replicating plasmids in covalently closed circular, linear, and ligated forms.

1. Grow strain to be transformed overnight on LA agar plus 1% insoluble starch.

2. Inoculate cells from the plate into 55 ml of YTA broth in a 250-ml flask to OD_{550} 0.1. Shake for 10–15 min at 37°C and then take half of the culture and put it in another flask (prewarmed at 37°C).

3. Grow to OD_{550} of 1.

4. Centrifuge culture in two sterile 50-ml Oak Ridge polycarbonate tubes, detergent free, in a SS34 rotor at room temperature for 10 min at 16,000 rpm.

5. Vacuum off media, paying attention to the fact that the pellet may not be very compact for some strains. Some media will most likely be left behind.

6. Resuspend the culture (both tubes) in 4.5 ml of P solution.

7. Add lysozyme (Boehringer Mannheim) to a final concentration of 2 mg/ml. The lysozyme needs to be prepared fresh at 20 mg/ml in P solution, and filter sterilized with a 0.45-μm filter. Add 500 μl lysozyme to 4.5 ml of cells.

8. Incubate at 37°C, no shaking, for 30–45 min and check protoplasting with a phase-contrast microscope. When >99% of cells are protoplasts, spin down protoplasts in Oak Ridge tubes (2000–4000 rpm for 20 min). The speed for pelleting depends on the stickiness of the strain used and will have to be determined. Wash once with 5 ml of P solution and resuspend cells in 1–5 ml of P solution. The resuspension volume may be varied depending on the number of protoplasts recovered and the number of transformations to be done.

9. Have 10–15 μl of the plasmid (high concentrations of plasmid should be used) to be transformed into an Eppendorf tube. Add 1 ml of cells and mix by gentle pipetting. Pipette the mixture into Oak Ridge tubes containing 1 ml 60% PEG 6000 (Serva or Fluka only). Mix gently for 3 min by rocking. Add 5 ml of ART liquid media and mix. Spin at 3000 rpm for 20 min at room temperature. Resuspend in 1 ml of ART and incubate for 90 min at 30°C, 100 rpm.

10. Microwave soft agar (ART medium) and place at 55°C (no antibiotic in soft agar). Distribute 10 ml of soft agar in 18-cm tubes, add transformed protoplasts (100 to 200 μl), and plate on large ART plates with 5 μg/ml of kanamycin or chloramphenicol. One to 10 plates can be used.

11. Dry for 15 min under hood, invert, and incubate at 30°C.

12. Colonies may take 3–8 days to appear, depending on a number of factors. When grown, pick to LA plates with antibiotic (for most bacillus plasmids, 10 μg/ml Kanamycin or chloramphenicol)

at 30°C. Some strains yield a large number of background colonies that will not grow on a LA antibiotic plate.

13. From 1 to 100 plasmid-bearing transformants can be expected.

Media and Solutions The indicated origin of some components was found to be crucial. Milli-Q H_2O was used during preparation.

ART Media Contains per Liter:

5 g NaCl
5 g yeast extract
5 g tryptone
5 g glucose
6.0577 g Tris
171.15 g sucrose
1 ml 0.5 *M* $CaCl_2$
1 ml 1.0 *M* $MgSO_4$
1 ml 0.5 *M* $MnSO_4$
pH 8.4
ART agar contains 1.8% (w/v) agar
ART soft agar contains 0.8% (w/v) agar

Detailed Preparation of 500 ml

1. Dissolve 3.03 g Tris (Boehringer Mannheim) and 85.575 sucrose (BBL) in 250 ml H_2O. Adjust pH to 8.4 with 5 *N* HCl. Bring to 300 ml. Autoclave for 15 min in a 500-ml bottle.
2. To prepare plate agar, use 9 g of BBL granulated agar in 100 ml H_2O, use 4 g for soft agar. Autoclave for 15 min in a 500-ml bottle with a stir bar inside.
3. After autoclaving, store bottles at 50°C.
4. To the cooled 300-ml Tris/sucrose solution, add 25 ml of the following sterile stocks: 10% NaCl (filter), 10% yeast extract (autoclave), 10% Bacto-Tryptone (autoclave), and 10% glucose (filter). Then add 1 ml of these filter-sterilized stocks: 0.5 *M* $CaCl_2$ (3.68 g of $CaCl_2 \cdot 2H_2O$ in 50 ml H_2O) , 1.0 *M* $MgSO_4$ (12.32 g of $MgSO_4 \cdot 7H_2O$ in 50 ml H_2O), and 0.5 *M* $MnSO_4$ (1.69 g of $MnSO_4 \cdot H_2O$ in 50 ml H_2O).
5. Add solution from step 4 to the 50°C agar bottle and stir. Add antibiotics, mix, and pour plates. Nine large plates can be obtained from 500 ml. For soft agar, divide into 100-ml sterile bottles.

P Solution Contains per Liter:

1.25 g NH_4Cl
15 g Tris (Boehringer Mannheim)
0.075 g NaCl
0.045 g KCl
0.375 g $Na_2SO_4 \cdot 10H_2O$
5.33 g $MgCl_2 \cdot 6H_2O$
171.15 g sucrose

Preparation: mix all components, adjust pH to 8.4, divide into bottles, and autoclave for 15 min.

PEG Solution

60% (w/v) PEG (polyethyleneglycol 6000, Serva or Fluka) in P solution
Stir and heat to dissolve, bring to volume, autoclave
Although the PEG tends to fall out of solution over time, it still works

YTA media Contains per Liter:

5 g NaCl
10 g Difco Bacto-tryptone
5 g Difco Bacto-yeast extract

LA Agar Contains per Liter:

10 g Bacto-tryptone
5 g Bacto-yeast extract
0.5 g NaCl
17.5 g agar

C. PBS1 Lysate Preparation and Transduction Protocols

First Day From a fresh overnight streak on TBAB (tryptose blood agar base; Difco) start a culture in PAB (Penassay antibiotic medium No. 3; Difco) at 37C in shaker at 200–250 rpm. At the end of the log phase or when cells reach motility, dilute the culture in Y medium (see below) to a concentration of $3–4 \times 10^7$. Prepare 10^{-4}, 10^{-5}, and 10^{-6} dilutions of any PBS1 lysate in Y medium. In three different tubes mix 0.5 ml of the strain with 1 ml of the phage dilutions. Incubate at 37°C for 10 min (without shaking). Place tubes with phage-bacteria mix in a 45°C bath and add 1.5 ml of TBAB soft agar kept melted at the same temperature. TBAB soft agar consists of 6.5 ml of TBAB mixed with 4 ml of water. Mix gently and pour on TBAB plates. Let soft agar solidify and incubate at 30°C for 16–20 hr.

Second Day With a sterile loop or a Pasteur pipette, pick two turbid plaques and inoculate in 40 ml of PAB in a 250-ml flask. Shake gently (about 150 rpm) at 37°C for 2 hr or until culture is slightly turbid and then place the flask in a thermostat at 37°C overnight.

Third Day Centrifuge the suspension to eliminate cellular debris. Add 25 μg/ml of DNAse and 0.05 *M* $MgCl_2$ and place at 37°C for 15–30 min. Filter sterilize with a 0.45-μm filter.

The lysate is ready. The titer of the phage should be about 10^8–10^9. To titer the lysate, use the same procedure used to prepare it. Although PBS1 is not a lysogenic phage, it makes only turbid plaques, which are not easy to spot.

Transduction Grow the cells as described in Day 1. When the cells appear motile, mix 1 ml of cells with 0.5 ml of phage, shake at 37°C gently for 20–30 min, spin down, wash with 1 ml of minimal medium, resuspend in 0.2 ml, and plate straight and 10^{-1} dilution on selective plates. About 300–500 transductants per plate should be obtained when selecting for an auxotrophic marker, a little less when selecting for an antibiotic resistance marker.

Y Medium

4 g NaCl
5 g K_2SO_4
1.5 g KH_2PO_4
3 g Na_2HPO_4
0.12 g $MgSO_4.7H_2O$
0.01 g $CaCl_2$
0.01 g $FeCl_3$
1 g yeast extract (Difco)
1 liter H_2O

The medium should be at pH 7.0 without adjustments. Sterilize at 121°C for 15 min.

References

Anagnostopoulos, C., and Spizizen, J. (1961). Requirements for transformation in *Bacillus subtilis*. *J. Bacteriol.* **81,** 741–746.

Aono, R., Ito, M., Joblin, K., and Horikoshi, K. (1994). Genetic recombination after cell fusion protoplasts from the facultative alkaliphile *Bacillus* sp. C-125. *Microbiology* **140,** 3085–3090.

Arbige, M. V., Bulthuis, B. A., Schultz, J., and Crabb, D. (1993). Fermentation of Bacillus. *In* "*Bacillus subtilis* and other Gram-positive Bacteria" (A. L. Sonenshein, J. A. Hoch, and R. Losick, eds.), pp. 871–895. Am. Soc. Microbiol., Washington, DC.

Barilla', D., Caramori, T., and Galizzi, A. (1994). Coupling of flagellin gene transcription to flagellar assembly in *Bacillus subtilis*. *J. Bacteriol.* *176*, 4558–4564.

Behnke, D. (1992). Protein export and the development of secretion vectors. *In* "Biology of Bacilli: Applications to Industry" (R. H. Doi and M. McGloughlin, eds.), pp. 143–188. Butterworth-Heinemann, Boston.

Biswas, I., Gruss, A., Ehrlich, D. S., and Maguin, E. (1993). High-efficiency gene inactivation and replacement system from Gram-positive bacteria. *J. Bacteriol.* **175,** 3628–3635.

Brehm, S. P., Staal, S. P., and Hoch, J. A. (1973). Phenotypes of pleiotropic-negative sporulation mutants of *Bacillus subtilis*. *J. Bacteriol.* **115,** 1063–1070.

Bron, S. (1990). Plasmids. *In* "Molecular Biological Methods for *Bacillus*" (C. C. R. Harwood and S. M. Cutting, eds.), pp. 75–174. Wiley, New York.

Canosi, U., Morelli, G., and Trautner, T. (1978). The relationship between molecular structure and transformation efficiency of some *Staphilococcus areus* plasmids isolated from *Bacillus subtilis*. *Mol. Gen. Genet.* **168,** 259–267.

Chang, S., and Cohen, S. N. (1979). High frequency transformation of *Bacillus subtilis* protoplasts by plasmid DNA. *Mol. Gen. Genet.* **168,** 111–115.

Chary, V. K., Amaya, E. I., and Piggot, P. J. (1997). Neomycin- and spectinomycin-resistance replacement vectors for *Bacillus subtilis*. *FEMS Microbiol. Lett.* **153,** 135–139.

Crawford, I. T., Greis, K. D., Parks L., and Streips, U. N. (1987). Facile autoplast generation and transformation in *Bacillus thuringiensis* subsp. *Kurstaki*. *J. Bacteriol.* 169, 5423–5428.

Cutting, S. M., and Van der Horn, P. B. (1990) Genetic analysis. *In* "Molecular Biological Methods for Bacillus" (C. C. R. Harwood and S. M. Cutting, eds.), pp. 27–74. Wiley, New York.

de Boer, A. S., Priest, F., and Diderichsen, B. (1994). On the industrial use of *Bacillus licheniformis*: A review. *Appl. Microbiol. Biotechnol.* **40,** 595–598.

DeRisi, J. L., Iyer, V. R., and Brown, P. (1997). Exploring the metabolic and genetic control of gene expression on a genomic scale. *Science* **270,** 680–686.

de Saizieu, A., Certa, U., Warrington, J., Gray, C., Keck, W., and Mous, J. (1998). Bacterial transcript imaging by hybridization of total RNA to nucleotide arrays. *Nat. Biotechnol.* **16,** 45–48.

de Vos, W. M., Kleerebezem, M., and Kuipers, O. P. (1997). Expression systems for industrial Gram-positive bacteria with low guanine and cytosine content. *Curr. Opin. Biotechnol.* **8,** 547–553.

Dod, B., and Balassa, G. (1978). Spore control (*sco*) mutations in *Bacillus subtilis*. III. Regulation of extracellular protease synthesis in the spore control mutations *scoC. Mol. Gen. Genet.* **163,** 57–63.

Dubnau, D. (1993). Genetic exchange and homologous recombination. *In "Bacillus subtilis* and other Gram-positive bacteria" (A. L. Sonenshein, J. A. Hoch, and R. Losick, eds.), pp. 555–584. Am. Soc, Microbiol., Washington, DC.

Dubnau, D., and Cirigliano, C. (1972). Fate of transforming DNA following uptake by competent *Bacillus subtilis*. III. Formation and properties of products isolated from transformed cells which are derived entirely from donor DNA. *J. Mol. Biol.* **64,** 9–29.

Eichler, K., Vockler, C., Carballido-Lopez, R. and Van Loon, A. P. G. M. (1997). Investigation of the biotechnological potential of folic acid biosynthesis in *Bacillus subtilis. Abstr. Int. Confer. Bacilli, 9th,* Lausanne, Switzerland, *1997,* p. 46.

Fahnestock, S. R., and Fisher, K. E. (1986). Expression of the staphylococcal protein A gene in *Bacillus subtilis* by gene fusions utilizing the promoter from a *Bacillus amyloliquefaciens* α-amylase gene. *J. Bacteriol.* **165,** 796–804.

Ferrari E., and Hoch, J. A. (1983). A single copy transducible system for complementation and dominance analyses in *Bacillus subtilis. Mol. Gen. Genet.* **189,** 321–325.

Ferrari E., and Hoch, J. A. (1989). Genetics. *In* "Bacillus" (C. R. Harwood, ed.), pp. 57–72. Plenum, New York.

Ferrari, E., Henner, D. J., Perego, M., and Hoch, J. A. (1988). Transcription of *Bacillus subtilis* subtilisin and expression of subtilisin in sporulation mutation. *J. Bacteriol.* **170,** 289–295.

Ferrari, E., Jarnagin, A. S., and Schmidt, B. F. (1993). Commercial production of extracellular enzymes. *In* "Bacillus subtilis and other Gram-positive Bacteria" (A. L. Sonenshein, J. A. Hoch, and R. Losick, eds.), pp. 917–937. Am. Soc. Microbiol., Washington, DC.

Fleming, A. B., Tangney, M., Jorgensen, P. L., Diderichsen, B., and Priest, F. G. (1995). Extracellular enzyme synthesis in a sporulation-deficient strain of *Bacillus licheniformis. Appl. Environ. Microbiol.* **61,** 3775–3780.

Gaur, N. K., Oppenheim, J., and Smith, I. (1991). The *Bacillus subtilis sin* gene, a regulator of alternate developmental processes, codes for a DNA-binding protein. *J. Bacteriol.* **173,** 678–686.

Haima, P., Bron, S., and Venema, G. (1987). The effect of restriction on shotgun cloning and plasmid stability in *Bacillus subtilis* Marburg. *Mol. Gen. Genet.* **209,** 335–342.

Hemphill, H. E., and Whiteley, H. R. (1975). Bacteriophages of *Bacillus subtilis. Bacteriol. Rev.* **39,** 257–315.

Henner, D.J., Ferrari, E., Perego, M., and Hoch, J. A. (1988a). Location of the target of the *hpr-97, sacU32(Hy),* and *sacQ36(Hy)* mutations in upstream region of the subtilisin promoter. *J. Bacteriol.* **170,** 296–300.

Henner, D. J., Yang, M., and Ferrari, E. (1988b). Localization of the *Bacillus subtilis sacU*(Hy) mutations to two linked genes with similarities to the conserved prokaryotic family of two component signaling systems. *J. Bacteriol.* **170,** 5102–5109.

Honjo, M., Nakayama, A., Iio, A., Kawamura, K., Sawakura, A., and Furutani, Y. (1987). Construction of a highly efficient host-vector system for secretion of heterologous proteins in *Bacillus subtilis. J. Biotechnol.* **6,** 191–204.

Itaya, M., and Tanaka, T. (1997). Experimental surgery to create subgenomes of *Bacillus subtilis* 168. *Proc. Natl. Acad. Sci. U.S.A.* **94,** 5378–5382.

Janniere, L., Bruand, C., and Ehrlich, S. D. (1990). Structurally stable *Bacillus subtilis* cloning vectors. *Gene* **87,** 53–61.

Janniere, L., Gruss, A., and Ehrlich, S. D. (1993). Plasmids. Fermentation of Bacillus. *In "Bacillus subtilis* and other Gram-positive Bacteria" (A. L. Sonenshein, J. A. Hoch, and R. Losick, eds.), pp. 625–645. Am. Soc. Microbiol., Washington, DC.

Jensen, K. K., and Hulett, M. F. (1989). Protoplast transformation of *Bacillus licheniformis* MC14. *J. Gen. Microbiol.* **135,** 2283–2287.

Kallio, P. T., Fagelson, J. E., Hoch, J. A. and Strauch, M. A. (1991). The transition state regulator Hpr of *Bacillus subtilis* is a DNA-binding protein. *J. Biol. Chem.* **266,** 13411–13417.

Kowalski, J. B., and Thorne, C. B. (1979). Genetic mapping of a bacteriophage resistance mutation, pha–1, and thi–1 mutation in *Bacillus licheniformis. J. Bacteriol.* **137,** 689–691.

Kunst, F., Pascal, M., Lepesant-Kejzlarova, J., Lepesant, J. A., Billault, A., and Dedonder, R. (1974). Pleiotropic mutations affecting sporulation conditions and the syntheses of extracellular enzymes in *Bacillus subtilis* 168. *Biochimie* **56,** 1481–1489.

Kunst, F., Debarbouillle, M., Msadek, T., Young, M., Mauel, C., Karamata, D., Klier, A., Rapoport, G., and Dedonder, R. (1988). Deduced polypeptides encoded by the *Bacillus subtilis sacU* locus share homology with two-component sensor regulator systems. *J. Bacteriol.* **170,** 5093–5101.

Kunst, F., Ogasawara, N., Moszer, I., and Danchin, A. (1997). The complete genome sequence of the gram-positive bacterium *Bacillus subtilis. Nature (London)* **390,** 249–256.

Lepesant-Kejzlarova, J., Lepesant, J. A., Walle, J., Billault, A., and Dedonder, R. (1975). Revision of the linkage map of *Bacillus subtilis* 168: Indication for circularity of the chromosome. *J. Bacteriol.* **121,** 823–834.

Makino, S.-I., Sasakawa, C., Uchida, I., Terakado, N., and Yoshikawa, M. (1987). Transformation of a cloning vector, pUB110, into *Bacillus anthracis. FEMS Microbiol. Lett.* 44, 45–48.

Mann, S. P., Orpin, C. G., and Hazlewood, G. P. (1986). Transformation of *Bacillus* spp.: And examination of the transformation of *Bacillus* protoplasts by plasmids pUB110 and pHV33. *Curr. Microbiol.* 13, 191–195.

McCuen, R. W., and Thorne, C. B. (1971). Genetic mapping of genes concerned with glutamyl polypeptide production by *Bacillus licheniformis* and a study of their relationship to the development of competence for transformation. *J. Bacteriol.* **107,** 636–645.

McGrath, M. E., Osawa, E. A., Barnes, M. G., Clark, J. M., Mortara, K. D., and Schmidt, B. F. (1997). Production of crystallizable human chymase from a *Bacillus subtilis* system. *FEBS Lett.* **414,** 486–488.

Mottes, M., Grandi, G., Sgaramella V., Canosi, U., Morelli, G., and Trautner, T. (1979). Different specific activities of the monomeric and oligomeric forms of plasmid DNA in transformation of *Bacillus subtilis* and *Escherichia coli*. *Mol. Gen. Genet.* **174,** 281–286.

Nakano, M. M., Dailly, Y. P., Zuber, P., and Clark, D. P. (1997). Characterization of anaerobic fermentative growth of *Bacillus subtilis*: Identification of fermentation end products and genes required for growth. *J. Bacteriol.* **179,** 6749–6755.

Nakazawa, K., Sasamoto, H., Shiraki, Y., Harada, S., Yanagi, K., and Yamame, K. (1991). Extracellular production of mouse interferon beta by the *Bacillus subtilis* alpha-amylase secretion vectors: Antiviral activity and deduced NH2-terminal amino acid sequences of the secreted proteins. *Intervirology* **32,** 216–227.

Novikov, A. A., Borukov, S. I., and Strongin, A. Y. (1990). Bacillus amyloliquefaciens -amylase signal sequence fused in frame with human proinsulin is properly processed by Bacillus subtilis cells. *Biochem. Biophys. Res. Commun.* **169,** 297–301.

Perego, M. (1993). Integrational vectors for genetic manipulations in *Bacillus subtilis*. *In* "*Bacillus subtilis* and other Gram-positive Bacteria" (A. L. Sonenshein, J. A. Hoch, and R. Losick, eds.), pp. 615–624. Am. Soc. Microbiol., Washington, DC.

Perego, M., and Hoch, J. A. (1988). Sequence analysis and regulation of the *hpr* locus, a regulatory gene for protease production and sporulation in *Bacillus subtilis*. *J. Bacteriol.* **170,** 2560–2567.

Perego, M., Spiegelman, G. B., and Hoch, J. A. (1988). Structure of the gene for the transition state regulator AbrB: Regulator synthesis is controlled by the Spo0A protein. *Mol. Microbiol.* **2,** 689–699

Perkins, J., and Pero, J. (1993). Biosynthesis of riboflavin, biotin, folic acid and cobalamin. *In* "*Bacillus subtilis* and other Gram-positive Bacteria" (A. L. Sonenshein, J. A. Hoch, and R. Losick, eds.), pp. 319–334. Am. Soc. Microbiol., Washington, DC.

Pero, J., and Sloma, A. (1993). Proteases. *In* "*Bacillus subtilis* and other Gram-positive Bacteria" (A. L. Sonenshein, J. A. Hoch, and R. Losick, eds.), pp. 939–952. Am. Soc. Microbiol., Washington, DC.

Pragai, Z., Holczinger, A., and Sik, T. (1994). Transformation of *Bacillus licheniformis* protoplasts by plasmid DNA. *Microbiology* **140,** 305–310.

Prestidge L., and Spizizen, J. (1969). Conditions for competence in the *Bacillus licheniformis* transformation system. *J. Bacteriol.* **99,** 70–77.

Priest, F. G. (1989). Products and applications. *In* "Bacillus" (C. R. Harwood, ed.), pp. 293–320. Plenum, New York.

Priest, F. G. (1993). Systematic and ecology of *Bacillus*. *In* "*Bacillus subtilis* and other Gram-positive Bacteria" (A. L. Sonenshein, J. A. Hoch, and R. Losick, eds.), pp. 3–16. Am. Soc. Microbiol., Washington, DC.

Saris, P., Taira, S., Airaksinen, U., Palva, A., Sarvas, M., Palva, I. and Nureberg-Nyman, K. (1990). production and secretion of pertussis toxin subunits in *Bacillus subtilis*. *FEMS Microbiol. Lett.* **68,** 143–148

Sauer, U., Hatzimanikatis, V., Hohman, H.-P., Manneberg, M., Van Loon, A. P. G. M., and Bailey, J. E. (1996). Physiology and metabolic fluxes of wild-type and riboflavin producing bacillus subtilis. *Appl. Environ. Microbiol.* **62,** 3687–3696.

Shafikani, S., Siegel, R. A., Ferrari, E., and Schellenberger, V. (1997). Generation of large libraries of random mutants in *Bacillus subtilis* by PCR-based plasmid multimerization. *BioTechniques* **23,** 304–310

Shariati, P., Mitchell, W. J., Boyd, A., and Priest, F. G. (1995). Anaerobic metabolism in *Bacillus licheniformis* NCIB 6346. *Microbiology* **141,** 1117–1124.

Simonen, M., and Palva, I. (1993). Protein secretion in *Bacillus* species. *Microbiol. Rev.* **57,** 109–137

Sloma, A., Pawlyk, D., and Pero, J. (1988). Development of and expression and secretion system in *Bacillus subtilis* utilizing *sacQ*. *In* "Genetics and Biotechnology of Bacilli" (A. T. Ganesan and J. A. Hoch, eds.), Vol. 2, pp. 23–26. Academic Press, San Diego CA.

Somma, S., and Polsinelli, M. (1970) Quantitative autoradiographic study of competence and deoxyribonucleic acid incorporation in *Bacillus subtilis*. *J. Bacteriol.* **101,** 851–855.

Stahl, M.L., and Ferrari, E. (1984). Replacement of the *Bacillus subtilis* subtilisin structural gene with an *in vitro*-derived deletion mutation. *J. Bacteriol.* **158,** 411–418.

Steinmetz, M. (1993) Carbohydrate catabolism: Pathways, enzymes, genetic regulation and evolution. *In "Bacillus subtilis* and other Gram-positive Bacteria" (A. L. Sonenshein, J. A. Hoch, and R. Losick, eds.), pp. 157–170. Am. Soc. Microbiol., Washington, DC.

Steinmetz, M., and Richter, R. (1994). Plasmids designed to alter the antibiotic resistance expressed by insertion mutations in *Bacillus subtilis*, through in vivo recombination. *Gene* **142,** 79–83.

Strauch, M. A., Spiegelman, G. B., Perego, M., Johnson, W. C., Burbulys, D., and Hoch, J. A. (1989). The transition state transcription regulator *abrB* of *Bacillus subtilis* is a DNA binding protein. *EMBO J.* **8,** 1615–1621.

Strauch, M. A., Webb, V., Spiegelman, G. B., and Hoch, J. A. (1990). The Spo0A protein of *Bacillus subtilis* is a repressor of the *abrB* gene. *Proc. Natl. Acad. Sci. U.S.A.* **87,** 1801–1805.

Takao, M., Morioka, T., Yamagata, H., Tsukagoshi, N., and Udaka, S. (1989). Production of swine pepsinogen by protein producing *Bacillus brevis*, carrying swine pepsinogen cDNA. *Appl. Microbiol. Biotechnol.* **30,** 75–80

Temeyer, K. B. (1987). Comparison of methods for protoplast formation in *Bacillus thuringiensis*. *J. Gen. Microbiol.* **133,** 503–506.

Tjalsma, H., Noback, M. A., Bron, S., Venema, G., Yamane, K., and van Dijl, J. M. (1997). *Bacillus subtilis* contains four closely related type I signal peptidases with overlapping substrate specificities. *J. Biol. Chem.* **272,** 25983–25992.

Udaka, S., and Yamagata, I. (1993a). Protein secretion in *Bacillus brevis*. *Antonie van Leeuwenhoek.* **64,** 137–143.

Udaka, S., and Yamagata, I. (1993b). High-level secretion of heterologous proteins by *Bacillus brevis*. *In* "Methods in Enzymology" (R. Wu, ed.), Vol. 217, pp. 23–33. Academic Press, San Diego, CA.

Udaka, S., Tsukagoshi, N., and Yamagata, I. (1989) *Bacillus brevis*, a host bacterium for efficient extracellular production of useful proteins. *In* "Biotechnology and Genetic Engineering Reviews" (G. E. Russel and M. P Tombs, eds.), pp.113–146. Intercept, Andover, MA.

van Leen, R. W., Bakhuis, J. G., van Beckoven, R. F. W. C., Burger, H., Dorssers, L. C. J., Hommes, R. W. J., Lemson, P. J., Noordam, B., Persoon, N. L. M., and Wagemaker, G. (1991). Production of human interleukin-3 using industrial microorganisms. *Bio/Technology* **9,** 47–52

Vary, P. S. (1994). Prime time for *Bacillus megaterium*. *Microbiology* **140,** 1001–1013

Vehmaanpera, J. (1988). Transformation of *Bacillus amyloliquefaciens* protoplasts with plasmid DNA. *FEMS Microbiol. Lett.* **49,** 101–105.

Wang, L.-F., and Doi, R. H. (1992). Heterologous gene expression in *Bacillus.* In "Biology of Bacilli: Applications to Industry" (R. H. Doi and M. McGloughlin, eds.), pp. 63–104. Butterworth-Heinemann, Boston.

Welker, N. E., and Campbell, L. L. (1967). Unrelatedness of *Bacillus amyloliquefaciens* and *Bacillus subtilis. J. Bacteriol.* **94,** 1124–1130.

Wong, S.-L. (1995). Advances in the use of *Bacillus subtilis* for the expression and secretion of heterologous proteins. *Curr. Opin. Biotechnol.* **6,** 517–522.

Yansura, D. G., and Henner, D. J. (1984). Use of the *Escherichia coli* lac repressor and operator to control gene expression in *Bacillus subtilis. Proc. Natl. Acad. Sci. U.S.A.* **81,** 439–443.

Yoshimura, K., Miyazaki, T., Nakahama, K., and Kikuchi, M. (1986) *Bacillus subtilis* secretes a foreign protein by the signal sequence of *Bacillus amyloliquefaciens* neutral protease. *Appl. Microbiol. Biotechnol.* **23,** 250–256.

Youngman, P. (1990). Use of transposons and integrational vectors for mutagenesis and construction of gene fusions in *Bacillus* species. *In* "Molecular Biological Methods for *Bacillus*" (C. C. R. Harwood and S. M. Cutting, eds.), pp. 221–266. Wiley, New York.

Zhang, M., Nakai, H., and Imanaka, T. (1988). Useful host-vector systems in *Bacillus stearothermophilus. Appl. Environ. Microbiol.* **54,** 3162–3164.

4

araB EXPRESSION SYSTEM IN *Escherichia coli*

Marc Better
XOMA Corporation, Santa Monica, California 90404

Introduction

Because of the relative simplicity of gene expression in prokaryotes, a bacterium such as *Escherichia coli* is often a first choice for recombinant expression of heterologous proteins. Yet even within the context of bacterial gene expression, a staggering number of options can confront both the novice and the experienced scientist. In the current era of commercially available expression systems, and literally thousands of published papers describing solutions (and failures) to specific expression challenges, making careful choices remains an important key to success.

Certain common elements are required for effective gene expression in bacteria. Obviously a bacterial host is required, and most often a plasmid is used to carry the gene of interest. The expression

Gene Expression Systems: Using Nature for the Art of Expression

plasmid must carry a selectable marker to identify transformants, and the gene of interest must be equipped with appropriate initiation signals. Transcription must be signaled from a promoter and a transcriptional terminator also can be employed. When necessary, appropriate accessory proteins can be included to properly regulate an expression system. Second in importance only to the gene of interest, however, is selection of the proper promoter. This key element can, in many instances, dictate success or failure to a particular expression project.

For mostly historic reasons, many expression projects in *E. coli* and related bacteria have incorporated only a limited set of bacterial promoters. The most popular have been the lactose [*lac*] (Yanisch-Perron *et al.*, 1985) and the tryptophan [*trp*] (Goeddel *et al.*, 1980) promoters and the hybrid promoters derived from these two [*tac*] and *trc*] (Brosius, 1984; Amann and Brosius, 1985). Other commonly used expression systems include the phage λ-promoters P_L and P_R (Elvin *et al.*, 1990), the phage T7 promoter (Tabor and Richardson, 1985), and the alkaline phosphatase promoter [*pho*] (Chang *et al.*, 1986). Although each of these promoters offers many desirable features, the ideal promoter for expression of a wide variety of heterologous proteins would offer certain features that may not all be found in the commonly used systems. Many recombinant products can be toxic to the expression host, and therefore it is often important for the promoter to tightly regulate gene expression during culture propagation when gene expression is undesirable. However, when gene expression is required, the promoter must be easily controlled and a high expression level is often ideal. The agent or environmental condition that initiates gene expression must be easy to use and preferably of low cost. In general, a tightly regulated system is most desirable—one where gene expression is tightly repressed in the absence of inducer and highly derepressed in its presence. One bacterial promoter system that meets these specific requirements is the *araB* promoter (P_{BAD}).

The *araB* promoters of the Enterobacteriaceae, such as *E. coli* and *Salmonella typhimurium*, are highly conserved and regulated in a similar manner (Horwitz *et al.*, 1981). Both systems have been studied in detail and considerable information is available about their mechanism of gene regulation. Although the tightly controlled and highly inducible expression of the native genes involved in arabinose catabolism regulated by the *araB* promoter has been recognized for many years, reports of heterologous gene expression under arabinose control have appeared somewhat sporadically. More recently, it has become clear that the *araB* system offers a number of features that are particularly amenable for expression of a wide variety of recombinant products in bacteria. In come cases, recombinant products have been expressed in *E. coli* and purified for potential use as human pharmaceuticals. In this chapter, features of the *araB* expression systems are explored with emphasis on examples of how the system has been exploited to produce a wide array of recombinant products.

The *araB* Promoter and How It Works

Arabinose is a five-carbon sugar that is found widely in nature and can serve as a sole carbon source in many bacteria. The protein products from three genes (*araB*, *araA*, and *araD*) are needed for arabinose degradation in members of the *Enterobacteriaceae* family, such as *E. coli* and *S. typhimurium*, and these genes form a cluster abbreviated *araBAD*. These genes are arranged in a single operon and encode the enzymes ribulokinase (AraB), L-arabinose isomerase (AraA), and L-ribulose-5-phosphate 4-epimerase (AraD). A schematic view of the gene organization of the *E. coli ara* operon is shown in Fig. 1. Adjacent to the *araBAD* operon is a complex

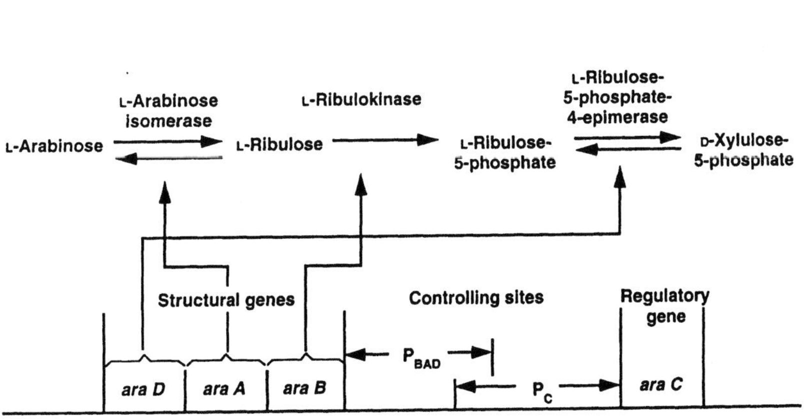

Figure 1 Map of the *S. typhimurium* or *E. coli ara* operon and the pathway for arabinose metabolism. P_{BAD} and P_C refer to the promoters for the *araBAD* and *araC* genes, respectively.

promoter region and the regulatory gene *araC*. The *araBAD* and *araC* genes are transcribed in opposite directions. Within and around the promoters for *araBAD* (P_{BAD}) and *araC* (P_C) lie binding sites for the AraC protein, the cyclic AMP (cAMP) receptor protein (CRP), and RNA polymerase. Alone or in combination, proteins bound to these regions in both the presence or the absence of the inducer, L-arabinose, tightly regulate expression from both promoters. Figure 2 illustrates the spatial arrangement of regulatory sites within this complex *araBAD* and *araC* promoter region.

Transcription from P_{BAD} is inducible with L-arabinose. In the presence of arabinose, AraC protein bound at the *araI* site immediately adjacent to the RNA polymerase binding site of the P_{BAD} promoter stimulates transcription of the *araBAD* operon (Huo *et al.*, 1988). In the absence of arabinose, however, the AraC protein represses mRNA synthesis from P_{BAD} by a mechanism requiring the formation of a DNA loop. Without arabinose, most copies of the *ara* regulatory region contain a DNA loop between the *araO*$_2$ and *araI* sites mediated by AraC protein bound to both of these sites. This loop constrains AraC protein bound at *araI* from entering the inducing state and holds the uninduced level of P_{BAD} expression low. Upon the addition of arabinose, the *araO*$_2$–*araI* loop opens, and arabinose bound to AraC protein on the *araI* site drives AraC into the inducing conformation, thereby inducing P_{BAD}. Regulation of this operon is also subject to catabolite repression, so even in the presence of arabinose, significantly less induction occurs when in-

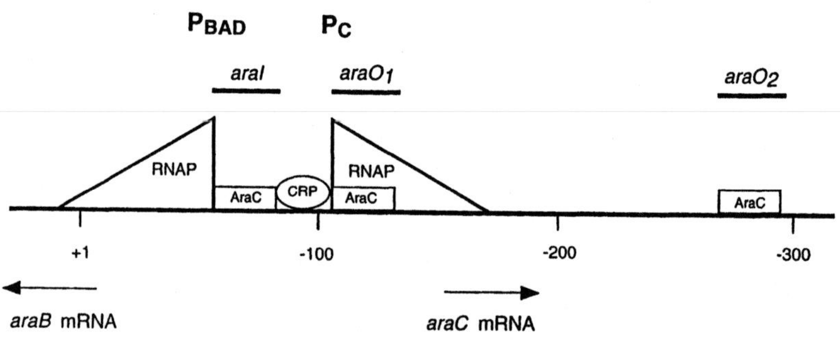

Figure 2 Regulatory control features of the *ara* operon. Sites for RNA polymerase (RNAP), AraC, and cyclic AMP repressor protein (CRP) binding within the *ara* control region are shown schematically. AraC protein binds to each of the three binding sites, *araI*, *araO*$_1$, and *araO*$_2$, as a protein dimer. +1 refers to the transcription start point for the *araBAD* promoter.

tracellular cAMP levels are low, such as when the cells are grown in the presence of glucose.

Transcription from P_C is also regulated by arabinose. In the absence of arabinose, access of RNA polymerase to P_C is limited by formation of the AraC-mediated *araI*–*araO*$_2$ loop, and *araC* expression remains low. In the presence of arabinose, AraC bound to *araI* in its inducing conformation relieves the steric constraint and the RNA polymerase has increased access to P_C. Transcriptional activity of P_C is transient, however, as AraC can bind to *araO*$_1$ and AraC-mediated DNA loops form between *araO*$_1$ and *araO*$_2$ that ultimately limit transcription from P_C. In addition, P_C is also subject to catabolite repression.

Although seemingly complex, these regulatory arrangements allow bacteria to produce enzymes for arabinose metabolism only when they are needed. The system has evolved to include two positive regulatory proteins (AraC + arabinose; CRP), one negative regulatory element (AraC – arabinose), a mechanism to control gene expression through DNA looping, and a system that can prevent or enhance an interaction with the RNA polymerase. All of the features of this entire regulatory region are conveniently contained within about 300 bp of DNA.

Use of the *ara* System for Expression of Recombinant Products

Two features of the *ara* system make it particularly well suited for the expression of recombinant products in bacteria such as *E. coli*. First, it is simple to exploit as only the 300-bp regulatory region is required, along with the full coding sequence for the *araC* gene. Second, regulation of the system has proven to be particularly tight. Table 1 illustrates that the induced/repressed level of gene expression from the *araB* promoter from multicopy expression plasmids is relatively high. Equally important, the uninduced level of protein expression is very low. This feature is especially important when proteins that are toxic to the host are to be expressed. In many reports, the level of product induction in *E. coli* is 1000-fold or more. Often it is difficult to quantify the full induction level because many assays for expressed proteins are not sensitive enough

Table 1
Inducibility of the *araB* Promoter

References	Expressed protein[a]	Expression ratio: Induced/repressed		
		P_{TRC}	P_{TAC}	P_{araB}
Lee *et al.* (1987)	AraA			1200[b]
Cagnon *et al.* (1991)	β-gal	9		240–700[c]
Guzman *et al.* (1995)	AP		50	200–1200[d]
Jacobs *et al.* (1989)	HMT			75,000
M. Better (unpublished)	Fab, enzymes, etc.			>1000

[a]AraA, L-arabinose isomerase; β-gal, β-galactosidase; AP, alkaline phosphatase; HMT, human metallothionine.
[b]Activation of the *araB* promoter was from the *E. coli* chromosome containing a single copy of the *araBAD* operon.
[c]Extent of induction depended on expression vector features.
[d]Extent of induction depended on cell culture medium.

to detect the extremely low expression levels in the uninduced state. Figure 3 illustrates a typical time course for the expression of a recombinant product after induction of the *araB* promoter in *E. coli* grown to a high cell density in a fermentor.

A number of individual expression vectors containing the *ara* system from *E. coli* or *S. typhimurium* have been described. In 1985, Johnston *et al.* described the vector pING1, which contained the *ara* regulatory region, the complete *araC* gene, and a portion of the *araB* gene from *S. typhimurium*. Restriction sites were introduced into the coding region of *araB* so that a gene fusion or a multigene transcription unit could be expressed under arabinose control. This system was used to express proteins that are normally expressed in *E. coli* under certain circumstances (homologous recombinant proteins), namely the M13 gene II (Johnston *et al.*, 1985) and gene 8 proteins (Kuhn and Wickner, 1985), and a similar vector was used to express the RepE protein from the *E. coli* F plasmid (Masson and Ray, 1986). Each of these proteins was produced in the cytoplasm of *E. coli* cells, and at least some of the expressed protein was soluble and could be detected in an active form either *in vivo* or in cell extracts.

An *araB* expression system derived from pING1 was subsequently engineered to regulate the expression of heterologous (nonbacterial) proteins in *E. coli*. Proteins produced successfully with this

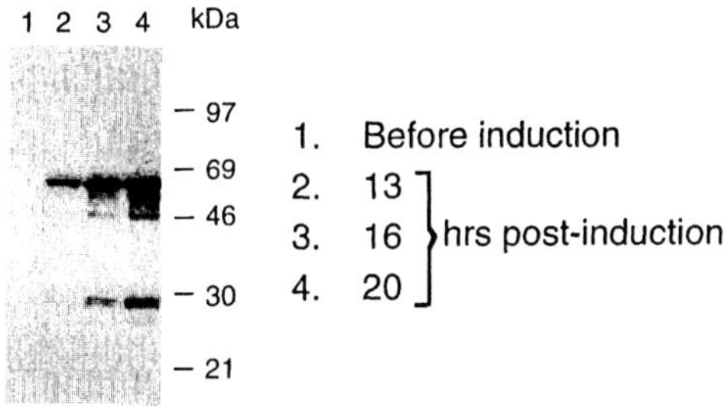

Figure 3 *E. coli* harboring a plasmid encoding a single chain antibody–gelonin fusion protein (SCA-gelonin) were grown to a high cell density (OD_{600} ~ 100) in glycerol minimal medium and induced with arabinose. After induction, proteins in the cell-free culture supernatant were separated by SDS–PAGE and transferred to a nylon membrane. Recombinant product was detected with antibody to gelonin. The fusion protein (~55 kDa) accumulated over time along with several lower molecular weight species. SCA-gelonin was purified directly from the cell-free culture supernatant by cation exchange and hydrophobic interaction chromatography (Better *et al.*, 1995).

system in *E. coli* included immunoglobulin Fab domains (Better *et al.*, 1988). In this case, the *araB* expression system was used to direct the production of polypeptides linked directly to hydrophobic signal sequences through the bacterial cytoplasmic membrane where Fab accumulated in the correctly folded, fully active configuration and could be recovered directly from the culture supernatant. Expression of Fab domains under the transcriptional control of P_{BAD} was the first demonstration that a native eukaryotic, heterodimeric molecule could be produced in *E. coli*.

Because each of the Fab molecules produced consisted of two similar polypeptide chains (κ and Fd), a dicistronic expression construct was assembled to include each gene under the transcriptional control of *araB* (Better *et al.*, 1988; Better and Horwitz, 1989). Each coding sequence, Fd and κ, was fused directly to the leader peptide segment of the bacterial *pelB* gene (encoding pectate lyase) from *Erwinia carotovora* (Lei *et al.*, 1987). Plasmids containing the dicistronic Fab expression vector under arabinose control were transformed into *E. coli* and grown in broth. The cultures were then induced by the addition of arabinose to the medium, and Fab accumulated directly into the culture medium (Fig. 4). The initially reported expression level was approximately 1–2 μg/ml, but it was subsequently demonstrated that the level of protein expression could be increased nearly 1000-fold by growing the bacteria to a high cell density in a fermentor (Better *et al.*, 1990). Fab yield from

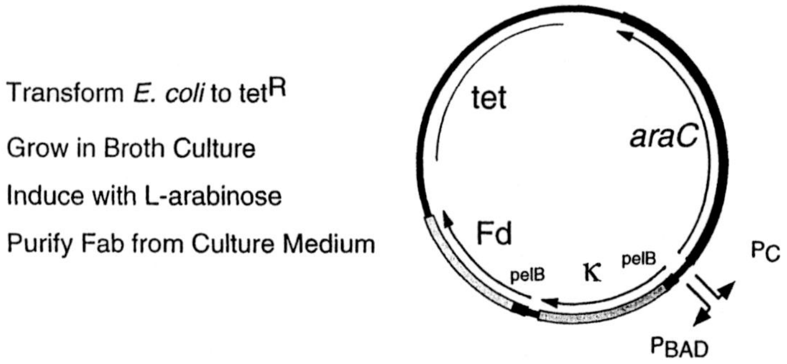

Figure 4 Schematic representation of an expression vector for Fab production under the transcriptional control of the *araB* promoter. The truncated heavy chain (Fd) and the κ light chain were each individually linked to the *pelB* leader sequence and cloned downstream of P_{BAD} as a dicistronic transcription unit. Also included in the expression plasmid is a gene for selection of tetracycline resistance in *E. coli* (tet).

bacterial cultures has been reported to achieve 0.5–2 g per liter from high cell density cultures (Better *et al.*, 1990; Carter *et al.*, 1992).

A relatively simple purification process has been developed for the purification of Fab molecules from bacteria (Gavit *et al.*, 1992). A series of ion-exchange or affinity chromatography steps can be used to separate bacterially produced Fab from cellular impurities. These processes can be very efficient, with overall recoveries of 80% or more. Greater than eight logs of bacterial endotoxin can be removed, and the bacterial DNA level can be reduced to less than 1 pg/mg of Fab (Gavit *et al.*, 1992). Fab protein suitable for pharmaceutical applications can be prepared easily from bacterial cultures.

The tight regulation of the arabinose expression system made it particularly well suited for the expression of Fab molecules in *E. coli*. In the example described earlier, the two similar polypeptide chains were each linked to the same signal sequence in a dicistronic message. In the absence of the inducer L-arabinose, no Fab could be detected in bacterial cultures. When a similar dicistronic expression system was placed under the transcriptional control of the *lac* promoter in a pUC expression vector (Yanisch-Perron *et al.*, 1985), a higher amount of background expression (in the absence of inducer) resulted in inhibition of cell growth and eventual deletion of antibody-encoding gene sequences between the two homologous leader sequences. Even with a *lacI^q* gene cloned directly onto the same pUC expression plasmid, the background expression of Fab genes that interfered with cell growth could not be completely prevented (data not shown). Clark *et al.* (1997) have also found that Fab genes under *lac* control can inhibit bacterial growth and that Fab expression from P_{BAD} can be more tightly repressed than that from P_{lac}.

The *araB* expression system used successfully for the production of Fab was also used for the expression of other proteins in *E. coli.* Several plant and fungal ribosome-inactivating proteins were expressed under the direct control of arabinose in *E. coli* (Nolan *et al.*, 1993; Better *et al.*, 1992; Bernhard *et al.*, 1994). One such protein, gelonin, was expressed as a secreted protein in *E. coli* and accumulated to greater than 1 g/liter as a fully active protein in the cell-free culture supernatant. Fusion proteins between antibody domains that can target antigens on human cells and cytotoxic molecules such as gelonin were also expressed under the transcriptional control of arabinose (Better *et al.*, 1995). These immunofusion proteins accumulated in the cell-free culture supernatant from arabinose-induced cells at greater than 400 mg/liter. Members of a family of immunofusion proteins expressed in *E. coli* under the transcriptional control of *araB* retained *in vitro* and *in vivo* biological activity comparable to that of chemically prepared immunoconjugates made from animal cell-produced whole antibodies and ribosome-inactivating proteins purified directly from plants.

Plasmids derived from pING1 have also been used to encode mammalian proteins that can become localized in the outer cell membrane of bacteria (Jacobs *et al.*, 1989; Romeyer *et al.*, 1990). In one example, a human metallothionein II gene was linked to the leader and membrane association fragment of the *E. coli* lipoprotein Lpp (Jacobs *et al.*, 1989). When induced with arabinose, bacteria carrying this expression vector directed an active metallothionein protein to the outer cell membrane. The recombinant protein was produced at ~75,000-fold over the uninduced level. This system was used to express an active heterologous protein that had previously been somewhat toxic and unstable in *E. coli.*

Several other groups have reported the construction of plasmid vectors that harness the *araB* expression system for the production of recombinant proteins. Cagnon *et al.* (1991) described a series of expression vectors that contained the *ara* expression system from pING1 in the vector pKK233.2 along with a number of other optional features. The promoter/operator region of *araB* was followed by a polylinker region for convenient gene cloning, and some vectors contained synthetic signal sequences, an f1 phage origin of replication, and mutated *araB* promoter sequences. The mutated *araB* promoter incorporated changes in the -10 region that made the promoter match more closely a consensus *E. coli* promoter. Interestingly, the promoter mutations resulted in a higher level of inducible expression (twofold) for a marker gene, but the uninduced expression level increased as well. Several recombinant proteins were expressed from this family of *ara* expression vectors, including the full-length Tat protein from the HIV virus (Armengaud *et al.*, 1991), the bacterial protein β-galactosidase (Cagnon *et al.*, 1991), the *Streptoalloteichus hindustanus* bleomycin-binding protein (Cagnon *et al.*, 1991), and the cholera toxin subunit B (CT-B) (Slos *et al.*, 1994). CT-B was linked to the *ompA* signal sequence and was expressed as a secreted protein. CT-B accumulated to approximately 60% of the total periplasmic protein, and at pilot scale CT-B was produced at about 1 g/liter. The majority of CT-B was released into the culture medium and could be recovered at greater than 80% efficiency from the cell-free culture medium.

Two additional reports of expression vectors have appeared. Perez-Perez and Gutierrez (1995) described an expression system in pACYC184 that remains compatible with ColE1-derived plasmids in an expression host. In addition, Guzman *et al.* (1995) described a series of versatile *araB* expression vectors that incorporate various selectable markers and multicloning sites. This latter series of vectors was studied extensively for the expression of native *E. coli* proteins, and it was demonstrated again that the *araB* system can be used "to achieve very low levels of uninduced expression, obtain moderately high levels of expression in the presence of inducer, and modulate expression over a wide range of inducer concentrations." These authors raise the possibility that the extent of arabinose induction can be regulated by the amount of inducer added to the culture, although Siegele and Hu (1997) suggest that intermediate expression levels in cultures simply reflect a population average of induced and uninduced cells.

Conclusions

In many cases, recombinant prokaryotic and eukaryotic proteins can be expressed readily in bacteria under the inducible control of the *araB* promoter. Some proteins have been efficiently expressed intracellularly where they can accumulate into inclusion bodies or fold into a native configuration. Alternatively, a number of proteins have been expressed under *araB* control that are well suited to fold correctly when they are directed across the cellular cytoplasmic membrane with a bacterial leader sequence. Numerous reports have highlighted that the arabinose promoter is tightly controlled and readily induced with arabinose. The *ara* system is easy to use both at laboratory scale and in large fermentors, and the inducing agent, L-arabinose, is widely available and relatively inexpensive. Compared to commonly used promoter systems such as *lac, tac,* and *trp,* the *ara* system is more tightly repressed in the absence of inducer and, compared to very strong promoters such as P_L, P_R, and T7, *ara* may be more adaptable and easier to use. Although some proteins may not be amenable to expression in bacteria at all, e.g., proteins that require posttranslational modification not possible in

Table 2
Advantages and Disadvantages of the *ara* Expression System for Bacterial Expression

Advantages	Disadvantages
Genes under *ara* control are tightly repressed in the absence of inducer	Regulated expression from the *araB* promoter requires the AraC protein
Upon induction, the resulting protein can be produced 1000-fold or more over the uninduced level	Bacterial systems cannot provide posttranslational modifications such as glycosylation
Arabinose in widely available and relatively inexpensive	
Very little arabinose is required for full induction	
The *ara* system is easy to implement both at laboratory scale and in large fermentors	
Expression yield of recombinant protein can be high	
The *ara* system can function in a variety of *E. coli* strains as well as in other bacterial species	
The *ara* system works well with proteins that are secreted as well as those that remain intracellular	
The control elements for the *ara* system are conveniently contained within about 300 bp of DNA	

prokaryotes, *ara* is an ideal and often overlooked system for controlling heterologous gene expression in bacteria (Table 2).

References

Amann, E., and Brosius, J. (1985). 'ATG vectors' for regulated high-level expression of cloned genes in *Escherichia coli. Gene* **40**, 183–190.

Armengaud, J., de Nuova Perez, L., Lemay, P., and Masson, J.-M. (1991). Production of a full length Tat protein in *E. coli* and its purification. *FEBS Lett.* **282**, 157–160.

Bernhard, S., Better, M., Fishwild, D. M., Lane, J. A., Orme, A. E., Garrison, D. A., Birr, C. A., Lei, S.-P., and Carroll, S.F. (1994). Cysteine analogs of recombinant barley ribosome-inactivating protein form antibody conjugates with enhanced stability and potency *in vitro. Bioconjugate Chem.* **5**, 126–132.

Better, M., and Horwitz, A. (1989). Expression of engineered antibodies and antibody fragments in microorganisms. *In "Methods in Enzymology"* (J.J. Langone, ed.), Vol. 178, pp. 476–496. Academic Press, San Diego, CA.

Better, M., Chang, C. P., Robinson, R. R., and Horwitz, A. (1988). *Escherichia coli* secretion of an active chimeric antibody fragment. *Science* **240**, 1041–1043.

Better, M., Weickmann, J., and Lin, Y.-L. (1990). Production and scale up of chimeric Fab fragments from bacteria. *ICSU Short Rep.* **10**, 105.

Better, M., Bernhard, S. L., Lei, S.-P., Fishwild, D. M., and Carroll, S. F. (1992). Activity of recombinant mitogillin and mitogillin immunoconjugates. *J. Biol. Chem.* **267**, 16712–16718.

Better, M., Bernhard, S. L., Williams, R. E., Leigh, S., Bauer, R. J., Kung, A. H. C., Carroll, S. F., and Fishwild, D. M. (1995). T cell-targeted immunofusion proteins from *E. coli. J. Biol. Chem.* **270**, 14951–14957.

Brosius, J. (1984). Plasmid vectors for the selection of promoters. *Gene* **27**, 161–172.

Cagnon, C., Valverde, V., and Masson, J.-M. (1991). A new family of sugar-inducible expression vectors for *Escherichia coli. Protein Eng.* **4**, 843–847.

Carter, P., Kelley, R. F., Rodrigues, M. L., Snedecor, B., Covarrubias, M., Velligan, M. D., Wong, W.-L. T., Rowland, A. M., Kotts, C. E., Carver, M. E., Yang, M., Bourell, J. H., Shepard, H. M., and Henner, D. (1992). High level *Escherichia coli* expression and production of a bivalent humanized antibody fragment. *Bio/Technology* **10**, 163.

Chang, C. N., Kuang, W.-J., and Chen, E. Y. (1986). Nucleotide sequence of the alkaline phosphatase gene of *Escherichia coli. Gene* **44**, 121–125.

Clark, A. A., Hammond, F. R., Papaioannou, A., Hawkins, N. J., and Ward, R. L. (1997). Regulation and expression of human Fabs under the control of the *Escherichia coli* arabinose promoter, P_{BAD}. *Immunotechnology* **3**, 217–226.

Elvin, C. M., Thompson, P. R., Argall, M. E., Hendry, P., Stamford, N. P., Lilley, P. E., and Dixon, N. E. (1990). Modified bacteriophage lambda promoter vectors for overproduction of proteins in *Escherichia coli. Gene* **37**, 123–126.

Gavit, P., Walker, M., Wheeler, T., Bui, P., Lei, S.-P., and Weickmann, J. (1992). Purification of a mouse-human chimeric Fab secreted from *E. coli. Bio. Pharm.* **5**, 28–58.

Goeddel, D. V., Yelverton, E., Ulrich, A., Heyneker, H. L., Miozzari, G., Holmes, W., Seeburg, P. H., Dull, T., May, L., Stebbin, N., Crea, R., Maeda, S., McCandliss,

R., Sloma, A, Tabor, J. M., Gross, M., Familletti, R. C., and Pestka, S. (1980). Human leukocyte interferon produced by *E. coli* is biologically active. *Nature (London)* **287**, 411–416.

Guzman, L.-M., Belin, D., Carson, M. J., and Beckwith, J. (1995). Tight regulation, modulation, and high-level expression by vectors containing the arabinose P_{BAD} promoter. *J. Bacteriol.* **177**, 4121–4130.

Horwitz, A. H., Heffernan, L., Morandi, C., Lee, J.-H., Timko, J., and Wilcox, G. (1981). DNA sequence of the *araBAD-araC* controlling region in *Salmonella typhimurium* LT2. *Gene* **14**, 309–319.

Huo, L., Martin, K. J., and Schleif, R. (1988). Alternative DNA loops regulate the arabinose operon in *Escherichia coli. Proc. Natl. Acad. Sci. U.S.A.* **85**, 5444–5448.

Jacobs, F. A., Romeyer, F. M., Beauchemin, M., and Brousseau, R. (1989). Human metallothionein-II is synthesized as a stable membrane-localized fusion protein in *Escherichia coli. Gene* **83**, 95–103.

Johnston, S., Lee, J.-H., and Ray, D. S. (1985). High-level expression of M13 gene II protein from an inducible polycistronic messenger RNA. *Gene* **34**, 137–145.

Kuhn, A., and Wickner, W. (1985). Isolation of mutants in M13 coat protein that affect its synthesis processing and assembly in phage. *J. Biol. Chem.* **260**, 15907–15913.

Lee, N., Francklyn, C., and Hamilton, E. (1987). Arabinose-induced binding of AraC protein to *araI*$_2$ activates the *araBAD* operon promoter. *Proc. Natl. Acad. Sci. U.S.A.* **84**, 8814–8818.

Lei, S.-P., Lin, H.-C., Wang, S.-S., Callaway, J., and Wilcox, G. (1987). Characterization of the *Erwinia carotovora pelB* gene and its product pectate lyase. *J. Bacteriol.* **169**, 4379–4383.

Masson, L., and Ray, D. S. (1986). Mechanism of autonomous control of the *Escherichia coli* F plasmid: Different complexes of the initiator/repression protein are bound to its operator and to an F plasmid replication origin. *Nucleic Acids Res.* **14**, 5693–5711.

Nolan, P. A., Garrison, D. A., and Better, M. (1993). Cloning and expression of a gene encoding gelonin, a ribosome-inactivating protein from *Gelonium multiflorum. Gene* **134**, 223–227.

Perez-Perez, J., and Gutierrez, J. (1995). An arabinose-inducible expression vector, pAR3, compatible with ColE1-derived plasmids. *Gene* **158**, 141–142.

Romeyer, F. M., Jacobs, F. A., and Brousseau, R. (1990). Expression of a *Neurospora crassa* metallothionein and its variants in *Escherichia coli. Appl. Environ. Microbiol.* **56**, 2748–2754.

Siegele, A., and Hu, J. C. (1997). Gene expression from plasmids containing the *araBAD* promoter at subsaturating inducer concentrations represents mixed populations. *Proc. Natl. Acad. Sci. U.S.A.* **94**, 8168–8172.

Slos, P., Speck, D., Accart, N., Kolbe, H. V. J., Schubnel, D., Bouchon, B., Bischoff, R., and Kieny, M.-P. (1994). Recombinant cholera toxin B subunit in *Escherichia coli:* High-level secretion, purification and characterization. *Protein Express. Purif.* **5**, 518–526.

Tabor, S., and Richardson, C. C. (1985). A bacteriophage T7 RNA polymerase/promoter system for controlled exclusive expression of specific genes. *Proc. Natl. Acad. Sci. U.S.A.* **82**, 1074–1078.

Yanisch-Perron, C., Vieira, J., and Messing, J. (1985). Improved M13 phage cloning vectors and host strains: Nucleotide sequences of the M13mp18 and pUC19 vectors. *Gene* **33**, 103–119.

Section II

EUKARYOTIC EXPRESSION SYSTEMS

5

ADENOVIRAL VECTORS FOR PROTEIN EXPRESSION

Dan J. Von Seggern and Glen R. Nemerow
Department of Immunology, The Scripps Research Institute, La Jolla, California 92037

Gene Expression Systems: Using Nature for the Art of Expression

Adenovirus Biology

Introduction

Adenoviruses (Ad) are a large family of nonenveloped, double-stranded DNA viruses. At least 47 serotypes of human adenovirus have been isolated, as well as related animal viruses including avian, canine, and bovine types. In humans, the most common serotypes generally cause mild, self-limiting infections of the respiratory tract or gut (Horwitz, 1996). Adenovirus has served as an important model system for studying many aspects of molecular biology, and a large body of knowledge has accumulated that allows it to be readily manipulated in the laboratory (for an extensive review of Ad biology, see Shenk, 1996). Many features of this viral system, including the ability to infect a wide variety of nondividing cells, a large capacity for insertion of DNA, ease of production and stability of the viral particles, and the high viral titers that can be produced, make Ad-based vectors especially useful in gene transfer. Vectors based on several Ad serotypes have been used for gene expression work both *in vitro* and *in vivo*.

Virus Structure and Receptor Interactions

The adenoviral particle consists of an icosahedral protein capsid encasing a double-stranded DNA molecule of approximately 36 kb (see Fig. 1). The most abundant viral protein is hexon, which makes up most of the outer shell of the virus (van Oostrum and Burnett, 1985). At each vertex is a complex composed of the penton base and fiber proteins, both of which interact with cellular receptors during the process of virus infection (Philipson *et al.*, 1968; Bergelson *et al.*, 1997; Tomko *et al.*, 1997; Wickham *et al.*, 1993). Several minor viral proteins also contribute to the capsid and may play roles in capsid assembly and in viral chromosome packaging. Ad DNA is packaged in a complex with several viral proteins (Mirza and Weber, 1982), and each end of the chromosome is covalently attached to a single molecule of terminal protein (TP) (Rekosh *et*

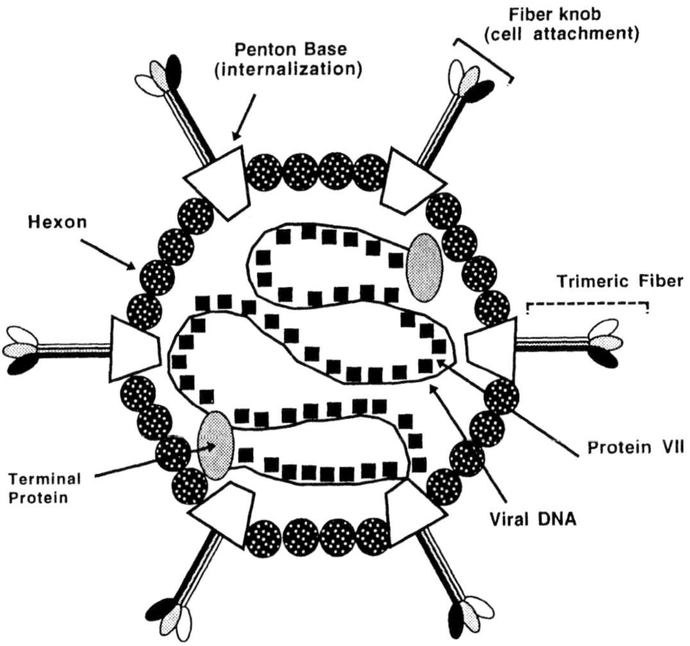

Figure 1 Idealized cross section of the adenovirus particle. Hexon protein makes up the majority of the capsid by mass. A complex of penton base and fiber proteins is located at each vertex. The 36-kb double-stranded DNA chromosome is packaged as a complex with protein VII, and one molecule of terminal protein is attached covalently at each end. For clarity, a number of minor proteins have been omitted from the diagram.

al., 1977). TP acts as a primer for DNA synthesis (Van der Vliet, 1995) and also serves to anchor the viral chromosome to the nuclear matrix (Schaack *et al.,* 1990) .

Genome Structure and Expression

Most gene transfer and expression experiments have been done with adenovirus serotypes 2 and 5. The entire genomic sequences of these viruses are known (Chroboczek *et al.,* 1992; Roberts *et al.,* 1984), and many years of work have been spent elucidating a well-defined program of viral gene expression and defining functions for most of the viral reading frames (Shenk, 1996). Historically, the Ad chromosome has been divided into 100 map units (1 m.u. is roughly equivalent to 360 bp). The positions of the major transcriptional units in the virus, which can be broadly divided into two classes, are shown in Fig. 2. Early genes, encoding mainly regulatory functions, are those transcribed shortly after infection. Late genes, including those encoding the major structural proteins, are highly expressed only after viral DNA replication begins. Both early and late Ad promoters have also been used to express heterologous genes.

Early Events in Infection Shortly after infection, transcription from viral early promoters begins. The most important regulator of viral growth is the immediate-early E1a protein, a promiscuous transcriptional activator that interacts with many cellular transcription factors (reviewed by Jones, 1995). E1a both activates further viral gene expression and modifies host cell regulation to drive the cell into S phase, which ensures that sufficient precursors for viral DNA and protein will be available. Deletion of the E1a gene blocks most viral gene expression as well as DNA synthesis and therefore renders Ad replication defective. Infection by E1-deleted Ad vectors often does not cause cytotoxic effects or grossly perturb the underlying cell biology; indeed, such vectors have been approved for administration to human subjects in many gene therapy protocols. These vectors are therefore useful in studies of protein function in intact cells, and most recombinant Ads now in use are E1 deleted.

E2 transcripts are spliced to produce several proteins that are involved in viral DNA synthesis. These include the viral DNA polymerase and terminal protein precursor encoded by E2b, and the

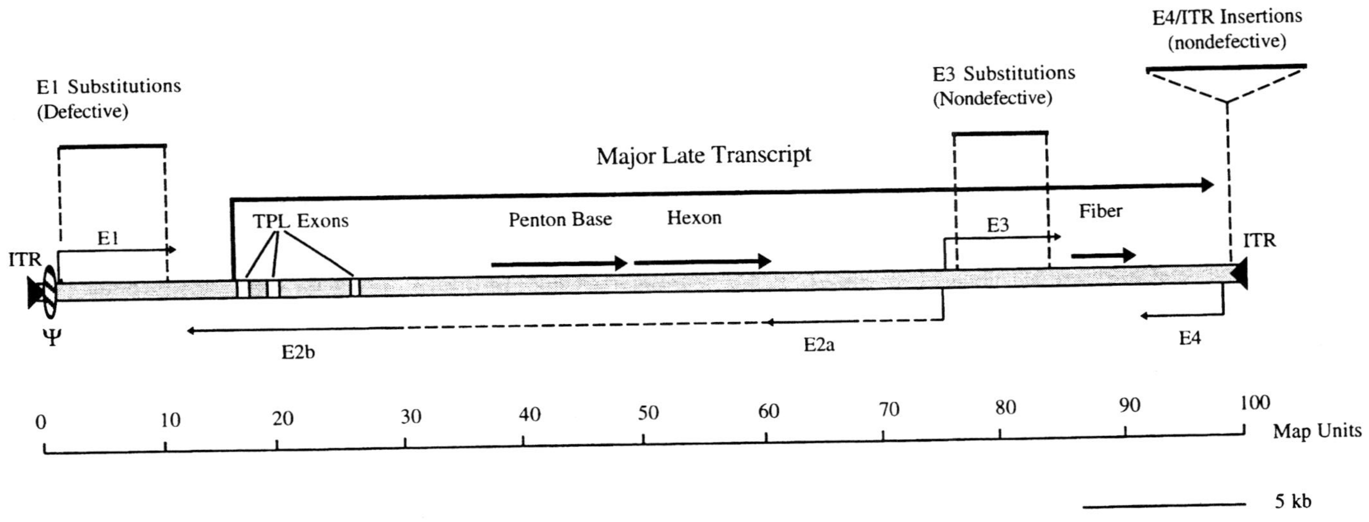

Figure 2 Diagram of the adenovirus type 5 (Ad5) genome. Locations of the E1, E2, E3, and E4 early transcriptional units are indicated. The major late transcript is represented by a heavy arrow, and the positions of the penton, hexon, and fiber open reading frames within this transcript are shown. The three small exons making up the tripartite leader (TPL) are shown as open boxes. The inverted terminal repeats (ITR) at each end of the chromosome are represented by black triangles, and the location of the essential packaging sequences is indicated by Ψ. The E1 and E3 regions can be substituted with foreign DNA as shown to generate defective and nondefective vectors, respectively. Nondefective vectors can also be produced by insertions between the right ITR and the E4 promoter. One map unit (m.u.) is equal to approximately 360 bp.

single-stranded DNA-binding protein produced from E2a. For a review of the roles of E2 products in DNA replication, see Van der Vliet (1995). E4 contains several open reading frames (ORFs) encoding proteins that regulate processes including viral transcription, the shut off of host cell macromolecular synthesis, and correct synthesis of unit length viral chromosomes (Falgout and Ketner, 1987; Weiden and Ginsberg, 1994; Halbert *et al.*, 1985).

Unlike other early regions, E3 products are not required for the replication of virus in cultured cells (Ginsberg *et al.*, 1989). Instead, these proteins are involved in evading the host antiviral immune response. E3 encodes a 19-kDa glycoprotein that downregulates the transport of major histocompatibility complex I molecules to the surface of an infected cell and reduces the inflammatory response generated against the vector (Ginsberg *et al.*, 1989). Other E3 products block tumor necrosis factor-induced lysis or apoptosis of infected cells (see Wold, 1993; Tollefson, 1998). Because the E3 ORF can be deleted without affecting viability, a foreign gene can be substituted into this region and expressed under control of the E3 promoter.

Late Events in Infection Viral DNA replication is initiated by about 8 hr postinfection and is followed by the activation of late viral gene expression. Genes encoding the major structural proteins, such as penton, hexon, and fiber, are all transcribed from the major late promoter (MLP), which initiates a long primary transcript spanning most of the virus genome. After the onset of viral DNA replication, the MLP is transcribed at very high levels and its transcripts are efficiently spliced into mRNAs encoding the various structural proteins (Shenk, 1996).

Three small exons totaling 200 nucleotides (the tripartite leader, TPL) are spliced onto the 5′ ends of all the late mRNAs (Chow *et al.*, 1997; Akusjärvi and Pettersson, 1979; Zain *et al.*, 1979). Late in infection, host cell protein synthesis is shut off due to inhibition of eIF-4E, a cellular translation initiation factor (Zhang *et al.*, 1994). The presence of the TPL confers eIF-4E independence and ensures that the viral mRNAs are translated preferentially (Dolph *et al.*, 1988, 1990). This allows viral proteins to accumulate to very high levels late in infection. These viral transcriptional and translational elements have also been used in recombinant Ad to express foreign proteins at high levels and are discussed further below.

As the late gene products accumulate, virus particles are assembled in the nucleus and released by lysis of the cell, which may involve virus-induced apoptosis (Tollefson *et al.*, 1996). An infected cell can produce 1000 or more progeny virus, allowing large numbers of virions to be isolated.

Adaptation of Adenovirus as a Gene Transfer Vector

Recombinant adenoviruses have been used to express foreign proteins for a number of years, and the recent explosion of interest in gene therapy has led to the development both of improved adenoviral vectors and of more efficient techniques for generating them. Adenovirus-mediated gene therapy is not the focus of this work and has previously been reviewed (Haddada *et al.*, 1995; Kozarsky and Wilson, 1993; Ali *et al.*, 1994). However, technology developed for gene therapy is, in many cases, applicable to the problem of protein expression in general, and improved vectors are likely to make the adenovirus system even more versatile.

Nondefective (Replication-Competent) Adenoviral Vectors

Provided that the total length of the recombinant DNA molecule does not exceed about 105% of wild type (Bett *et al.*, 1993), DNA can be inserted at several sites in the adenovirus chromosome to create a nondefective recombinant (see Fig. 2). Such vectors have been used to express foreign proteins in many studies, particularly for vaccine development studies where the ability of the virus to replicate postadministration may be advantageous. In addition to the commonly used serotypes 2 and 5, adenoviruses of other subgroups such as Ad4 and Ad7 have been used for protein expression (Imler, 1995). A property of nondefective vectors that may be useful in some situations is their ability to shut off host cell macromolecular synthesis. This effectively enriches the expression of a recombinant protein regulated by late viral transcriptional and translational elements.

In many studies, the nonessential E3 region has been replaced by a foreign sequence of interest (Schneider *et al.*, 1989; Morin *et al.*, 1987; Dewar *et al.*, 1989; Johnson *et al.*, 1988), and the plasmid-based systems now available for construction of recombinant viruses (see later) contain unique restriction sites to simplify cloning into the E3 locus. However, it may be advantageous to retain these sequences in viruses that are to be used for experiments *in vivo* because of the role of E3 in reducing the antiviral immune response (Wold, 1993; Ginsberg *et al.*, 1989; Tollefson, 1998). Helper-independent viruses can also be created by inserting DNA between the E4 promoter and the inverted terminal repeat at the right end of the chromosome (Saito *et al.*, 1985; Mason *et al.*, 1990). However, unless another viral sequence (such as E3) is deleted, the packaging size constraint limits such insertions to about 1.8 kb.

Defective (Replication-Incompetent) Adenoviral Vectors

A replication-defective adenovirus is produced by the removal of essential sequences (such as E1a or E4) from the viral chromosome. As it is possible for cells to be infected simultaneously with two different adenoviruses, defective mutants can be grown by coinfection with a nondefective helper that *trans* complements the missing function. Pure stocks of defective vectors can also be propagated by growth in cell lines that provide the missing viral function(s). The first such cell line, 293, was constructed by the transformation of human embryonic kidney cells with sheared Ad5 DNA and contains an integrated copy of the E1 region (Graham *et al.*, 1977). This allows growth of viral mutants in which the E1a region has been replaced by foreign sequences. Complementation by 293 cells is very efficient and allows growth of E1-deleted adenoviruses to near wild-type levels. A subline of 293, which was adapted for growth in suspension culture, has been described (Graham, 1987). This should be of use in producing large quantities of recombinant protein in high-density cultures.

Regions of the viral chromosome required for replication and packaging of DNA have been identified and make up only a small part of the chromosome. The ends of the Ad chromosome are composed of 105-bp inverted terminal repeat (ITR) sequence. Binding sites for a number of proteins within the ITR are necessary and sufficient for the initiation of viral DNA synthesis (Bernstein *et al.*, 1986; Hay, 1985). A 164 nucleotide sequence adjacent to the left ITR is also required for the packaging of viral DNA into particles (Hearing *et al.*, 1987). In principle, these are the only elements required in *cis*, and the rest of the 36-kb chromosome could be replaced with foreign DNA.

Defective vectors lacking essentially all of the Ad genome have been described (Haecker *et al.*, 1996; Parks *et al.*, 1996; Kumar-Singh and Chamberlain, 1997; Kochanek *et al.*, 1997; Fisher *et al.*, 1996). These vectors are propagated by coinfection with a helper virus and have very high capacities to accept DNA inserts. In one example of this "empty" vector strategy, the Cre/Lox recombinase system was used to remove the packaging signals from the helper virus chromosome during vector growth, rendering it unpackageable (Parks *et al.*, 1996). However, so far there are no such vectors that are truly helper independent, and it is currently impossible to remove all the helper virus from a preparation of vector. This may be a disadvantage, depending on the intended application of the vector. As helper viruses themselves are usually replication defective, their presence at low levels may not be an issue for *in vitro* experiments. It also appears that the minimum allowable size of DNA for efficient packaging is about 27 kb, as shorter chromosomes are not propagated stably (Parks and Graham, 1997).

Because E1a products are essential for progression through the viral life cycle, E1-deleted vectors are almost completely defective for growth in most cells and can generally be expected to have only minimal effects on host cell function. This has made them valuable tools for delivering genes in cell biology studies. However, the E1 deletion may not completely block the expression of viral genes following administration *in vivo*. This leads to an antivector immune response and contributes to the short-term gene expression seen in most gene therapy experiments to date (Yang *et al.*, 1994, 1995). This has prompted the development of vector systems deleted for additional viral genes such as E4 or the E2a DNA-binding protein, which are expected to be less immunogenic and safer *in vivo* (Wang *et al.*, 1995; Brough *et al.*, 1996; Yeh *et al.*, 1996; Krougliak and Graham, 1995; Zhou *et al.*, 1996; Brough *et al.*, 1996; Gorziglia *et al.*, 1996). These multiply deleted vector systems also have higher capacities for insertion of a gene of interest and should be more versatile transfer vectors for protein expression.

Construction of Adenovirus Expression Vectors

Adenovirus DNA is infectious, in that transfection of viral DNA alone into an appropriate cell line results in its replication, viral protein synthesis, and production of complete viral particles. This process is much more efficient if the DNA–TP complex is isolated intact from virions, probably due to the role of TP in priming viral DNA synthesis (Sharp *et al.*, 1976). A number of methods for generating recombinant virus by the manipulation of viral DNA and retransfection have been developed (for an excellent recent review of recombinant Ad construction techniques, including protocols, see Graham and Prevec, 1995). The large size of the adenovirus chromosome means that it contains few convenient unique restriction sites for insertion of DNA. Most protocols for the generation of recombinant viruses involve manipulation of the vector in two or more fragments, followed either by ligation *in vitro* or by cotransfection and recombination in host cells to regenerate a full-length chromosome. Use of the most common methods to construct an Ad recombinant with a cDNA inserted in place of the E1a gene is illustrated in Fig. 3.

Generation of Adenoviruses by Manipulation of Viral DNA

Recombinant adenoviruses were originally created by isolating fragments of the viral chromosome, followed by ligation or recombination to regenerate a complete viral backbone. Generally, DNA isolated from virions is digested with a restriction enzyme that removes the left end of the chromosome, including the viral packaging signals. This renders the large DNA fragment noninfectious. The unique ClaI site located at 2.6 m.u. is commonly used for this purpose (Fig. 3A). Undigested viral DNA and the smaller DNA fragments are often removed by sucrose gradient ultracentrifugation. The large fragment is ligated *in vitro* to a DNA fragment containing the left end of the chromosome and containing the desired modifications (Stow, 1981). If the viral DNA contained in this fragment has a deletion of the E1a region, the resulting recombinant Ad will be a replication-defective E1 mutant. Inclusion of unique restriction sites in the left end-containing plasmid (generally in the place of the E1a gene) simplifies insertion of a cDNA. The ligation products are then transfected into host cells to regenerate infectious virus (Fig. 3A).

In a variation of this method (Fig. 3B), the purified large fragment is transfected into host cells along with a plasmid that contains overlapping viral DNA, including the left end (Gluzman et al., 1982; Graham and Prevec, 1995; Haj-Ahmad and Graham, 1986).

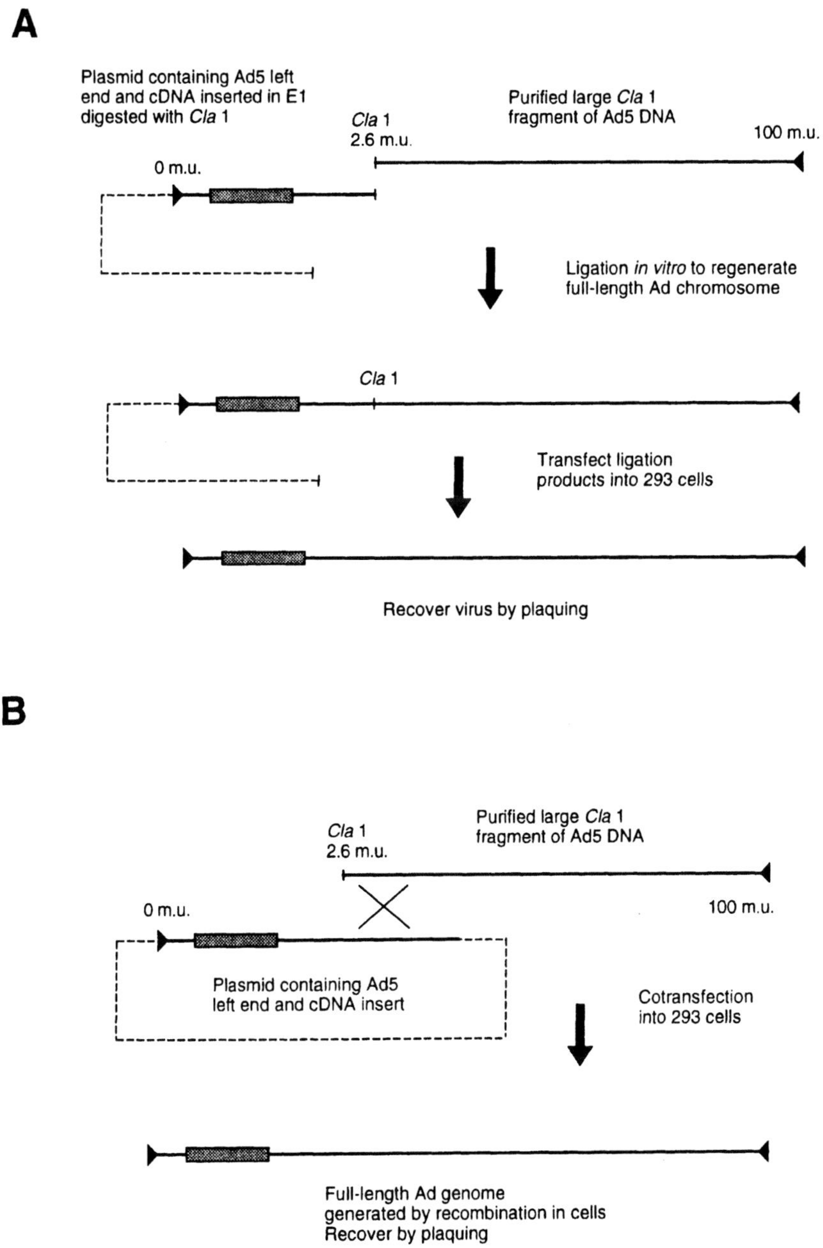

Figure 3 Commonly used methods for the construction of recombinant Ad. The use of each method to construct a replication-defective recombinant with a cDNA inserted into the E1 region is shown. Adenoviral DNA and plasmid sequences are indicated as solid and dashed lines, respectively. (A) *In vitro* ligation of a purified restriction fragment of Ad DNA containing most of the genome (here, the *Cla*I fragment containing the Ad5 sequence from 2.6 to 100 m.u.) to plasmid DNA representing the remaining part of the genome and modified by the insertion of the desired sequence. Ligation products are then transfected to 293

(continues on next page)

C

D

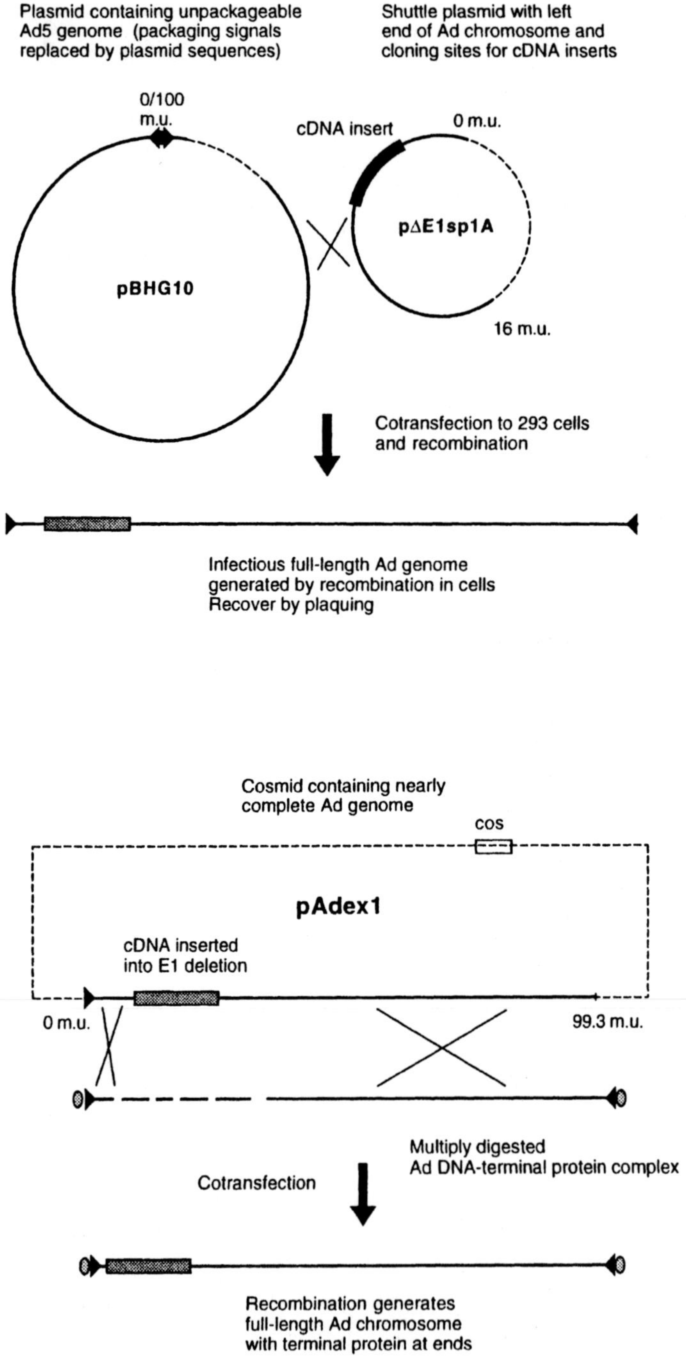

Plasmid containing unpackageable Ad5 genome (packaging signals replaced by plasmid sequences)
Shuttle plasmid with left end of Ad chromosome and cloning sites for cDNA inserts
0/100 m.u.
cDNA insert
0 m.u.
pBHG10
pΔE1sp1A
16 m.u.
Cotransfection to 293 cells and recombination
Infectious full-length Ad genome generated by recombination in cells
Recover by plaquing
Cosmid containing nearly complete Ad genome
cos
pAdex1
cDNA inserted into E1 deletion
0 m.u.
99.3 m.u.
Multiply digested Ad DNA-terminal protein complex
Cotransfection
Recombination generates full-length Ad chromosome with terminal protein at ends

An infectious Ad chromosome can then be generated by homologous recombination between the viral fragment and the plasmid sequences, resulting in the production of infectious virus. Similar methods have been used to create viral genomes with modifications or insertions near the right end of the chromosome (e.g., insertions into the E3 region).

These methods of generating recombinant Ads may produce a background of wild-type virus due to contamination with undigested viral DNA or religation of the wild-type fragments. This increases the number of plaques that must be isolated and screened to identify the desired recombinant. Imler *et al.* (1995) found that the background can be reduced if the donor virus from which the large restriction fragment is derived contains a copy of the HSV thymidine kinase gene inserted into the E1 region. This parental virus cannot form plaques in the presence of gancyclovir. Growth of any residual donor virus is therefore suppressed by the drug selection, and isolation of the desired recombinant is simplified. Schaack *et al.* (1995a) have described a donor virus containing a β-galactosidase gene near the left end of the genome. Plaques resulting from undigested parental DNA or religation of the two fragments in the host cells will express β-galactosidase and can be identified by incorporation of X-Gal (a chromogenic β-galactosidase substrate) into the agarose overlay.

Figure 3 *(continued)*
cells to generate replicating virus. (B) Overlap recombination in 293 cells between the purified adenoviral fragment as in (A) and a plasmid containing the left part of the viral genome following cotransfection into cells. (C) Use of the two-plasmid system described in the text. The cDNA of interest is inserted into one of the unique cloning sites in the plasmid pΔE1sp1a (which contains Ad DNA corresponding to 0–16 m.u. with an E1 deletion), and the resulting construct is cotransfected with a plasmid such as pBHG10, which contains an unpackageable Ad genome. A complete, packageable Ad chromosome is generated only after recombination takes place between the two DNAs. (D) The cosmid-terminal protein system. pAdex1 is a cosmid containing most (0–99.3 m.u.) of the Ad5 genome and is modified to contain a unique cloning site in place of the E1a gene. Following insertion of the desired cDNA into pAdex1, it is transfected into cells along with an Ad5 DNA–terminal protein complex that has been digested by a restriction enzyme that cuts at several sites in the genome. Recombination between pAdex1 and the terminal fragments of the Ad5 DNA generates a full-length chromosome with a molecule of terminal protein covalently attached at each end, which is efficiently packaged as virus.

Generation of Adenoviruses by Recombination of Plasmid DNA

An adenovirus chromosome, ligated into a circular *Escherichia coli* plasmid vector, can be excised from the plasmid and propagated as virus after transfection into an appropriate host cell line (Graham, 1984; Hanahan and Gluzman, 1984). Berkner and Sharp (1983) also demonstrated that two or more plasmids containing overlapping adenoviral DNA fragments could undergo recombination in cells to regenerate a complete viral chromosome.

Taking advantage of these findings, a two-plasmid system (Fig. 3C) has been developed to simplify the construction of recombinant viral genomes (McGrory *et al.*, 1988; Bett *et al.*, 1994). An essentially complete but unpackageable viral genome is cloned into one plasmid, such as pJM17 or pBHG10. A construct such as pJM17 is noninfectious due to the insertion of plasmid sequences into the viral region, rendering it too large to package into viral particles (McGrory *et al.*, 1988). Deletion of the viral packaging sequences from plasmids such as pBHG10, as shown in Fig. 3C, prevents replication of the nonrecombinant plasmid (Bett *et al.*, 1994). A second plasmid contains the left end of the viral chromosome, including an intact packaging sequence. The E1a region is generally deleted from this second plasmid (pΔE1Sp1A in Fig. 3C) and replaced by a linker with one or more unique restriction sites. The presence of unique cloning sites, and the smaller size of this second plasmid, greatly simplifies the process of inserting a gene of interest into the viral backbone. Another unique restriction site has been engineered into the E3 deletion of pBHG10, which allows insertion of foreign sequence here as well. After cloning of the desired sequences into either or both plasmids, they are cotransfected into a packaging cell line such as 293. Only after recombination takes place between the two plasmids will a packageable viral chromosome be generated.

This method allows propagation of all viral DNA segments in *E. coli,* which is technically much simpler than isolating and purifying fragments from virion DNA. The background of nonrecombinant virus produced following plasmid transfection can be very low. Using the pJM17 system, background plaques arising from rearrangement of the oversized vector plasmid have been reported (McGrory *et al.*, 1988; Graham and Prevec, 1995). Using a pBHG10-type plasmid (which lacks packaging signals) reduces the background of nonrecombinant plaques to essentially zero.

Use of an Adenovirus DNA–Terminal Protein Complex

As noted earlier, the Ad DNA–TP complex is much more infectious on a per microgram basis than plasmid DNA or proteinase-treated viral DNA alone. A method developed by Saito and co-workers (Fig. 3D) takes advantage of this to simplify the recovery of recombinant viruses (Miyake *et al.*, 1996). Most of the Ad genome is contained in the cosmid pAdex1, which has been modified to contain a unique restriction site for the insertion of a foreign sequence in place of the viral E1 region. Cosmid DNA is transfected into host cells along with an adenovirus DNA–TP complex that has been digested by a restriction enzyme that cuts the viral backbone many times. Fragmentation of the viral backbone reduces background due to the carryover of donor virus or to religation of its fragments. Homologous recombination between the pAdex cosmid and the terminal fragments of the donor viral DNA regenerates a full-length chromosome that has a molecule of TP covalently attached at each end. These workers report that recovery of the desired recombinant from this system is very efficient and much simpler than using purified viral DNA alone.

Other Systems for Adenovirus Vector Construction

Two methods have been described for generating and modifying full-length adenoviral chromosomes in lower organisms. This strategy has the theoretical advantage that regeneration of the desired Ad chromosome does not depend on any viral functions. This might simplify the recovery of poorly growing Ad mutants, which may be difficult to recover using the recombination strategies outlined earlier.

Ketner *et al.*, (1994) have succeeded in constructing a yeast artificial chromosome (YAC) containing the entire Ad2 genome. The Ad chromosome can be excised from the YAC by restriction digestion and transfected to mammalian cells. This system allows any desired alterations of the viral chromosome to be performed using the very powerful homologous recombination techniques available in the yeast system, without the need to isolate partial adenovirus clones. Methods for virus construction using recombination in *E. coli* have recently been described (Chartier *et al.*, 1996; He *et al.*, 1998). As in the yeast system, DNA isolated from bacteria after recombination is infectious and can be used to produce virus by transfection of mammalian cells. These vector systems have not yet come into widespread use, however.

Purification of Recombinant Adenoviruses

The presence of recombinant Ad following transfection of cells by appropriate DNA is readily detectable by the presence of plaques or cytopathic effect (CPE). Regardless of the method used for generating a recombinant Ad, the recovered virus should be plaque purified before use. Nondefective virus will form plaques on monolayers of essentially any cell type that is permissive for viral growth, whereas defective vectors must be grown on a complementing cell line (such as 293 for E1-deleted vectors). If the cell monolayers are overlaid with culture medium containing agarose following transfection, the resulting plaques can be picked directly for further purification (for Ad plaquing and propagation protocols, see Graham and Prevec, 1995).

Alternatively, transfected cells can be maintained under liquid medium and the wells or plates observed for the spreading cell death characteristic of an active viral infection (Zhang *et al.*, 1993). The virus present in a crude extract from the infected plate is then purified by plaquing. It is sometimes easier to detect the production of virus using this method, as detection of the relatively few plaques that result from transfection is dependent on the cell monolayer maintaining good morphology throughout the experiment. This is especially important if the recombinant virus grows slowly.

Production of Viral Stocks

A major advantage of the adenovirus system is the ease with which large quantities of high-titer viral stocks can be produced. Nondefective vectors can be propagated in a number of cell lines, most commonly epithelial cells such as HeLa. Defective vectors are either grown as mixed stocks, including a helper virus, or as pure stocks using complementing cell lines. For propagation of E1-deleted Ads, the standard has been the human embryonic kidney-derived cell line 293 (Graham *et al.*, 1977). Additional E1-complementing lines based on retinal epithelial cells or A549 lung carcinoma cells have been reported (Gorziglia *et al.*, 1996; Fallaux *et al.*, 1996). The newer generation of vectors deleted for viral genes such as E2, E4, or fiber are grown in packaging lines expressing the corresponding viral proteins (Yeh *et al.*, 1996; Gorziglia *et al.*, 1996; Wang *et al.*, 1995; Brough *et al.*, 1996; Zhou *et al.*, 1996; Langer and Schaack, 1996; Amalfitano *et al.*, 1996; Caravokyri and Leppard, 1995; Schaack *et al.*, 1995b; Weinberg and Ketner, 1983; Krougliak and Graham, 1995; Von Seggern *et al.*, 1998). In most cases, high-titer stocks of these viruses can be prepared readily.

Growth of Ad vectors requires no special facilities beyond those normally used for mammalian tissue culture (NIH Biosafety Level 2). The majority of virions produced remain associated with the cell, allowing concentration by pelleting infected cells prior to spontaneous lysis. A simple freeze–thaw lysis of infected cells is then performed, followed by banding of virus on CsCl gradients (Graham and Prevec, 1995; Everitt *et al.*, 1977). For some uses, such as infection of cells *in vitro* to produce a stock of recombinant protein, CsCl purification may be unnecessary. Purified virus is dialyzed into buffer containing 10% glycerol and is quite stable for long-term storage at −70 °C (Graham and Prevec, 1995). Storing the purified virus in 10 m*M* Tris, pH 8.1, 0.9% NaCl, and 10% glycerol has been found to improve stability. Titers of a standard E1-deleted adenovirus in excess of 10^{11} plaque-forming units (PFU)/ml can be obtained readily by this method. Chromatographic methods for virus isolation have been developed that may further simplify production and scaleup (Huyghe *et al.*, 1995).

Virus concentration has been determined by a number of methods. For a detailed study of several methods for Ad quantitation and of factors that affect results, see Mittereder *et al.* (1996). The number of virus particles in solution is often determined by absorbance at 280 nm or by a standard protein assay (1 μg of Ad2 proteins corresponds to 4×10^9 viral particles). An HPLC-based method for quantifying Ad particles has also been described (Shabram *et al.*, 1997). However, measuring the virus particle number does not provide information about the biological activity of a preparation, and infectious titers are generally reported in PFU. Plaque assays are straightforward to carry out, but require from 10 to 14 days from start to finish, depending on the rate of growth of an individual virus. A more rapid assay is the measurement of fluorescent focus units (FFU) (Thiel and Smith, 1967). Results of the FFU assay are available 1 to 2 days after infection of cells. Because of their high level of synthesis, viral structural proteins are good targets for immunostaining of infected cells. The authors have used polyclonal antibodies raised against either penton base or fiber to measure FFU.

Time Line for Vector Construction

The steps involved in producing a recombinant Ad vector with the standard two-plasmid system (Fig. 3C) are (1) cloning a foreign gene or genes into the appropriate Ad plasmid(s), (2) transfection into host cells followed by recombination, and (3) purification of the resulting virus. As shown in Fig. 4, a standard first-generation vector (e.g., insertion of a new expression cassette into the El region) can be generated in approximately 7–10 weeks. Much of the time involved is spent waiting for plaques or CPE to appear in the transfected cells and in the successive rounds of plaque purification, which do not require much "hands-on" work.

Applications of Adenoviral Vectors in Gene Expression

Proteins successfully produced from recombinant Ad (see below) have included heterologous viral proteins, intracellular enzymes, cell surface proteins, and secreted proteins such as cytokines and

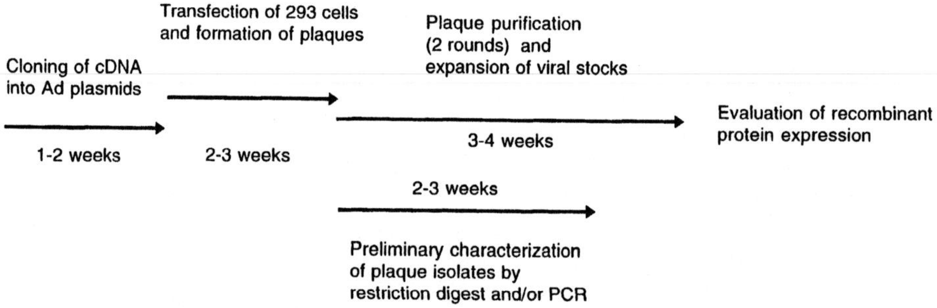

Figure 4 Representative time line for construction of a standard E1-deleted Ad recombinant. Note that after the primary plaques appear, both plaque purification and characterization of the recombinant virus (by restriction digestion or PCR analysis) can be carried out simultaneously.

growth factors. An area in which Ad-mediated expression is particularly useful is the study of protein function in the cell. Because of the interest in Ad-mediated gene therapy, much of this work has been aimed at expression in clinically relevant cell types rather than at producing large amounts of purified protein. Because of this, relatively few workers have reported the mass yields of their adenovirus-produced protein. A number of examples where the yields of expressed protein were reported are summarized in Table 1.

Schaffhausen *et al.* (1987) purified polyomavirus middle T antigen from cells infected with their Ad5 vector and reported yields of up to 100 μg protein/100-mm dish of infected 293 cells. Zhao *et al.* (1993) inserted the gene encoding a protein tyrosine phosphatase (PTP1C), driven by the MLP and TPL sequences, into the E1 region of an Ad5 vector. They purified the recombinant protein from infected 293 cells and reported isolating more than 100 μg of purified enzyme per 150-mm dish of cells (Zhao *et al.*, 1993). Garnier *et al.* (1994) used the same recombinant virus to explore scaleup of protein production using a 293 suspension cell line. They were able to produce the PTP1C protein at levels of 15% of the total cellular protein content (equivalent to 90 mg protein/liter of culture) by growing cells at high density (2×10^6 cells/ml). This type of strategy should be generally applicable for the large-scale production of other proteins.

Adenovirus-Mediated Expression of Heterologous Viral Proteins

An important application for recombinant adenoviruses has been the expression of heterologous viral proteins (Berkner, 1992), either for structure/function studies or for their possible use as vaccines. A discussion of Ad-based vaccines is outside the scope of this chapter, and this topic has previously been reviewed (Imler, 1995). However, this work illustrates some of the myriad potential strategies for Ad-mediated protein expression.

Hepatitis B virus surface antigen (HBsAg) has been produced from Ad-infected cells using a wide variety of strategies. In one case the HBsAg gene, driven by the MLP either with or without inclusion of TPL sequences, was inserted in place of the Ad5 E1 region (Davis *et al.*, 1985), and addition of the TPL resulted in approximately a 70-fold increase in protein synthesis. As expected for expression driven by late viral regulatory elements, the appearance of HBsAg in the medium was observed at late times (10–20 hr

Table 1
Recombinant Proteins Expressed Using Ad Vectors and Yields Obtained

Protein expressed	Position	Serotype	Promoter	Cell line[a]	Yield[b]	Reference
Human α_1-antitrypsin[c]	E1	Ad5	MLP + TPL	HeLa	6 μg/ml/10^6 cells	Gilardi *et al.* (1990)
Human α_1-antitrypsin[c]	E1	Ad5	MLP	HUVEC	0.3–0.6 μg/10^6 cells/day	Lemarchand *et al.* (1992)
β-Galactosidase	E1	Ad5	CMV	MRC5	27 μg/6 cm dish[d]	Wilkinson and Akrigs (1992)
Luciferase	E3	Ad5[e]	SV40 early	HeLa	20 μg/10^6 cells	Mittal *et al.* (1993)
PTP1C protein tyrosine phosphatase	E1	Ad5	MLP + TPL	293 S	90 mg/liter	Garnier *et al.* (1994)
PTP1C protein tyrosine phosphatase	E1	Ad5	MLP + TPL	293	100 μg/150 mm dish	Zhao *et al.* (1993)
HBsAg[c]	E1	Ad5	E1a	Vero	0.5–1 μg/10^6 cells	Ballay *et al.* (1985)
HBsAg	E4/ITR	Ad4[e], Ad7[e]	MLP + TPL	A549	3–4 μg/10^6 cells	Mason *et al.* (1990)
HbsAg[c] and IL-6[c]	E4/ITR, E3	Ad7[e]	MLP + TPL, E3	A549	8–10 μg/10^6 cells, 0.1–0.2 μg/10^6 cells	Lindley *et al.* (1994)
Hepatitis B precore antigen,[c] HbSAg[c]	E3, E4/ITR	Ad7[e]	E3, MLP + TPL	A549	1.4 μg, 6 μg /10^6 cells	Ye *et al.* (1991)
Measles virus N protein	E1	Ad5	CMV	MRC5	20% of soluble cell protein[b]	Fooks *et al.* (1995)
Porcine respiratory coronavirus spike glycoprotein	E3	Ad5	E3	ST	33 μg/10^6 cells	Callebaut and Pensaert (1995)

HIV-1 env	E3	Ad5[e]	E3	HeLa S	1 mg/liter of culture	Dewar *et al.* (1989)
HIV-1 env	E4/ITR	Ad7[e]	MLP + TPL	A549	4 μg/10^6 cells	Chanda *et al.* (1990)
Polyomavirus middle T antigen	E1	Ad5	MLP + TPL	293	100 μg/100-mm dish	Schaffhausen *et al.* (1987)
Murine IL-12 (both subunits)[c]	E1, E3	Ad5	CMV, CMV	293	42,000 U/10^6 cells/day	Bramson *et al.* (1996)
Soluble ciliary neurotrophic factor[c]	E1	Ad5	RSV LTR	1° rat astrocytes	120 pg/10^6 cells/hr	Smith *et al.* (1996)
Neurotrophin-3[c]	E1	Ad5	RSV LTR	1° rat astrocytes	350 pg/10^6 cells/hr	Smith *et al.* (1996)
Fibroblast growth factor 4	E1	Ad5	Sr α	Dami	2–3 μg/10^6 cells	Konishi *et al.* (1996)
Human IL–2[c]	E1	Ad5	CMV	Various human	1–2 μg/10^6 cells	Addison *et al.* (1995)
VEGF[c]	E1	Ad5	CMV	HUVEC	2.2 μg/10^6 cells/day	Mühlhauser *et al.* (1995)
GM-CSF[c]	E1	Ad5	CMV	293	8 μg/10^6 cells	Xing *et al.* (1996)

[a]HeLa, human cervical carcinoma cells; A549, human lung carcinoma; HUVEC, primary human vascular endothelial cells; MRC5, human fibroblasts; 293, Ad5-transformed human embryonic kidney cell line; 293S, suspension subline of 293; Vero, monkey kidney cell line; ST, swine testicle cell line; Dami, human megakaryocytic cell line.

[b]Yields reported by the authors were converted to μg recombinant protein/10^6 cells.

[c]Secreted to medium.

[d]Forskolin induced.

[e]Nondefective vector.

postinfection). HBsAg has also been placed into the E1 region downstream of the Ad5 E1a promoter (Ballay *et al.*, 1985) and inserted into the E3 region of a nondefective Ad5 (Morin *et al.*, 1987). In the latter case, the hepatitis B gene was inserted without a promoter, and its expression appeared to be driven by both early (probably E3) and late (possibly the MLP) Ad promoters.

Mason *et al.* (1990) inserted a cassette consisting of the Ad4 MLP, TPL, and HBsAg coding sequences between the right ITR and the E4 promoter of a nondefective Ad4. In addition to the TPL exons, their constructs retained varying amounts of the intronic sequence located between the first and the second TPL segments. As had been noted previously for the analogous Ad5 intron (Mansour *et al.*, 1986), these workers reported that sequences within this intron could dramatically upregulate mRNA levels driven by the MLP. HBsAg levels produced by infected A549 cells reached levels of almost 4 $\mu g/10^6$ cells in these experiments. Saito *et al.* (1985) also created a nondefective vector by inserting the HBsAg gene (this time driven by the HBV promoter) between E4 and the ITR of Ad5. This construct lacked a TPL and although HBV mRNA was readily detectable, HBsAg was expressed only at very low levels (less than 1 $ng/10^6$ cells), emphasizing the importance of translational regulation after infection with nondefective Ad vectors.

Finally, the ability of Ad vectors to accommodate multiple DNA insertions has been used to coexpress HBsAg along with other proteins. A nondefective Ad7 vector was constructed with MLP, TPL (including the first intron), and HBsAg sequences inserted between the E4 and right ITR, and either the HBV core or precore protein gene inserted in place of the E3 region and driven by the E3 promoter (Ye *et al.*, 1991). HBsAg was produced by these viruses with the late timing expected for MLP transcription and reached levels of 6 $\mu g/10^6$ cells in the medium. The E3-driven core and precore antigens were expressed by 12 hr after infection (as expected for Ad early genes) and were reported to reach levels of 1.46 and 0.33 $\mu g/10^6$ cells, respectively.

Fooks and co-workers (1995) used the CMV promoter to direct expression of the measles nucleocapsid (N) protein from an E1-deleted Ad5 vector. By treating infected cells with forskolin (to increase CMV promoter transcription), they were able to produce recombinant N protein at levels representing 20% of total soluble cellular protein (Fooks *et al.*, 1995). These workers compared the adenovirus-produced protein to recombinant N protein produced using a baculovirus system or expressed in *E. coli* (Warnes *et al.*,

1994). The approximate level of expression of the baculovirus- and adenovirus-expressed N proteins was found to be similar and considerably higher than in the *E. coli* system (the yields reported in this case were 40, 25, and 3% of total cellular protein, respectively).

The *env* gene of HIV-1 has been expressed using nondefective Ad5 (Dewar *et al.*, 1989) and Ad7 (Chanda *et al.*, 1990) vectors, with proper cleavage of the gp160 protein into the gp120 and gp41 fragments and syncitia formation by infected cells. Dewar *et al.* (1989) found that with their Ad5-based vector, gp160 was produced in HeLa suspension cultures at approximately 1 mg/liter. Chanda *et al.* (1990) reported production of very high levels (2–5 mg/liter of culture) of gp160 in A549 cells infected with their Ad7 vector if the HIV-1 *rev* gene product was provided, either by coinfection with a rev-expressing Ad vector or from a *rev* gene fragment expressed from the Ad E3 promoter. A genomic fragment of SIV containing the overlapping *env* and *rev* genes was inserted into the E3 region of Ad5 (Cheng *et al.*, 1992), and proper splicing of the mRNA to produce both SIV proteins was observed. The presence of the *rev* gene upregulated *env* protein production in this context as well, indicating that a complex foreign transcript incorporated into an Ad vector can be spliced into two different functional mRNAs that produce appropriately interacting proteins.

Adenovirus-Mediated Expression of Cytokines and Growth Factors

The emphasis on potential therapeutic uses of adenovirus has led a number of investigators to insert cytokine or growth factor genes into Ad vectors. Biologically active interleukin (IL)-6 was expressed using a recombinant Ad that also contained the HBsAg gene (Lindley *et al.*, 1994). A CMV-driven human IL-2 gene was inserted into the E1 region of an Ad5 vector, and expression was evaluated in several human cell lines (Addison *et al.*, 1995). Secretion of IL-2 was measured at 25–110 ng/10^6 cells over a period of several days. The recombinant protein was active, as measured by the ability to stimulate growth of an IL-2-dependent cell line. Smith and coworkers (1996) inserted the genes for secretable ciliary neurotrophic factor (sCNTF) or for neurotrophin-3 into replication-defective Ad and reported that both viruses allowed secretion of their respective proteins into the medium at concentrations sufficient to promote neuronal survival *in vitro*. Other growth factors that have been expressed using Ad vectors include fibroblast growth factor-4 (HST-1/FGF-4) and vascular endothelial growth factor (VEGF) (Konishi *et al.*, 1996; Mühlhauser *et al.*, 1995).

Expression of heterodimeric proteins such as IL-12 is complicated by the need to deliver the genes encoding both subunits to the same cells in approximately equal dosages. Here, the ability to insert more than one foreign DNA segment into the adenovirus chromosome is valuable. Bramson *et al.* (1996) inserted a gene encoding the p35 subunit of murine IL-12 into the E1 region and that encoding the p40 subunit into E3, both driven by the strong CMV promoter. This strategy ensured that all infected cells received both genes at the same copy number. Infection of 293 cells or the human fibroblast line MRC5 with the IL-12-bearing Ad vector resulted in the secretion of biologically active IL-12 into the medium at levels up to 40,000 units/10^6 cells (Bramson *et al.*, 1996).

Adenovirus-Mediated Protein Expression for Cell Biology Studies

A major advantage of the adenovirus system is that highly efficient infection of a population of cells can be achieved in a relatively synchronous way. A protein of interest can be introduced readily into most or all cells in an experimental setting, and infection of primary cultures, biopsy samples, or cell lines that are difficult to transfect using standard methods has been reported. By using a replication-defective Ad vector, it is often possible to infect cells without grossly perturbing normal cell functions, thus allowing analysis of the functional properties of the recombinant protein.

Askanas *et al.* (1996) used an Ad vector to express the β-amyloid precursor protein (β-APP) in primary cells derived from biopsy specimens of normal human muscle in order to determine whether β-APP could cause degenerative changes seen in inclusion body myositisis, a degenerative muscle disease. By using the strong CMV immediate-early promoter, this group was able to express high levels of the recombinant β-APP in essentially all cells and to demonstrate that its overexpression caused morphological changes similar to those seen in the disease state.

Huang *et al.* (1997) recently investigated the role of different MAP kinase isoforms in Fas-induced apoptosis by expressing several different MAP kinase molecules, as well as dominant negative, constitutively active, and chimeric forms of these proteins, in the

Jurkat T-cell line. This line is refractory to transfection by standard methods, and the ability of Ad to synchronously transfect essentially all of the cells was instrumental in defining the role of a specific MAP kinase molecule in the apoptosis pathway.

In a study of glucose metabolism, an Ad vector allowed expression of hexokinase 1 in the majority of cells of isolated (but intact) rat islets of Langerhans (Becker *et al.*, 1994). Here, the Ad system allowed transfection of a much higher percentage of cells (60–70%) than could be transfected by nonviral methods. These investigators were able to gather information about hexokinase overexpression without the use of transgenic animals.

In endothelial cells, Ad-mediated expression of a nuclear-localized form of the NF-κB inhibitor, Iκ-Bα, inhibited NF-κB activity and blocked expression of several NF-κB-dependent markers of cell activation in response to lipopolysaccharide. NF-κB-independent markers such as secretion of von Willebrand's factor or of prostacyclin were unaffected (Wrighton *et al.*, 1996). In another study, Ad vectors were used to investigate signal transduction by the type II transforming growth factor-β (TGF-β) receptor (Yamamoto *et al.*, 1996). Expression of a truncated receptor in cultured cells completely blocked both the antiproliferative and the transcriptional activation effects of TGF-β treatment, without affecting the response to two other growth factors.

The ability of more than one adenovirus to infect the same cell has also been used to coexpress multiple protein subunits. Herpes simplex virus 1 encodes a heterodimeric immunoglobulin receptor that is expressed on the surface of an infected cell. After cloning the genes for the two subunits into separate Ad vectors, Hanke *et al.* (1990) expressed either or both subunits simply by infecting target cells with either or both recombinant Ads. This allowed determination of the IgG-binding properties of the individual subunits as well as of the heterodimeric complex on the surface of the infected cells.

In Vivo Expression of Recombinant Adenoviral-Encoded Proteins

Ad vectors can efficiently transfect a large number of organs or tissues *in vivo,* including liver (Jaffe *et al.*, 1992), kidney (Moullier *et al.*, 1994), skin (Setoguchi *et al.*, 1994b), brain (Akli *et al.*, 1993; Doran *et al.*, 1997), skeletal muscle (Ragot *et al.*, 1993), heart (Stratford-Perricaudet *et al.*, 1992), lung epithelium (Rosenfeld *et al.*,

1991), and several ocular tissues (Mashhour *et al.*, 1994) (for further review, see Ali *et al.*, 1994, or Kozarsky and Wilson, 1993). This property allows protein function to be investigated *in vivo* without generating transgenic animals.

Adenoviral gene delivery is particularly useful if a large number of variant proteins are to be evaluated. For example, to identify the domains of the closely related lipoprotein lipase and hepatic lipase proteins responsible for determining their substrate specificity *in vivo*, Kobayashi *et al.* (1996) made several chimeric proteins by exchanging domains between the two and cloned the resulting genes into E1-deleted Ad vectors. They systemically administered these viruses to hepatic lipase-deficient mice and examined changes in serum lipids. These experiments showed that the greater ability of hepatic lipase to hydrolyze phospholipids and to reduce total serum cholesterol is largely determined by the "lid" domain of the proteins, which is thought to restrict access to the active site of the enzyme.

Cytokine and chemokine functions *in vivo* have been studied using Ad vectors to locally produce factors, including GM-CSF, RANTES, and IL-6 in rat lung (Xing *et al.*, 1994, 1996; Braciak *et al.*, 1996). Expression of these molecules in a defined tissue has allowed investigation of their roles on inflammatory processes and recruitment of cells such as monocytes and T lymphocytes. Local expression of Ad-delivered genes may be more relevant to their normal physiological roles than systemic administration of the purified protein.

Systemically administered Ad vectors can also provide high-level temporary expression of a protein that is secreted into the bloodstream. Much of this work has focused on preclinical applications, as in expression of factor VIII for the treatment of hemophilia A (Connelly *et al.*, 1996) or of erythropoietin to stimulate erythropoiesis (Setoguchi *et al.*, 1994a), but this property can also be useful in understanding the *in vivo* roles of secreted proteins. For example, rats treated intravenously with a viral vector directing expression of the rat leptin protein exhibited reduced food intake and weight gain, as well as a disappearance of fat deposits (Chen *et al.*, 1996). Intraperitoneal administration of an Ad vector containing the gene encoding HST-1/FGF-4 to mice stimulated platelet production (Sakamoto *et al.*, 1994). This group then used the same viral vector in an *in vitro* study to demonstrate that the increased platelet count was due to FGF-4 stimulation of megakaryocyte maturation (Konishi *et al.*, 1996).

Miscellaneous Applications

As noted previously, the large size of the Ad chromosome should allow for the transfer of very large genes. Vectors containing only the minimum Ad packaging sequences have been used to express full-length human (Haecker *et al.*, 1996) or murine (Kochanek *et al.*, 1997; Kumar-Singh and Chamberlain, 1997) dystrophin in cultured myotubes and *in vivo*. The 14-kb dystrophin cDNA far exceeds the capacity of other viral vectors now in widespread use.

Several groups have constructed Ad vectors that express the Cre recombinase protein of bacteriophage P1. This enzyme catalyzes recombination between 34-bp target sites (*LoxP*) with excision of the intervening DNA. Expression of Cre from an Ad recombinant could remove an inactivating sequence flanked by *LoxP* sites and switch on expression of a reporter gene either located on a second Ad coinfected along with the Ad–Cre recombinant or stably integrated into the host cell chromosome (Anton and Graham, 1995; Miyake *et al.*, 1996; Sakai *et al.*, 1995). Wang *et al.* (1996) extended these findings *in vivo* and demonstrated that the Ad–Cre system was capable of causing efficient recombination between *LoxP* sites on the chromosomes of transgenic mice. This approach should be useful in conditionally expressing proteins or in producing gene knockouts that are spatially and/or temporally restricted.

Another application for Ad-mediated gene delivery is modulation of the receptor repertoire expressed on the surface of infected cells. Lieber *et al.* (1995) used an Ad vector to deliver the amphotrophic retrovirus receptor (RAM) cDNA to cells that are normally resistant to infection with the amphotropic retroviral vectors commonly used in gene therapy work. Following Ad infection, target cells were markedly more infectible with a recombinant retrovirus. An Ad recombinant was used to express human CD4 on the surface of a number of CD4-negative cell lines (Yasukawa *et al.*, 1997). In several of these lines, CD4 expression allowed infection by the herpesvirus HHV–7, confirming the role of CD4 in HHV-7 entry.

As replication-defective Ad vectors can be introduced to a relatively small area of tissue by local injection or application, they have utility as markers for the study of cell migration and development. When *LacZ*-containing Ad vectors were administered to the external surface of early stage chicken hearts, β-galactosidase-expressing cells in some, but not all, regions of the heart were detectable at later times (Fisher and Watanabe, 1996), providing information about the origin of the cells making up various heart structures. Similar experiments have demonstrated that Ad can be used to mark cells in the developing rat central nervous system and in preimplantation mouse embryos (Lisovoski *et al.*, 1994; Tsukui *et al.*, 1995).

Considerations in the Use of Adenovirus Expression Vectors

Choice of Insertion Site: E1 vs E3

Insertion of foreign DNA into either or both of the E1 and E3 regions is compatible with high-titer virus production. The ability to insert two genes into the same vector chromosome has obvious advantages if a heterodimeric protein (Bramson *et al.*, 1996) or two proteins designed to work together (Hanke *et al.*, 1990) are to be expressed in the same cells. There are functional differences in the use of E3 vs E1 insertions, however, and one site may be preferable over the other for a particular application.

Substitution of the nonessential E3 region by foreign DNA has the advantage that replication-competent viruses can be generated readily. This property is of special interest for vaccine development, and many E3 substitution vectors have been designed for this purpose (Imler, 1995) . The most common method of generating E3 insertions in Ad5 vectors has been replacement of the *Xba*I fragment between 78.8 and 85.7 m.u. by foreign sequence. This removes most E3 coding sequences but leaves the E3 promoter intact, which allows it to direct expression of the transgene. A number of studies using such substitutions have reported that transgene expression was largely E3 driven, even when the inserted DNA contained its own promoter sequences (Both *et al.*, 1993; Schneider *et al.*, 1989; Johnson *et al.*, 1988). Expression of genes inserted into the E3 region of a nondefective vector may also be affected by late viral promoters, probably the MLP, late in infection (Dewar *et al.*, 1989; Morin *et al.*, 1987). This viral regulation of transgene expression may be a disadvantage if cell type or temporally restricted expression is desired.

In contrast, the E1 deletions generally used remove the E1a promoter sequences. In many cases, cellular promoters incorporated into the E1 region of adenoviral vectors have been shown to be regulated in a manner similar to their endogenous genes, allowing expression of a gene of interest in a cell-type or tissue-specific manner. As E1-deleted vectors are almost completely replication defective and can have minimal effects on normal cellular processes, they have been chosen for most work involving cell biology studies or *in vivo* experiments.

Use of Viral Regulatory Elements

Because viral proteins are translated preferentially at the expense of host proteins late in infection, it can be advantageous to use viral late gene control elements for applications where very high level expression is desired (for a detailed discussion of this topic, see Berkner, 1992). As noted earlier, expression of a foreign gene placed under the control of the viral MLP and TPL can be enhanced along with that of bona fide viral gene products (Berkner *et al.*, 1987; Alkhatib and Briedis, 1988; Berkner, 1992; Davis *et al.*, 1985; Garnier *et al.*, 1994). Sequences in the first intron of the TPL have also been found to increase MLP transcription in infected cells, contributing to the very high level expression seen in some of these studies (Mansour *et al.*, 1986; Mason *et al.*, 1990). The highest yields of recombinant protein reported from Ad vectors (Table 1) have been from constructs using both the MLP and the TPL for transgene expression in cells that support productive viral infections (e.g., E1-deleted vectors in 293 cells or nondefective vectors in other cell types).

Inclusion of the TPL in expression constructs has been reported previously to boost the production of a variety of proteins, even in uninfected cells (Sheay *et al.*, 1993), and TPL can increase greatly the expression of the Ad5 fiber protein in 293-based cell lines in the absence of any viral infection (Von Seggern *et al.*, 1998). These sequences may therefore improve protein yield even in the context of a nonpermissive cell type infected by a defective virus and may have general value in protein expression work.

As noted earlier, the E3 promoter has been used (sometimes inadvertently) to express many foreign genes. E3 is transcribed in many cell types, and transcription is activated within a few hours postinfection.

Use of Heterologous Promoters

Many different promoters have been inserted into Ad vectors and in most cases have retained their activity. Widely used viral promoters include the long terminal repeat of RSV, the SV40 early promoter, and the cytomegalovirus (CMV) major immediate-early promoter. The CMV promoter provides very strong expression of a transgene in many cell types, at least in the short term (Jiang *et al.*, 1996; Guo *et al.*, 1996), and has become perhaps the most widely used promoter in adenoviral vectors. Bartlett *et al.* (1996) have reported that the constitutively active U1 small nuclear RNA promoter, which is transcribed in essentially all cell types, has activity similar to CMV and is active when placed into an E1-deleted Ad vector.

When placed in the adenovirus context, a number of tissue-specific promoters have been shown to be regulated much like their cellular counterparts. Friedman and co-workers (1986) found that the liver-specific rat albumin promoter in an Ad5 vector was expressed in rat or human hepatoma cells or in primary hepatocytes, but not in a myeloma cell line, whereas the B-cell-specific immunoglobulin heavy chain promoter was expressed strongly in myeloma but not hepatoma cells. More recently, studies of a modified mouse albumin promoter driving expression of the human factor VIII gene demonstrated specificity both for hepatocytes *in vitro* and for liver *in vivo* (Connelly *et al.*, 1996).

A number of groups have sought to target adenovirus-based gene expression to tumor cells. Hepatoma cells often express elevated levels of α-fetoprotein (AFP) relative to normal liver. A recombinant Ad carrying AFP promoter and enhancer sequences driving the herpes simplex virus thymidine kinase (TK) gene expressed high levels of TK in AFP-expressing tumor cell lines such as HuH-7 or HepG2, but not in non-AFP-expressing liver cell lines such as HLF (Kanai *et al.*, 1996). Other workers have reported that expression of foreign genes placed under the control of the carcinoembryonic antigen (CEA), osteocalcin (OC), or DF-3 promoters in Ad vectors is specific for CEA-, OC-, or DF-3-expressing tumor cells, respectively (Lan *et al.*, 1996; Ko *et al.*, 1996; Chen *et al.*, 1995). Hashimoto *et al.* (1996) found that Ad-mediated expression using two different neural cell-specific promoters was restricted to the expected cell types both in culture and after administration of the respective Ad vectors to rat brain *in vivo*.

The ability to control the level or timing of transgene expression may be desirable in a particular experimental setting. A number of inducible or regulated promoters have been placed into the Ad backbone. For example, a synthetic construct consisting of minimal thymidine kinase promoter sequences and several retinoic acid response elements was inserted into the E1 region of an Ad5 vector. Transcription of a reporter gene in infected cells could be induced by the addition of retinoids, both in culture and *in vivo* (Hayashi *et al.*, 1994). The mouse metallothionine promoter also retained its zinc-inducible expression in an Ad context (Yajima *et al.*, 1996). Varley *et al.* (1995) made Ad constructs containing a luciferase gene driven by either of two inflammation-responsive promoters. Cells infected with either virus were found to express high levels of luciferase in response to the cytokines present in conditioned culture medium. *In vivo* studies demonstrated that luciferase expression was upregulated in tissues of mice infected with these vectors and challenged with inflammatory agents such as lipopolysaccharide toxins or turpentine (Varley *et al.*, 1995).

The pattern that emerges from these studies is that cellular promoters in an Ad chromosome often behave as would be predicted from their normal activities, although it may be worth noting that all of these specific promoter activities were generated using insertions into the viral E1 region. The ability to reproduce cell-specific patterns of gene expression after adenovirus-mediated transfer will be invaluable in targeting transgene function to specific cells and may be useful in optimizing expression of a recombinant protein in the cell type of interest.

Advantages and Disadvantages of Adenoviral Vectors

The unique advantage of the adenoviral expression system is its versatility. Once constructed, a single vector can be used for *in vitro* protein expression and purification, studies of the effect of the gene product on cell biology, or *in vivo* studies in many tissue types and a number of different species. For these reasons, adenovirus is becoming the system of choice for work involving both *in vitro* and *in vivo* studies.

The investigator has the option of using either well-characterized viral regulatory sequences or a heterologous promoter, which should allow expression to be tailored to the requirements of the particular system of interest. The large (and growing) capacity of Ad vectors allows expression of large genes or of a combination of genes, as in expression of both subunits of a heterodimeric protein or of an antigenic protein together with a cytokine. Finally, the wide tropism and ability to infect nondividing cell types allows Ad vectors to be used in many cases where other vectors, such as retroviruses, cannot be used.

The major disadvantage of this system is the complexity of vector construction. A recombinant vector for Ad-based expression of a protein can take considerably longer to produce than a plasmid for use in a prokaryotic expression system or simple transfection into eukaryotic cells. However, techniques for adenovirus construction have improved dramatically. The plasmid-based systems now available for construction of standard E1- or E3-deleted vectors are quite straightforward, and the effort required for the construction of recombinant adenoviruses is similar to that needed to construct recombinant baculoviruses.

As Ad DNA does not normally integrate into host cell chromosomes, the expression of the transgene is transient. This is of little consequence in short-term or *in vitro* experiments, but may limit some long-term *in vivo* studies.

Although Ad vectors are being used successfully *in vivo* to study protein function, there are still problems associated with the host response to the vector itself. Following the administration of currently available Ad vectors, antivector immune responses are generated that can lead to inflammation, elimination of the infected cells, and short duration of transgene expression. This limits long-term studies in immunocompetent animals. Infection even by replication-defective Ads also affects the endogenous signal transduction pathways (Li *et al.*, (1998) and can lead to responses in the infected cells, such as the production of cytokines. This can potentially complicate the interpretation of studies of protein expression *in vivo*. This type of effect can often be controlled for by administration of an Ad vector lacking an insert.

Finally, although Ad vectors have been found to infect a wide variety of cells in several species, including human, mouse, rat, and chicken, not all cells of interest may express the appropriate integrin and CAR proteins to allow infection. The repertoire of integrin and fiber receptor expression on the target cell type should therefore be considered when planning *in vivo* work using Ad vectors (Table 2).

Table 2
Advantages and Disadvantages of Adenoviral Vectors

Advantages	Disadvantages
Versatility; recombinant protein expressed from same vector both *in vitro* and *in vivo*	Complexity of recombinant Ad construction
High capacity for (multiple) DNA insertion	Expression is generally transient
Ability to use wide variety of cellular promoters in Ad recombinants	*In vivo* antivector immune response
High-level protein expression *via* late Ad regulatory elements	Current Ad vectors cannot be specifically targeted *in vivo*
Ease of producing high-titer viral stocks	
Wide tropism of Ad includes many nondividing cells	

Future Directions

As interest in adenovirus as a gene therapy vector remains high, further improvements in vector design and techniques for their construction can be expected. Packaging systems for the complementation of additional Ad genes such as E2, E4, terminal protein, or fiber are now available (Krougliak and Graham, 1995; Yeh *et al.*, 1996; Caravokyri and Leppard, 1995; Schaack *et al.*, 1995b; Gorziglia *et al.*, 1996; Wang *et al.*, 1995; Brough *et al.*, 1996; Weinberg and Ketner, 1983; Amalfitano *et al.*, 1996; Zhou *et al.*, 1996, Von Seggern *et al.*, 1998). Viral vectors deleted for the corresponding sequences will have a higher capacity for insertion of DNA and should be useful for expressing large proteins such as dystrophin or combinations of genes (perhaps multisubunit enzyme complexes). As they are expected to be less immunogenic and to persist longer *in vivo*, they should also improve the prospects for the use of Ad to study protein function in the context of an intact animal.

Ongoing work is also extending the already impressive list of cell types that can be infected. Some cell types (most notably many hematopoietic cells) are difficult to infect using Ad. This is largely due to these cells lacking expression of either the CAR protein, which serves as the fiber receptor (Bergelson *et al.*, 1997; Tomko *et al.*, 1997), or integrins appropriate for the internalization of virus (Wickham *et al.*, 1993). For example, in human airway organ culture, regenerating cells were found to be much more infectible than normal cells (Dupuit *et al.*, 1995). Another study showed that immature cells in the airway were much more susceptible to infection by Ad and that these cells expressed higher levels of the integrins needed for virus internalization (Goldman and Wilson, 1995). In skeletal muscle, the infectibility of different cell populations has also been shown to correlate with α_v integrin expression (Acsadi *et al.*, 1994). Huang and colleagues have demonstrated that lymphoid cells can be infected efficiently if they are treated with growth factors that upregulate the synthesis of α_v integrins (Huang *et al.*, 1995) and that there is an alternate pathway of infection using α_m integrins in at least some lymphoid cell types (Huang *et al.*, 1996).

Several groups are now developing adenoviruses that have modified tropism by altering the fiber or penton base proteins that interact with cellular receptors. Wickham *et al.* (1996) have described Ads with epitope tags incorporated into the penton base, allowing the virus to be redirected using a bispecific antibody approach. The modified virus was shown to infect endothelial and smooth muscle cells, which are normally difficult to infect due to low CAR expression (Wickham *et al.*, 1996).

Replacement of the fiber gene in a vector of one serotype by that of another has been shown to confer the cell-binding specificity associated with the new fiber serotype (Gall *et al.*, 1996). Chimeric fiber genes containing the receptor-binding domain of a different Ad serotype have also been used to alter viral tropism (Stevenson *et al.*, 1997; Krasnykh *et al.*, 1996). Michael and colleagues added a short peptide sequence from gastrin-releasing peptide to the fiber (Michael *et al.*, 1995) and demonstrated that this peptide epitope is accessible for antibody binding at the surface of the fiber protein.

A high-resolution structure for the receptor-binding knob domain of the fiber protein is now available, which should lead to improved strategies for redirecting fiber binding (Xia *et al.*, 1994). This type of work should lead to Ad vectors that can be targeted to particular cell types, extending the utility of the system for studies *in vivo* or using tissue samples.

Conclusions

Adenoviral gene transfer vectors are extremely versatile and are becoming widely used, both for protein expression *in vitro* and for studies of protein function *in vivo*. The high insert capacity of these vectors means that large genes or combinations of different genes can be transferred to cells. By using transcriptional and translational control elements from adenoviral late genes, a very high level expression of recombinant proteins has been demonstrated. Many tissue-specific or regulatable promoters have been shown to retain their function when placed on Ad chromosomes, allowing foreign genes to be expressed in a predictable way in the desired cell type.

As Ad vectors are becoming more widely used in gene therapy work, there has been much interest in improving the technology. Plasmid-based systems for vector construction have simplified the generation of recombinant Ad greatly, and the new Ad vector/packaging cell systems becoming available will provide further increases in capacity and in the effectiveness of *in vivo* gene transfer. Targetable vectors are being developed that will eventually allow expression of a protein of interest in specific cell types *in vivo*.

Acknowledgments

The authors thank members of the Nemerow Laboratory for their assistance, Colleen McKiernan for comments on the manuscript, and Joan Gausepohl for her help with the manuscript.

References

Acsadi, G., Jani, A., Massie, B., Simoneau, M., Holland, P., Blaschuk, K., and Karpati, G. (1994). A differential efficiency of adenovirus-mediated *in vivo* gene transfer into skeletal muscle cells of different maturity. *Hum. Mol. Genet.* **3,** 579–584.

Addison, C. L., Braciak, T., Ralston, R., Muller, W. J., Gauldie, J., and Graham, F. L. (1995). Intratumoral injection of an adenovirus expressing interleukin 2 induces regression and immunity in a murine breast cancer model. *Proc. Natl. Acad. Sci. U.S.A.* **69,** 8522–8526.

Akli, S., Caillaud, C., Vigne, E., Stratford-Perricaudet, L. D., Poenaru, L., Perricaudet, M., Kahn, A., and Peschanski, M. R. (1993). Transfer of a foreign gene into the brain using adenovirus vectors. *Nat. Genet.* **3,** 224–228.

Akusjärvi, G., and Pettersson, U. (1979). Sequence analysis of adenovirus DNA: Complete nucleotide sequence of the spliced 5′ noncoding region of adenovirus 2 hexon messenger RNA. *Cell (Cambridge, Mass.)* **16,** 841–850.

Ali, M., Lemoine, N. R., and Ring, C. J. A. (1994). The use of DNA viruses as vectors for gene therapy. *Gene Ther.* **1,** 367–384.

Alkhatib, G., and Briedis, D. J. (1988). High-level eucaryotic in vivo expression of biologically active measles virus hemagglutinin by using an adenovirus Type 5 helper-free vector system. *J. Virol.* **62,** 2718–2727.

Amalfitano, A., Begy, C. R., and Chamberlain, J. S. (1996). Improved adenovirus packaging cell lines to support the growth of replication-defective gene-delivery vectors. *Proc. Natl. Acad. Sci. U.S.A.* **93,** 3352–3356.

Anton, M., and Graham, F. L. (1995). Site-specific recombination mediated by an adenovirus vector expressing the Cre recombinase protein: A molecular swiitch for control of gene expression. *J. Virol.* **69,** 4600–4606.

Askanas, V., McFerrin, J., Baque, S., Alvarez, B., Sarkozi, E., and Engel, W. K. (1996). Transfer of β-amyloid precursor protein gene using adenovirus vector causes mitochondrial abnormalities in cultured normal human muscle. *Proc. Natl. Acad. Sci. U.S.A.* **93,** 1314–1319.

Ballay, A., Levrero, M., Buendia, M.-A., Tiollais, P., and Perricaudet, M. (1985). *In vitro* and *in vivo* synthesis of the hepatitis B virus surface antigen and of the receptor for polymerized human serum albumin from recombinant human adenoviruses. *EMBO J.* **4,** 3861–3865.

Bartlett, J. S., Sethna, M., Ramamurthy, L., Gowen, S. A., Samulski, R. J., and Marzluff, W. F. (1996). Efficient expression of protein coding genes from the murine U1 small nuclear RNA promoter. *Proc. Natl. Acad. Sci. U.S.A.* **93,** 8852–8857.

Becker, T. C., BeltrandelRio, H., Noel, R. J., Johnson, J. H., and Newgard, C. B. (1994). Overexpression of hexokinase I in isolated islets of Langerhans *via* recombinant adenovirus. *J. Biol. Chem.* **269,** 21234–21238.

Bergelson, J. M., Cunningham, J. A., Droguett, G., Kurt-Jones, E. A., Krithivas, A., Hong, J. S., Horwitz, M.S., Crowell, R. L., and Finberg, R. W. (1997). Isolation of a common receptor for coxsackie B viruses and adenoviruses 2 and 5. *Science* **275,** 1320–1323.

Berkner, K. L. (1992). Expression of heterologous sequences in adenoviral vectors. *Curr. Top. Microbiol. Immunol.* **158,** 39–66.

Berkner, K. L., and Sharp, P. A. (1983). Generation of adenovirus by transfection of plasmids. *Nucleic Acids Res.* **11,** 6003–6020.

Berkner, K. L., Schaffhausen, B. S., Roberts, T. M., and Sharp, P. A. (1987). Abundant expression of polyomavirus middle T antigen and dihydrofolate reductase in an adenovirus recombinant. *J. Virol.* **61,** 1213–1220.

Bernstein, J. A., Porter, J. M., and Challberg, M. D. (1986). Template requirements for *in vivo* replication of adenovirus DNA. *Mol. Cell. Biol.* **6,** 2115–2124.

Bett, A. J., Prevec, L., and Graham, F. L. (1993). Packaging capacity and stability of human adenovirus type 5 vectors. *J. Virol.* **67,** 5911–5921.

Bett, A. J., Haddara, W., Prevec, L., and Graham, F. L. (1994). An efficient and flexible system for construction of adenovirus vectors with insertions or deletions in early regions 1 and 3. *Proc. Natl. Acad. Sci. U.S.A.* **91,** 8802–8806.

Both, G. W., Lockett, L. J., Janardhana, V., Edwards, S. J., Bellamy, A. R., Graham, F. L., Prevec, L., and Andrew, M. E. (1993). Protective immunity to rotavirus-induced diarrhoea is passively transferred to newborn mice from naive dams vaccinated with a single dose of a recombinant adenovirus expressing rotavirus VP7sc. *Virology* **193,** 940–950.

Braciak, T. A., Bacon, K., Xing, Z., Torry, D. J., Graham, F. L., Schall, T. J., Richards, C.D., Croitoru, K., and Gauldie, J. (1996). Overexpression of RANTES using a recombinant adenovirus vector induces the tissue-directed recruitment of monocytes to the lung. *J. Immunol.* **157,** 5076–5084.

Bramson, J., Hitt, M., Gallichan, W. S., Rosenthal, K. L., Gauldie, J., and Graham, F. L. (1996). Construction of a double recombinant adenovirus vector expressing a heterodimeric cytokine: *In vitro* and *in vivo* production of biologically active interleukin-12. *Hum. Gene Ther.* **7,** 333–342.

Brough, D. E., Lizonova, A., Hsu, C., Kulesa, V. A., and Kovesdi, I. (1996). A gene transfer vector-cell line system for complete functional complementation of adenovirus early regions E1 and E4. *J. Virol.* **70,** 6497–6501.

Callebaut, P. and Pensaert, M. (1995). Expression and immunogenicity of the spike glycoprotein of porcine respiratory coronavirus encoded in the E3 region of adenovirus. *In* "Corona- and Related Viruses" (P.J. Talbot and G.A. Levy, eds.), pp. 265–270. Plenum, New York.

Caravokyri, C. and Leppard, K. N. (1995). Constitutive episomal expression of polypeptide IX (pIX) in a 293-based cell line complements the deficiency of pIX mutant adenovirus type 5. *J. Virol.* **69,** 6627–6633.

Chanda, P. K., Natuk, R. J., Mason, B. B., Bhat, B. M., Greenberg, L., Dheer, S. K., Molnar-Kimber, K. L., Mizutani, S., Lubeck, M. D., Davis, A. R., and Hung, P. P. (1990). High level expression of the envelope glycoproteins of the human immunodeficiency virus type 1 in presence of *rev* gene using helper-independent adenovirus Type 7 recombinants. *Virology* **175,** 535–547.

Chartier, C., Degryse, E., Gantzer, M., Dieterlé, A., Pavirani, A., and Mehtali, M. (1996). Efficient generation of recombinant adenovirus vectors by homologous recombination in *Escherichia coli*. *J. Virol.* **70,** 4805–4810.

Chen, G., Koyama, K., Yuan, Lee, Y., Zhou, Y.-T., O'Doherty, R. O., Newgard, C. B., and Unger, R. H. (1996). Disappearance of body fat in normal rats induced by adenovirus-mediated leptin gene therapy. *Proc. Natl. Acad. Sci. U.S.A.* **93,** 14795–14799.

Chen, L., Chen, D., Manome, Y., Dong, Y., Fine, H. A., and Kufe, D. W. (1995). Breast cancer selective gene expression and therapy mediated by recombinant adenoviruses containing the DF3/MUC1 promoter. *J. Clin. Invest.* **96,** 2775–2782.

Cheng, S.-M., Lee, S.-G., Ronchetti-Blume, M., Virk, K. P., Mizutani, S., Eichberg, J. W., Davis, A., Hung, P. P., Hirsch, V. M., Chanock, R. M., Purcell, R. H., and Johnson, P. R. (1992). Coexpression of the simian immunodeficiency virus Env and Rev proteins by a recombinant human adenovirus host range mutant. *J. Virol.* **66,** 6721–6727.

Chow, L. T., Gelinas, R. E., Broker, T. R., and Roberts, R .J. (1997). An amazing sequence arrangement at the 5′ ends of adenovirus 2 messenger RNA. *Cell (Cambridge, Mass.)* **12,** 1–8.

Chroboczek, J., Bieber, F., and Jacrot, B. (1992). The sequence of the genome of adenovirus type 5 and its comparison with the genome of adenovirus type 2. *Virology* **186,** 280–285.

Connelly, S., Gardner, J.M., McClelland, A., and Kaleko, M. (1996). High-level tissue-specific expression of functional human factor VIII in mice. *Hum. Gene Ther.* **7,** 183–195.

Davis, A. R., Kostek, B., Mason, B. B., Hsiao, C. L., Morin, J., Dheer, S. K., and Hung, P. P. (1985). Expression of hepatitis B surface antigen with a recombinant adenovirus. *Proc. Natl. Acad. Sci. U.S.A.* **82,** 7560–7564.

Dewar, R. L., Natarajan, V., Vasudevacchari, M. B., and Salzman, N. P. (1989). Synthesis and processing of human immunodeficiency virus type 1 envelope proteins encoded by a recombinant human adenovirus. *J. Virol.* **63,** 129–136.

Dolph, P. J., Racaniello, V., Villamarin, A., Palladino, F., and Schneider, R.J. (1988). The adenovirus tripartite leader may eliminate the requirement for cap-binding protein complex during translation initiation. *J. Virol.* **62,** 2059–2066.

Dolph, P. J., Huang, J., and Schneider, R.J. (1990). Translation by the adenovirus tripartite leader: Elements which determine independence from cap-binding protein complex. *J. Virol.* **64,** 2669–2677.

Doran, S. E., Ren, X. D., Betz, A. L., Pagel, M. A., Neuwelt, E. A., Roessler, B. J., and Davidson, B. L. (1997). Gene expression from recombinant viral vectors in the central nervous system after blood-brain barrier disruption. *Neurosurgery* **36,** 965–970.

Dupuit, F., Zahm, J.-M., Pierrot, D., Brezillon, S., Bonnet, N., Imler, J.-L., Pavirani, A., and Puchelle, E. (1995). Regenerating cells in human airway surface epithelium represent preferential targets for recombinant adenovirus. *Hum. Gene Ther.* **6,** 1185–1193.

Everitt, E., Meador, S. A., and Levine, A. S. (1977). Synthesis and processing of the precursor to the major core protein of adenovirus type 2. *J. Virol.* **21,** 199–214.

Falgout, B., and Ketner, G. (1987). Adenovirus early region 4 is required for efficient virus particle assembly. *J. Virol.* **61,** 3759–3768.

Fallaux, F. J., Kranenburg, O., Cramer, S. J., Houweling, A., Van Ormondt, H., Hoeben, R. C., and van der Eb, A. J. (1996). Characterization of 911: A new helper cell line for the titration and propagation of early region 1-deleted adenoviral vectors. *Hum. Gene Ther.* **7,** 215–222.

Fisher, K. J., Choi, H., Burda, J., Chen, S.-J., and Wilson, J. M. (1996). Recombinant adenovirus deleted of all viral genes for gene therapy of cystic fibrosis. *Virology* **217,** 11–22.

Fisher, S. A., and Watanabe, M. (1996). Expression of exogenous protein and analysis of morphogenesis in the developing chicken heart using an adenoviral vector. *Cardiovasc. Res.* **31,** E86-E95.

Fooks, A. R., Schadekc, E., Liebert, U. G., Dowsett, A. B., Rima, B. K., Steward, M., Stephenson, J.R ., and Wilkinson, G. W. G. (1995). High-level expression of the measles virus nucleocapsid protein by using a replication-deficient adenovirus vector: Induction of an MHC–1-restricted CTL response and protection in a murine model. *Virology* **210,** 456–465.

Friedman, J. M., Babiss, L. E., Clayton, D. F., and Darnell, J. E., Jr. (1986). Cellular promoters incorporated into the adenovirus genome: Cell specificity of albumin and immunoglobulin expression. *Mol. Cell. Biol.* **6,** 3791–3797.

Gall, J., Kass-Eisler, A., Leinwand, L., and Falck-Pedersen, E. (1996). Adenovirus type 5 and 7 capsid chimera: Fiber replacement alters receptor tropism without affecting primary immune neutralization epitopes. *J. Virol.* **70,** 2116–2123.

Garnier, A., Coté, J., Nadeau, I., Kamen, A., and Massie, B. (1994). Scale-up of the adenovirus expression system for the production of recombinant protein in human 293S cells. *Cytotechnology* **15,** 145–155.

Gilardi, P., Courtney, M., Pavirani, A., and Perricaudet, M. (1990). Expression of human α_1-antitrypsin using a recombinant adenovirus vector. *FEBS Lett* **1,** 60–62.

Ginsberg, H. S., Lundholm-Beauchamp, U., Horswood, R. L., Pernis, B., Wold, W. S. M., Chanock, R. M., and Prince, G. A. (1989). Role of early region 3 (E3) in pathogenesis of adenovirus disease. *Proc. Natl. Acad. Sci. U.S.A.* **86,** 3823–3827.

Gluzman, Y., Reichl, H., and Solnick, D. (1982). Helper-free adenovirus type-5 vectors. *In* "Eukaryotic Viral Vectors" (Y. Gluzman, ed.), pp. 187–192. Cold Spring Harbor Lab., Cold Spring Harbor, NY.

Goldman, M. J., and Wilson, J. M. (1995). Expression of αvβ5 integrin is necessary for efficient adenovirus-mediated gene transfer in the human airway. *J. Virol.* **69**, 5951–5958.

Gorziglia, M. I., Kadan, M. J., Yei, S., Lim, J., Lee, G. M., Luthra, R., and Trapnell, B. C. (1996). Elimination of both El and E2a from adenovirus vectors further improves prospects for *in vivo* human gene therapy. *J. Virol.* **70**, 4173–4178.

Graham, F. L. (1984). Covalently closed circles of human adenovirus DNA are infectious. *EMBO J.* **3**, 2917–2922.

Graham, F. L. (1987). Growth of 293 cells in suspension culture. *J. Gen. Virol.* **68**, 937–940.

Graham, F. L., and Prevec, L. (1995). Methods for construction of adenovirus vectors. *Mol. Biotechnol.* **3**, 207–220.

Graham, F. L., Smiley, J., Russell, W. C., and Nairn, R. (1977). Characteristics of a human cell line transformed by DNA from human adenovirus type 5. *J. Gen. Virol.* **36**, 59–72.

Guo, Z. S., Wang, L.-H., Eisensmith, R. C., and Woo, S. L. C. (1996). Evaluation of promoter strength for hepatic gene expression *in vivo* following adenovirus-mediated gene transfer. *Gene Therapy* **3**(9), 802–810.

Haddada, H., Cordier, L., and Perricaudet, M. (1995). Gene therapy using adenovirus vectors. *Curr. Top. Microbiol. Immunol.* **199**(3), 297–306.

Haecker, S. E., Stedman, H. H., Balice-Gordon, R. J., Smith, D. B. J., Greelish, J. P., Mitchell, M. A., Wells, A., Sweeney, H. L., and Wilson, J. M. (1996). *In vivo* expression of full-length human dystrophin from adenoviral vectors deleted of all viral genes. *Hum. Gene Ther.* **7**, 1907–1914.

Haj-Ahmad, Y. and Graham, F. L. (1986). Development of a helper-independent human adenovirus vector and its use in the transfer of the herpes simplex virus thymidine kinase gene. *J. Virol.* **57**, 267–274.

Halbert, D. N., Cutt, J. R., and Shenk, T. (1985). Adenovirus early region 4 encodes functions required for efficient DNA replication, late gene expression, and host cell shutoff. *J. Virol.* **56**, 250–257.

Hanahan, D., and Gluzman, Y. (1984). Rescue of functional replication origins from embedded configurations in a plasmid carrying the adenovirus genome. *Mol. Cell. Biol.* **4**, 302–309.

Hanke, T., Graham, F. L., Lulitanond, V., and Johnson, D. C. (1990). Herpes simplex virus IgG receptors induced using recombinant adenovirus vectors expressing glycoproteins E and I. *Virology* **177**, 437–444.

Hashimoto, M., Aruga, J., Hosoya, Y., Kanegae, Y., Saito, I., and Mikoshiba, K. (1996). A neural cell-type-specific expression system using recombinant adenovirus vectors. *Hum. Gene Ther.* **7**, 149–158.

Hay, R. T. (1985). The origin of adenovirus DNA replication: Minimal DNA sequence requirement in vivo. *EMBO J.* **4**, 421–426.

Hayashi, Y., DePaoli, A. M., Burant, C. F., and Refetoff, S. (1994). Expression of a thyroid hormone-responsive recombinant gene introduced into adult mouse liver by replication-defective adenovirus can be regulated by endogenous thyroid hormone receptor. *J. Biol. Chem.* **269**, 23872–23875.

He, T. C., Zhou, S., da Costa, L. T., Yu, J., Kinzler, K. W., and Vogelstein, B. (1988). A simplified system for generating recombinant adenoviruses. *Proc. Natl. Acad. Sci. U.S.A.* **95**, 2509–2514.

Hearing, P., Samulski, R. J., Wishart, W. L., and Shenk, T. (1987). Identification of a repeated sequence element required for efficient encapsidation of the adenovirus type 5 chromosome. *J. Virol.* **61,** 2555–2558.

Horwitz, M. S. (1996). Adenoviruses. In "Virology"(B. N. Fields, D. M. Knipe, and P. M. Howley, eds.), pp. 2149–2170. Lippincott-Raven, Philadelphia.

Huang, S., Endo, R. I., and Nemerow, G. R. (1995). Upregulation of integrins αvβ3 and αvβ5 on human monocytes and T lymphocytes facilitates adenovirus-mediated gene delivery. *J. Virol.* **69,** 2257–2263.

Huang, S., Kamata, T., Takada, Y., Ruggeri, Z. M., and Nemerow, G. R. (1996). Adenovirus interaction with distinct integrins mediates separate events in cell entry and gene delivery to hematopoietic cells. *J. Virol.* **70,** 4502–4508.

Huang, S., Jiang, Y., Li, Z., Nishida, E., Mathias, P., Lin, S., Ulevitch, R. J., Nemerow, G. R., and Han, J. (1997). Apoptosis signaling pathway in T cells is composed of ICE/Ced–3 family proteases and MAP kinase kinase 6b. *Immunology* **6,** 739–749.

Huyghe, B. G., Liu, X., Sutjipto, S., Sugarman, B.J., Horn, M. T., Shepard, M., Scandella, C. J., and Shabram, P. (1995). Purification of type 5 recombinant adenovirus encoding human p53 by column chromatography. *Hum. Gene Ther.* **6,** 1403–1416.

Imler, J.-L. (1995). Adenovirus vectors as recombinant viral vaccines. *Vaccine* **13,** 1143–1151.

Imler, J.-L., Chartier, C., Dieterlé, A., Dreyer, D., Mehtali, M., and Pavirani, A. (1995). An efficient procedure to select and recover recombinant adenovirus vectors. *Gene Ther.* **2,** 263–268.

Jaffe, H. A., Danel, C., Longenecker, G., Metzger, M., Setoguchi, Y., Rosenfeld, M. A., Gant, T. W., Thorgeirsson, S. S., Stratford-Perricaudet, L. D., Perricaudet, M., Pavirani, A., Lecocq, J.-P., and Crystal, R. G. (1992). Adenovirus-mediated *in vivo* gene transfer and expression in normal rat liver. *Nat. Genet.* **1,** 372–378.

Jiang, C., O'Connor, S.P., Armentano, D., Berthelette, P. B., Schiavi, S. C., Jefferson, D.M., Smith, A.E., Wadsworth, S.C., and Cheng, S.H. (1996). Ability of adenovirus vectors containing different CFTR transcriptional cassettes to correct ion transport defects in CF cells. *Am. J. Physiol.* **4** (Pt. 1), L527–L537.

Johnson, D. C., Ghosh-Choudhury, G., Smiley, J. R., Fallis, L., and Graham, F. L. (1988). Abundant expression of herpes simplex virus glycoprotein gB using an adenovirus vector. *Virology* **164,** 1–14.

Jones, N. (1995). Transcriptional modulation by the adenovirus E1A gene. *Curr. Top. Microbiol. Immunol.* **199**(3), 59–80.

Kanai, F., Shiratori, Y., Yoshida, Y., Wakimoto, H., Hamada, H., Kanegae, Y., Saito, I., Nakabayashi, H., Tamaoki, T., Tanaka, T., Lan, K.-H., Kato, N., Shina, S., and Omata, M. (1996). Gene therapy for α-fetoprotein-producing human hepatoma cells by adenovirus-mediated transfer of the herpes simplex virus thymidine kinase gene. *Hepatology* **23,** 1359–1368.

Ketner, G., Spencer, F., Tugendreich, S., Connelly, C., and Hieter, P. (1994). Efficient manipulation of the human adenovirus genome as an infectious yeast artificial chromosome clone. *Proc. Natl. Acad. Sci. U.S.A.* **91,** 6186–6190.

Ko, S.-C., Choen, J., Kao, C., Gotoh, A., Shirakawa, T., Sikes, R.A., Karsenty, G., and Chung, L. W. K. (1996). Osteocalcin promoter-based toxic gene therapy for the treatment of osteosarcoma in experimental models. *Cancer Res.* **56,** 4614–4619.

Kobayashi, J., Applebaum-Bowden, D., Dugii, K. A., Brown, D. R., Kashyap, V. S.,

Parrott, C., Duarte, C., Maeda, N., and Santamarina-Fojo, S. (1996). Analysis of protein structure-function *in vivo*. *J. Biol. Chem.* **271**, 26296–26301.

Kochanek, S., Clemens, P. R., Mitani, K., Chen, H.-H., and Chan, S. (1997). A new adenoviral vector: Replacement of all viral coding sequences with 28 kb of DNA independently expressing both full-length dystrophin and β-galactosidase. *Proc. Natl. Acad. Sci. U.S.A.* **93**, 5731–5736.

Konishi, H., Ochiya, T., Yasuda, Y., Sakamoto, H., Muto, T., Sugimura, T., and Terada, M. (1996). HST-1/FGF-4 stimulates proliferation of megakaryocyte progenitors synergistically and promotes megakaryocyte maturation. *Oncogene* **13**, 9–19.

Kozarsky, K. F., and Wilson, J. M. (1993). Gene therapy: Adenovirus vectors. *Curr. Opin. Genet. Dev.* **3**, 499–503.

Krasnykh, V. N., Mikheeva, G. V., Douglas, J. T., and Curiel, D. T. (1996). Generation of recombinant adenovirus vectors with modified fibers for altering viral tropism. *J. Virol.* **70**, 6839–6846.

Krougliak, V., and Graham, F. L. (1995). Development of cell lines capable of complementing E1, E4, and protein IX defective adenovirus type 5 mutants. *Hum. Gene Ther.* **6**, 1575–1586.

Kumar-Singh, R., and Chamberlain, J. S. (1997). Encapsidated adenovirus minichromosomes allow delivery and expression of a 14 kb dystrophin cDNA to muscle cells. *Hum. Mol. Genet.* **5**, 913–921.

Lan, K.-H., Kanai, F., Shiratori, Y., Okabe, S., Yoshida, Y., Wakimoto, H., Hamada, H., Tanaka, T., Ohashi, M., and Omata, M. (1996). Tumor-specific gene expression in carcinoembryonic antigen-producing gastric cancer cells using adenovirus vectors. *Gastroenterology* **111**, 1241–1251.

Langer, S. J., and Schaack, J. (1996). 293 cell lines that inducibly express high levels of adenovirus type 5 precursor terminal protein. *Virology* **221**, 172–179.

Lemarchand, P., Jaffe, H. A., Danel, C., Cid, M. D., Kleinman, H. K., Stratford-Perricaudet, L. D., Perricaudet, M., Pavirani, A., Lecocq, J.-P., and Crystal, R. G. (1992). Adenovirus-mediated transfer of a recombinant human α_1-antitrypsin cDNA to human endothelial cells. *Proc. Natl. Acad. Sci. U.S.A.* **89**, 6482–6486.

Li, E., Stupack, D., Klemke, R., Cheresh, D. A., and Nemerow, G. R. (1998). Adenovirus endocytosis via α_v integrins requires phosphoinositide-3-OH kinase. *J. Virology* **72**, 2055–2061.

Lieber, A., Vrancken Peeters, M. J., and Kay, M. A. (1995). Adenovirus-mediated transfer of the amphotropic retrovirus receptor cDNA increases retroviral transduction in cultured cells. *Hum. Gene Ther.* **6**, 5–11.

Lindley, T., Virk, K. P., Ronchetti-Blume, M., Goldberg, K., Lee, S.-G., Eichberg, J. W., Hung, P. P., and Cheng, S.-M. (1994). Construction and characterization of adenovirus co-expressing hepatitis B virus surface antigen and interleukin-6. *Gene* **138**, 165–170.

Lisovoski, F., Cadusseau, J., Akli, S., Caillaud, C., Vigne, E., Poenaru, L., Stratford-Perricaudet, L., Perricaudet, M., Kahn, A., and Peschanski, M. (1994). In vivo transfer of a marker gene to study motoneuronal development. *NeuroReport* **5**, 1069–1072.

Mansour, S. L., Grodzicker, T., and Tjian, R. (1986). Downstream sequences affect transcription initiation from the adenovirus major late promoter. *Mol. Cell. Biol.* **6**, 2684–2694.

Mashhour, B., Couton, D., Perricaudet, M., and Briand, P. (1994). *In vivo* adenovirus-mediated gene transfer into ocular tissues. *Gene Ther.* **1**, 122–126.

Mason, B. B., Davis, A. R., Bhat, B. M., Hengalvala, M., Lubeck, M. D., Zandle, G., Kostek, B., Cholodofsky, S., Dheer, S., Molnar-Kimber, K., Mizutani, S., and Hung, P. P. (1990). Adenovirus vaccine vectors expressing hepatitis B surface antigen: Importance of regulatory elements in the adenovirus major late intron. *Virology* **177,** 452–461.

McGrory, W. J., Bautista, D. S., and Graham, F. L. (1988). A simple technique for the rescue of early region I mutations into infectious human adenovirus type 5. *Virology* **163,** 614–617.

Michael, S. I., Hong, J. S., Curiel, D. T., and Engler, J. A. (1995). Addition of a short peptide ligand to the adenovirus fiber protein. *Gene Ther.* **2,** 660–668.

Mirza, M. A., and Weber, J. (1982). Structure of adenovirus chromatin. *Biochim. Biophys. Acta* **696,** 76–86.

Mittal, S. K., McDermott, M. R., Johnson, D. C., Prevec, L., and Graham, F. L. (1993). Monitoring foreign gene expression by a human adenovirus-based vector using the firefly luciferase gene as a reporter. Virus Res. **28,** 67–90.

Mittereder, N., March, K. L., and Trapnell, B. C. (1996). Evaluation of the concentration and bioactivity of adenovirus vectors for gene therapy. *J. Virol.* **70,** 7498–7509.

Miyake, S., Makimura, M., Kanegae, Y., Harada, S., Sato, Y., Takamori, K., Tokuda, C., and Saito, I. (1996). Efficient generation of recombinant adenoviruses using adenovirus DNA-terminal protein complex and a cosmid bearing the full-length virus genome. *Proc. Natl. Acad. Sci. U.S.A.* **93,** 1320–1324.

Morin, J. E., Lubeck, M. D., Barton, J. E., Conley, A. J., Davis, A. R., and Hung, P. P. (1987). Recombinant adenovirus induces antibody response to hepatitis B virus surface antigen in hamsters. *Proc. Natl. Acad. Sci. U.S.A.* **84,** 4626–4630.

Moullier, P., Friedlander, G., Calise, D., Ranco, P., Perricaudet, M., and Ferry, N. (1994). Adenoviral-mediated gene transfer to renal tubular cells *in vivo*. *Kidney Int.* **45,** 1220–1225.

Mühlhauser, J., Merrill, M. J., Pili, R., Maeda, H., Bacic, M., Bewig, B., Passaniti, A., Edwards, N. A., Crystal, R. G., and Capogrossi, M. C. (1995). $VEGF_{165}$ expressed by a replication-deficient recombinant adenovirus vector induces angiogenesis in vivo. *Circ. Res.* **77,** 1078–1086.

Parks, R. J., and Graham, F. L. (1997). A helper-dependent system for adenovirus vector production helps define a lower limit for efficient DNA packaging. *J. Virol.* **71,** 3293–3298.

Parks, R. J., Chen, L., Anton, M., Sankar, U., Rudnicki, M. A., and Graham, F. L. (1996). A helper-dependent adenovirus vector system: Removal of helper virus by Cre-mediated excision of the viral packaging signal. *Proc. Natl. Acad. Sci. U.S.A.* **93,** 13565–13570.

Philipson, L., Lonberg-Holm, K., and Petterson, U. (1968). Virus-receptor interaction in an adenovirus system. *J. Virol.* **2,** 1064–1075.

Ragot, T., Vincent, N., Chafey, P., Vigne, E., Gilgenkrantz, H., Conton, D., Cartand, J., Briand, P., Kaplan, J.-C., Perricaudet, M., and Kahn, A. (1993). Efficient adenovirus-mediated transfer of a human minidystrophin gene to skeletal muscle of *mdx* mice. *Nature (London)* **361,** 647–650.

Rekosh, D. M. K., Russell, W. C., and Bellet, A. J. D. (1977). Identification of a protein linked to the ends of adenovirus DNA. *Cell (Cambridge, Mass.)* **11,** 283–295.

Roberts, R. J., O'Neill, K. E., and Yen, C. T. (1984). DNA sequences from the adenovirus 2 genome. *J. Biol. Chem.* **259,** 13968–13975.

Rosenfeld, M. A., Siegfried, W., Yoshimura, K., Yoneyama, K., Fukayama, M., Stier, L. E., Pääkkö, P. K., Gilardi, P., Stratford-Perricaudet, L. D., Perricaudet, M., Jallat, S., Pavirani, A., Lecocq, J.-P., and Crystal, R. G. (1991). Adenovirus-mediated transfer of a recombinant α1-antitrypsin gene to the lung epithelium in vivo. *Science* **252,** 431–434.

Saito, I., Oya, Y., Yamamoto, K., Yuasa, T., and Shimojo, H. (1985). Construction of nondefective adenovirus type 5 bearing a 2.8-kilobase hepatitis B virus DNA near the right end of its genome. *J. Virol.* **54,** 711–719.

Sakai, K., Mitani, K., and Miyazaki, J. (1995). Efficient regulation of gene expression by adenovirus vector-mediated delivery of the CRE recombinase. *Biochem. Biophys. Res. Commun.* **217,** 393–401.

Sakamoto, H., Ochiiya, T., Sato, Y., Tsukamoto, M., Konishi, H., Saito, I., Sugimura, T., and Terada, M. (1994). Adenovirus-mediated transfer of the *HST-1 (FGF4)* gene induces increased levels of platelet count *in vivo. Proc. Natl. Acad. Sci. U.S.A.* **91,** 12368–12372.

Schaack, J., Yew-Wai Ho, W., Freimuth, P., and Shenk, T. (1990). Adenovirus terminal protein mediates both nuclear matrix association and efficient transcription of adenovirus DNA. *Genes Dev.* **4,** 1197–1208.

Schaack, J., Langer, S., and Guo, X. (1995a). Efficient selection of recombinant adenoviruses by vectors that express β-galactosidase. *J. Virol.* **69,** 3920–3923.

Schaack, J., Guo, X., Yew-Wai Ho, W., Karlok, M., Chen, C., and Ornelles, D. (1995b). Adenovirus type 5 precursor terminal protein-expressing 293 and HeLa cell lines. *J. Virol.* **69,** 4079–4085.

Schaffhausen, B. S., Bockus, B. J., Berkner, K. L., Kaplan, D., and Roberts, T. M. (1987). Characterization of middle T antigen expressed by using an adenovirus expression system. *J. Virol.* **61,** 1221–1225.

Schneider, M., Graham, F. L., and Prevec, L. (1989). Expression of the glycoprotein of vesicular stomatitis virus by infectious adenovirus vectors. *J. Gen. Virol.* **70,** 417–427.

Setoguchi, Y., Danel, C., and Crystal, R. G. (1994a). Stimulation of erythropoiesis by in vivo gene therapy: Physiologic consequences of transfer of the human erythropoietin gene to experimental animals using an adenovirus vector. *Blood* **84,** 2946–2953.

Setoguchi, Y., Jaffe, H. A., Danel, C., and Crystal, R. G. (1994b). *Ex vivo* and *in vivo* gene transfer to the skin using replication-deficient recombinant adenovirus vectors. *J. Invest. Dermatol.* **102,** 415–421.

Shabram, P. W., Giroux, D. D., Goudreau, A. M., Gregory, R. J., Horn, M. T., Huyghe, B. G., Liu, X., Nunnally, M. H., Sugarman, B. J., and Sutjipto, S. (1997). Analytical anion-exchange HPLC of recombinant type-5 adenoviral particles. *Hum. Gene Ther.* **8,** 453–465.

Sharp, P. A., Moore, C., and Haverty, J. L. (1976). The infectivity of adenovirus 5 DNA-protein complex. *Virology* **75,** 442–456.

Sheay, W., Nelson, S., Martinez, I., Chu, T.-H. T., Bhatia, S., and Dornburg, R. (1993). Downstream insertion of the adenovirus tripartite leader sequence enhances expression in universal eukaryotic vectors. *BioTechniques* **15,** 856–862.

Shenk, T. (1996). Adenoviridae: The viruses and their replication. *In* "Virology" (B.N. Fields, D.M. Knipe, and P.M. Howley, eds.), pp. 2111–2148. Lippincott-Raven, Philadelphia.

Smith, G. M., Hale, J., Pasnikowski, E. M., Lindsay, R. M., Wong, V., and Rudge, J. S. (1996). Astroytes infected with replication-defective adenovirus containing a secreted form of CNTF or NT3 show enhanced support of neuronal populations *in vitro*. *Exp. Neurol.* **139,** 156–166.

Stevenson, S. C., Rollence, M., Marshall-Neff, J., and McClelland, A. (1997). Selective targeting of human cells by a chimeric adenovirus vector containing a modified fiber protein. *J. Virol.* **71,** 4782–4790.

Stow, N. D. (1981). Cloning of a DNA fragment from the left-hand terminus of the adenovirus type 2 genome and its use in site-directed mutagenesis. *J. Virol.* **37,** 171–180.

Stratford-Perricaudet, L. D., Makeh, I., Perricaudet, M., and Briand, P. (1992). Widespread long-term gene transfer to mouse skeletal muscles and heart. *J. Clin. Invest.* **90,** 626–630.

Thiel, J. F. and Smith, K. O. (1967). Fluorescent focus assay of viruses on cell monolayers in plastic petri plates. *Proc. Soc. Exp. Biol. Med.* **125,** 892–895.

Tollefson, A. E., Ryerse, J. S., Scaria, A., Hermiston, T. W., and Wold, W. S. M. (1996). The E3 11.6 KDa adenovirus death protein (ADP) is required for efficient cell death: Characterization of cells infected with *adp* mutants. *Virology* **220,** 152–162.

Tollefson, A. E., Hermiston, T. W., Lichtenstein, D. L., Colle, C. F., Tripp, R. A., Dimitrov, T., Toth, K., Wells, C. E., Doherty, P. C., and Wold, W. S. N. (1998). Forced degradation of Fas inhibits apoptosis in adenovirus-infected cells. *Nature* **392,** 726–730.

Tomko, R. P., Xu, R., and Philipson, L. (1997). HCAR and MAR: The human and mouse cellular receptors for subgroup C adenoviruses and group B coxsackieviruses. *Proc. Natl. Acad. Sci. U.S.A.* **94,** 3352–3356.

Tsukui, T., Miyake, S., Azuma, S., Ichise, H., and Saito, I. (1995). Gene transfer and expression in mouse preimplantation embryos by recombinant adenovirus vector. *Mol. Reprod. Develop.* **42,** 291–297.

Van der Vliet, P. C. (1995). Adenovirus DNA replication. *Curr. Top. Microbiol. Immunol.* **199**(2), 1–30.

van Oostrum, J. and Burnett, R. M. (1985). Molecular composition of the adenovirus type 2 virion. *J. Virol.* **56,** 439–448.

Varley, A. W., Coulthard, M.G., Meidell, R. S., Gerard, R. D., and Munford, R. S. (1995). Inflammation-induced recombinant protein expression *in vivo* using promoters from acute-phase protein genes. *Proc. Natl. Acad. Sci. U.S.A.* **92,** 5346–5350.

Von Seggern, D. J., Kehler, J., Endo, R. I., and Nemerow, G. R. (1998). Complementation of a fiber mutant adenovirus by packaging cell lines stably expressing the Ad5 fiber protein. *J. Gen. Virol.* **79,** 1461–1468.

Wang, Q., Jia, X.-C., and Finer, M. H. (1995). A packaging cell line for propagation of recombinant adenovirus vectors containing two lethal gene-region deletions. *Gene Ther.* **2,** 775–783.

Wang, Y., Krushel, L. A., and Edelman, G. M. (1996). Targeted DNA recombination *in vivo* using an adenovirus carrying the *cre* recombinase gene. *Proc. Natl. Acad. Sci. U.S.A.* **93,** 3932–3936.

Warnes, A., Fooks, A. R., and Stephenson, J. R. (1994). Production of measles nucleoprotein in different expression systems and its use as a diagnostic reagent. *J. Virol. Methods* **49,** 257–268.

Weiden, M. D., and Ginsberg, H. S. (1994). Deletion of the E4 region of the genome produces adenovirus DNA concatemers. *Proc. Natl. Acad. Sci. U.S.A.* **91,** 153–157.

Weinberg, D. H., and Ketner, G. (1983). A cell line that supports the growth of a defective early region 4 deletion mutant of human adenovirus type 2. *Proc. Natl. Acad. Sci. U.S.A.* **80,** 5383–5386.

Wickham, T .J., Mathias, P., Cheresh, D. A., and Nemerow, G. R. (1993). Integrins $a_v\beta_3$ and $a_v\beta_5$ promote adenovirus internalization but not virus attachment. *Cell (Cambridge, Mass.)* **73,** 309–319.

Wickham, T. J., Segal, D. M., Roelvink, P. W., Carrion, M. E., Lizonova, A., Lee, G.M ., and Kovesdi, I. (1996). Targeted adenovirus gene transfer to endothelial and smooth muscle cells by using bispecific antibodies. *J. Virol.* **70,** 6831–6838.

Wilkinson, G. W. G., and Akrigg, A. (1992). Constitutive and enhanced expression from the CMV major IE promoter in a defective adenovirus vector. *Nucleic Acids Res.* **20,** 2233–2239.

Wold, W. S. M. (1993). Adenovirus genes that modulate the sensitivity of virus-infected cells to lysis by TNF. *J. Cell. Biochem.* **53,** 329–335.

Wrighton, C. J., Hofer-Warbinek, R., Moll, T., Eytner, R., Bach, F. H., and deMartin, R. (1996). Inhibition of endothelial cell activation by adenovirus-mediated expression of IkBα, an inhibitor of the transcription factor NF-kB. *J. Exp. Med.* **183,** 1013–1022.

Xia, D., Henry, L. J., Gerard, R. D., and Deisenhofer, J. (1994). Crystal structure of the receptor-binding domain of adenovirus type 5 fiber protein at 1.7 Å resolution. *Structure* **2,** 1259–1270.

Xing, Z., Braciak, T., Jordana, M., Croitoru, K., Graham, F. L., and Gauldie, J. (1994). Adenovirus-mediated cytokine gene transfer at tissue sites. *J. Immunol.* **153,** 4059–4069.

Xing, Z., Ohkawara, Y., Jordana, M., Graham, F. L., and Gauldie, J. (1996). Transfer of granulocyte-macrophage colony-stimulating factor gene to rat lung induces eosinophilia, monocytosis, and fibrotic reactions. *J. Clin. Invest.* **97,** 1102–1110.

Yajima, H., Kosukegawa, A., Hoque, M. M., Shimojima, T., Ishizu, K., Takayama, M., Sasaki, Y., Sakai, H., Otsuka, M., Inokuchi, S., and Handa, H. (1996). Construction and characterization of a recombinant adenovirus vector carrying the human preproinsulin gene under the control of the metallothionein gene promoter. *Biochem. Biophys. Res. Commun.* **229,** 778–787.

Yamamoto, H., Ueno, H., Ooshima, A., and Takeshita, A. (1996). Adenovirus-mediated transfer of a truncated transforming growth factor-β (TGF-β) Type II receptor completely and specifically abolished diverse signaling by TGF-β in vascular wall cells in primary culture. *J. Biol. Chem.* **271,** 16253–16259.

Yang, Y., Nunes, F., Berencsi, K., Furth, E. E., Gönczöl, E., and Wilson, J. M. (1994). Cellular immunity to viral antigens limits E1-deleted adenoviruses for gene therapy. *Proc. Natl. Acad. Sci. U.S.A.* **91,** 4407–4411.

Yang, Y., Li, Q., Ertl, H. C., and Wilson, J. M. (1995). Cellular and humoral immune responses to viral antigens create barriers to lung-directed gene therapy with recombinant adenoviruses. *J. Virol.* **69,** 2004–2015.

Yasukawa, M., Inoue, Y., Ohminami, H., Sada, E., Miyake, K., Tohyama, T., Shimada, T., and Fujita, S. (1997). Human herpesvirus 7 infection of lymphoid and myeloid cell lines transduced with an adenovirus vector containing the CD4 gene. *J. Virol.* **71,** 1708–1712.

Ye, W. W., Mason, B. B., Chengalvala, M., Cheng, S.-M., Zandle, G., Lubeck, M. D., Lee, S.-G., Mizutani, S., Davis, A. R., and Hung, P. P. (1991). Co-expression of hepatitis B virus antigens by a non-defective adenovirus vaccine vector. *Arch. Virol.* **118,** 11–27.

Yeh, P., Dedieu, J.-F., Orsini, C., Vigne, E., Denefle, P., and Perricaudet, M. (1996). Efficient dual transcomplementation of adenovirus E1 and E4 regions from a

293-derived cell line expressing a minimal E4 functional unit. *J. Virol.* **70,** 559–565.

Zain, S., Sambrook, J., Roberts, R. J., Keller, W., Fried, M., and Dunn, A. R. (1979). Nucleotide sequence analysis of the leader segments in a cloned copy of adenovirus 2 fiber mRNA. *Cell (Cambridge, Mass.)* **16,** 851–861.

Zhang, W.-W., Fang, X., Branch, C. D., Mazur, W., French, B. A., and Roth, J. A. (1993). Generation and identification of recombinant adenovirus by liposome-mediated transfection and PCR analysis. *BioTechniques* **15,** 868–872.

Zhang, Y., Feigenblum, D., and Schneider, R. J. (1994). A late adenovirus factor induces eIF-4E dephosphorylation and inhibition of cell protein synthesis. *J. Virol.* **68,** 7040–7050.

Zhao, Z., Bouchard, P., Diltz, C. D., Shen, S.-H., and Fischer, E. H. (1993). Purification and characterization of a protein tyrosine phosphatase containing SH2 domains. *J. Biol. Chem.* **268,** 2816–2820.

Zhou, H., O'Neal, W., Morral, N., and Beaudet, A. L. (1996). Development of a complementing cell line and a system for construction of adenovirus vectors with E1 and E2a deleted. *J. Virol.* **70,** 7030–7038.

6

EXPRESSION IN THE METHYLOTROPHIC YEAST *Pichia pastoris*

James M. Cregg

Department of Biochemistry and Molecular Biology, Oregon Graduate Institute of Science and Technology, Portland, Oregon 97291

Introduction

Selection of an Expression System

Foreign gene expression systems based on yeasts have proven to be efficient and economical sources of a variety of higher eukaryotic proteins of commercial and academic interest (Romanos *et al.*, 1992). Yeasts combine the microbial growth and genetic manipulation advantages of *Escherichia coli* along with a eukaryotic environment and the ability to perform many eukaryote-specific post-translational protein modifications such as proteolytic processing, folding, disulfide bridge formation, and glycosylation. Although *E. coli* typically produces eukarotic foreign proteins at high levels, the absence of these eukaryotic features results in proteins that are often insoluble and inactive. Thus, for eukaryotic proteins that are needed in a biologically active and/or native form and for which *in vitro* refolding procedures have proven to be inefficient, a eukaryotic expression system is required.

Yeasts offer significant advantages over higher eukaryotic tissue culture-based systems such as those utilizing Chinese hamster ovary or baculovirus-infected insect cell lines. Their rapid growth rate (1–3 hr/generation) relative to tissue culture cells (~1 day/generation) greatly reduces production times. The cost of media, equipment, and facilities needed to culture yeasts is substantially less than that needed for tissue culturing, especially on a commercial scale. For secreted recombinant proteins, the absence of serum components in yeast media substantially reduces the time and effort needed for their purification.

However, yeast expression systems also have limitations that make them inappropriate for some proteins. The best documented of these involves the glycosylation of secreted proteins. Yeasts add oligosaccharides composed primarily of mannose residues to proteins at O-serine/threonine- and N-asparagine-linked sites rather than the complex oligosaccharides added at these sites by higher eukaryotes (Goochee *et al.*, 1991; see later for details). As a result, yeast-secreted human glycoproteins are not suitable for intravenous therapeutic use due to their antigenicity, immunogenicity, and rapid blood clearance rates. The length of the oligosaccharide chains added at N-linked sites by yeasts is also an important consideration for the expression of certain foreign proteins. Whereas individual oligosaccharide structures added to proteins by higher organisms are relatively small (<20 sugar residues), yeasts often add much longer mannose chains (>50), a phenomenon termed hyperglycosylation. These large oligosaccharide structures can interfere with the folding, activity, or antigenicity of some proteins. Whether a specific foreign glycoprotein will be hyperglycosylated when expressed in a yeast system or whether hyperglycosylation will interfere with the desired biological function of a foreign glycoprotein is difficult to predict.

A second disadvantage of yeasts is that some higher eukaryotic proteins are not compatible with the yeast secretory apparatus. Instead of being secreted, these proteins remain trapped at some point along the secretory pathway. In some instances, the failure of yeasts to secrete a foreign protein appears to be due to the inability of the yeast secretory apparatus to utilize the native higher eukaryotic signal sequence, a problem that can often be corrected by the substitution of a yeast signal sequence. However, for other proteins, substitution of a yeast signal sequence does not result in efficient secretion, and the cause of their failure to be secreted remains largely unknown. At present, the only means of examining questions of secretion and carbohydrate structure is to express the gene of interest in a yeast system and directly analyze the protein product.

Why *Pichia pastoris* Instead of *Saccharomyces cerevisiae*?

The methylotrophic yeast *Pichia pastoris* has two key advantages over *Saccharomyces cerevisiae* as a host for the production of foreign proteins (Cregg *et al.*, 1993). The first is the promoter used to transcribe foreign genes that is derived from the methanol-regulated *P. pastoris* alcohol oxidase I gene (*AOX1*), which is ideally suited for this purpose. In cells exposed to methanol as the sole carbon source, transcription initiation at the *AOX1* promoter (*AOX1p*) is highly efficient and comparable to that of promoters derived from highly expressed glycolytic pathway genes (Waterham *et al.*, 1997).

However, unlike glycolytic promoters, *AOX1p* is tightly regulated and highly repressed under nonmethanolic growth conditions. Because most foreign proteins are at least somewhat deleterious to the cell when expressed at high levels, the ability to maintain cultures in a repressed or "expression off" mode is important to minimize selection for nonexpressing mutant strains during cell growth. *AOX1p* transcription appears to virtually require methanol and remains substantially repressed under growth conditions that result in substantial derepression of other promoters (e.g., carbon limitation/starvation). With other more easily derepressed promoters, the opportunity exists for the selective growth of nonproducing mutants, especially during the period required to generate biomass in high volume and high-cell density fermenter cultures. Because *AOX1p* is controlled by manipulating the carbon source added to the culture medium, growth and induction of *P. pastoris* expression strains are easily performed at all scales, from a shake flask to large fermenters.

The second key advantage of *P. pastoris* is that it is not a strong fermenter like *S. cerevisiae.* Yeast fermentation generates ethanol, which, in high-density cultures, can rapidly build to toxic levels (the Crabtree effect). Especially for secreted proteins, where the concentration of a foreign protein in the medium is roughly proportional to the concentration of cells, high-cell density fermenter culturing is desirable but not easily achievable with *S. cerevisiae.* In contrast, *P. pastoris* expression strains are relatively easy to culture at cell densities of ~100 g/liter, dry cell weight, or greater (Fig. 1) (Siegel and Brierley, 1989).

Purpose of This Chapter

The purpose of this chapter is to provide readers with a basic understanding of the *P. pastoris* expression system: How the system works. What proteins are most likely to work well in the system. What problems to anticipate in using the system and potential solutions. Other aspects of the system have been described in previous reviews (Romanos *et al.*, 1992; Cregg *et al.*, 1993; Romanos, 1995). For detailed protocols describing the construction of *P. pastoris* expression strains and the expression of foreign genes in the system, readers are referred to Higgins and Cregg (1998) or the Invitrogen *Pichia* expression manual. A list of heterologous proteins that have been produced in *P. pastoris* and reported in the literature is presented in Table 1.

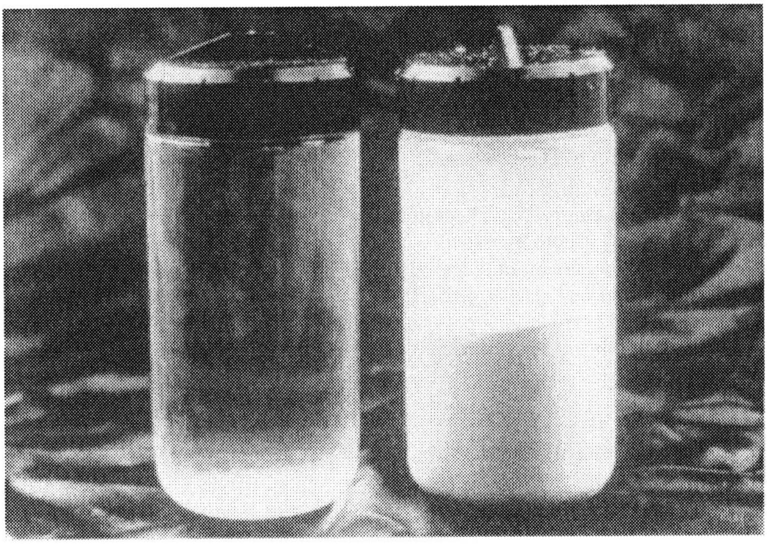

Figure 1 Centrifuge bottles containing *P. pastoris* cultures. Bottle on left contains culture at density of 1.0 OD_{600} units. Bottle on right contains culture at approximately 130 g/liter dry cell weight (with permission from Cregg and Higgins, 1995).

Background Information

Methanol Metabolism

The ability of certain yeast species to grow on methanol as sole carbon and energy sources was first described by Koichi Ogata (Ogata *et al.*, 1969). The methylotrophs attracted immediate attention as potential sources of single-cell protein (SCP), unusual enzymes (e.g., alcohol oxidase, formate dehydrogenase), and metabolites (e.g., ATP, aldehydes, amino acids) (Wegner, 1990). Biochemical studies revealed that methanol utilization requires a novel metabolic pathway involving several unique enzymes (Veenhuis *et al.*, 1983). The pathway begins with the oxidation of methanol to formaldehyde and hydrogen peroxide, catalyzed by the enzyme alcohol oxidase (AOX). AOX is sequestered within an organelle called the peroxisome along with catalase, which degrades hydrogen peroxide to oxygen and water. A portion of the formaldehyde generated by AOX leaves the peroxisome and is further oxidized to formate and carbon dioxide by two cytoplasmic dehydrogenases,

Table 1
Foreign Proteins Expressed in *P. pastoris*

Protein (organism)	Mode[a]	Reference
Bacteria		
β-Galactosidase (*Escherichia*)	I	Tschopp *et al.* (1987a)
β-Lactamase (*Escherichia*)	I	Waterham *et al.* (1997)
Pertussis pertactin protein (*Bordetella*)	I	Romanos *et al.* (1991)
Streptokinase (*Streptomyces*)	I	Hagenson *et al.* (1989)
Tetanus toxin C fragment (*Clostridium*)	I	Clare *et al.* (1991a)
D-Alanine carboxypeptidase (*Bacillus*)	S	Despreaux and Manning (1993)
Subtilisin inhibitor (*Streptomyces*)	S	Markaryan *et al.* (1996)
Fungi		
Alt a 1 allergen (*Alternaria*)	S	de Vouge *et al.* (1996)
β-Cryptogein (*Phytophthora*)	S	O'Donohue *et al.* (1996)
Dipeptidyl-peptidase V (*Aspergillus*)	S	Beauvais *et al.* (1997)
Glucoamylase (*Aspergillus*)	S	Fierobe *et al.* (1997)
β-Glucosidase (*Candida*)	S	Skory *et al.* (1996)
Invertase (*Saccharomyces*)	S	Tschopp *et al.* (1987b)
α1,2-Mannosyltransferase (*Saccharomyces*)	S	Romero *et al.* (1997)
Pectate lyase (*Fusarium*)	S	Guo *et al.* (1996)
Plants		
Glycolate oxidase (spinach)	I	M. S. Payne *et al.* (1995)
Nitrate reductase (*Arabidopsis*)	I	Su *et al.* (1996)
Phosphoribulokinase (spinach)	I	Brandes *et al.* (1996)
Phytochromes A and B (potato)	I	Ruddat *et al.* (1997)
α-Amylase isozymes (barley)	S	Juge *et al.* (1996)
Cyn d 1 allergen (Bermuda grass)	S	Smith *et al.* (1996)
α-Galactosidase (coffee bean)	S	Zhu *et al.* (1995)
Invertebrates		
Green fluorescent protein (jellyfish)	I	Monosov *et al.* (1996)
Bm86 antigen (tick)	I	Rodriguez *et al.* (1994)
Dragline silk protein (spider)	I	Fahnestock and Bedzyk (1997)
Luciferase (fire fly)	I	McCollum *et al.* (1993)
Angiotensin-converting enzyme (fruit fly)	S	Williams *et al.* (1996)
Anticoagulant peptide (tick)	S	Laroche *et al.* (1994)
Ghilanten (leech)	S	Brandkamp *et al.* (1995)
Hirudin (leech)	S	Rosenfeld *et al.* (1996b)
Mammals (other than humans)		
Carnitine palmitoyltransferases (rat)	I	de Vries *et al.* (1997)
17 α-Hydroxylase/C17,20-lyase (shark)	I	Trant (1996)
Leukocyte 12-lipoxygenase (pig)	I	Reddy *et al.* (1994)
Multifunctional enzyme (rat)	I	Qin *et al.* (1997)
NO synthase reductase domain (rat)	I	Gachhui *et al.* (1996)
α-*N*-Acetylgalactosaminidase (chicken)	S	Zhu *et al.* (1996)
Angiotensin converting enzyme (rabbit)	S*	Sadhukhan *et al.* (1996)

Table 1 *(continued)*

Protein (organism)	Mode[a]	Reference
β-Casein (cow)	S	Choi and Jiménez-Flores (1996)
Cholesteryl ester transfer protein (rabbit)	S	Kotake *et al.* (1996)
Enterokinase catalytic domain (cow)	S	Vozza *et al.* (1996)
Epidermal growth factor (mouse)	S	Clare *et al.* (1991b)
Gelatinase B (mouse)	S	Masure *et al.* (1997)
Interferon-τ (goat)	S	van Heeke *et al.* (1996)
Lysozyme (cow)	S	Digan *et al.* (1989)
Major urinary protein complex (mouse)	S	Ferrari *et al.* (1997)
5-HT5A serotonin receptor (mouse)	S*	Weiss *et al.* (1995)
Single-chain Fv fragments (mouse)	S	Luo *et al.* (1995)
Humans		
Cytomegalovirus ppUL44 antigen	I	Battista *et al.* (1996)
Hepatitis B surface antigen	I	Cregg *et al.* (1987)
Hepatitis B surface antigen-HIV gp41 epitope chimera	I	Eckhart *et al.* (1996)
α-Mannosidase	I	Liao *et al.* (1996)
Protein kinase C	I	Lima *et al.* (1996)
Proteinase inhibitor 6	I	Sun *et al.* (1995)
Tumor necrosis factor	I	Sreekrishna *et al.* (1989)
Amyloid precursor protein	S	Ohsawa *et al.* (1995)
Amyloid β-protein precursor-like protein-2 (Kunitz-type proteinase inhibitor domain)	S	van Nostrand *et al.* (1994)
Angiostatin (kringles 1–4 of plasminogen)	S	Sim *et al.* (1997)
Aprotinin analogue	S	Vedvick *et al.* (1991)
Cathepsin E	S	Yamada *et al.* (1994)
Cathepsin L propeptide	S	Carmona *et al.* (1996)
Epidermal growth factor	S	Siegel *et al.* (1990)
Fas ligand (soluble form)	S	Tanaka *et al.* (1997)
Fibroblast collagenase	S	Rosenfeld *et al.* (1996a)
Human serum albumin	S	Barr *et al.* (1992)
Interleukin-6 (cytokine receptor domain)	S	Vollmer *et al.* (1996)
Lymphocyte surface antigen CD38	S	Fryxell *et al.* (1995)
MHC class II heterodimers (soluble form)	S	Kalandadze *et al.* (1996)
Monocyte chemotactic protein	S	Masure *et al.* (1995)
μ-Opioid recepter	S*	Talmont *et al.* (1996)
Procathepsin B	S	Illy *et al.* (1997)
Tissue-type plasminogen activator (kringle 2 domain)	S	Nilsen *et al.* (1997)
Thrombomodulin	S	White *et al.* (1995)
Transferrin (N-terminal domain)	S	Steinlein *et al.* (1995)
Urokinase-type plasminogen activator-anexin v chimeras	S	Okabayshi *et al.* (1996)

[a]I, intracellular; S, secreted; S*, secreted plasma membrane protein.

reactions that are the sole source of energy for cells growing on methanol.

The remaining formaldehyde is assimilated to form cellular constituents by a cyclic pathway that starts with the condensation of formaldehyde with xylulose 5-monophosphate, a reaction catalyzed by a third peroxisomal enzyme dihydroxyacetone synthase (DHAS). The products of this reaction, glyceraldehyde 3-phosphate and dihydroxyacetone, leave the peroxisome and enter a cytoplasmic cyclic pathway that regenerates xylulose 5-monophosphate and, for every three cycles, one net molecule of glyceraldehyde 3-phosphate. Two methanol pathway enzymes, AOX and DHAS, are present at high levels in cells grown on methanol but are absent in cells grown on most other carbon sources (e.g., glucose, glycerol, ethanol). Alcohol oxidase concentrations can be especially high at >30% of total soluble protein in methanol-grown cells (Couderc and Baratti, 1980).

High-Cell Density Fermentation

The parent strain from which all *P. pastoris* expression strains are derived was developed during the 1970s by Phillips Petroleum Co. for the large-scale production of SCP, yeast biomass primarily intended for use as a high protein animal feed (Wegner, 1990). These investigations resulted in the definition of media and culture conditions that allow *P. pastoris* to be grown in continuous culture for long periods and at high-cell densities in large fermenter cultures. Much of the SCP process has been adapted for the growth and induction of *P. pastoris* expression strains (Siegel *et al.*, 1990). Culture media are inexpensive and defined, consisting of pure carbon sources (glycerol and methanol), biotin, salts, trace elements, and water. Media are free of undefined ingredients that can be a source of pyrogens or toxins, an important consideration in the production of human pharmaceuticals. *P. pastoris* is cultured in media with a relatively low pH and methanol, a carbon source that cannot be utilized by most microorganisms. Thus, *P. pastoris* cultures are less susceptible than most to contamination.

The *AOX1* Promoter

Two essential elements are required for any expression system: a strong promoter to drive the expression of foreign genes and methods to manipulate foreign DNA in the host. These elements were developed for *P. pastoris* during the 1980s under contracts from Phillips Petroleum Co. to the Salk Institute Biotechnology/Industrial Associates (SIBIA). For the reasons described earlier, the promoter selected to drive the expression of foreign genes in *P. pastoris* was *AOX1p*, and it has proven to be well suited for this task. The *AOX1* gene was isolated from a cDNA library constructed with RNA derived from methanol-grown *P. pastoris* cells by screening for clones that hybridized to RNAs specific to methanol-grown cells (Ellis *et al.*, 1985). An AOX encoding cDNA was identified and used as a probe to isolate *AOX* encoding genomic DNA fragments. Detailed studies of these clones revealed that the *P. pastoris* genome harbors two functional *AOX* genes; *AOX1* and *AOX2* (Cregg and Madden, 1987). The two AOX proteins are >90% identical at the primary sequence level and have similar specific activities (Cregg *et al.*, 1989; Koutz *et al.*, 1989). However, *AOX1* is responsible for all but a minor fraction of the total *AOX* message and protein in methanol-grown cells. *AOX1* message levels are estimated to constitute ~5% of total poly(A)$^+$ RNA from methanol-grown cells but are undetectable in cells utilizing most other carbon sources (Cregg and Madden, 1988).

Molecular Genetic Manipulation of *P. pastoris*

Molecular genetic methods for *P. pastoris* such as DNA-mediated transformation, gene targeting, gene replacement, and cloning by functional complementation are similar to those described for *S. cerevisiae*. The yeast can be transformed using either the spheroplast generation method or whole cell methods such as those involving LiCl, polyethylene glycol$_{1000}$, or electroporation (Cregg *et al.*, 1985; Liu *et al.*, 1992; Waterham *et al.*, 1996). As in *S. cerevisiae*, cleavage of a *P. pastoris* vector within a sequence shared by the host genome stimulates homologous recombination events that efficiently target integration of the vector to that genomic locus (Cregg and Madden, 1987). However, gene replacements appear to require longer terminal flanking sequences to efficiently direct integration than does *S. cerevisiae*. Replacement frequencies of greater than 20% typically require flanking segments of at least 0.5 kb at each terminus (Cregg and Russell, 1998). Specific examples of the integration of expression vectors into the *P. pastoris* genome are presented in the following section.

P. pastoris is an ascomycetous homothallic yeast that can also be manipulated by classical genetic methods (Cregg *et al.*, 1989). Unlike homothallic strains of *S. cerevisiae,* which are diploid, *P. pastoris* remains haploid unless forced to mate. To initiate a mating cycle, strains with complementary markers are mixed and subjected to a nitrogen-limited medium. After a period on this medium, cells are shifted to a standard minimal medium supplemented with a combination of nutrients designed to allow complementing diploid cells to grow but not self-mated or nonmated parental cells. The resulting diploids are stable as long as they are not subject to nutritional stress. To obtain spore products, diploids are returned to the nitrogen-limited medium, which stimulates them to proceed through meiosis and sporulation. Relative to *S. cerevisiae, P. pastoris* asci are small, and spores stick together tenaciously. Therefore, spore products are handled by random spore techniques instead of micromanipulation. With random spore techniques, most standard classical genetic manipulations can be accomplished with *P. pastoris,* including mutant isolation, complementation analysis, backcrossing, strain construction, and spore analysis.

Construction of Expression Strains

Expression Vectors

A list of commonly used *P. pastoris* expression vectors and their features is presented in Table 2. All *P. pastoris* expression vectors are shuttle plasmids designed for transformation into both *E. coli* and *P. pastoris*. For maintenance in *E. coli,* the plasmids contain an origin of replication and an ampicillin resistance gene. As a selectable marker for *P. pastoris,* most vectors contain the *P. pastoris* histidinol dehydrogenase (*HIS4*) gene, which is used in combination with *P. pastoris his4* host strains (Fig. 2) (Cregg *et al.*, 1985). A family of *P. pastoris* vectors containing the *Sh ble* gene, which confers resistance to the drug Zeocin, has been developed (Fig. 3) (Higgins *et al.*, 1998). In these vectors, the *Sh ble* or *Zeo* gene is the selectable marker for both *E. coli* and *P. pastoris,* substantially reducing the size of these plasmids.

For the expression of foreign genes, vectors contain an expression cassette composed of a fragment of ~0.9 kb containing sequences 5′ of the *AOX1* gene (*AOX1p*) and a fragment of ~0.3 kb containing

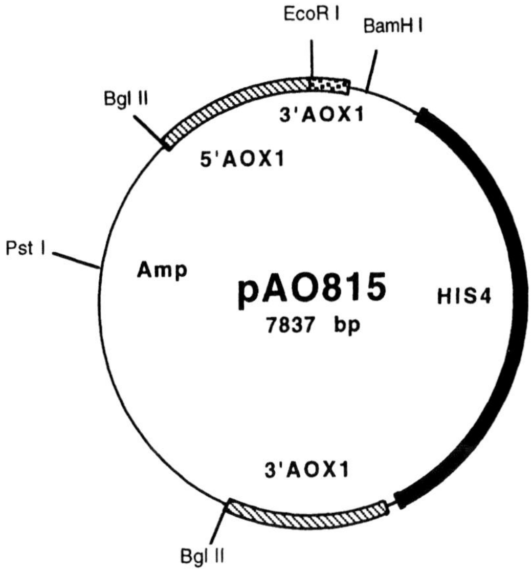

Figure 2 *P. pastoris* expression vector pAO815. The thin line represents sequences from *E. coli* plasmid pBR322 (with permission from Cregg and Higgins, 1995).

Table 2
Common *P. pastoris* Expression Vectors

Vector name	Selectable markers	Feature	Reference
Intracellular			
pHIL-D2	HIS4	NotI sites for *AOX1* gene replacement	K. Sreekrishna (personal communication)
pAO815	*HIS4*	Expression cassette bounded by *Bam*HI and *Bgl*II sites for generation of multicopy expression vector	Thill *et al.* (1990)
pPIC3K	*HIS4* and *kan*^{r}	Multiple cloning sites for insertion of foreign genes; G418 selection for multicopy strains	Scorer *et al.* (1993b)
pPICZ	*ble*^{r}	Multiple cloning sites for insertion of foreign genes;	Higgins *et al.* (1998)

(continues)

Table 2 *(continued)*

Vector name	Selectable markers	Feature	Reference
		Zeocin selection for multicopy strains; potential for fusion of foreign protein to His_6 and *myc* epitope tags	
pHWO10	*HIS4*	Expression controlled by constitutive *GAPp*	Waterham *et al.* (1997)
pGAPZ	*ble*^{r}	Expression controlled by constitutive *GAPp*; multiple cloning site for insertion of foreign genes; Zeocin selection for multicopy strains; potential for fusion of foreign protein to His_6 and *myc* epitope tags	Invitrogen, Carlsbad, CA
Secretion			
pHIL-S1	*HIS4*	*AOX1p* fused to *PHO1* secretion signal; *Xho*I, *Eco*RI, and *Bam*HI sites available for insertion of foreign genes	K. Sreekrishna (personal communication); Invitrogen
pPIC9K	*HIS4* and *kan*^{r}	*AOX1p* fused to α-MF prepro signal sequence; *Xho*I (not unique), *Eco*RI, *Not*I, *Sna*BI, and *Avr*II sites available for insertion of foreign genes; G418 selection for multicopy strains	Scorer *et al.* (1993b)
pPICZα	*ble*^{r}	*AOX1p* fused to α-MF prepro signal sequence; multiple cloning site for insertion of foreign genes; Zeocin selection for multicopy strains; potential for fusion of foreign protein to His_6 and *myc* epitope tags	Higgins *et al.* (1998)
pGAPZα	*ble*^{r}	Expression controlled by constitutive *GAPp*; *GAPp* fused to α-MF prepro signal sequence; multiple cloning site for insertion of foreign genes; Zeocin selection for multicopy strains; potential for fusion of foreign protein to His_6 and *myc* epitope tags	Invitrogen, Carlsbad, CA

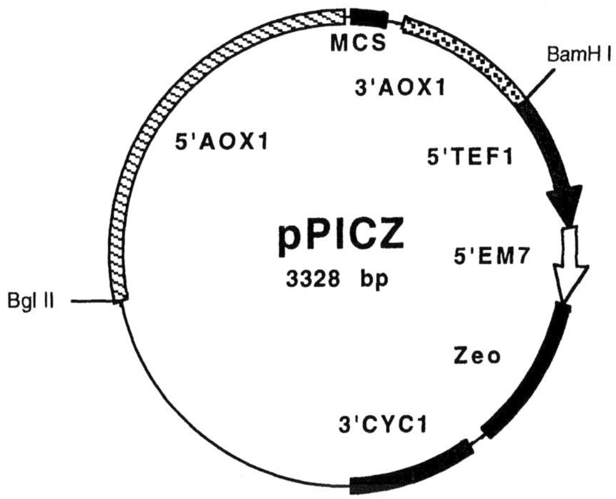

Figure 3 *P. pastoris* vector pPICZ. MCS, multiple cloning site. *Sh ble* is the gene from *Streptoalloteichus hindustanus* conferring resistance to Zeocin. 5′*TEF1* contains promoter from the *S. cerevisiae TEF1* gene. 5′*EM7* is the *E. coli EM7* promoter. 3′CYC1 contains the transcriptional terminator from the *S. cerevisiae CYC1* gene.

sequences just 3′ of the gene for transcription termination (Koutz *et al.*, 1989). The promoter and terminator fragments are separated by one or more unique restriction sites. In addition, for the secretion of foreign proteins, vectors are available in which sequences encoding the secretion signals of *S. cerevisiae* α mating factor prepro (α-MF prepro) or *P. pastoris* acid phosphatase (PHO1) have been properly fused to *AOX1p*.

Expression vectors with an alternative promoter have become available (Waterham *et al.*, 1997; Invitrogen, Carlsbad, CA); these vectors contain *GAPp*, a strong constitutive promoter derived from the glyceraldehyde 3-phosphate dehydrogenase (*GAP*) gene. In glucose-grown cultures, expression levels provided by *GAPp* are similar to those seen with the *AOX1p* in shake-flask methanol-grown cultures. An advantage of *GAPp* relative to *AOX1p* is that, because it is constitutive, it is not necessary to shift cultures from one medium to another to induce expression. However, use of *GAPp* is appropriate only for genes whose products are not deleterious to the cell. In addition, expression levels from *AOX1p* are often enhanced by feeding methanol to cultures at a growth-limiting rate in fermenter cultures. A similar phenomenon is not observed with *GAPp*.

Integration of Expression Vectors into the *P. pastoris* Genome

To maximize the stability of expression strains, *P. pastoris* expression vectors are integrated into the host genome. The most common means of insertion is to linearize vectors prior to transformation by cutting at one of several unique restriction sites in either the *HIS4* or *AOX1p* sequences. The free DNA termini stimulate homologous recombination events that result in single crossover-type integration events into these loci at a high frequency (50–80% of His$^+$ transformants) (Fig. 4A). Most of the remaining transformants appear to have undergone gene conversion events in which only the *HIS4* gene from the vector has integrated into the mutant host *his4* locus without other vector sequences.

An alternative integration strategy is available with *P. pastoris* expression vectors that contain an additional fragment derived from sequences 3′ of the *AOX1* gene (Fig. 2) (Cregg *et al.*, 1987). These vectors can be cut with selected restriction enzymes that release the expression cassette (i.e., the fragment containing the promoter, foreign gene, and terminator) and *HIS4* gene on a DNA fragment flanked by *AOX1* 5′ and 3′ terminal sequences. Approximately 10–20% of His$^+$ transformation events with these vector fragments are the consequence of a gene replacement event in which the *AOX1* gene is deleted and replaced by the expression cassette and *HIS4* gene (Fig. 4B). The resulting strains are forced to rely on the transcriptionally weak *AOX2* gene for growth on methanol and, as a result, metabolize methanol at a greatly reduced rate. These gene replacement strains are easily identified among His$^+$-transformed colonies by replica plating them to methanol and selecting those with reduced ability to grow on methanol (Mut^s phenotype). Mut^s expression strains can also be constructed by utilizing a Mut^s host strain (such as KM71) and the single crossover integration strategy. The potential advantage of Mut^s strains is that they sometimes express higher levels of foreign protein than wild-type (Mut$^+$) strains, especially in shake-flask cultures [e.g., hepatitis B surface antigen (Cregg *et al.*, 1987)].

Strains with Multiple Expression Cassettes

Methods to construct *P. pastoris* strains with multiple copies of a heterologous gene expression cassette have been described. Early *P. pastoris* expression strains were made with only a single expression

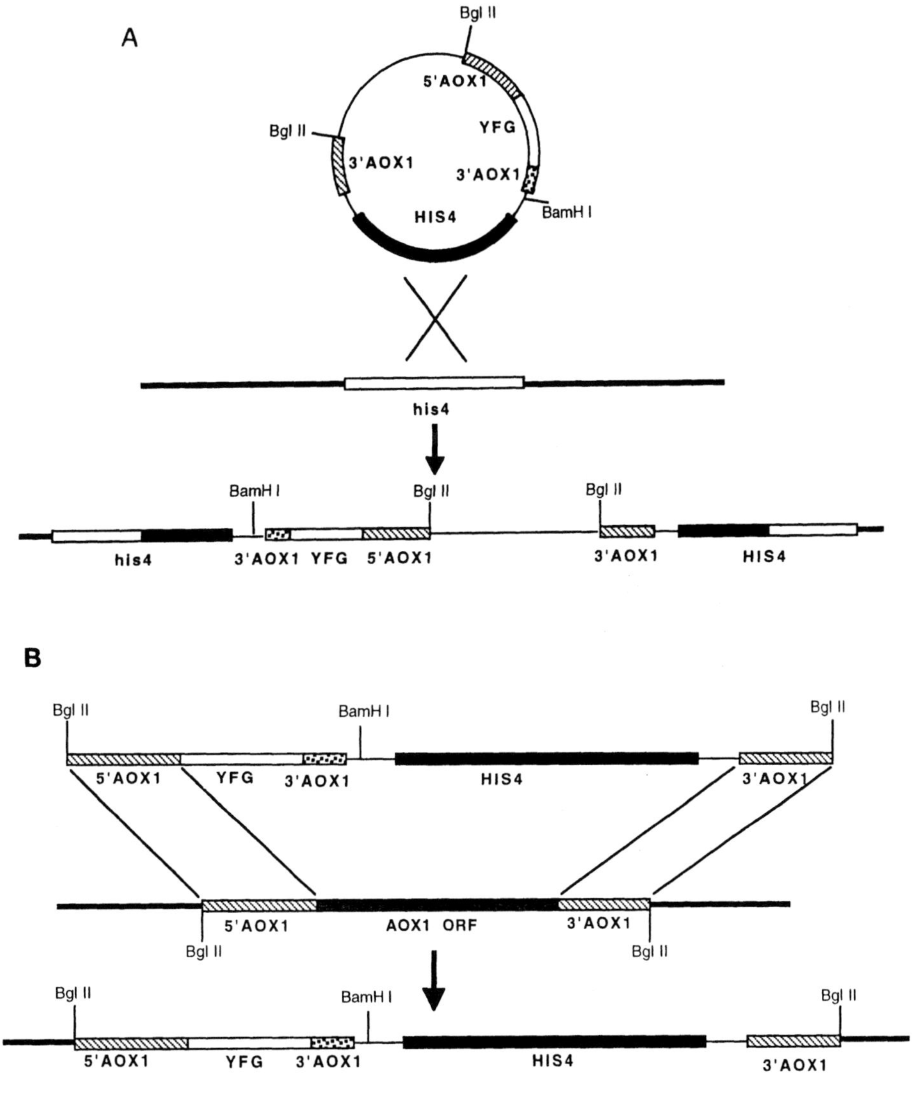

Figure 4 Integration of expression vectors into the *P. pastoris* genome. (A) Single crossover integration into the AOX1p locus. (B) Integration of vector fragment by replacement of *AOX1* gene.

cassette copy [e.g., Invertase (Tschopp *et al.*, 1987b)]. Although several of these strains produced high levels of foreign proteins, strains harboring multiple copies often (but not always; see Thill *et al.*, 1990) synthesize significantly higher levels of protein. Four methods of obtaining multicopy expression strains have evolved. The first method is to construct vectors that contain multiple expression cassette copies (Vedvick *et al.*, 1991). Starting with an expression vector such as pAO815 (Fig. 2), the foreign gene is first inserted under the control of *AOX1p*. The resulting vector containing a single copy of the gene is then digested with *Bgl*II and *Bam*HI. These sites are located just 5′ of *AOX1p* and 3′ of the *AOX1* terminator, so that cleavage results in excision of the expression cassette from the vector. The cassette is then reinserted into the *Bam*HI site of the same single-copy vector to create a vector with two cassettes in a direct repeat orientation. Importantly, the two cassettes are joined at a *Bgl*II/*Bam*HI junction that cannot be cleaved further by either enzyme. The process of *Bgl*II and *Bam*HI cleavage of a resulting multicopy copy vector and reinsertion of additional cassettes is repeated to generate a series of vectors that contain increasing numbers of cassettes. An advantage of this approach for the production of human pharmaceuticals is that the precise number of expression cassettes is known and can be recovered from the production strain for verification by direct DNA sequencing.

The second method involves transforming *P. pastoris* with vectors that contain a single expression cassette and then screening for strains that contain more than one expression vector integrated into their genomes. Such multicopy strains exist naturally within transformed cell populations at a frequency of a few percent. These multicopy strains can be identified by screening large numbers of individual transformants for high product levels using sodium dodecylsulfate–polyacrylamide gel electrophoresis (SDS–PAGE) (Sreekrishna *et al.*, 1989), immunoblotting (Clare *et al.*, 1991a), or colony immunoblotting techniques (Wung and Gascoigne, 1996) or by screening transformants for ones with multiple copies of the foreign gene by colony DNA dot blot hybridization (Romanos *et al.*, 1991).

The third method utilizes expression vectors that contain the bacterial *Tn903kan*^{r} gene, which confers resistance to G418 (Scorer *et al.*, 1993b). The level of resistance conferred by this gene is roughly proportional to the vector copy number. The *kan*^{r} vectors are transformed into *P. pastoris* by selection for His$^+$ prototrophy and then screened by replica plating to G418-containing plates.

This method results in a subset of colonies enriched for those that contain multiple copies of the expression vector. However, the vector copy number varies greatly. Thus, several His⁺- and G418-resistant colonies must be selected and subjected to further analysis of copy number and expression level. Using these approaches, strains harboring up to 30 copies of an expression cassette have been isolated (Clare *et al.*, 1991b).

The fourth method utilizes Zeocin resistance vectors such as pPICZ (Fig. 3) (Higgins *et al.*, 1998). The presence of Zeocin in plates at a concentration of 50 μg/ml results in a population of transformants that mostly contain a single copy of a Zeo vector. However, increasing the concentration of Zeocin to between 0.5 and 1 mg/ml significantly enriches for multicopy transformants. As with the G418 selection method, transformants vary in copy number and must be examined further to determine which have the highest number and which are the most productive. However, with the Zeocin vectors, multicopy strains can be selected directly during transformation without the intermediate step of selecting for His⁺ colonies. Furthermore, these vectors also have a *Bgl*II and *Bam*HI site-bounded expression cassette for the construction of multicopy vectors. Thus, it is possible to construct a vector with multiple expression cassettes and then further select for strains that have multiple copies of a multiexpression cassette vector.

Strains derived by each of the four methods have proven to be stable, even under the selective pressure of production in fermenter cultures.

Host Strain Selection

A list of *P. pastoris* host strains commonly employed for the expression of foreign genes is presented in Table 3. To allow for transformation with *HIS4*-based vectors, most expression strains are *his4* auxotrophs. When using Zeocin resistance-based vectors, the host strain should be prototrophic or made prototrophic by transformation with a DNA fragment containing the *HIS4* gene or a vector containing *HIS4*. A prototrophic expression strain avoids potential problems caused by the depletion of histidine pools in cells that may occur with a histidine-requiring strain.

Three types of *P. pastoris* host strains are available that vary with regard to their ability to grow on methanol. In addition to wild-type methanol-utilizing (Mut⁺) and *AOX1* deletion (Mut^s) strains men-

Table 3
P. pastoris Expression Host Strains

Strain name	Genotype	Phenotype	Reference
Y-11430		Wild type	NRRL[a]
GS115	*his4*	Mut^+ His^-	Cregg *et al.* (1985)
KM71	*aox1Δ::SARG4 his4 arg4*	Mut^s His^-	Cregg and Madden (1987)
MC100-3	*aox1Δ::SARG4 aox2Δ::Phis4 his4 arg4*	Mut^- His^-	Cregg *et al.* (1989)
SMD1168	*pep4Δ his4*	Mut^+ His^+ Protease deficient	White *et al.* (1995)

[a]Northern Regional Research Laboratories (Peoria, IL).
Note: Superscript s refers to slow phenotype.

tioned earlier, a strain in which both *AOX1* and *AOX2* genes have been deleted has been described for expression (Chiruvola *et al.*, 1997). Because this strain (MC100–3) does not grow on methanol at all (Mut^-) and is *his4* marked, it can be transformed by *HIS4*-based expression vectors. Foreign genes under control of the *AOX1* promoter in the Mut^- strain are still induced by methanol to levels comparable to those in the other hosts. Because methanol is used solely for induction and not for growth by this strain, the amount of methanol used is reduced greatly. The Mut^- strain may be an attractive host for certain large-scale protein production facilities, if the presence of the large volumes of methanol needed to grow and induce a Mut^+ expression strain is considered a potential fire hazard. A second potential application for the Mut^- expression strain is in the production of recombinant proteins for use by the food products industry where feedstocks for growth of microorganisms must come from nonpetrochemical sources. Methanol is available from agricultural sources, but is more expensive than methanol generated from natural gas. Thus, the use of a Mut^- expression strain could significantly lower the production cost of recombinant proteins needed by the food industry.

A third host strain option is a protease-deficient strain such as SMD1168 (White *et al.*, 1995). In this strain, *PEP4*, a gene encoding a key hydrolytic enzyme, has been deleted. Although normally compartmentalized within the vacuole of the cell, the PEP4 protein and other proteases activated by the PEP4 protein are present in *P. pastoris* culture media due to the lysis of a small percentage of cells during culturing. Because the concentration of these proteases in the medium increases along with culture density, the protein stability problem is exacerbated at high cell densities. Certain secreted proteins have been found to be particularly sensitive to proteases, and their production in *P. pastoris* is aided significantly by the use of this protease-deficient strain (e.g., White *et al.*, 1995; Brierley, 1998). However, protease-deficient strains are not as "healthy" as strains that are wild type with respect to *PEP4* (e.g., lower viability, lower transformation efficiency, slower growth rate); therefore, their use is recommended as an alternative in situations where proteases may be the cause of low yields and where other measures to reduce proteolysis, such as those discussed in the following section, have not proven to be sufficient.

Important Considerations in the Use of the Expression System

General Considerations

Because *P. pastoris* is a eukaryotic organism, the production of foreign proteins in this system is a more complex process than in *E. coli*. Even for proteins that are eventually produced at >1 g/liter concentration in *P. pastoris*, initial shake-flask levels can be disappointingly low (<1 mg/liter). The generation of a productive strain often requires significant effort in investigating factors that may be limiting protein yield and testing improvement strategies. Because initial levels may be low, a key reagent for the development of a *P. pastoris* expression strain is a preparation of antibodies specific for a foreign protein so that it can be detected at these initial low levels and examined for characteristics such as intactness and glycosylation.

System users should also be aware that, for high expression levels, it is usually necessary to grow their *P. pastoris* expression strain in fermenter culture. One obvious advantage of fermentation, especially for secreted proteins, is the potential for high-cell density culturing. A second advantage is that *AOX1p* is induced to levels three to five times higher in cells fed methanol at growth-limiting rates relative to cells growing in excess methanol. The controlled feeding of methanol is practical only in a fermenter where the methanol feed rate and the dissolved oxygen level can be monitored.

For a list of the advantages and disadvantages of the *P. pastoris* expression system, see Table 4.

Table 4
Advantages and Disadvantages of the *P. pastoris* Expression System

Advantages	Disadvantages
Culturing	
Rapid growth rate	
High-cell density: >100 g (dcw)/liter	
Clean medium composed of salts, biotin, and carbon source	
Easily scaled up to large-volume, high-density fermentor cultures	Fermentor culturing often needed to achieve high level of foreign protein
Molecular genetics	
Classical genetic methods available	
Molecular methods similar to *S. cerevisiae*	Range of vectors limited
Stable integrated expression vectors	
Promoters	
AOX1p—strong, tightly regulated, and easily controlled	
GAPp—strong constitutive	
Expression	
Eukaryotic environment aids folding of higher eukaryotic foreign proteins	
High expression levels	
Secretion	Improper posttranslational modifications
Proper posttranslational modifications	Native signals not always processed
Sulfhydril bond formation	Some proteins misfold and become stuck in secretory pathway
Signal sequence processing	Lower eukaryotic (high mannose)-type glycosylation
Folding	Proteases in medium degrade some foreign proteins
Glycosylation	
High levels—g/liter fermentor cultures	
Few yeast proteins in medium—high initial purity of foreign protein	

Preparation of the Foreign Gene

Most *P. pastoris* expression vectors contain a unique *Eco*RI site into which foreign genes are inserted that is positioned just 5′ of the methionine initiator ATG of *AOX1*. In preparing DNA for insertion at this site, 5′ nontranslated sequences from the foreign gene (cDNA) should be removed, as these sequences can greatly reduce expression levels (Sreekrishna, 1993). 3′ untranslated sequences do not appear to affect the expression of foreign genes and can be left in place or removed, whichever is more convenient.

The expression of foreign genes with a high A + T content can be affected by premature transcription termination. Although the transcription termination signal for *P. pastoris* is not known, it appears to be an AT-rich motif, and termination within AT-rich sequences located within the coding region of mammalian genes has occurred in yeasts and *P. pastoris* specifically [e.g., tetanus toxin C fragment (Clare *et al.*, 1991a), HIV gp120 (Scorer *et al.*, 1993a)]. Fortunately, this problem can be corrected. In expressing HIV–1 gp120 protein in *P. pastoris,* Scorer et al. (1993a) observed the presence of multiple partial transcription termination points within the gene. These sites were removed successfully by increasing the GC content of that region of the gene through chemical synthesis.

Many *P. pastoris* expression vectors now contain additional cloning sites further 3′ of the standard *Eco*RI site into which a foreign gene can be inserted. Caution should be taken when using these sites as expression levels from these positions relative to *Eco*RI have not been determined and may be significantly lower.

Signal Sequence Selection

Secretion requires an amino-terminal signal sequence to target the proteins into the secretory pathway. For *P. pastoris,* the native signal sequence of the foreign protein has been utilized successfully in some instances [e.g., HSA (Sreekrishna, 1993)]. However, for other proteins, *P. pastoris* has not been able to efficiently utilize the native signal sequence, e.g., invertase (Tschopp *et al.*, 1987b). The most generally applicable signal for the secretion of foreign proteins has been the 89 amino acid *S. cerevisiae* α-MF prepro leader, and sequences encoding this signal have been incorporated into several of the standard *P. pastoris* expression vectors (Table 2). In *S. cerevisiae,* processing of the α-MF prepro leader requires three proteolytic activities: (1) a signal peptidase that cleaves the 19 amino acid pre region; (2) a dibasic endopeptidase, the product of the *KEX2* gene, that cleaves on the carboxy-terminal side of Lys–Arg at the junction between α-MF prepro and the foreign protein; and (3) a dipeptidyl amino peptidase, the *STE13* gene product, that removes Glu–Ala spacer residues from the amino terminus of the foreign protein. Using the α-MF prepro leader, a number of foreign proteins have been secreted at high levels from *P. pastoris* [e.g., mEGF (Clare *et al.*, 1991b)]. For most of these, processing of the pro leader at the *KEX2* and *STE13* sites has occurred efficiently, generating proteins with a uniform authentic amino terminus. Thus, the *P. pastoris* secretory pathway has ample quantities of these proteases for recognition and processing of the *S. cerevisiae* α-MF prepro leader.

An α-MF prepro leader may not be appropriate for all secreted proteins. Vedvick *et al.* (1991) expressed aprotinin (bovine pancreatic trypsin inhibitor) fused to the α-MF prepro leader at the Lys–Arg processing site (i.e., without a Glu–Ala spacer). Although nearly 1 g/liter of active aprotinin was secreted, amino-terminal sequencing showed that the product was a mixture of two peptides with amino-terminal extensions of either 11 or 4 amino acids with little properly processed aprotinin. A second strain that included a Glu–Ala spacer between the Lys–Arg site and the first amino acid of mature aprotinin secreted a product that was efficiently cleaved after Lys–Arg, but which had retained Glu–Ala at the amino terminus. Similar results have been obtained with other α-MF prepro foreign protein fusions and suggest that the amino termini of these proteins fold into a conformation that is inaccessible to the *KEX2* and *STE13* proteins. To achieve proper processing in these instances, it may be necessary to use a standard pre-type signal such as that from *P. pastoris PHO1* (W. E. Payne *et al.*, 1995), *S. cerevisiae SUC2* (Tschopp *et al.*, 1987b), or other genes. In contrast to the α-MF pro leader, which is removed in the Golgi after the protein has folded, pre-type signal sequences are cleaved during translocation into the endoplasmic reticulum (ER) while nascent polypeptides are in an extended conformation (Lyman and Schekman, 1996). Unfortunately, and for unknown reasons, success in secreting foreign proteins using pre-type signal sequences has been variable. Thus, if an authentic amino terminus is essential and the α-MF prepro leader is not processed properly, it may be necessary to empirically test several signal sequences to find one that results in the efficient secretion and processing of the foreign protein.

Glycosylation

P. pastoris is capable of adding both O- and N-linked carbohydrate to secreted proteins (Goochee *et al.*, 1991). O-glycosylation involves the attachment of carbohydrate to polypeptides via the hydroxyl group of serine and threonine. In mammals, O-linked oligosaccharides are composed of a variety of sugars, including *N*-acetylgalactosamine, galactose (Gal), and sialic acid (NeuAc). In contrast, lower eukaryotes, including *P. pastoris*, add O-oligosaccharides composed solely of mannose (Man) residues. The number of Man residues per chain, their manner of linkage, and the frequency and specificity of O-glycosylation in *P. pastoris* have yet to be determined. It should not be assumed that because a protein is not glycosylated by its native host, *P. pastoris* will not glycosylate it. *P. pastoris* added O-linked mannose to ~15% of human IGF-1 protein, although this protein is not glycosylated in humans (Brierley, 1998). Furthermore, it should not be assumed that the specific Ser and Thr residue(s) selected for O-glycosylation by *P. pastoris* will be the same as the native host.

N-glycosylation begins in the ER with the transfer of a lipid-linked oligosaccharide unit, $Glc_3Man_9GlcNAc_2$ (Glc, glucose; GlcNAc, *N*-acetylglucosamine), to asparagine at the recognition sequence Asn–X–Ser/Thr. This oligosaccharide core unit is subsequently trimmed to $Man_8GlcNAc_2$ (Fig. 5). Beyond this point, lower and higher eukaryotic glycosylation patterns differ substantially. The mammalian Golgi apparatus performs a series of trimming and addition reactions that generate oligosaccharides composed of $Man_{5-6}GlcNAc_2$ (high-mannose type), a mixture of several different sugars (complex type), or a combination of both (hybrid type) (Goochee *et al.*, 1991). In *S. cerevisiae*, N-linked core units are elongated in the Golgi through the addition of mannose outer chains. These chains vary greatly in length, but typically are long, containing more than 50 mannose residues. As a result, *S. cerevisiae* N-glycosylated proteins are heterogeneous in size. Mammalian glycoproteins secreted from *S. cerevisiae* also receive these long mannose outer chains, a condition referred to as hyperglycosylation.

Two distinct patterns of N-glycosylation have been observed on foreign proteins secreted by *P. pastoris*. Some proteins such as *S. cerevisiae* invertase are secreted with carbohydrate structures similar in size and structure to the core unit (Tschopp *et al.*, 1987b;

Grinna and Tschopp, 1989). The lack of outer mannose chains on *P. pastoris*-secreted invertase is surprising given the large amount of outer chain addition that occurs during the secretion of invertase by its native host *S. cerevisiae*. Trimble *et al.* (1991) determined the structures of the most abundant species of oligosaccharide molecules on *P. pastoris*-secreted invertase and found that most were similar to those of *S. cerevisiae* core oligosaccharides.

Figure 5 shows the major species identified along with the percentage of the total oligosaccharide molecules each species represented. Almost all were $Man_{8-11}GlcNAc_2$ and, with the exception of the $Man_{11}GlcNAc_2$ species, were identical to *S. cerevisiae* core structures. The most abundant $Man_{11}GlcNAc_2$ oligosaccharides contained an α1,2-linked mannose at one of three terminal positions, whereas in *S. cerevisiae* an α1,3 mannose is the most frequently observed $Man_{11}GlcNAc_2$ species (Fig. 5, bottom left). Based on the similarity of *P. pastoris* and *S. cerevisiae* core structures, Trimble *et al.* (1991) hypothesized that processing of core oligosaccharides occurs via similar pathways in both yeasts. Interestingly, *P. pastoris* does not appear to be capable of adding α1,3-terminal mannose to oligosaccharides, as α1,3-linked mannose is absent from total *P. pastoris* oligosaccharide preparations and activity for α1,3 mannosyl transferase is also absent (R. B. Trimble, personal communication). This contrasts with *S. cerevisiae* oligosaccharides where α1,3-linked terminal mannose is common.

Other foreign proteins secreted from *P. pastoris* receive much more carbohydrate and appear by SDS–PAGE and western blotting to be hyperglycosylated [e.g., HIV gp120 (Scorer *et al.*, 1993a)]. Aside from the probable absence of α1,3-linked mannose, little is known regarding the structure of *P. pastoris* outer-chain oligosaccharides. Furthermore, it is also not clear why outer chains are added to some *P. pastoris* secreted proteins and not to others or how outer chain addition may be prevented.

Growth in Fermenter Cultures

Although a few foreign proteins have been expressed well in *P. pastoris* shake-flask cultures, expression levels in shake flasks are more typically relatively low and are improved greatly by fermenter culturing. One reason fermenter culturing is necessary is that only in the controlled environment of a fermenter is it possible to grow the organism to high-cell densities (>100 g/liter dry cell

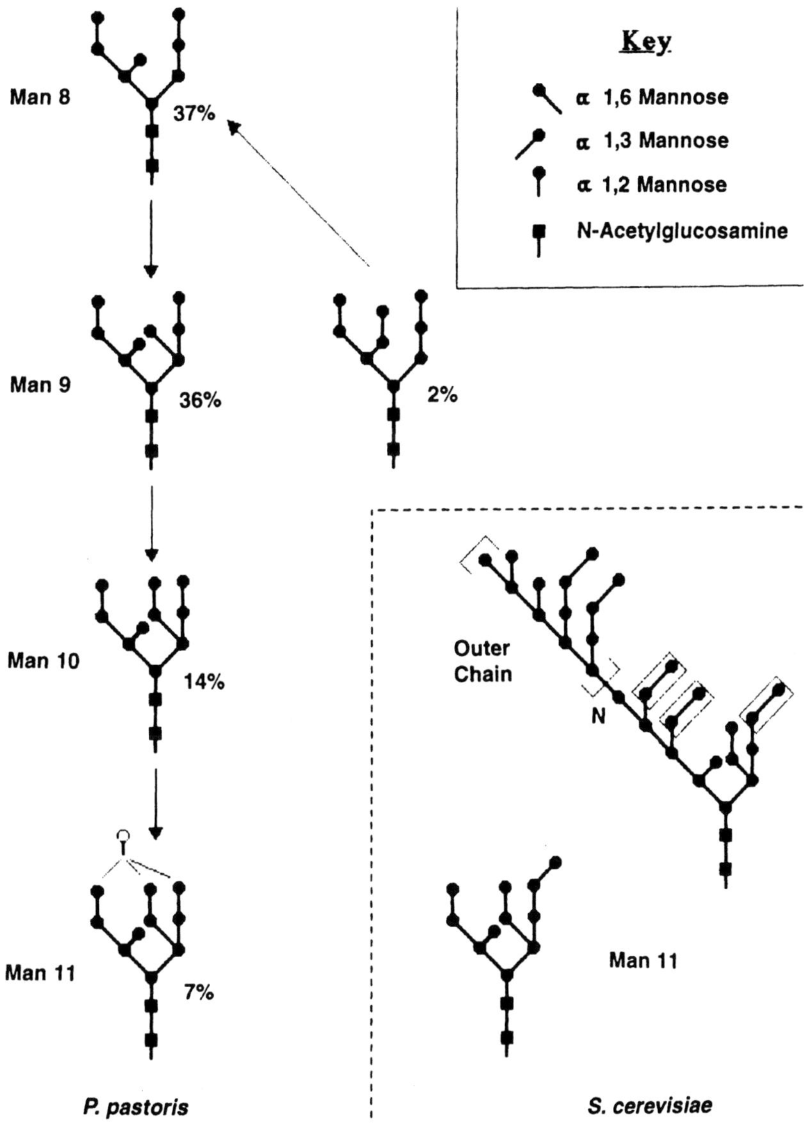

Figure 5 Structures of selected N-linked oligosaccharides on *S. cerevisiae* invertase secreted by *P. pastoris* (Trimble *et al.*, 1991). Sugar and linkage type are indicated in the upper right-hand corner. Open circles with dashed lines represent the most frequent position observed for the last mannose on *P. pastoris* oligosaccharides. Terminal α1,3-linked mannoses on the *S. cerevisiae* core structure are shown enclosed in rectangles (with permission from Cregg and Higgins, 1995).

weight or 500 OD_{600} units/ml). A second reason is that the level of transcription initiated from *AOX1p* is greater in *P. pastoris* cells fed methanol at growth-limiting rates in fermenter culture relative to cells grown in excess methanol (Cregg and Madden, 1988). Thus, even for intracellularly expressed proteins, product yields from a given strain as a percentage of total cellular proteins are significantly higher from fermenter cultured cells. A third reason is that methanol metabolism utilizes oxygen at a high rate and the expression of foreign genes is affected negatively by oxygen limitation. Only in the controlled environment of a fermenter is it feasible to accurately monitor and adjust oxygen levels in the culture medium. Thus, many users of the *P. pastoris* expression system will need to produce their foreign protein in fermenters.

Considerable effort has gone into the optimization of high cell density fermentation techniques for expression strains and, as a result, a variety of fed-batch and continuous culture schemes are available (see Higgins and Cregg, 1998). To varying degrees, each takes advantage of the ability to separately control cell growth and production in *P. pastoris*. All schemes involve the initial growth of strains in a defined medium on glycerol. During this period, growth is rapid, but heterologous gene expression is fully repressed. Upon depletion of glycerol, a transition phase is initiated in which additional glycerol is fed to cultures at a growth-limiting rate. Finally, methanol or a mixture of glycerol and methanol is fed to cultures to induce expression. The methanol feed rate is increased by steps, and the ability of cultures to utilize methanol is monitored by observing changes in dissolved oxygen levels. The time of harvest, typically the peak concentration of a foreign protein, is determined empirically for each protein.

Unfortunately, high-cell density fermenation also increases the concentrations of other cellular materials, particularly proteases. In this environment, some secreted proteins are quite stable, whereas others are significantly degraded. Three strategies have proven effective in minimizing the proteolytic instability of foreign proteins secreted into the *P. pastoris* culture medium. One is the addition of amino acid-rich supplements, such as peptone or casamino acids to the culture medium, which appear to reduce product degradation by acting as excess substrates for problem proteases (Clare *et al.*, 1991b). A second is changing the culture medium pH (Clare *et al.*, 1991b). *P. pastoris* is capable of growing across a relatively broad range of pH from 3.0 to 7.0, which allows considerable leeway in adjusting the pH to one that is not optimal for a problem protease.

A third is the use of a protease-deficient *P. pastoris* host strain. Finally, it is possible to combine these proteolytic-stabilizing strategies with an additive effect.

Conclusion

The *P. pastoris* expression system has gained acceptance as an important host organism for the production of foreign proteins. One protein, HSA, synthesized in *P. pastoris* has passed clinical trials for use as part of a serum replacement product and is expected to receive final approval for release in the near future. Another protein, hepatitis B surface antigen, is currently on the market as a subunit vaccine against the hepatitis B virus in South America. Starting in 1994, the system was made available to both academic and industrial researchers and, as a result, the number and variety of proteins expressed in *P. pastoris* have increased dramatically. These reports have provided new reagents and methods for the system and insights into its use. For example, Laroche et al. (1994) described the use of *P. pastoris* to efficiently incorporate ^{13}C and ^{15}N isotopes into proteins for the determination of protein structures via nuclear magnetic resonance techniques, thus extending the usefulness of the system to researchers investigating protein structure/function relationships.

Despite the success of the *P. pastoris* system, opportunities exist to increase the range of proteins that can be expressed in the system. Alternative promoters with a range of transcriptional strengths and regulatory properties would be useful along with additional selectable markers for the transformation of *P. pastoris*. These tools would allow investigators to express heterooligomeric proteins and to further increase the potential number of expression cassettes that can be introduced into a single strain. Little is known about *AOX1p* regulation at the molecular level. Such studies could lead to modified *AOX1* promoters with increased transcriptional strength or to the identification and overexpression of factors that limit transcription of the *AOX1p*.

Studies are also needed to address problems associated with the secretion of mammalian proteins from *P. pastoris*. It is not clear why some native signal sequences work and others do not or why heterologous signal sequences only work on some proteins and not on others. The nonmammalian character of yeast glycosylation remains a major stumbling block for the system. Methods to significantly reduce outer chain addition to N-linked sites or to make *P. pastoris* N-linked oligosaccharides more mammalian like in structure would be most useful. Finally, a better understanding of *P. pastoris* proteases and, in particular, the identification of those responsible for the degradation of some foreign proteins would aid in the development of strategies to eliminate or circumvent their effects.

Acknowledgments

I thank Terrie Hadfield for help in preparing this manuscript. Its preparation was supported in part by grants from the National Institutes of Health (DK43698) and the National Science Foundation (MCB 9514289).

References

Barr, K. A., Hopkins, S. A., and Sreekrishna, K. (1992). Protocol for efficient secretion of HSA developed from *Pichia pastoris*. *Pharm. Eng.* **12,** 48–51.

Battista, M. C., Bergamini, G., Campanini, F., Landini, M. P., and Ripaldi, A. (1996). Intracellular production of a major cytomegalovirus antigenic protein in the methylotrophic yeast *Pichia pastoris*. *Gene* **176,** 197–201.

Beauvais, A., Monod, M., Debeaupuis, J. P., Diaquin, M., Kobayashi, H., and Latgé, J. P. (1997). Biochemical and antigenic characterization of a new didpeptidyl-peptidase isolated from *Aspergillus fumigatus*. *J. Biol. Chem.* **272,** 6238–3244.

Brandes, H. K., Hartman, F. C., Lu, T. Y., and Larimer, F. W. (1996). Efficient expression of the gene for spinach phosphoribulokinase in *Pichia pastoris* and utilization of the recombinant enzyme to explore the role of regulatory cysteinyl residues by site-directed mutagenesis. *J. Biol. Chem.* **271,** 6490–6796.

Brandkamp, R. G., Sreekrishna, K., Smith, P. L., Blankenship, D. T., and Cardin, A. D. (1995). Expression of a synthetic gene encoding the anticoagulant-antimetastatic protein ghilanten by the methylotropic yeast *Pichia pastoris*. *Protein Express. Purif.* **6,** 813–820.

Brierley, R. A. (1998). Secretion of recombinant human insulin-like growth factor I. *In* "*Pichia* Protocols: Methods in Molecular Biology" (D. R. Higgins and J. M. Cregg, eds.), pp. 149–177, Chapter 11. Humana Press, Totowa, NJ.

Carmona, E., Dufour, E., Plouffe, C., Takebe, S., Mason, P., Mort, J. S., and Ménard, R. (1996). Potency and selectivity of the cathepsin L propeptide as an inhibitor of cysteine proteases. *Biochemistry* **35,** 8149–8157.

Chirulova, V., Cregg, J. M., and Meagher, M. M. (1997). Recombinant protein production in an alcohol oxidase-defective strain of *Pichia pastoris* in fed-batch fermentations. *Enzyme Microb. Technol.* **21,** 277–283.

Choi, B.-K., and Jiménez-Flores, R. (1996). Study of putative glycosylation sites in bovine β-casein introduced by PCR-based site-directed mutagenesis. *J. Agric. Food Chem.* **44,** 358–364.

Clare, J. J., Rayment, F. B., Ballantine, S. P., Sreekrishna, K., and Romanos, M. A. (1991a). High-level expression of tetanus toxin fragment C in *Pichia pastoris* strains containing multiple tandem integrations of the gene. *Bio/Technology* **9,** 455–460.

Clare, J. J., Romanos, M. A., Rayment, F. B., Rowedder, J. E., Smith, M. A., Payne, M. M., Sreekrishna, K., and Henwood, C. A. (1991b). Production of mouse epidermal growth factor in yeast: High-level secretion using *Pichia pastoris* strains containing multiple gene copies. *Gene* **105,** 205–212.

Couderc, R., and Baratti, J. (1980). Oxidation of methanol by the yeast *Pichia pastoris*: Purification and properties of alcohol oxidase. *Agric. Biol. Chem.* **44,** 2279–2289.

Cregg, J. M., and Higgins, D. R. (1995). Production of foreign proteins in the yeast *Pichia pastoris. Can. J. Bot.* **73**(Suppl. 1), S891–S897.

Cregg, J. M., and Madden, K. R. (1987). Development of yeast transformation systems and construction of methanol-utilization-defective mutants of *Pichia pastoris* by gene disruption. *In* "Biological Research on Industrial Yeasts" (G. G. Stewart, I. Russell, R. D. Klein, and R. R. Hiebsch, eds.), Vol. 2, pp. 1–18. CRC Press, Boca Raton, FL.

Cregg, J. M., and Madden, K. R. (1988). Development of the methylotrophic yeast, *Pichia pastoris,* as a host system for the production of foreign proteins. Dev. *Ind.* Microbiol. **29,** 33–41.

Cregg, J. M., and Russell, K. A. (1997). Transformation. *In* "*Pichia* Protocols: Methods in Molecular Biology" (D. R. Higgins and J. M. Cregg, eds.), pp. 27–39, Chapter 3. Humana Press, Totowa, NJ.

Cregg, J. M., Barringer, K. J., Hessler, A. Y., and Madden, K. R. (1985). *Pichia pastoris* as a host system for transformations. *Mol. Cell. Biol.* **5,** 3376–3385.

Cregg, J. M., Tschopp, J. F., Stillman, C., Siegel, R., Akong, M., Craig, W. S., Buckholz, R. G., Madden, K. R., Kellaris, P. A., Davis, G. R., Smiley, B. L., Cruze, J., Torregrossa, R., Velicelebi, G., and Thill, G. P. (1987). High-level expression and efficient assembly of hepatitis B surface antigen in the methylotrophic yeast, *Pichia pastoris. Bio/Technology* **5,** 479–485.

Cregg, J. M., Madden, K. R., Barringer, K. J., Thill, G. P., and Stillman, C. A. (1989). Functional characterization of the two alcohol oxidase genes from the yeast *Pichia pastoris. Mol. Cell. Biol.* **9,** 1316–1323.

Cregg, J. M., Vedvick, T. S., and Raschke, W. C. (1993). Recent advances in the expression of foreign genes in *Pichia pastoris. Bio/Technology* **11,** 905–910.

Despreaux, C. W., and Manning, R. F. (1993). The dacA gene of Bacillus stearothermophilus coding for D-alanine carboxypeptidase: Cloning, structure and expression in *Escherichia coli* and *Pichia pastoris. Gene* **131,** 35–41.

de Vouge, M. W., Thaker, A. J., Curran, I. H., Zhang, L., Muradia, G., Rode, H., and Vijay, H. M. (1996). Isolation and expression of a cDNA clone encoding an Alternaria alternata Alt a 1 subunit. *Int. Arch. Allergy Immunol.* **111,** 385–395.

de Vries, Y., Arvidson, D. N., Waterham, H. R., Cregg, J. M., and Woldegiorgis, G. (1997). Functional characterization of mitochondrial carnitine palmitoyltransferases I and II expressed in the yeast *Pichia pastoris. Biochemistry* **36,** 5285–5292.

Digan, M. E., Lair, S. V., Brierley, R. A., Siegel, R. S., Williams, M. E., Ellis, S. B., Kellaris, P. A., Provow, S. A., Craig, W. S., Velicelebi, G., Harpold, M. M., and Thill, G. P. (1989) Continuous production of a novel lysozyme via secretion from the yeast, *Pichia pastoris*. *Bio/Technology* **7**, 160–164.

Eckhart, L., Raffelsberger, W., Ferko, B., Klima, A., Purtscher, M., Katinger, H., and Ruker, F. (1996). Immunogenic presentation of a conserved gp41 epitope of human immunodeficiency virus type 1 on recombinant surface antigen of hepatitis B virus. *J. Gen. Virol.* **77**, 2001–2008.

Ellis, S. B., Brust, P. F., Koutz, P. J., Waters, A. F., Harpold, M. M., and Gingeras, T. R. (1985) Isolation of alcohol oxidase and two other methanol regulatable genes from the yeast *Pichia pastoris*. *Mol. Cell. Biol.* **5**, 1111–1121.

Fahnestock, S. R., and Bedzyk, L. A. (1997). Production of synthetic spider dragline silk protein in *Pichia pastoris*. *Appl. Microbiol. Biotechnol.* **47**, 33–39.

Ferrari, E., Lodi, T., Sorbi, R. T., Tirindelli, R., Cavaggioni, A., and Spisni, A. (1997). Expression of a lipocalin in *Pichia pastoris*: Secretion, purification and binding activity of a recombinant mouse major urinary protein. *FEBS Lett.* **401**, 73–77.

Fierobe, H.-P., Mirgorodskaya, E., Frandsen, T. P., Roepstorff, P., and Svensson, B. (1997). Overexpression and characterization of *Aspergillus awamori* wild-type and mutant glucoamylase secreted by the methylotrophic yeast *Pichia pastoris*: Comparison with wild-type recombinant glucoamylase produced using *Saccharomyces cerevisiae* and *Aspergillus niger* as hosts. *Protein Express. Purif.* **9**, 159–170.

Fryxell, K. B., O'Donoghue, K., Graeff, R. M., Lee, H. C., and Branton, W. D. (1995). Functional expression of soluble forms of human CD38 in *Escherichia coli* and *Pichia pastoris*. *Protein Express. Purif.* **6**, 329–336.

Gachhui, R., Presta, A., Bentley, D. F., Abu Soud, H. M., McArthur, R., Brudvig, G., Ghosh, D. K., and Strehr, D. J. (1996). Characterization of the reductase domain of rat neuronal nitric oxide synthase generated in the methyotrophic yeast *Pichia pastoris*. Calmodulin response is complete within the reductase domain itself. *J. Biol. Chem.* **271**, 20594–20602.

Goochee, C. F., Gramer, M. J., Andersen, D. C., Bahr, J. B., and Rasmussen, J. R. (1991). The oligosaccharides of glycoproteins: Bioprocess factors affecting oligosaccharide structure and their effect on glycoprotein properties. *Bio/Technology* **9**, 1347–1355.

Grinna, L. S., and Tschopp, J. F. (1989). Size distribution and general structural features of N-linked oligosaccharides from the methylotrophic yeast, *Pichia pastoris*. *Yeast* **5**, 107–115.

Guo, W., González-Candelas, L., and Kolattukudy, P. E. (1996). Identification of a novel *pelD* gene expressed uniquely in planta by *Fusarium solani* f. sp. pisi (*Nectria haematococca*, mating type VI) and characterization of its protein product as an endo-pectate lyase. *Arch. Biochem. Biophys.* **332**, 305–312.

Hagenson, M. J., Holden, K. A., Parker, K. A., Wood, P. J., Cruze, J. A., Fuke, M., Hopkins, T. R., and Stroman, D. W. (1989). Expression of streptokinase in *Pichia pastoris* yeast. *Enzyme Microb. Technol.* **11**, 650–656.

Higgins, D. R., and Cregg, J. M. (1998). *Pichia* Protocols: Methods in Molecular Biology, Human Press, Totawa, N.J.

Higgins, D. R., Busser, K., Comiskey, J., Whittier, D. S., Purcell, T. J., and Hoeffler, J. P. (1997). Small vectors for expression based on dominant drug resistance with direct multicopy selection. *In* "*Pichia* Protocols: Methods in Molecular Biology" (D. R. Higgins and J. M. Cregg, eds.), pp. 41–53, Chapter 4. Humana Press, Totowa, NJ.

Illy, C., Quraishi, O., Wang, J., Purisima, E., Vernet, T., and Mort, J. S. (1997). Role of the occluding loop in cathepsin B activity. *J. Biol. Chem.* **272,** 1197–1202.

Juge, N., Andersen, J. S., Tull, D., Roepstorff, P., and Svensson, B. (1996). Overexpression, purification, and characterization of recombinant barley α-amylases 1 and 2 secreted by the methylotrophic yeast *Pichia pastoris. Protein Express. Purif.* **8,** 204–214.

Kalandadze, A., Galleno, M., Foncerrada, L., Strominger, J. L., and Wucherpfennig, K. W. (1996). Expression of recombinant HLA-DR2 molecules. Replacement of the hydrophobic transmembrane region by a leucine zipper dimerization motif allows the assembly and secretion of soluble DR αβ heterodimers. *J. Biol. Chem.* **271,** 20156–20162.

Kotake, H., Li, Q., Ohnishi, T., Ko, K. W., Agellon, L. B., and Yokoyama, S. (1996). Expression and secretion of rabbit plasma cholesteryl ester transfer protein by *Pichia pastoris. J. Lipid Res.* **37,** 599–605.

Koutz, P., Davis, G. R., Stillman, C., Barringer, K., Cregg, J., and Thill, G. (1989). Structural comparison of the *Pichia pastoris* alcohol oxidase genes. *Yeast* **5,** 167–177.

Laroche, Y., Storme, V., De Meutter, J., Messens, J., and Lauwereys, M. (1994). High-level secretion and very efficient isotopic labeling of tick anticoagulant peptide (TAP) expressed in the methylotrophic yeast, *Pichia pastoris. Bio/Technology* **12,** 1119–1124.

Liao, Y. F., Lal, A., and Moremen, K. W. (1996). Cloning, expression, purification, and characterization of the human broad specificity lysosomal acid α-mannosidase. *J. Biol. Chem.* **271,** 28348–28358.

Lima, C. D., Klein, M. G., Weinstein, I. B., and Hendrickson, W. A. (1996). Three-dimensional structure of human protein kinase C interacting protein 1, a member of the HIT family of proteins. *Proc. Natl. Acad. Sci. U.S.A.* **93,** 5357–5362.

Liu, H., Tan, X., Veenhuis, M., McCollum, D., and Cregg, J. M. (1992). An efficient screen for peroxisome-deficient mutants of *Pichia pastoris. J. Bacteriol.* **174,** 4943–4951.

Luo, D., Mah, N., Krantz, M., Wilde, K., Wishart, D., Zhang, Y., Jacobs, F.. and Martin, L. (1995). Vl-linker-Vh orientation-dependent expression of single chain Fv-containing an engineered disulfide-stabilized bond in the framework regions. *J. Biochem. (Tokyo)* **118,** 825–831.

Lyman, S. K., and Schekman, R. (1996). Polypeptide translocation machinery of the yeast endoplasmic reticulum. *Experientia* **52,** 1042–1049.

Markaryan, A., Beall, C. J., and Kolattukudy, P. E. (1996). Inhibition of *Aspergillus serine* proteinase by *Streptomyces subtilisin* inhibitor and high-level expression of this inhibitor in *Pichia pastoris. Biochem. Biophys. Res. Commun.* **220,** 372–376.

Masure, S., Paemen, L., Proost, P., Van Damme, J., and Opdenakker, G. (1995). Expression of a human mutant monocyte chemotactic protein 3 in *Pichia pastoris* and characterization as an MCP-3 receptor antagonist. *J. Interferon Cytokine Res.* **15,** 955–963.

Masure, S., Paemen, L., van Aelst, I., Fiten, P., Proost, P., Billiau, A., van Damme, J., and Opdenakker, G. (1997). Production and characterization of recombinant active mouse gelatinase B from eukaryotic cells and *in vivo* effects after intravenous administration. *Eur. J. Biochem.* **244,** 21–30.

McCollum, D., Monosov, E., and Subramani, S. (1993). The *pas8* mutant of *Pichia pastoris* exhibits the peroxisomal protein import deficiencies of Zellweger syndrome—the PAS8 protein binds to the COOH-terminal tripeptide peroxisomal targeting signal, and is a member of the TPR protein family. *J. Cell Biol.* **121,** 761–774.

Monosov, E. Z., Wenzel, T. J., Luers, G. H., Heyman, J. A., and Subramani, S. (1996). Labeling of peroxisomes with green fluorescent protein in living *P. pastoris* cells. *J. Histochem. Cytochem.* **44,** 581–589.

Nilsen, S. L., DeFord, M. E., Prorok, M., Chibber, B. A., Bretthauer, R. K., and Castellino, F. J. (1997). High-level secretion in *Pichia pastoris* and biochemical characterization of the recombinant kringle 2 domain of tissue-type plasminogen activator. *Biotechnol. Appl. Biochem.* **25,** 63–74.

O'Donohue, M. J., Boissy, G., Huet, J. C., Nespoulous, C., Brunie, S., and Pernollet, J. C. (1996). Overexpression in *Pichia pastoris* and crystallization of an elicitor protein secreted by the phytopathogenic fungus, *Phytophthora cryptogea. Protein Express. Purif.* **8,** 254–261.

Ogata, K., Nishikawa, H., and Ohsugi, M. (1969). A yeast capable of utilizing methanol. *Agric. Biol. Chem.* **33,** 1519–1520.

Ohsawa, I., Hirose, Y., Ishiguro, M., Imai, Y, Ishiura, S., and Kohsaka, S. (1995). Expression, purification, and neurotrophic activity of amyloid precursor protein-secreted forms produced by yeast. *Biochem. Biophys. Res. Commun.* **213,** 52–58.

Okabayashi, K., Tsujikawa, M., Morita, M., Einaga, K., Tanaka, K., Tanabe, T., Yamanouchi, K., Hirama, M., Tait, J. F., and Fujikawa, K. (1996). Secretory production of recombinant urokinase-type plaminogen activator-annexin V chimerase in *Pichia pastoris. Gene* **177,** 69–76.

Payne, M. S., Petrillo, K. L., Gavagan, J. E., Wagner, L. W., DiCosimo, R., and Anton, D. L. (1995). High-level production of spinach glycolate oxidase in the methylotrophic yeast *Pichia pastoris*: engineering a biocatalyst. *Gene* **167,** 215–2 19.

Payne, W. E., Gannon, P. M., and Kaiser, C. A. (1995). An inducible acid phosphatase from the yeast *Pichia pastoris*: Characterization of the gene and its product. *Gene* **163,** 19–26.

Qin, Y. M., Poutanen, M. H., Helander, H. M., Kvist, A. P., Siivari, K. M., Schmitz, W., Conzelmann, E., Hellman, U., and Hiltunen, J. K. (1997). Peroxisomal multifunctional enzyme of β-oxidation metabolizing D–3-hydroxyacyl-CoA esters in rat liver: Molecular cloning, expression and characterization. *Biochem. J.* **321,** 21–28.

Reddy, R. G., Yoshimoto, T., Yamamoto, S., and Marnett, L. J. (1994). Expression, purification, and characterization of porcine leukocyte 12-lipoxygenase produced in the methylotrophic yeast, *Pichia pastoris. Biochem. Biophys. Res. Commun.* **205,** 381–388.

Rodriguez, M., Rubiera, R., Penichet, M., Montesinos, R., Cremata, J., and de la Fuente, J. (1994). High level expression of the *B. microplus* Bm86 antigen in the yeast *Pichia pastoris* forming highly immunogenic particles for cattle. *J. Biotechnol.* **33,** 135–146.

Romanos, M. (1995). Advances in the use of *Pichia pastoris* for high-level gene expression. *Curr. Opin. Biotechnol.* **6,** 527–533.

Romanos, M. A., Clare, J. J., Beesley, K. M., Rayment, F. B., Ballantine, S. P., Makoff, A. J., Dougan, G., Fairweather, N. F., and Charles, I. G. (1991). Recombinant *Bordetella pertussis* pertactin (P69) from the yeast *Pichia pastoris*: High-level production and immunological properties. *Vaccine* **9,** 901–906.

Romanos, M. A., Scorer, C. A., and Clare, J. J. (1992). Foreign gene expression in yeast: A review. *Yeast* **8,** 423–488.

Romero, P. A., Lussier, M., Sdicu, A. M., Bussey, H., and Herscovics, A. (1997). Ktr1p is an α-1,2-mannosyltransferase of *Saccharomyces cerevisiae.* Comparison of the enzymic properties of soluble recombinant Ktr1p and Kre2p/Mnt1p produced in *Pichia pastoris. Biochem. J.* **321,** 289–295.

Rosenfeld, S. A., Ross, O. H., Hillman, M. C., Corman, J. I., and Dowling, R. L. (1996a). Production and purification of human fibroblast collagenase (MMP-1) expressed in the methylotrophic yeast *Pichia pastoris. Protein Express. Purif.* **7,** 423–430.

Rosenfeld, S. A., Nadeau, D., Tirado, J., Hollis, G. F., Knabb, R. M., and Jia, S. (1996b). Production and purification of recombinant hirudin expressed in the methylotrophic yeast *Pichia pastoris. Protein Express. Purif.* **8,** 476–482.

Ruddat, A., Schmidt, P., Gatz, C., Braslavsky, S. E., Gartner, W., and Schaffner, K. (1997). Recombinant type A and B phytochromes from potato. Transient absorption spectroscopy. *Biochemistry* **36,** 103–111.

Sadhukhan, R., Sen, G. C., and Sen, I. (1996). Synthesis and cleavage—secretion of enzymatically active rabbit angiotensin-converting enzyme in *Pichia pastoris. J. Biol. Chem.* **271,** 18310–18313.

Scorer, C. A., Buckholz, R. G., Clare, J. J., and Romanos, M. A. (1993a). The intracellular production and secretion of HIV-1 envelope protein in the methylotrophic yeast *Pichia pastoris. Gene* **136,** 111–119.

Scorer, C. A., Clare, J. J., McCombie, W. R., Romanos, M. A., and Sreekrishna, K. (1993b). Rapid selection using G418 of high copy number transformants of *Pichia pastoris* for high-level foreign gene expression. *Bio/Technology* **12,** 181–184.

Siegel, R. S., and Brierley, R. A. (1989) Methylotrophic yeast *Pichia pastoris* produced in high-cell-density fermentations with high cell yields as vehicle for recombinant protein production. *Biotechnol. Bioeng.* **34,** 403–404.

Siegel, R. S., Buckholz, R. G., Thill, G. P., and Wondrack, L. M. (1990). Production of epidermal growth factor in methylogrophic yeast cells. Eur. Pat. App. W090/10697.

Sim, B. K., O'Reilly, M. S., Liang, H., Fortier, A. H., He, W., Madsen, J. W., Lapcevich, R., and Nacy, C. A. (1997). A recombinant human angiostatin protein inhibits experimental primary and metastic cancer. *Cancer Res.* **57,** 1329–1334.

Skory, C. D., Freer, S. N., and Bothast, R. J. (1996). Expression and secretion of the *Candida wickerhamii* extracellular β-glucosidase gene, *bglB,* in *Saccharomyces cerevisiae. Curr. Genet.* **30,** 417–422.

Smith, P. M., Suphioglu, C., Griffith, I. J., Theriault, K., Knox, R. B., and Singh, M. B. (1996). Cloning and expression in yeast *Pichia pastoris* of a biologically active form of Cyn d 1, the major allergen of Bermuda grass pollen. *J. Allergy Clin. Immunol.* **98,** 331–343.

Sreekrishna, K. (1993). Strategies for optimizing protein expression and secretion in the methylotrophic yeast *Pichia pastoris. In* "Industrial Microorganisms: Basic and Applied Molecular Genetics" (R. H. Baltz, G. D. Hegeman, and P. L. Skatrud, eds.), Chapter 16, pp. 119–126. Am. Soc. Microbiol., Washington, DC.

Sreekrishna, K., Nelles, L., Potenz, R., Cruze, J., Mazzaferro, P., Fish, W., Fuke, M., Holden, K., Phelps, D., Wood, P., and Parker, K. (1989). High-level expression, purification, and characterization of recombinant human tumor necrosis factor synthesized in the methylotrophic yeast *Pichia pastoris*. *Biochemistry* **28,** 4117–4125.

Steinlein, L. M., Graf, T. N., and Ikeda, R. A. (1995). Production and purification of N-terminal half-transferrin in *Pichia pastoris*. *Protein Express. Purif.* **6,** 619–624.

Su, W., Huber, S. C., and Crawford, N. M. (1996). Identification *in vitro* of a post-translational regulatory site in the hinge 1 region of Arabidopsis nitrate reductase. *Plant Cell (Cambridge, Mass.)* **8,** 519–527.

Sun, J., Coughlin, P., Salem, H. H., and Bird, P. (1995). Production and characterization of recombinant human proteinase inhibitor 6 expressed in *Pichia pastoris*. *Biochim. Biophys. Acta* **1252,** 28–34.

Talmont, F., Sidobre, S., Demange, P., Milon, A., and Emorine, L. J. (1996). Expression and pharmacological characterization of the human μ-opioid receptor in the methylotrophic yeast *Pichia pastoris*. *FEBS Lett.* **394,** 268–272.

Tanaka, M., Suda, T., Yatomi, T., Nakamura, N., and Nagata, S. (1997). Lethal effect of recombinant human Fas ligand in mice pretreated with Propionibacterium acnes. *J. Immunol.* **158,** 2303–2309.

Thill, G. P., Davis, G. R., Stillman, C., Holtz, G., Brierley, R., Engel, M., Buckholz, R., Kenney, J., Provow, S., Vedvick, T., and Siegel, R. S. (1990). Positive and negative effects of multi-copy integrated expression vectors on protein expression in *Pichia pastoris*. *In* "Proceedings of the 6th International Symposium on Genetics of Microorganisms (H. Heslot, J. Davies, J. Florent, L. Bobichon, G. Durand, and L. Penasse, eds.), Vol. 2, pp. 477–490. Soc. Fr. Microbiol., Paris.

Trant, J. M. (1996). Functional expression of recombinant spiny dogfish shark (*Squalus acanthias*) cytochrome P450c17 (17 α-hydroxylase/C17,20-lyase) in yeast (*Pichia pastoris*). *Arch. Biochem. Biophys.* **326,** 8–14.

Trimble, R. B., Atkinson, P. H., Tschopp, J. F., Townsend, R. R., and Maley, F. (1991). Structure of oligosaccharides on *Saccharomyces SUC2* invertase secreted by the methylotrophic yeast *Pichia pastoris*. *J. Biol. Chem.* **266,** 22807–22817.

Tschopp, J. F., Brust, P. F., Cregg, J. M., Stillman, C. A., and Gingeras, T. R. (1987a). Expression of the *lacZ* gene from two methanol-regulated promoters in *Pichia pastoris*. *Nucleic Acids Res.* **15,** 3859–3876.

Tschopp, J. F., Sverlow, G., Kosson, R., Craig, W., and Grinna, L. (1987b). High-level secretion of glycosylated invertase in the methylotrophic yeast, *Pichia pastoris*. *Bio/Technology* **5,** 1305–1308.

van Heeke, G., Ott, T. L., Strauss, A., Ammaturo, D., and Bazer, F. W. (1996). High yield expression and secretion of ovine pregnancy recognition hormone interferon-τ by *Pichia pastoris*. *J. Interferon Cytokine Res.* **16,** 119–126.

van Nostrand, W. E., Schmaier, A. H., Neiditch, B. R., Siegel, R. S., Raschke, W. C., Sisodia, S. S., and Wagner, S. L. (1994). Expression, purification, and characterization of the Kunitz-type proteinase inhibitor domain of the amyloid β-protein precursor-like protein–2. *Biochim. Biophys. Acta* **1209,** 165–170.

Vedvick, T., Buckholz, R. G., Engel, M., Urcan, M., Kinney, J., Provow, S., Siegel, R. S., and Thill, G. P. (1991). High-level secretion of biologically active aprotinin from the yeast *Pichia pastoris*. *J. Ind. Microbiol.* **7,** 197–202.

Veenhuis, M., van Dijken, J. P., and Harder, W. (1983). The significance of peroxisomes in the metabolism of one-carbon compounds in yeasts. *Adv. Microb. Physiol.* **24,** 1–82.

Vollmer, P., Peters, M., Ehlers, M., Yagame, H., Matsuba, T., Kondo, M., Yasukawa,

K., Buschenfelde, K. H., and Rose-John, S. (1996). Yeast expression of the cytokine receptor domain of the soluble interleukin-6 receptor. *J. Immunol. Methods* **199,** 47–54.

Vozza, L. A., Wittwer, L., Higgins, D. R., Purcell, T. J., Bergseid, M., Collins-Racie, L. A., LaVallie, E. R., and Hoeffler, J. P. (1996). Production of a recombinant bovine enterokinase catalytic subunit in the methylotrophic yeast *Pichia pastoris. Bio/Technology* **14,** 77–81.

Waterham, H. R., de Vries, Y., Russell, K. A., Xie, W., Veenhuis, M., and Cregg, J. M. (1996). The *Pichia pastoris PER6* gene product is a peroxisomal integral membrane protein essential for peroxisome biogenesis and has sequence similarity to the Zellweger syndrome protein PAF-1. *Mol. Cell. Biol.* **16,** 2527–2536.

Waterham, H. R., Digan, M. E., Koutz, P. J., Lair, S. L., and Cregg, J. M. (1997). Isolation of the *Pichia pastoris* glyceraldehyde-3-phosphate deydrogenase gene and regulation and use of its promoter. *Gene* **186,** 37–44.

Wegner, G. (1990). Emerging applications of methylotrophic yeasts. *FEMS Microbiol. Rev.* **87,** 279–284.

Weiss, H. M., Haase, W., Michel, H., and Reilander, H. (1995). Expression of functional mouse 5-HT5A serotonin receptor in the methylotrophic yeast *Pichia pastoris*: Pharmacological characterization and localization. *FEBS Lett.* **377,** 451–456.

White, C. E., Hunter, M. J., Meininger, D. P., White, L. R., and Komives, E. A. (1995). Large-scale expression, purification and characterization of small fragments of thrombomodulin: The roles of the sixth domain and of methionine 388. *Protein Eng.* **8,** 1177–1187.

Williams, T. A., Michaud, A., Houard, X., Chauvet, M. T., Soubrier, F., and Corvol, P. (1996). *Drospholia melanogaster* angiotensin I-converting enzyme expressed in *Pichia pastoris* resembles the C domain of the mammalian homologue and does not require glycosylation for secretion and enzymic activity. *Biochem. J.* **318,** 125–131.

Wung, J. L., and Gascoigne, N. R. (1996). Antibody screening for secreted proteins expressed in *Pichia pastoris. BioTechniques* **21,** 808, 810, 812.

Yamada, M., Azuma, T., Matsuba, T., Iida, H., Suzuki, H., Yamamoto, K., Kohli, Y., and Hori, H. (1994). Secretion of human intracellular aspartic proteinase cathepsin E expressed in the methylotrophic yeast, *Pichia pastoris* and characterization of produced recombinant cathepsin E. *Biochim. Biophys. Acta* **1206,** 279–285.

Zhu, A., Monohan, C., Zhang, Z., Hurst, R., Leng, L., and Goldstein, J. (1995). High-level expression and purification of coffee bean α-galactosidase produced in the yeast *Pichia pastoris. Arch. Biochem. Biophys.* **324,** 65–70.

Zhu, A., Monohan, C., Wang, Z. K., and Goldstein, J. (1996). Expression, purification and characterization of recombinant α-N-acetylgalactosaminidase produced in the yeast *Pichia pastoris. Protein Express. Purif.* **8,** 456–462.

7

RECOMBINANT PROTEIN EXPRESSION IN *Pichia methanolica*

Christopher K. Raymond

Department of Protein Expression Technology
ZymoGenetics, Inc., Seattle, Washington 98102

Background

Methylotrophic yeasts have a well-established niche in the expression of recombinant proteins. Several species are homothallic, hap-

Gene Expression Systems: Using Nature for the Art of Expression

loid organisms with readily manipulable genetics. The transcriptional activity of the alcohol oxidase promoters used to control foreign protein expression can be both tightly regulated by carbon source and strongly induced by the addition of methanol. Recombinant expression cassettes are integrated into the host genome, and thus expression strains have the favorable characteristic of being genetically stable. Furthermore, expression cassettes occasionally integrate in multiple, tandem copies, giving rise in at least some instances to strains with dramatically increased productivity (Clare *et al.*, 1991). Methylotrophic yeasts possess the capacity to secrete large quantities of correctly folded proteins, and carbohydrate modifications, although "nonmammalian" in structure, are often modest in size.

Taxonomic investigation revealed that methylotrophic yeasts *Pichia pinus (MH4)*, *Pichia aganobii*, and *Pichia methanolica* are a single species that is now unified under the species name *Pichia methanolica* (Jong and Birmingham, 1996; Tolstorukov and Burkett, 1991; Urakami and Michimi, 1977). It normally exists as a homothallic haploid organism, but it can be induced by nutritional starvation to mate and undergo a sexual cycle (Tolstorukov *et al.*, 1982). Auxotrophic mutants of *P. methanolica* have been identified, and the complementing wild-type genes from *P. methanolica* or *Saccharomyces cerevisiae* were used to develop transformation procedures (Hiep *et al.*, 1993a,b). *P. methanolica* was investigated for its utility in the production of single cell protein and was found to grow well in continuous fermentations using a synthetic salts medium (Urakami *et al.*, 1983). It was also found to utilize methanol efficiently as a carbon and energy source, and the crude protein content of the resulting biomass was greater than in the other methylotrophic yeast species examined. These observations motivated the development of *P. methanolica* as a host for protein expression.

DNA Transformation

P. methanolica type strain CBS6515 was mutagenized with UV irradiation, and auxotrophic mutants were enriched with nystatin (Raymond *et al.*, 1998). Two complementation groups of pink, Ade^- auxotrophs were identified. The wild-type *P. methanolica ADE2* gene was cloned by cross-species complementation of an *S. cerevisiae ade2* strain (Hiep *et al.*, 1993b; Raymond *et al.*, 1998). Transformation with this gene was used to identify one group of pink, Ade^- mutants as *ade2* strains, and it was used to develop an efficient transformation protocol using electroporation (Fig. 1). The cells are prepared for transformation by a rather generic "yeast" protocol (Faber *et al.*, 1994; Meilhoc *et al.*, 1990), and parameters for efficient transformation were determined empirically (Fig. 1; Raymond *et al.*, 1998).

The *ade2/ADE2* transformation system is complicated by the fact that the *P. methanolica ADE2* gene has a strong autonomously replicating sequence (ARS; Hiep *et al.*, 1993a). Moreover, transformed linear fragments undergo efficient repair to become circular plasmids, and therefore the majority of Ade^+ transformants (~98%) harbor unstable extrachromosomal episomes. For the purposes of

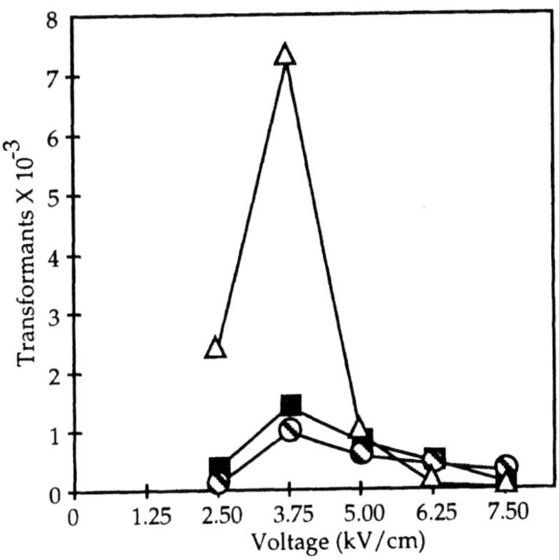

Figure 1 Optimal conditions for electrotransformation of *P. methanolica*. Electrocompetent PMAD11 cells (Table 1) were mixed with 1 μg of *Not*I-digested pCZR134 (see text) and pulsed at the indicated voltage settings with resistance settings of 200 Ω (⊘), 600 Ω (■), or "infinite" Ω (Δ). Cells were then plated on selective media. Data shown are the average of three separate transformation experiments. Reproduced with permission from Raymond, C. K., Bukowski, T., Holderman, S. D., Ching, A. F. T., Vanaja, E., and Stamm, M. R. (1998). Development of the methylotrophic yeast, *Pichia methanolica*, for the expression of the 65-kilodalton isoform of human glutamate decarboxylase. *Yeast* **14,** 11–23. Copyright © John Wiley & Sons Limited.

evaluating protein expression, stable integration of the transforming DNA into the host genome is desired; unstable transformants generate little recombinant protein relative to stable expression strains (C. K. Raymond, unpublished observation). Fortunately, unstable transformant colonies turn pink by virtue of the fact that they are a heterogeneous collection of Ade^+ and Ade^- cells. Stable transformant colonies remain white in color. Furthermore, it was discovered empirically that stable transformants grow much more rapidly than unstable transformants on transformation plates that contain 1.2 *M* sorbitol. In the routine generation of an expression strain, *P. methanolica ade2* cells are transformed with an *ADE2*-marked protein expression fragment and plated onto adenine-minus plates that contain 1.2 *M* sorbitol. Rapidly growing, white transformant colonies are then picked and subsequently screened for recombinant protein expression.

Expression Vector and Identification of an Expression Strain

Polymerase chain reaction (PCR) techniques were used to identify two alcohol oxidase-encoding genes in *P. methanolica*: *AUG1* and *AUG2* (alcohol utilization gene; Raymond *et al.*, 1998). The alcohol oxidases encoded by these genes share 83% identity with one another. Aug1p shares 84% identity with Mox1p from *Hansenula polymorpha*, 72% identity with Aod1p from *Candida boidini*, and 69% identity with Aox1p from *P. pastoris*. Three lines of evidence indicate that Aug1p is the dominant alcohol oxidase in *P. methanolica* and therefore the *AUG1* promoter is appropriate choice to drive recombinant protein expression. First, amino-terminal sequence analysis of the alcohol oxidase induced by methanol in a high-cell density fermentation demonstrated that only Aug1p was present. Second, an *aug1Δ* mutant grows poorly in minimal methanol broth whereas an *aug2Δ* mutant has a wild-type growth phenotype (an *aug1Δ aug2Δ* double mutant cannot grow at all). Finally, the levels of recombinant protein expression driven by the *AUG1* promoter in *P. methanolica* are comparable to the expression levels achieved with the *AOX1* promoter in *P. pastoris* (see later).

A protein expression vector, pCZR134, that utilizes the *AUG1* transcriptional control elements and the *ADE2* selectable marker was assembled into a pUC19 vector (Fig. 2). Protein-coding sequences can be cloned into unique *Eco*RI, *Bam*HI, or *Spe*I sites. The *Eco*RI site is positioned at nucleotide +1 of the alcohol oxidase-coding region. In limited studies, placement of the cloning sites 5′ of the alcohol oxidase-coding region at position -50 or -100 relative to the alcohol oxidase start codon had no noticeable influence on protein expression levels (C. K. Raymond, unpublished). For transformation into *P. methanolica*, the expression cassette is liberated from the pUC backbone as a linear DNA fragment by digestion with either *Not*I or *Sfi*I.

To identify an expression strain, stable Ade⁺ transformants are routinely gridded to minimal methanol plates, overlayed with nitrocellulose, and grown for 2–3 days. To detect intracellular proteins, colonies that adhere to the filter can be lysed using a simple protocol (Wuestehube *et al.*, 1996). Following lysis, or in the case of secreted proteins, the filters are then probed with an antibody that recognizes the recombinant protein using standard Western blot techniques. To date, most transformants have been *AUG1*, consistent with a mechanism whereby expression cassettes integrate into the host genome via circularization followed by integration. Disruption of the *AUG1* gene by direct, double crossover

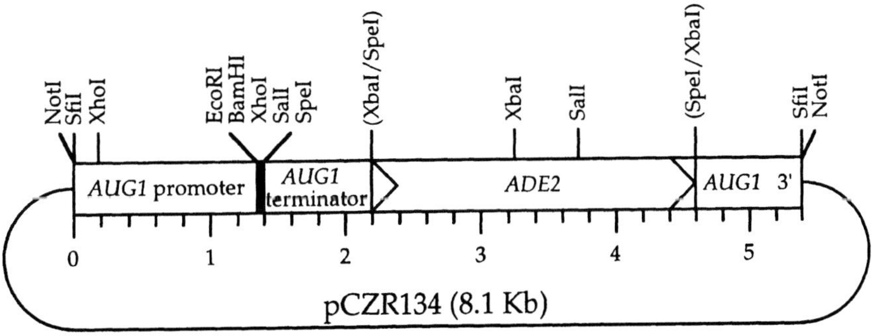

Figure 2 The *P. methanolica* protein expression vector, pCZR134, was made as described in Raymond *et al.* (1998). cDNAs to be expressed can be cloned into unique *Eco*RI, *Bam*HI, and/or *Spe*I sites. The scale shown is in kilobases, and the empty expression vector is 8.1 kb. Reproduced with permission from Raymond, C. K., Bukowski, T., Holderman, S. D., Ching, A. F. T., Vanaja, E., and Stamm, M. R. (1998). Development of the methylotrophic yeast, Pichia methanolica, for the expression of the 65-kilodalton isoform of human glutamate decarboxylase. *Yeast* **14,** 11–23. Copyright © John Wiley & Sons Limited.

transplacement with the expression cassette appears to be a relatively rare event. Most (~90%) transformants appear to make some level of recombinant protein. Strategies to evaluate protein expression and to identify high-yield expression strains are considered next.

Gene Disruptions and Generation of a Protease-Deficient Strain

The *ade2/ADE2* genetic system has the advantage of having a forward selection for Ade$^+$ transformants and a simple visual screen for colonies that have lost the *ADE2* marker. This facilitated site-directed gene disruptions by loop-in/loop-out mutagenesis (Raymond *et al.*, 1998; Rothstein, 1991). Sequential rounds of mutagenesis were used to generate the strains shown in Table 1. PCR techniques were used to clone the *P. methanolica PEP4* and *PRB1* genes. A *pep4Δ* single mutant, PMAD15, appeared wild type for vacuolar protease activity, but the *pep4Δ prb1Δ* double mutant, PMAD16, possessed a vacuolar protease-deficient phenotype (Jones, 1991; Raymond *et al.*, 1998). This strain has no noticeable vegetative growth phenotype and it ferments to high-cell density providing that an adequate supply of nitrogen is maintained. Unlike the "wild-type" *ade2* strain, PMAD11, it has not yet been possible to reconstitute efficient transformation frequencies from *frozen* electrocompetent PMAD16 cells (C. K. Raymond, unpublished).

Table 1
Isogenic *Pichia methanolica* Strains

Strain	Genotype
PMAD11	*ade2-11*
PMAD12	*ade2-11 aug1Δ*
PMAD13	*ade2-11 aug2Δ*
PMAD14	*ade2-11 aug1Δ aug2Δ*
PMAD15	*ade2-11 pep4Δ*
PMAD16	*ade2-11 pep4Δ prb1Δ*
PMAD17	*ade2-11 aug1Δ aug2Δ pep4Δ prb1Δ*

Fermentation

A simple, fed-batch fermentation protocol has been developed for protein expression in *P. methanolica* (Raymond *et al.*, 1998; Appendix). The recipe was formulated based on the elemental composition of yeasts (Jones and Greenfield, 1984; Zabriskie *et al.*, 1980) and on the notion of "theoretical excess factors" for the supplementation of trace elements (Egli and Fiechter, 1981). For *P. methanolica*, the preferred carbon source is glucose. Cells are grown initially for 24–36 hr in batch, and an additional biomass is generated during a 24-hr carbon-limited glucose feed. In *P. methanolica*, partial derepression of the *AUG1* promoter is evident during the glucose fed-batch phase (Fig. 3). Full induction of the recombinant protein synthesis is achieved by a simple 36-hr transition to a mixed feed of glucose and methanol. The rationale for the mixed feed induction is based on empirical studies that demonstrated full induction of alcohol oxidase activity yet increased growth rates for *Hansenula polymorpha* cells grown in continuous culture on mixed glucose/methanol feeds (Egli *et al.*, 1982, 1986). For *P. methanolica*, mixed feed induction allows for rapid induction of recombinant protein synthesis while significantly reducing the amount of methanol required to induce protein synthesis. The entire fermentation process requires 3–4 days and typically yields 70–80 g/liter dry cell weight.

Evaluation of Protein Expression

As typically conducted, an investigation of protein expression in *P. methanolica* involves three phases. These are (a) plasmid construction, (b) transformation and primary screening, and fermentation concomitant with secondary screening (Fig. 4). Given a ready source of cDNA to be expressed and PCR primers, plasmid assembly generally requires less than 1 week. Intracellular constructs generally utilize their own translational initiation codons whereas secreted constructs are most often successfully expressed when fused C-terminal to the *S. cerevisiae* α factor prepro segment. Epitope tagging of the recombinant proteins may be desirable for sub-

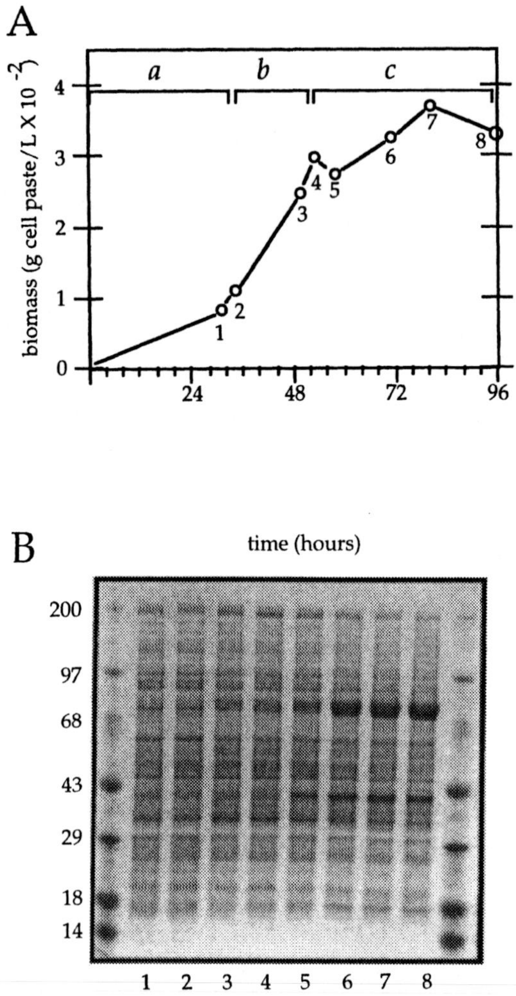

Figure 3 High-cell density fermentation of *P. methanolica.* (A) Wet cell biomass as a function of fermentation time. The indicated intervals are (*a*) batch mode with an initial carbon load of 5% glucose; (*b*) fed-batch mode, 0.4% glucose/hr; and (c) mixed feed methanol fed-batch mode, 0.1% glucose + 0.4% methanol/hr. (B) Total intracellular protein profile of *P. methanolica* cells as a function of fermentation time. Samples 1–8 were withdrawn at the times indicated in (A), and 10 μg of total protein was loaded in each lane. Induction of the 70-kDa *AUG1*-encoded alcohol oxidase is most evident in lanes 6, 7, and 8.

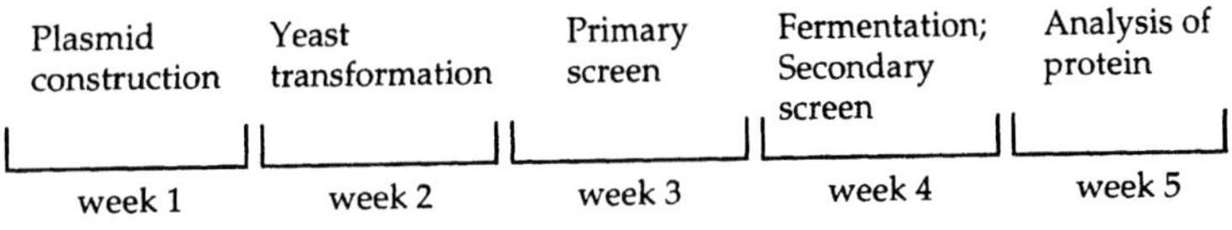

Figure 4 Time line for evaluation of recombinant protein expression in *P. methanolica*.

sequent detection and purification. An expression cassette consisting of the *AUG1* transcriptional control elements, the cDNA under investigation, and the *ADE2* selectable marker is generated as a *Not*I (or *Sfi*I) restriction fragment and is used to transform freshly prepared electrocompetent cells of the *pep4*Δ *prb1*Δ strain PMAD16 to Ade^+. Under ideal conditions, 1 μg of transforming DNA yields about 100 *stable* transformants (described earlier) that are readily apparent after 2–3 days of growth at 30°C. Stable transformants are routinely picked to minimal methanol plates, and an antibody-based primary colony screen is generally used to identify promising expression strains (described earlier). In cases where an antibody to the recombinant product is not available, PCR techniques and Southern blotting have been used to identify transformant colonies that harbor the recombinant expression cassette.

Once potential expression strains have been identified, secondary screening involves assessment of the quality and quantity of the recombinant protein. A high priority is generally placed on assessing the quality of the recombinant protein. A promising strain from the primary screen is fermented to high-cell density, protein expression is induced, and the recombinant protein is analyzed for posttranslational modifications. The major concerns involve appropriate proteolytic removal of the α factor prepro segment from secreted proteins, inappropriate proteolytic cleavage within the recombinant protein, and the extent of glycosyl modifications added to secreted proteins with N-linked glycosylation sites. Assuming that a recombinant protein is expressed appropriately, it is straightforward to rescreen for potential high-expressing strains in a secondary screen. Promising clones are grown in shake-flask cultures of minimal methanol broth, and the highest expressing strains are detected by semiquantitive protein analysis, i.e., Western blotting or ELISA assays. This "quality then quantity" secondary screening strategy is based on the (generally true) assumption that the posttranslational processing of proteins is the same in strains that vary in protein productivity.

Expression of the 65-kDa Isoform of Human Glutamate Decarboxylase in *P. pastoris* and *P. methanolica*

Human glutamate decarboxylase (GAD_{65}) is expressed exclusively in pancreatic islet β cells and is thought to be an important autoantigen in the development of type I diabetes (Baekkeskov *et al.*, 1990). GAD_{65} was expressed as an intracellular protein in *P. pastoris* and *P. methanolica* (Raymond *et al.*, 1998). For *P. pastoris*, the GAD_{65} cDNA (Karlsen *et al.*, 1991) was inserted into pHIL-D2, several hundred His^+ transformants were screened for GAD_{65} expression, and a high-expressing strain identified in a secondary expression screen was fermented (see Chapter 6). For *P. methanolica*, the GAD_{65} cDNA inserted into pCZR134 was used to transform PMAD16 cells to Ade^+, and screening yielded an *aug1Δ* strain, PGAD4-2, that harbored about eight copies of the GAD_{65} cDNA (Raymond *et al.*, 1998). Strain PGAD4-2 was used to express GAD_{65} in a fermentor as described in the Appendix. Protein-normalized cell extracts were prepared from samples of GAD_{65} expressing *P. pastoris* and *P. methanolica* cells, and GAD_{65} expression was compared by Western blotting and enzymatic activity measurements (Fig. 5). The expression levels and enzymatic activity of GAD_{65} produced in the two yeast systems are essentially identical. GAD_{65} was purified to greater than 90% purity from *P. methanolica* cell extracts by hydrophobic partitioning into Triton X-114 (Moody *et al.*, 1995) followed by column chromatography steps (Fig. 5; Raymond *et al.*, 1998). Purified GAD_{65} was used to estimate GAD_{65} expression levels in fermentation samples. Semiquantitative Western blotting and enzymatic activity indicate that GAD_{65} expression levels were 2% of total cell protein, or about 500 mg/liter.

Protein Secretion

The *P. methanolica* expression system is still relatively new, and there are very few nonproprietary examples of recombinant protein *secretion* from this system. Two examples are provided to il-

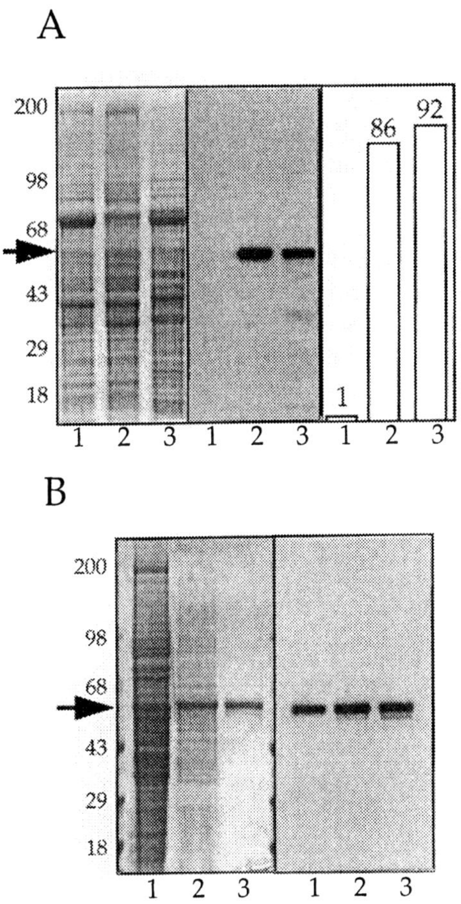

Figure 5 (A) Expression of human GAD_{65} in *P. methanolica* and *P. pastoris*. Cell extracts were prepared from fermentation samples of a control strain of *P. methanolica* (lane 1), *P. methanolica* strain PGAD4-2 (lane 2), and a GAD_{65} high-expressing strain of *P. pastoris* (lane 3). (Left) A Coomassie blue-stained gel of 10 μg of total protein, (middle) a GAD_{65} Western blot of 0.5 μg of total protein, and (right) GAD_{65} enzymatic activities in units of nmol/min/50 mg total protein. The arrow indicates the position of the GAD_{65} protein. (B) Purification of GAD_{65}. A total cell extract from a fermented sample of PGAD4-2 (lane 1), a sample of the Triton X-114 detergent phase (lane 2), and the final hydroxyapetite-eluted GAD_{65} sample (lane 3) were analyzed for total protein (left) and for GAD_{65} by Western blotting (right). Total proteins were loaded to display 0.5 μg of GAD_{65} and diluted 50-fold for GAD_{65} Western analysis. Reproduced with permission from Raymond, C. K., Bukowski, T., Holderman, S. D., Ching, A. F. T., Vanaja, E., and Stamm, M. R. (1998). Development of the methylotrophic yeast, *Pichia methanolica,* for the expression of the 65-kilodalton isoform of human glutamate decarboxylase. *Yeast* **14,** 11–23. Copyright © John Wiley & Sons Limited.

lustrate the potential for high levels of secretion on the one hand and for secretion of a complex protein on the other. A fusion protein consisting of the *S. cerevisiae* α factor prepro sequence fused to the N terminus of human leptin (Zhang *et al.*, 1994) was secreted at levels of about 500 mg/liter under fermentation conditions (Fig. 6). The Kex2p cleavage site in this protein, Lys–Arg–**Val–Pro**, was not cleaved in *S. cerevisiae* or in *P. methanolica*, and the resulting fusion protein was secreted in a hyperglycosylated form from both yeast systems (C. K. Raymond, unpublished). Treatment with endoglycosidase H reveals that the major secreted species is the full-length fusion protein. Minor amounts of alternatively cleaved species are evident, possibly as a result that the protein was expressed in a Pep^+ host strain. A second example illustrates the capacity of the *P. methanolica* system to secrete a complex protein. A cDNA fusion between α factor pre-

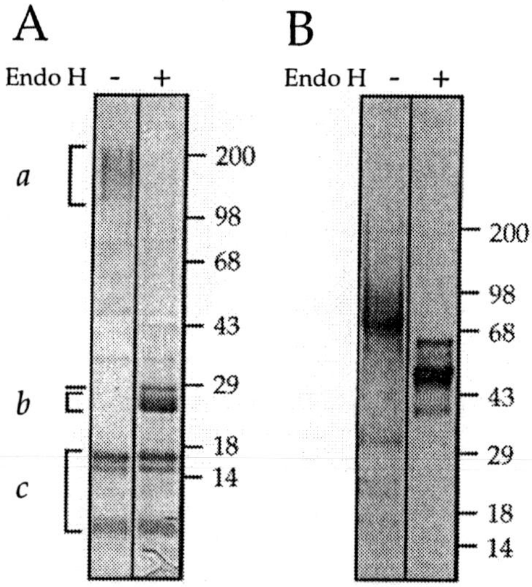

Figure 6 (A) Expression of an *S. cerevisiae* α factor prepro–human leptin fusion protein. Five microliters of fermentation broth was analyzed by SDS–PAGE and Coomassie blue staining. (a) Hyperglycosylated fusion protein, (*b*) deglycosylated fusion protein, and (c) partial proteolytic cleavage products of the fusion protein. Recombinant Endo H used to deglycosylate the sample is indicated by the line at 29 kDa. (B) Western blot of a human cytokine receptor expressed in *P. methanolica*. Treatment with Endo H reveals a major species of ~52 kDa and a minor species of ~60^+ kDa that is presumed to be an uncleaved α factor prepro–receptor fusion protein.

pro and the extracellular domain of a human cytokine receptor was introduced into PMAD16 cells and expressed under fermentation conditions. The extracellular domain is a 52-kDa complex structure with six N-linked glycosylation sites and the potential to form seven disulfide bonds. A Western blot of secreted protein probed with an antireceptor antibody is shown in Fig. 6. Expression levels of receptor fragment were estimated to be ~10 mg/liter. In the absence of endoglycosidase H treatment, a relatively discrete species of ~ 80 kDa was apparent. Treatment with endoglycosidase H predominantly yielded a ~52-kDa protein species. These data indicate three key points. First, by virtue of the fact that it was secreted, the extracellular receptor domain was folded into a secretion-competent conformation; ligand-binding studies are underway. Second, the Kex2p cleavage site in this construct (Lys–Arg–**Glu–Glu–Glu**) appears to have been utilized efficiently by an endogenous *KEX2*-like protease in *P. methanolica*. Finally, the relatively compact mobility of the glycosylated receptor suggests that *P. methanolica*, as with other methylotrophic yeasts, exhibits modest elaboration of carbohydrate structures on some recombinant proteins.

Benefits and Liabilities of the *P. methanolica* Expression System

Data generated so far indicate that the *P. methanolica* expression system shares the same attractive features as expression systems in other methylotrophic yeasts (Chapter 6; Gellisen *et al.*, 1995; Table 2). These systems provide for the rapid evaluation of protein expression in genetically malleable systems. They possess the capacity to secrete large quantities of relatively complex proteins. Scale-up to fermentation conditions is relatively straightforward. In the case of *P. methanolica*, a fermentation strategy is described that utilizes glucose as an inexpensive carbon source for the generation of biomass and that requires modest amounts of methanol for the induction of recombinant protein synthesis.

The main limitation of the *P. methanolica* expression system is that it is not as well characterized as other yeast expression systems. Optimal parameters for transformation, protein expression,

Table 2
Advantages and Disadvantages of the *P. methanolica* Expression System

Advantages	Disadvantages
Same advantages as other methylotrophic yeast expression systems	System less intensively investigated compared to other yeast systems
Tractable genetics	Nonmammalian glycosylation of secreted proteins
Rapid evaluation of recombinant protein expression	Fermentation equipment necessary to fully evaluate protein expression
Simple, rapid, high-biomass fermentation	Limited availability of system
	No plasmid systems as in *S. cerevisiae*

and fermentation have not been investigated in any significant detail. Differences in the repertoire of posttranslational modifications between yeasts and mammalian cells will undoubtedly preclude the use of the *P. methanolica* expression system for the synthesis of complex mammalian proteins. In particular, proteins bearing yeast glycosyl modifications are probably not suitable as injectable therapeutics. Furthermore, in the author's opinion, protein expression in methylotrophic yeasts can only be evaluated adequately under fermentation conditions, and acquisition and operation of fermentation equipment can sometimes be a limiting resource. In summary, it is anticipated that the *P. methanolica* expression system will share the same, well-documented strengths and weaknesses as other yeast expression systems.

Summary

A fully functional recombinant protein expression system has been developed in the methylotrophic yeast *P. methanolica*. The system is simple to use and is capable of generating large amounts of recombinant protein. Future improvements will focus on procedures for generating and identifying strains that produce high levels of recombinant protein and on fermentation technologies that are amenable to large-scale protein production.

Appendix

Fermentation of *P. methanolica*

Fed-batch fermentations are performed in a BioFlo 3000 fermentor (New Brunswick) fitted with a 5-liter vessel. The inoculum is grown in 250 ml of YPB broth at 30°C. The fermentation vessel is charged with 2.5 liters of media containing 57.8 g $(NH_4)_2SO_4$, 46.6 g KCl, 30.8 g $MgSO_4 \cdot 7H_2O$, 8.6 g $CaSO_4 \cdot SO_4$, and 2.0 g NaCl. After autoclaving and cooling of the vessel to a working temperature of 29°C, 350 ml of 50% glucose, 210 ml of 30% sodium hexametaphosphate (phosphate glass), and 250 ml of 10 × trace elements are added. 10 × trace elements contain per liter 27.8 g $FeSO_4 \cdot 7H_2O$, 0.5 g $CuSO_4 \cdot 5H_2O$, 1.09 g $ZnCl_2$, 1.35 g $MnSO_4 \cdot H_2O$, 0.48 g $CoCl_2 \cdot H_2O$, 0.24 g $Na_2MoO_4 \cdot 2H_2O$, 0.5 g H_3BO_3, 0.26 g $NiSO_4 \cdot 6H_2O$, 0.08 g KI, 5 mg biotin, 0.5 g thiamine, and 2.5 ml H_2SO_4. The pH of the fermentor is adjusted to 5.0 and is controlled automatically with 10% NH_4OH and 10% H_3PO_4. Aeration is provided initially as compressed air provided at a flow rate of 5 liters/min and an impeller agitation rate of 300 rpm. The dissolved oxygen control is set to be maintained at 30% of saturation within an agitation range of 300–800 rpm. Oxygen demand above 800 rpm activates automatic supplementation with pure oxygen. The batch phase requires 24–36 hr. Following the exhaustion of glucose, a glucose feed consisting of 50% (w/v) glucose, 300 m*M* $(NH_4)SO_4$, and 1× trace elements solution is initiated at a rate of 60 ml/hr for 24 hr. Recombinant protein expression in wild-type cells is induced by a mixed feed of 20 ml/hr of glucose feed and 25 ml/hr pure methanol for 36 hr. The induction of recombinant GAD_{65} expression in the *aug1Δ* PGAD4-2 strain was achieved with a mixed feed of 40 ml/hr glucose feed and 12.5 ml/hr pure methanol for 36 hr. Biomass yields are typically about 250–300 g of washed cell paste/liter or 60–75 g/liter dry cell weight.

References

Baekkeskov, S., Aanstoot, H.-J., Christagau, S., Reets, A., Solimena, M., Cascalho, M., Folli, F., Richter-Olesen, H., and DeCamilli, P. (1990). Identification of the 64K autoantigen in insulin-dependent diabetes as the GABA-synthesizing enzyme glutamic acid decarboxylase. *Nature (London)* **347**, 151–156.

Clare, J. J., Rayment, F. B., Ballantine, S. P., Sreekrishna, K., and Romanos, M.

(1991). High-level expression of tetanus toxin fragment C in *Pichia pastoris* strains containing multiple tandem integrations of the gene. *Bio/Technology* **9**, 455–460.

Egli, T., and Fiechter, A. (1981). Theoretical analysis of media used in the growth of yeasts on methanol. *J. Gen. Microbiol.* **123**, 365–369.

Egli, T., Kappeli, O., and Fiechter, A. (1982). Regulatory flexibility of methylotrophic yeasts in chemostat cultures: Simultaneous assimilation of glucose and methanol at a fixed dilution rate. *Arch. Microbiol.* **131**, 1–7.

Egli, T., Bosshard, C., and Hamer, G. (1986). Simultaneous utilization of methanol-glucose mixtures by *Hansenula polymorpha* in chemostat: Influence of dilution rate and mixture composition on utilization pattern. *Biotechnol. Bioeng.* **28**, 1735–1741.

Faber, K. N., Haima, P., Harder, W., Veenhuis, M., and AB, G. (1994). Highly-efficient electrotransformation of the yeast *Hansenula polymorpha. Curr. Genet.* **25**, 305–310.

Gellisen, G., Hollenberg, C. P., and Janowicz, Z. A. (1995). Gene expression in methylotrophic yeasts. *Bioprocess Technol.* **22**, 195–239.

Hiep, T. T., Noskov, V. N., and Pavlov, Y. I. (1993a). Transformation in the methylotrophic yeast *Pichia methanolica* utilizing homologous *ADE1* and heterologous *Saccharomyces cerevisiae* ADE2 and LEU2 genes as genetic markers. *Yeast* **9**, 1189–1197.

Hiep, T. T., Kulikov, V. N., Noskov, V. N., Sizonenko, G. I., Chernoff, Y. O., and Pavlov, Y. I. (1993b). The 5-aminoimidazole ribonucleotide-carboxylase structural gene of the methylotrophic yeast *Pichia methanolica:* Cloning, sequencing and homology analysis. *Yeast* **9**, 1251–1258.

Jones, E. W. (1991). Tackling the protease problem in *Saccharomyces cerevisiae. In* "Methods in Enzymology" (C. Guthrie and G. R. Fink, eds.), Academic Press, San Diego, CA. Vol. 194, pp. 428–453.

Jones, R. P., and Greenfield, P. F. (1984). A review of yeast ionic nutrition. *Process Biochem.* 48–60.

Jong, S. C., and Birmingham, J. M. (1996). Methylotrophic yeasts as transformation hosts for the expression of foreign genes. *Soc. Ind. Microbiol. NEWS* **46**, 199–203.

Karlsen, A. E., Hagopian, W. A., Grubin, C. E., Dube, S., Disteche, C. M., Adler, D. A., Barmeier, H., Mathewes, S., Grant, F. J., Foster, D., and Lernmark, A. (1991). Cloning and primary structure of a human islet isoform of glutamic acid decarboxylase form chromosome 10. *Proc. Natl. Acad. Sci. U.S.A.* **88**, 8337–8341.

Meilhoc, E., Masson, J.-M., and Teissie, J. (1990). High efficiency transformation of intact yeast cells by electric field pulses. *Bio/Technology* **8**, 223–227.

Moody, A. J., Hejnaes, K. R., Marshall, M. O., Larsen, F. S., Boel, E., Svendsen, I., Mortensen, E., and Dyrberg, T. (1995). Isolation by anion-exchange of immunologically and enzymatically active human islet glutamic acid decarboxylase 65 overexpressed in Sf9 insect cells. *Diabetologia* **38**, 14–23.

Raymond, C. K., Bukowski, T., Holderman, S. D., Ching, A. F. T., Vanaja, E., and Stamm, M. R. (1997). Development of the methylotrophic yeast, *Pichia methanolica,* for the expression of the 65-kilodalton isoform of human glutamate decarboxylase. *Yeast* **14,** 11–23.

Rothstein, R. (1991). Targeting, disruption, replacement, and allele rescue: Integrative DNA transformation in yeast. *In* "Methods in Enzymology" (C. Guthrie and G. R. Fink, eds.), Vol. 194, pp. 281–301.

Tolstorukov, I. I., and Burkett, T. (1991). Comparative analysis of genomic structures of methylotrophic yeasts. *Abstr. Int. Spec. Symp. Yeasts 15th,* Riga (USSR), *1991,* p. 141.

Tolstorukov, I. I., Benevolensky, S. V., and Efremov, B. D. (1982). Genetic control of cell type and complex organization of the mating type locus in the yeast *Pichia pinus. Curr. Genet.* **5**, 137–142.

Urakami, T., and Michimi, R. (1977). Process for producing yeast cells. U. S. Pat. 4,033,821.

Urakami, T., Terao, I., and Nagai, I. (1983). Isolation, identification and cultivation of methanol-utilizing yeasts. *J. Ferment. Technol.* **61**, 221–231.

Wuestehube, L. J., Duden, R., Eun, A., Hamamoto, S., Korn, P., Ram, R., and Schekman, R. (1996). New mutants of *Saccharomyces cerevisiae* affected in the transport of proteins from the endoplasmic reticulum to the golgi complex. *Genetics* **142**, 393–406.

Zabriskie, D. W., Armiger, W. B., Phillips, D. H., and Albano, P. A. (1980). "Traders' Guide to Fermentation Media Formulation." Traders Protein, Inc., Memphis, TN.

Zhang, Y., Proenca, R., Maffei, M., Barone, M., Leopold, L., and Friedman, J. M. (1994). Positional cloning of the mouse *obese* gene and its human homologue. *Nature (London)* **372**, 425–427.

8

CYTOMEGALOVIRUS PROMOTER FOR EXPRESSION IN MAMMALIAN CELLS

Mark F. Stinski
Department of Microbiology, University of Iowa, Iowa City, Iowa 52242

Gene Expression Systems: Using Nature for the Art of Expression

Introduction

In the context of the human cytomegalovirus (CMV) genome, the CMV promoter drives the expression of two regulatory proteins important for productive viral replication. It is paradoxical that CMVs should contain an extremely strong enhancer-containing promoter relative to other animal viruses because CMVs replicate slowly and to relatively low titers in cell culture or in the host. For example, peak titers of infectious virus are not reached until approximately 3 and 12 days after high multiplicity infection of permissive human fibroblast or monocyte-derived macrophage, respectively (Fish *et al.*, 1995; Stinski, 1977). In a host that is immunocompromised because of organ transplantation, clinical symptoms due to viral infection generally do not appear until 3–4 weeks after exposure to the virus (Alford and Britt, 1990; Ho, 1991).

The human CMV major immediate early enhancer-containing promoter, which was the first of the CMV promoters to be cloned and characterized, was originally shown to drive the synthesis of a spliced 1.95-kb mRNA. The viral RNA is detectable in the infected cell by Northern blot analysis within 2 hr after infection. Synthesis of this viral mRNA does not require *de novo* protein synthesis, which suggests that either cellular transcription factors or incoming virion-associated proteins drive the synthesis of the viral RNA. The human CMV promoter was initially mapped to a large viral genomic DNA fragment of approximately 20 kb (Wathen and Stinski, 1982) and was then more accurately mapped to approximately 0.751 map units on the human CMV genome (Stinski *et al.*, 1983). The region, originally referred to as the major immediate-early (IE) promoter-regulatory region, was subjected to DNA sequence analysis and a consensus TATA box was identified (Thomsen *et al.*, 1984). The region upstream of the basal promoter was found to contain numerous repeat motifs. This region was demonstrated to be an enhancer that could substitute for and rescue an enhancer-less SV40 virus (Boshart *et al.*, 1985). Using nuclear extracts prepared from HeLa cells, *in vitro* transcription assays demonstrated that the upstream sequence motifs affected the level of downstream transcription (Thomsen *et al.*, 1984). The complete sequencing of the human CMV genome (AD169 strain) placed the CMV enhancer-containing promoter between 173,731 and 174,281 bp on the viral genome (Chee *et al.*, 1990). The transcription start site of the human CMV promoter and the exons and introns of the downstream viral RNA were mapped

(Stenberg *et al.*, 1984). Because of the high level of transcription from the human CMV promoter in the absence of *de novo* protein synthesis, the downstream gene was designated IE1 or the major IE gene (also referred to as IE1-72 or IE1 491aa). Subsequently, Stenberg *et al.* (1985) discovered that there are two genes downstream of the promoter as a result of alternative splicing events and separate 3′ end polyadenylation signals. The second gene was referred to as IE2 (also referred to as IE2-86 or IE2 579aa).

With the identification of eukaryotic transcription factors and their DNA-binding sites, it became apparent that the human CMV enhancer contains different types of *cis*-acting elements in both repetitive and nonrepetitive DNA sequence motifs (Faisst and Meyer, 1992). *cis*-acting elements are critical for the activity of the CMV promoter because both 5′ end truncations and internal deletions reduced the level of downstream transcription (Stinski and Roehr, 1985).

The human CMV major immediate-early enhancer-containing promoter (~−550 to +8) is usually referred to as the "CMV promoter." Two types of recombinant plasmids were constructed for mammalian cellular gene expression using the human CMV promoter. The first plasmid, referred to as pLink760, contained approximately 760 bp (−760 to +8 bp) of the entire enhancer and promoter region flanked at both the 5′ and 3′ ends with several different restriction endonuclease sites in a pAT153 vector backbone. This plasmid facilitated the efficient transfer of the human CMV enhancer–promoter upstream of several different types of reporter genes (Stinski and Roehr, 1985). The second type of expression plasmids, referred to as pCMV1, 2, 3, 4, and 5, were constructed to contain the CMV enhancer–promoter (−760 to +3 bp) upstream of different polylinker restriction endonuclease sites for the insertion of different eukaryotic genes upstream of the human growth hormone (hGH) transcription termination and polyadenylation signal (Andersson *et al.*, 1989; Brewer, 1994). Because these plasmids also contained the SV40 origin of DNA replication, plasmid DNAs could be amplified by DNA replication in SV40 T antigen-containing cells (Andersson *et al.*, 1989; Gluzman, 1981; Stenberg and Stinski, 1985). One of these expression plasmids (pCMV1) was modified further to contain either the neomycin gene for G418 selection or the hygromycin B resistant gene, and the resultant constructs were designated pCB6 and pCB7, respectively (Brewer, 1994).

Between 1985 and 1990, these human CMV promoter expression plasmids were distributed as research reagents to hundreds of laboratories around the world. In 1991, human CMV promoter expression plasmids became available commercially, such as pRcCMV from Invitrogen (San Diego, CA) and pBKCMV from Stratagene (La Jolla, CA). This chapter reviews the structure and function of the human CMV enhancer-containing promoter for the expression of various eukaryotic gene products in different mammalian cell types. The cellular factors that either upregulate or downregulate transcription from the human CMV promoter will be discussed.

The Cytomegalovirus Enhancer-Containing Promoter

cis-Acting Sites in the Enhancer

The CMV major immediate-early enhancers contain an array of constitutive and inducible transcriptional regulatory elements. Of the mammalian CMV enhancers sequenced to date, which include those of human, simian, murine, and rat CMVs, all contain repeat motifs with consensus binding sites for eukaryotic transcription factors, but the arrangements of the elements differ among species (Alcendor *et al.*, 1993; Boshart *et al.*, 1985; Chang *et al.*, 1990; Dorsch-Hasler *et al.*, 1985; Fickensher *et al.*, 1989; Hunninghake *et al.*, 1989; Niller and Hennighausen, 1990; Sandford and Burns, 1996; Stinski and Roehr, 1985; Thomsen *et al.*, 1984). In addition, the number of elements or type of elements may differ. For example, the simian CMV enhancer has several serum response elements whereas the human CMV has only one (Chan *et al.*, 1996). Human CMV has primarily CREB/ATF elements, but murine CMV has primarily AP-1 elements (Dorsch-Hasler *et al.*, 1985; Hunninghake *et al.*, 1989). These differences may reflect an evolution for the efficient expression in mammalian cells of different species. It is proposed that the different *cis*-acting elements act individually and synergistically to stabilize the RNA polymerase II transcription initiation complex on the promoter. While the following section emphasizes the location and the effects of positive *cis*-acting elements, some negative *cis*-acting elements have also been identified and will be discussed later in the chapter.

Figure 1 is a diagram of the human CMV enhancer-containing

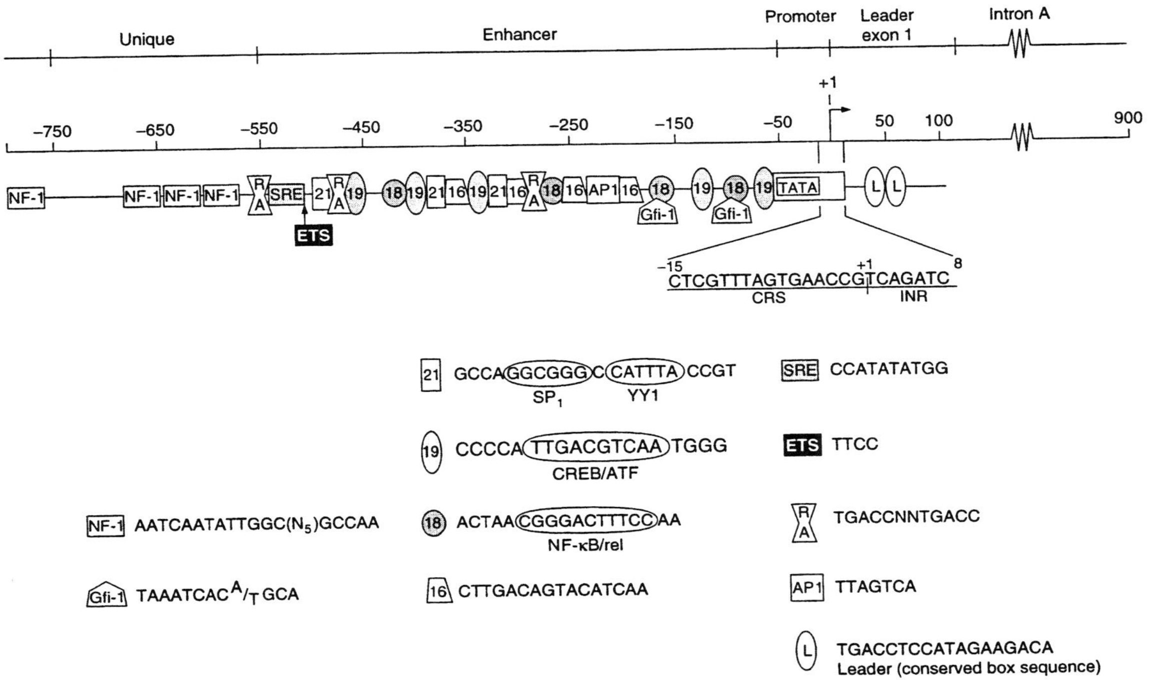

Figure 1 Diagram of the human CMV enhancer-containing promoter with downstream exon 1/intron A. The relative locations of various individual repeat elements are indicated. Below are the designation and consensus sequence for the various *cis*-acting elements involved in either activation or repression of the CMV promoter. Elements with reiterated DNA are identified according to the number of base pairs repeated. The promoter contains the crs for IE2 protein binding. An initiator (INR)-like sequence is downstream of the transcription start site (arrow).

promoter. The region between –550 and –50 relative to the transcription start site was designated the enhancer and the region between –50 and +8 the promoter (Stinski *et al.*, 1991, 1993). The downstream region is referred to as the exon 1 leader and intron A sequences (Chapman *et al.*, 1991; Stenberg *et al.*, 1984). The enhancer has repeat elements of 19, 18, and 21 bp that contain CREB/ATF, NF-κB/rel, and both SP-1 and YY1-binding sites, respectively (Boshart *et al.*, 1985; Stinski *et al.*, 1991, 1993; Thomsen *et al.*, 1984). A role for the 16-bp repeat elements in the enhancer region has not been established. Functional consensus and aberrant SP-1 sites are present in both repetitive and nonrepetitive DNA sequence (Lang *et al.*, 1992). Among the nonrepetitive sequences are AP-1, serum response (SRE), and ETS elements. Several retinoic acid (RA) elements have also been identified (Fig. 1) (Angulo *et al.*, 1995a; Chan *et al.*, 1996; Ghazal *et al.*, 1992). The cellular physiology may influence which of these elements plays a major role in affecting transcription.

Positive Regulation

Figure 2 summarizes the multiple pathways for activation of the CMV enhancer-containing promoter. Ca^{2+}/calmodulin, protein A, and protein C kinase pathways activate the CMV promoter via the CREB/ATF sites (Chan *et al.*, 1996; Hunninghake *et al.*, 1989; Liu and Stinski, 1992). The AP-1 and CREB/ATF elements have a role in constitutive expression from the promoter and respond to stimulation by various protein kinase pathways (Fig. 2) (Chang *et al.*, 1990; Hunninghake *et al.*, 1989; Niller and Hennighausen, 1990). The NF-κB/rel elements increase transcription after activation by a variety of stimuli, including cellular differentiation (Fig. 2) (Boldogh *et al.*, 1990; Cherrington and Mocarski, 1989; Laegreid *et al.*, 1994; Sambucetti *et al.*, 1989). Heat shock, stress, and tumor necrosis factor α (TNF-α) activate the CMV promoter via the NF-kB/rel sites (Chan *et al.*, 1996; Laegreid *et al.*, 1994). Virions of human CMV contain proteins in the viral tegument that activate transcription (Stinski and Roehr, 1985). Tegument proteins encoded by the viral UL26 (T. Stamminger, personal communication), UL69 (Winkler *et al.*, 1994), and UL82 (Liu and Stinski, 1992) genes are activators of transcription from the human CMV promoter. For

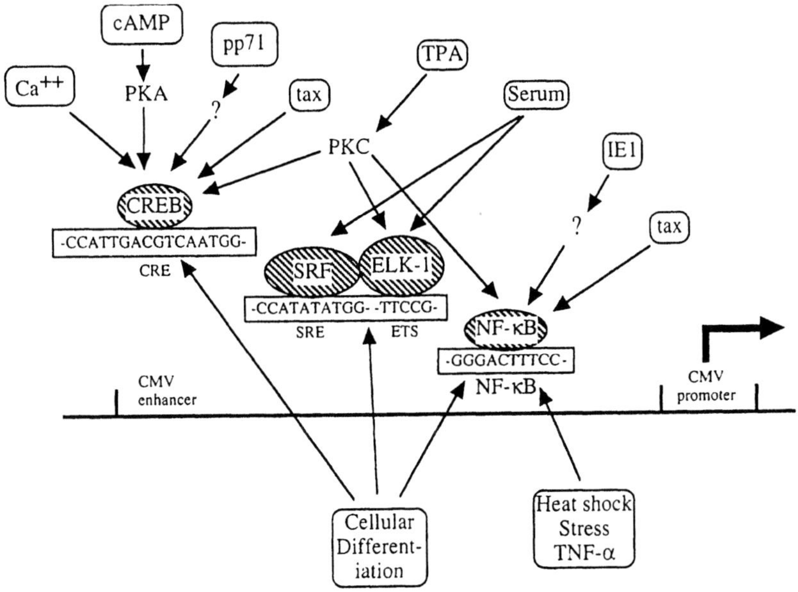

Figure 2 Diagram of the multiple pathways for activation of transcription from the CMV promoter. The activating factors are described in the text. The locations of the various *cis*-acting response elements are diagrammed in Fig. 1.

example, the gene product of UL82 (pp71) activates the CMV promoter via the CREB/ATF and AP-1 sites (Liu and Stinski, 1992). Phorbol esters such as phorbol 12-myristate 13-acetate (PMA) and 12-*O*-tetradecanoyl-phorbol-13-acetate (TPA) and mitogens such as phytohemagglutinin, lipopolysaccharide, and hydrocortisone activate the CMV promoter via the CREB/ATF and NF-κB/rel *cis*-acting sites (Chang *et al.*, 1990; Fickensher *et al.*, 1989; Hunninghake *et al.*, 1989; Laegreid *et al.*, 1994; Lathey and Spector, 1991; Niller and Hennighausen, 1990; Sambucetti *et al.*, 1989). Serum response factor (SRF) activates the CMV promoters via the SRE, and Elk-1 activates via the ETS element (Fig. 2) (Chan *et al.*, 1996). Finally, the CMV promoter responds to other viral transcription factors. For example, the CMV promoter is activated moderately by the human CMV IE1 protein via the NF-κB/rel motifs (Cherrington and Mocarski, 1989) and is strongly activated by the heterologous human T-cell leukemia virus tax protein via the CREB/ATF and NF-κB/rel motifs (Moch *et al.*, 1992). YY1 sites are frequently adjacent to SP-1 sites and, consequently, YY1 may interact with SP-1 to enhance downstream transcription in some cell types (Shrivastava and Calame, 1994). RA elements are functional as physiological levels of RA, a derivative of vitamin A, can activate human CMV replication in normally nonpermissive N-teratocarcinoma-2D1 (Tera-2) stem cells (Angulo *et al.*, 1995b; Ghazal *et al.*, 1992). The retinoic acid receptor (RAR) homodimer, which belong to a superfamily of steroid-thyroid hormone receptors, can activate the human CMV promoter (Angulo *et al.*, 1995a). RA sites within the human CMV enhancer may function as activators or repressors of transcription, depending on the physiology of the cell (Kurokawa *et al.*, 1995).

In summary, multiple transcription factors can bind to the CMV enhancer and influence expression from the promoter, depending on the physiological state of the cell. DNA-binding transcription factors presumably act in concert to facilitate high levels of transcription from the CMV promoter.

cis-Acting Sites in the Promoter

In addition to the TATA box and the CAAT box, two other regions influence transcription from the CMV promoter. These are the initiator-like sequence (INR) and the *cis*-repression sequence (crs) (Fig. 1). The initiator-like sequence from +2 to +8 relative to the adjacent transcription start site fits the consensus initiator sequence of PyPyANT/APyPy, where Py represents pyrimidines and N represents any nucleotide (Javahery *et al.*, 1994; Lo and Smale, 1996). A cellular protein of approximately 150 kDa, found in transcriptionally active extracts of HeLa cells, binds to the initiator-like sequence of the human CMV promoter (Macias *et al.*, 1996). Site-specific mutation of the initiator-like sequence adjacent to the transcription start site eliminates the binding of the 150-kDa cellular protein and significantly reduces the efficiency of transcription from the CMV promoter (Macias *et al.*, 1996). Footprinting analysis demonstrated that the binding site of the cellular protein overlaps the crs. This cellular protein may be similar to the human cellular initiation factor designated CIF 150 and described by Kaufman *et al.*, 1998. Human CMV IE2 homodimers can bind to the crs and potentially interfere with the binding of the 150-kDa cellular protein (Macias *et al.*, 1996). The physical blockage at the crs/initiator-like sequence by the viral IE2 protein may explain the repression of transcription from the CMV promoter, which is discussed in more detail later.

Experimental Procedures

Time Line

The CMV promoter has been used in expression plasmids for transient transfection experiments and in viral vectors for infection experiments. For transient transfection experiments, as little as 0.1 μg of expression plasmid per 10^6 cells has been used (Liu and Stinski, 1992). Limiting determinations of input DNA levels are necessary when measuring the effect of a *trans*-acting factor on the CMV promoter because of the relatively high basal activity of the promoter. A high DNA input amount of 2–5 μg per 10^6 cells is useful when protein synthesis is measured by immunological techniques. Although RNA produced by the CMV promoter from expression plasmids not containing the SV40 origin (ori) can be detected by Northern blot or RNase protection assay (Hermiston *et al.*, 1990; Hunninghake *et al.*, 1989), detection of gene products by either immunoprecipitation or Western blot may require the SV40 virus ori of DNA replication and cells (COS) expressing the SV40 virus T antigen for plasmid DNA amplification (Malone *et al.*, 1990; Stenberg and Stinski, 1985).

With the CMV promoter upstream of reporter genes such as the chloramphenicol acetyltransferase (CAT), β-galactosidase (β-gal), or luciferase (luc), the gene products can be detected from expression plasmids without the SV40 ori within 4–6 hr after transfection (Hunninghake *et al.*, 1989). However, most experiments use 24–48 hr of incubation to allow the gene product to accumulate.

The CMV promoter competes strongly for transcription factors in the cell. The activity of other promoters (e.g., early SV40 or RSVLTR) driving expression of other gene products or internal controls may be reduced significantly in the presence of the CMV promoter. One important control in a series of cotransfection experiments is to test the effect of the CMV promoter alone on another promoter driving the expression of a reporter gene or a transactivator of transcription. In *in vitro* transcription experiments, internal controls are particularly difficult to include when using the CMV promoter because of competition for transcription factors. Generally, 100–150 ng of CMV promoter template per 10 ng of nuclear extract is used for *in vitro* runoff transcription reactions, but the template concentrations are dependent on the quality of the nuclear extract. Peak levels of RNA are synthesized within 15 min at 37°C. Because of the high level of *in vitro* transcription, the CMV promoter is frequently used as a standard to test the competence of a cellular extract (Promega Corp., Madison, WI).

RNAs, Proteins, and Yields

Initial experiments to evaluate the strength of the CMV promoter used the enhancer and promoter without the downstream exon 1/intron A and the following three different reporter genes: thymidine kinase (TK), CAT, and ovalbumin (OV) (Stinski and Roehr, 1985; Thomsen *et al.*, 1984). With *in vitro* runoff transcription assays, the level of RNA synthesized from the CMV promoter was approximately 5- and 3-fold higher than that from the SV40 and late adenovirus promoters, respectively. Because previous studies had shown that measurements of the enzyme level expressed from chimeric genes containing the CAT gene correlated with the corresponding mRNA level, the majority of the experiments used CAT enzyme activity to compare mRNA levels. The CAT enzyme activity from the CMV promoter driving expression of the CAT gene was usually 5- to 7-fold higher than that of the SV40, RSVLTR, and HIVLTR promoters (Liu *et al.*, 1991). CAT enzyme activity from expression plasmids containing the CMV enhancer and promoter plus exon 1/intron A (see Fig. 1) was as much as 15-fold higher than the SV40 promoter (Chapman *et al.*, 1991). It was possible to express as much as 450 ng of a secretory glycoprotein per milliliter of COS cell tissue culture fluid (Chapman *et al.*, 1991). However, in stably transformed cells such as human 143 TK^+, the human CMV promoter is downregulated. The promoter can be strongly induced by either infectious or noninfectious human CMV virions or by the human CMV tegument protein designated pp71 (UL82) (Liu and Stinski, 1992; Stinski and Roehr, 1985). In cells not stimulated by human CMV virions, the CMV enhancer and promoter without exon 1/intron A expressed only 4–5 ng of OV per milligram of protein, but in cells stimulated with human CMV virions, 14–26 ng per milligram of protein was detectable (Stinski and Roehr, 1985).

After high multiplicity infection of cells permissive for an adenovirus vector containing truncated versions of the enhancer upstream of the CMV promoter and without the presence of exon 1/intron A, as much as 100 μg of a reporter gene product per 10^6 cells can be produced (Addison *et al.*, 1997). In summary, RNAs, proteins, and yields of gene products from the CMV enhancer containing promoter are influenced by the presence or absence of exon 1/intron A and whether the method for expression is transiently transfected cells, stably transformed cells, or viral vector-infected cells.

Cellular Effects

It is difficult to compare the strength of a promoter in different cell types because of the variability in transfection procedures that are optimal for various cells. With the popular calcium phosphate precipitation method (Graham and van der Eb, 1973), expression of the CAT gene from the CMV promoter in descending order is SV40 T antigen-containing COS, Chinese hamster ovary, HeLa, monkey CV-1, human fibroblast, and NIH 3T3 mouse cells (M. F. Stinski and T. J. Roehr, unpublished data). Expression of the CAT gene was consistently low in undifferentiated (Tera-2) human cells and high in their differentiated counterparts (Shelbourn *et al.*, 1989; Sinclair *et al.*, 1992). Likewise, expression was low in unstimulated human Jurkat T lymphoblastoid cells sensitive to mitogenic stimulation and high in the stimulated counterparts (Hunninghake *et al.*, 1989).

Effects Upstream of the Cytomegalovirus Enhancer

NF-1 Region

Immediately upstream of the human CMV enhancer between -610 and -780 bp relative to the transcription start site is a region containing four NF-1-binding sites (Fig. 1). NF-1-binding sites are also upstream of the enhancers of baboon, African green monkey, rhesus, and chimpanzee CMVs (Alcendor *et al.*, 1996). In addition, simian CMVs have a A:T-rich region also referred to as the bent domain (Chan *et al.*, 1996). The effects of these regions on the CMV promoter require further investigation. Although the effects of these regions in transient transfection experiments are negligible, the effects in viral vectors have not been investigated. In the context of a viral genome, these regions may influence nucleosome structure, which may have an effect on the transcription from the CMV promoter (Nelson and Groudine, 1986).

Modulator

The modulator was defined as a region between -750 and -1150 bp upstream of the human CMV promoter transcription start site (Nelson and Groudine, 1986; Nelson *et al.*, 1987). Although the modulator has a repressive effect on the CMV promoter in transient transfection experiments (Huang *et al.*, 1996; Shelbourn *et al.*, 1989; Stein *et al.*, 1993), deletion of the modulator from the human CMV genome did not affect transcription from the CMV promoter in permissive human fibroblast cells or undifferentiated Tera-2 or monocytic THP-1 cells (Meier and Stinski, 1997). In the context of the viral genome, the modulator may have a role in regulating the transcription of genes that overlap the modulator or are adjacent to it. Although further research is necessary, this region may be useful in insulating transcription units from the effects of the strong CMV enhancer.

Effects Downstream of the Cytomegalovirus Promoter

Exon 1 and Intron A

The region downstream of the initiator-like sequence, referred to as exon 1 leader and intron A, has a positive effect on transcription. There is a DNA region in exon 1 referred to as the "conserved box sequence" that binds unknown cellular proteins in transcription competent nuclear extracts (Fig. 1) (Ghazal and Nelson, 1991). Exon 1 alone can enhance the level of expression approximately 6-

fold (Ghazal and Nelson, 1991). In addition, the region referred to as intron A, the intron downstream of exon 1 leader, also has a positive effect on transcription (Fig. 1) (Chapman *et al.*, 1991). Exon 1/intron A sequences can increase expression from the CMV promoter by as much as 50-fold (Chapman *et al.*, 1991). Reasons for the enhanced level of expression are unknown. Intron A has a NF-1-binding site, but mutation of this site does not explain the strong level of enhanced expression (Chapman *et al.*, 1991). The leader and intron A sequences could affect transcriptional initiation and elongation or stabilize the mRNA in eukaryotic cells. However, these sequences are absent in most CMV promoter expression plasmids and have not been incorporated into viral vectors. Exon 1 adds an additional 121 bp and intron A adds 792 bp and, therefore, it has been difficult to incorporate these sequences into viral vectors because of virion packaging constraints.

Negative Regulation

Gfi-1

The CMV promoter in the context of the viral genome is downregulated by cellular and viral proteins in certain cell types. For example, in granulocyte–macrophage progenitors and blood monocytes, the human CMV genome is present, but there is little to no transcription from the CMV major immediate-early promoter (Kondo *et al.*, 1994; Taylor-Wiedeman *et al.*, 1991). In contrast, transcription is active in the monocyte-derived macrophage (Fish *et al.*, 1995). Little is known about the cellular or viral proteins that downregulate the CMV promoter. One candidate is the human homologue of Gfi-1, a murine protoncogene. Gfi-1 binds upstream of the CMV promoter region at approximately –100 and –157 bp relative to the transcription start site (Fig. 1) (Zweidler-McKay *et al.*, 1996). Ectopic expression of Gfi-1 in cotransfected murine NIH 3T3 fibroblast represses the CMV promoter (Zweidler-McKay *et al.*, 1996). In undifferentiated cells, the CMV promoter is repressed and Gfi-1 is highly expressed. In differentiated myelomonocytic cells, the CMV promoter is activated and Gfi-1 expression is downregulated. Hence, there is an inverse relationship between Gfi-1 expression and CMV promoter activity. However, the function of these Gfi-1 *cis*-acting sites in the context of the viral genome or in a viral vector has not been determined.

YY1 and Retinoic Acid

Within the enhancer region there are several YY1 and RA responsive *cis*-acting sites. The direct or indirect recruitment of histone deacetylases by YY1 or RAR could affect the CMV promoter by maintaining a repressive chromatin structure (Yang *et al.*, 1996). Cellular differentiation reverses the repressive effects on the human CMV promoter (Kothari *et al.*, 1991; Liu *et al.*, 1994; Sinclair *et al.*, 1992).

Viral Proteins

The second gene downstream of the human CMV major immediate-early promoter, IE2, negatively regulates transcription from the CMV promoter in *in vitro* transcription reactions (Macias and Stinski, 1993) or in transiently transfected cells (Cherrington *et al.*, 1991; Hermiston *et al.*, 1990; Liu *et al.*, 1991; Pizzorno and Hayward, 1990). The viral IE2 protein binds as a homodimer to a site that overlaps the transcription initiation site, referred to as the *cis*-repression sequence. Chemical cleavage DNA footprinting reactions mapped the IE2 protein-binding site between –15 and +2 relative to the transcription start site (Lang and Stamminger, 1994; Macias *et al.*, 1996). A cellular factor binds between –16 and +7 (Macias *et al.*, 1996). Although the binding of the viral IE2 protein represses transcription from the CMV promoter, the binding of the cellular factor enhances transcription (Macias *et al.*, 1996). It has been reported that components associated with TFII-D bind to an initiator sequence and facilitate transcription initiation complex assembly (Kaufman *et al.*, 1998; Kaufman and Smale, 1994; Martinez *et al.*, 1994; Verrijzer *et al.*, 1995). The structural context of the DNA template around the crs and the Inr plays a critical role in determining promoter activity.

Human CMV persistently infects 60–80% of the population and viral reactivation from latency occurs frequently. The vast majority of viral vectors using the human CMV promoter have the IE2 protein-binding site. Viral vectors using the CMV promoter for gene therapy are repressed if the human CMV IE2 protein is present in the target cells. Site-specific mutations that inhibit IE2 protein binding without affecting cellular initiation-like factor binding have been identified (Macias *et al.*, 1996), but these constructs have not been tested in viral vectors used for gene therapy.

Conclusions

Improvements

The advantages and disadvantages of the CMV promoter for eukaryotic gene expression are listed in Table 1. The majority of the CMV expression plasmids for transient transfection experiments contain the human CMV major immediate-early enhancer element to approximately –550 bp and the promoter region to approximately +8 relative to the transcription start site. A few laboratories use the human CMV enhancer-containing promoter with exon 1 and intron A sequences. The later constructs are the strongest for downstream gene expression in transient transfection experiments.

Table 1
Advantages and Disadvantages of the CMV Promoter for Gene Expression

Advantages	Disadvantages
High-level downstream gene expression in differentiated cell strains and cell lines	Low-level downstream gene expression in undifferentiated cell strains and cell lines
Strong competition by the enhancer-containing promoter for eukaryotic transcription factors	The CMV promoter competes with other promoters driving expression of internal controls or other test genes
The enhancer-containing promoter is activated by a wide variety of cellular transcription factors	The enhancer-containing promoter is activated by serum, various hormones, and cellular stress
The enhancer-containing promoter can be activated by a variety of chemical and biological inducers	The enhancer-containing promoter cannot be repressed completely for on- and off-type experiments
The CMV promoter is very strong for transient transfection experiments	The CMV promoter is downregulated when stably integrated into cellular genomes

Inefficient transcription from the human CMV promoter occurs in undifferentiated cells. Although some repressor elements within the enhancer element have been proposed, further research is necessary to determine the effects of these repressor elements in the context of the viral genome or in viral vectors.

In viral vectors, the 5′ end of the enhancer is frequently truncated to accommodate nucleic acid packaging constraints. As anticipated, the strength of the CMV promoter is reduced with 5′-end truncations. The human CMV promoter has been the promoter of choice for gene therapy or for pathogen immunization by direct DNA injection. The murine CMV promoter is expected to be efficient in transgenic mouse systems, but in the mouse system the human CMV promoter has been widely used and transcription correlates with cellular differentiation and the target tissues normally associated with human CMV infection *in utero* or in the adult (Baskar *et al.*, 1996a, b; Koedood *et al.*, 1995).

In stably transformed cells or in cells weakly permissive for human CMV replication, the human CMV promoter is downregulated. Sodium butyrate, which may affect the phasing of nucleosomes bound to the human CMV major immediate-early enhancer-containing promoter, activates and increases the level of transcription from the CMV promoter, which increases the level of viral replication (Radsak *et al.*, 1989; Tanaka *et al.*, 1991). Little is known about chromatin structural effects on the CMV enhancer-containing promoter in different cell types.

Future Directions

The CMV promoter has been used to express a plethora of eukaryotic gene products. Prior to commercialization of the CMV promoter as a research reagent in 1991, the author's laboratory, in collaboration with David Russell (University of Texas Southwestern, Dallas), distributed the CMV expression plasmids as a reagent for expression of the eukaryotic genes listed in Table 2. The CMV promoter is used frequently for specialty protein production, gene therapy, or DNA-based vaccination. Gene guns or viral vectors are being used to introduce the CMV promoter driving transcription of genes of interest. DNA injection is being used for vaccination against pathogens or for the correction of genetic errors. Although these procedures are still in experimental phases of development, they present new approaches toward solving problems involving

Table 2
Examples of Eukaryotic Gene Products Expressed Using the CMV Promoter[a]

Type of eukaryotic gene product			
Virus	Ligand–receptor	Enzyme	Other
α herpes	LDL	Steroid hydroxylase	GTP binding
β herpes	Complement	Tyrosine kinase	MHC class I
γ herpes	type 2	Kidney endopeptidase	Cytochrome
Retroviral	β-adrenergic	Epoxide hydrolase	Synapsins
packaging	Insulin	Glycogen synthetase	Hox, Pax, Pou
HTLV I and II	Insulin-like	Phosphatase	Annexin II
Adeno	growth factor	Phosphatase inhibitor	S100 protein
Hepatitis B	Androgen	Flavin monooxygenase	Amelogenin
Japanese	Human nerve	Galactosyltransferase	Chromogranin
encephalitis	growth factor	Sulfotransferase	Secretogranin
Parvovirus B19	Dopaminergic	Glucuronosyltransferase	Coagulation
Vesicular	Interleukin (IL)-2Ra	Adenyl cyclase	Antithrombin
stomatitis	Thyroid	Mitochondrial synthetase	Cadherins
Influenza HA	Estrogen	Ca/calmodulin kinase	Myosin heavy
Polyoma	Progesterone		
middle T	Rhodopsin		
v-rel	TGF-β		
	IL-6		
	Activin type II		
	Peroxisome		
	proliferator		

[a]Investigations using CMV promoter expression plasmids prior to commercialization of the expression plasmids.

pathogens or genetic errors. Methods to completely downregulate the CMV promoter for on- and off-type experiments require further development. Although the tet operator-repressor system can greatly repress expression from the CMV promoter, there is some expression. Placement of the tet operator between the TATA box and the Inr similar to the crs can significantly repress the CMV promoter in the presence of the tet repressor (F. Yao, personal communication), but the level of repression is not as strong as the interaction between the crs and the viral IE2 protein (Liu *et al.*, 1991; Macias and Stinski, 1993). Expression of cytotoxic proteins makes it difficult to isolate cell lines when the CMV promoter is not completely repressed.

Acknowledgments

I thank the members of the laboratory for helpful discussions and reading of the manuscript and P. Lashmit for assistance in preparation of the manuscript. The author is supported by a grant from the U.S. Public Health Service, AI-13562.

References

Addison, C. L., Hitt, M., Kunsken, D., and Graham F. L. (1997). Comparison of the human versus murine cytomegalovirus immediate early gene promoters for transgene expression in adenoviral vectors. *J. Gen. Virol.* **78**, 1653–1661.

Alcendor, D. J., Barry, P. A., Pratt-Lowe, E., and Luciw, P. A. (1993). Analysis of the rhesus cytomegalovirus immediate-early gene promoter. *Virology* **194**, 815–821.

Alcendor, D. J., Gu, W., Zong, J., Waheed, I., Peng, R. S., Chan, Y. J., and Hayward, G. S. (1996). Conservation and divergence among control elements and binding sites in the major immediate-early enhancer of five primate cytomegaloviruses: Likely antagonistic role of the YY1 repressor protein *Herpesvirus Workshop, 21st*, p. 70.

Alford, C. A., and Britt, W. J. (1990). Cytomegalovirus. *In* B. N. Fields, D. M. Knipe, and P. M. Howley, (eds.), pp. 1981–2010. Raven Press, New York.

Andersson, S., Davis, D. N., Dahlback H., Jornvall, H., and Russell, D. W. (1989). Cloning, structure, and expression of the mitochondrial cytochrome P-450 sterol 26-hydroxylase, A bile acid biosynthetic enzyme. *J. Biol. Chem.* **264**, 8222–8229.

Angulo, A., Suto, C., Heyman, R. A., and Ghazal, P. (1995a). Characterization of the sequences of the human cytomegalovirus enhancer that mediate differential regulation by natural and synthetic retinoids. *Mol. Endocrinol.* **10**, 781–793.

Angulo, A., Suto, C., Boehm, M. F., Heyman, R. A., and Ghazal, P. (1995b). Retinoid activation of retinoic acid receptors but not of retinoid X receptors promotes cellular differentiation and replication of human cytomegalovirus in embryonal cells. *J. Virol.* **69**, 3831–3837.

Baskar, J. F., Smith, P. P., Nilauer, G., Jupp, R. A., Hoffmann, S., Peffer, N. J., Tenney, D. J., Colberg-Poley, A. M., Ghazal, P., and Nelson, J. A. (1996a). The enhancer domain of the human cytomegalovirus major immediate-early promoter determines cell type-specific expression in transgenic mice. *J. Virol.* **70**, 3207–3214.

Baskar, J. F., Smith, P. P., Ciment, G. S., Hoffmann, S., Tucker, C., Tenney, D. J., Colberg-Poley, A. M., Nelson, J. A., and Ghazal, P. (1996b). Developmental analysis of the cytomegalovirus enhancer in transgenic animals. *J. Virol.* **70**, 3215–3226.

Boldogh, I., AbuBakar, S., and Albrecht, T. (1990). Activation of proto-oncogenes: an immediate early event in human cytomegalovirus infection. *Science* **247**, 561–564.

Boshart, M., Weber, F., Jahn, G., Dorsch-Hasler, K., Fleckenstein, B., and Schaffner, W. (1985). A very strong enhancer is located upstream of an immediate early gene of human cytomegalovirus. *Cell (Cambridge, Mass.)* **41**, 521–530.

Brewer, C. B. (1994). Cytomegalovirus plasmid vectors for permanent lines of polarized epithelial cells. *Methods Cell Biol.* **43**, 233–245.

Chan, Y.-J., Chiou, C.-J., Huang, Q., and Hayward, G. S. (1996). Synergistic interactions between overlapping binding sites for the serum response factor and ELK-1 proteins mediate both basal enhancement and phorbol ester responsiveness of primate cytomegalovirus major immediate-early promoters in monocyte and T-lymphocyte cell types. *J. Virol.* **70**, 8590–8605.

Chang, Y. N., Crawford, S., Stall, J., Rawlins, D. R., Jeang, K. T., and Hayward, G. S. (1990). The palindromic series I repeats in the simian cytomegalovirus major immediate-early promoter behave as both strong basal enhancers and cyclic-AMP response elements. *J. Virol.* **64**, 264–277.

Chapman, B. S., Tayer, R. M., Vincent, K. A., and Haigwood, N. L. (1991). Effect of intron A from human cytomegalovirus (Towne) immediate-early gene on heterologous expression in mammalian cells. *Nucleic Acids Res.* **19**, 3979–3986.

Chee, M. A., Bankier, A. T., Beck, S., Bohni, R., Brown, C. M., Cerny, R., Horsnell, T., Hutchison, C. A., Kouzarides, T., Martignetti, J. A., Preddie, E., Satchwell, S. C., Tomlinson, P., Weston, K. M., and Barrell, B. G. (1990). Analysis of the protein-coding content of the sequence of human cytomegalovirus strain AD169. *Curr. Top. Microbiol. Immunol.* **154**, 125–169.

Cherrington, J. M., and Mocarski, E. S. (1989). Human cytomegalovirus ie1 transactivates the α promoter-enhancer via an 18-base pair repeat element. *J. Virol.* **63**, 1435–1440.

Cherrington, J. M., Khoury, E. L., and Mocarski, E. S. (1991). Human cytomegalovirus ie2 negatively regulates α gene expression via a short target sequence near the transcription start site. *J. Virol.* **65**, 887–896.

Dorsch-Hasler, K., Keil, G. M., Weber, F., Schaffner, J. M., and Koszinowski, U. H. (1985). A long and complex enhancer activates transcription of the gene coding for the highly abundant immediate early mRNA in murine cytomegalovirus. *Proc. Natl. Acad. Sci. U.S.A.* **82**, 8325–8329.

Faisst, S., and Meyer, S. (1992). Compilation of vertebrate-encoded transcription factors. *Nucleic Acids Res.* **20**, 3–26.

Fickensher, H., Stamminger, T., Ruger, R., and Fleckenstein, B. (1989). The role of a repetitive palindromic sequence element in the human cytomegalovirus major immediate early enhancer. *J. Gen. Virol.* **70**, 107–123.

Fish, K. N., Depto, A. S., Moses, A. V., Britt, W., and Nelson, J. A. (1995). Growth kinetics of human cytomegalovirus are altered in monocyte-derived macrophages. *J. Virol.* **69**, 3737–3743.

Ghazal, P., and Nelson, J. A. (1991). Enhancement of RNA polymerase II initiation complexes by a novel DNA control domain downstream from the cap site of the cytomegalovirus major immediate-early promoter. *J. Virol.* **65**, 2299–2307.

Ghazal, P., MeDattei, C., Giulietti, E., Kliewer, S. A., and Umesono, K. (1992). Retinoic acid receptors initiate induction of the cytomegalovirus enhancer in embryonal cells. *Proc. Natl. Acad. Sci. U.S.A.* **89**, 7630–7634.

Gluzman, Y. (1981). SV40-tranformed simian cells support the replication of early SV40 mutants. *Cell (Cambridge, Mass.)* **23**, 175–182.

Graham, F. L., and van der Eb, A. J. (1973). A new technique for the assay of infectivity of adenovirus 5 DNA. *Virology* **52**, 456–467.

Hermiston, T. W., Malone, C. L., and Stinski, M. F. (1990). Human cytomegalovirus immediate-early two protein region involved in negative regulation of the major immediate-early promoter. *J. Virol.* **64**, 3532–3536.

Ho, M. (1991). "Cytomegalovirus: Biology and Infection." Plenum, New York.

Huang, T. H., Oka, T., Asai, T., Okada, T., Merrills, B. W., Gerston, P. N., Whitson,

R. N., and Itakura, K. (1996). Repression by a differentiation-specific factor of the human cytomegalovirus enhancer. *Nucleic Acids Res.* **24**, 1695–1701.

Hunninghake, G. W., Monick, M. M., Liu, B., and Stinski, M. F. (1989). The promoter-regulatory region of the major immediate-early gene of human cytomegalovirus responds to T-lymphocyte stimulation and contains functional cyclic AMP-response elements. *J. Virol.* **63**, 3026–3033.

Javahery, R. A., Khachi, A., Lo, K. Z.-G., B., and Smale, S. T. (1994). DNA sequence requirements for transcriptional initiator activity in mammalian cells. *Mol. Cell. Biol.* **14**, 116–127.

Kaufman, J., and Smale, S. T. (1994). Direct recognition of initiator elements by a component of the transcription factor IID complex. *Genes Dev.* **8**, 821–829.

Kaufman, J., Ahrens, K., Koop, R., Smale, S. T., and Muller, R. (1998). CIF 150, a human cofactor for transcription factor IID-dependent initiator function. *Mol. Cell. Biol.* **18**, 233–239.

Koedood, R., Fichtel, A., Meier, P., and Mitchell, P. J. (1995). Human cytomegalovirus (HCMV) immediate-early enhancer/promoter specificity during embryogenesis defines target tissues of congenital HCMV infection. *J. Virol.* **69**, 2194–2207.

Kondo, K., Kaneshima, H., and Mocarski, E. S. (1994). Human cytomegalovirus latent infection of granulocyte-macrophage progenitors. *Proc. Natl. Acad. Sci. U.S.A.* **91**, 11879–11883.

Kothari, S. K., Baillie, J., Sissons, J. G. P., and Sinclair, J. H. (1991). The 21 bp repeat element of the human cytomegalovirus major immediate early enhancer is a negative regulator of gene expression in undifferentiated cells. *Nucleic Acids Res.* **19**, 1767–1771.

Kurokawa, R., Soderstron, M., Horlein, A., Halachmi, S., Brown, M., Rosenfeld, M. C., and Glass, C. C. (1995). Polarity-specific activities of retinoic acid receptors determined by a co-repressor. *Nature (London)* **377**, 451–454.

Laegreid, A., Medvedev, A., Nonstad, U., Bombara, M. P., Ranges, G., Sundan, A., and Espevik, T. (1994). Tumor necrosis factor receptor p75 mediates cell-specific activation of nuclear factor κB and induction of human cytomegalovirus enhancer. *J. Biol. Chem.* **269**, 7785–7791.

Lang, D., and Stamminger, T. (1994). Minor groove contacts are essential for an interaction of the human cytomegalovirus IE2 protein with its DNA target. *Nucleic Acids Res.* **22**, 3331–3338.

Lang, D., Fickenscher, H., and Stamminger, T. (1992). Analysis of proteins binding to the proximal promoter region of the human cytomegalovirus IE-1/2 enhancer/promoter reveals both consensus and aberrant recognition sequences for transcription factors SP-1 and CREB. *Nucleic Acids Res.* **20**, 3287–3295.

Lathey, J. L., and Spector, S. A. (1991). Unrestricted replication of human cytomegalovirus in hydrocortisone-treated macrophages. *J. Virol.* **65**, 6371–6375.

Liu, B., and Stinski, M. F. (1992). Human cytomegalovirus contains a tegument protein that enhances transcription from promoters with upstream ATF and AP-1 cis-acting elements. *J. Virol.* **66**, 4434–4444.

Liu, B., Hermiston, T. W., and Stinski, M. F. (1991). A cis-acting element in the major immediate early (IE) promoter of human cytomegalovirus is required for negative regulation by IE2. *J. Virol.* **65**, 897–903.

Liu, R., Baillie, J., Sissons, J. G. P., and Sinclair, J. H. (1994). The transcription factor YY1 binds to negative regulatory elements in the human cytomegalovirus major immediate early enhancer/promoter and mediates repression in nonpermissive cells. *Nucleic Acids Res.* **22**, 2453–2459.

Lo, K., and Smale, S. T. (1996). Generality of a functional initiator consensus sequence. *Gene* **182**, 13–22.

Macias, M. P., and Stinski, M. F. (1993). An in vitro system for human cytomegalovirus immediate early 2 protein (IE2)-mediated site-dependent repression of transcription and direct binding of IE2 to the major immediate early promoter. *Proc. Natl. Acad. Sci. U.S.A.* **90**, 707–711.

Macias, M. P., Huang, L., Lashmit, P. E., and Stinski, M. F. (1996). Cellular and viral protein binding to a cytomegalovirus promoter transcription initiation site: Effects on transcription. *J. Virol.* **70**, 3628–3635.

Malone, C. L., Vesole, D. H., and Stinski, M. F. (1990). Transactivation of a human cytomegalovirus early promoter by gene products from the immediate-early gene IE2 and augmentation by IE1: Mutational analysis of the viral proteins. *J. Virol.* **64**, 1498–1506.

Martinez, E., Chiang, C.-M., Ga, H., and Roeder, R. G. (1994). TATA-binding protein-associated factor(s) in TFIID function through the initiator to direct basal transcription from a TATA-less class II promoter. *EMBO J.* **13**, 3115–3126.

Meier, J. L., and Stinski, M. F. (1997). Effect of a modulator deletion on transcription of the human cytomegalovirus major immediate-early genes in infected undifferentiated and differentiated cells. *J. Virol.* **71**, 1246–1255.

Moch, H., Lang, D., and Stamminger, T. (1992). Strong transactivation of the human cytomegalovirus major immediate-early enhancer by p40tax of human T-cell leukemia virus type I via two repetitive tax-responsive sequence elements. *J. Virol.* **66**, 7346–7354.

Nelson, J. A., and Groudine, M. (1986). Transcriptional regulation of the human cytomegalovirus major immediate-early gene is associated with induction of DNase I-hypersensitive sites. *Mol. Cell. Biol.* **6**, 452–461.

Nelson, J. A., Reynolds-Kohler, C., and Smith, B. (1987). Negative and positive regulation by a short segment in the 5′-flanking region of the human cytomegalovirus major immediate-early gene. *Mol. Cell. Biol.* **7**, 4125–4129.

Niller, H. H., and Hennighausen, L. (1990). Phytohemagglutinin-induced activity of cyclic AMP (cAMP) response elements from cytomegalovirus is reduced by cyclosporine and synergistically enhanced by cAMP. *J. Virol.* **64**, 2388–2391.

Pizzorno, M. C., and Hayward, G. S. (1990). The IE2 gene products of human cytomegalovirus specifically down-regulate expression from the major immediate-early promoter through a target located near the cap site. *J. Virol.* **64**, 6154–6165.

Radsak, E., Fuhrmann, R., Franke, R. P., Schneider, D., Kollert, A., Brucher, K. H., and Drenckhahn, D. (1989). Induction by sodium butyrate of cytomegalovirus replication in human endothelial cells. *Arch. Virol.* **107**, 151–158.

Sambucetti, L. C., Cherrington, J. M., Wilkinson, G. W. G., and Mocarski, E. S. (1989). NF-kappa B activation of the cytomegalovirus enhancer is mediated by a viral transactivator and by T cell stimulation. *EMBO, J.* **8**, 4251–4258.

Sandford, G. R., and Burns, W. H. (1996). Rat cytomegalovirus has a unique immediate early gene enhancer. *Virology* **222**, 310–317.

Shelbourn, S. L., Kothari, S. K., Sissons, J. G. P., and Sinclair, J. H. (1989). Repression of human cytomegalovirus gene expression associated with a novel immediate early regulatory region binding factor. *Nucleic Acids Res.* **17**, 9165–9171.

Shrivastava, A., and Calame, K. (1994). An analysis of genes regulated by the multifunctional transcription regulator Yin-Yang-1. *Nucleic Acids Res.* **22**, 5151–5155.

Sinclair, J. H., Baillie, J., Bryant, L. A., Taylor-Wiedeman, J. A., and Sissons, J. G. P. (1992). Repression of human cytomegalovirus major immediate early gene expression in a monocytic cell line. *J. Gen. Virol.* **73**, 433–435.

Stein, J., Volk, H., Liebenthal, C., Kruger, D. H., and Prosch, S. (1993). Tumor necrosis factor α stimulates the activity of the human cytomegalovirus major immediate early enhancer/promoter in immature monocytic cells. *J. Gen. Virol.* **74**, 2333–2338.

Stenberg, R. M., and Stinski, M. F. (1985). Autoregulation of the human cytomegalovirus major immediate early gene. *J. Virol.* **56**, 676–682.

Stenberg, R. M., Thomsen, D. R., and Stinski, M. F. (1984). Structural analysis of the major immediate early gene of human cytomegalovirus. *J. Virol.* **49**, 190–199.

Stenberg, R. M., Witte, P. R., and Stinski, M. F. (1985). Multiple spliced and unspliced transcripts from human cytomegalovirus immediate-early region 2 and evidence for a common initiation site within immediate-early region 1. *J. Virol.* **56**, 665–675.

Stinski, M. F. (1977). Synthesis of proteins and glycoproteins in cells infected with human cytomegalovirus. *J. Virol.* **23**, 751–767.

Stinski, M. F., and Roehr, T. J. (1985). Activation of the major immediate early gene of human cytomegalovirus by cis-acting elements in the promoter-regulatory sequence and by virus-specific trans-acting components. *J. Virol.* **55**, 431–441.

Stinski, M. F., Thomsen, D. R., Stenberg, R. M., and Goldstein, L. M. (1983). Organization and expression of the immediate early genes of human cytomegalovirus. *J. Virol.* **46**, 1–14.

Stinski, M. F., Malone, C. L., Hermiston, T. W., and Liu, B. (1991). Regulation of human cytomegalovirus transcription. *In* "Herpesvirus Transcription and its Control" (E. K. Wagner, ed.), pp. 246–260. CRC Press, Boca Raton, FL.

Stinski, M. F., Macias, M. P., Malone, C. L., Thrower, A. R., and Huang, L. (1993). Regulation of transcription from the cytomegalovirus major immediate early promoter by cellular and viral proteins. *In* "Multidisciplinary Approach to Understanding Cytomegalovirus" (S. Michelson and S. A. Plotkin, eds.). Elsevier, Amsterdam, pp. 3–12.

Tanaka, J., Sadanari, H., Sato, H., and Fukuda, S. (1991). Sodium butyrate-inducible replication of human cytomegalovirus in a human epithelial cell line. *Virology* **185**, 271–280.

Taylor-Wiedeman, J., Sissons, J. G., Borysiewicz, L. K., and Sinclair, J. H. (1991). Monocytes are a major site of persistence of human cytomegalovirus in peripheral blood mononuclear cells. *J. Gen. Virol.* **72**, 2059–2064.

Thomsen, D. R., Stenberg, R. M., Goins, W. F., and Stinski, M. F. (1984). Promoter-regulatory region of the major immediate early gene of human cytomegalovirus. *Proc. Natl. Acad. Sci. U.S.A.* **81**, 659–663.

Verrijzer, C. P., Chen, J.-L., Yokomori, K., and Tjian, R. (1995). Binding of TAFs to core elements directs promoter selectivity by RNA polymerase II. *Cell (Cambridge, Mass.)* **81**, 1115–1125.

Wathen, M. W., and Stinski, M. F. (1982). Temporal patterns of human cytomegalovirus transcription: Mapping the viral RNAs synthesized at immediate early, early, and late times after infection. *J. Virol.* **41**, 462–477.

Winkler, M., Rice, S. A., and Stamminger, T. (1994). UL69 of human cytomegalovirus, an open reading frame with homology to ICP27 of herpes simplex virus, encodes a transactivator of gene expression. *J. Virol.* **68**, 3943–3954.

Yang, W. M., Inouye, C., Zeng, Y. Y., Bearss, D., and Seto, E. (1996). Transcriptional repression by YY1 is mediated by interaction with mammalian homolog of the yeast global regulator RPD3. *Proc. Natl. Acad. Sci. U.S.A.* **93**, 12845–12850.

Zweidler-McKay, P. A., Grimes, H. L., Flubacher, M. M., and Tsichlis, P. N. (1996). Gfi-1 encodes a nuclear zinc finger protein that binds DNA and functions as a transcriptional repressor. *Mol. Cell. Biol.* **16**, 4024–4034.

9

INDUCIBLE MAMMALIAN EXPRESSION SYSTEMS

Marijane Russell
Invitrogen Corporation, Carlsbad, California 92008

Background: Development of Inducible Expression Systems

The tightly regulated expression of foreign proteins of interest in mammalian cells has been difficult to achieve, but is vital when studying proteins that are potentially toxic to the host cell. There have been numerous attempts to develop tightly regulated inducible expression systems (Yarranton, 1992; Gossen *et al.*, 1993; Shockett and Schatz, 1996). The ideal inducible expression system would possess several key features, including low basal expression and a high level of induction of the protein of interest that is reversible (i.e., ability to regulate and turn both on and off). In addition the ideal system would induce protein synthesis rapidly and the inducer would exhibit no pleiotropic effects on host cells. Finally the optimal system would utilize vectors that are user friendly (small vector size, good multiple cloning site, resistance marker for selection of stable clones). In a perfect world, a one vector system containing all the regulatory elements required that would allow one to express their gene of interest would be ideal. Many systems have been developed and optimized to try and address these key features. So far the perfect inducible expression system has yet to be discovered, but this chapter discusses the advantages and disadvantages of systems available currently that have many of the desired elements just described.

This chapter first gives a brief history of the original attempts at making inducible mammalian expression systems. Achieving tight regulation of expression *in vitro* and *in vivo* with a single promoter or control element has been a very difficult problem for researchers to solve. The majority of the chapter compares and contrasts the three systems that have been characterized and made available commercially over the years. These three systems are the *Lac*-inducible expression system, the tetracycline-regulated expression system (of which there are several different versions available, each with different features), and the most recent advancement in this field, the ecdysone-inducible expression system.

Initial attempts to develop inducible expression systems utilized a variety of promoters. However, in general, these early systems relied on inducers that elicited pleiotropic effects on the cells, which made interpreting results of the experiments more difficult (depending on the pathways being studied). One of the first systems studied utilized the heat shock promoter, which was induced by a temperature shift from 37° to 42°C (Schweinfest *et al.*, 1988). Another promoter that demonstrated some level of utility, depending on the host cell system, is the metallothionine promoter, which can be induced by a variety of heavy metal ions, including zinc, copper, and cadmium. The temporal effects of a variety of proteins on cellular physiology have been studied using the metallothionine promoter to control expression in mammalian cells. These include v-src (Cox and Maness, 1991), trk (Mulherkar and Coulier, 1991), heat shock protein (Suzuki and Watanabe, 1994), and cAMP-dependent protein kinase (Tasken *et al.*, 1994). Other systems tried to use steroid regulatory promoters that require induction with glucocorticoids such as dexamethasone (Israel and Kaufman, 1889; Ko *et al.*, 1989; Planas-Silva and Means, 1992). Yet another system utilized the interferon type 1-inducible murine Mx promoter to regulate the expression of phospholipase Cγ (Totzke *et al.*, 1992). Using expressed human growth hormone under the control of the β-globin promoter, an inducible system was described by Needham *et al.* (1992). A major disadvantage of all of these early systems was that the "generic"-inducing agents led to an induction of a variety of cellular genes, in addition to the gene of interest. These systems also tended to have high basal levels of expression, which in turn led to lower induction levels of the protein of interest.

Major problems with early inducible expression systems included (1) inducers caused pleiotropic effects, (2) basal activity was too high, (3) the off in many systems was not really off and the increased basal activities resulted in only modest induction factors, (4) when working with toxic proteins, high basal expression can inhibit the ability to select stable clones, and (5) cloning vectors had limited sites and did not always contain selectable markers.

In order to overcome one of the major disadvantages of early inducible systems, which is having to use generic inducing agents, researchers used regulatory elements from distant species to create a more useful regulated expression system. By using prokaryotic regulatory elements to control gene expression in higher eukaryotic cells, the effectors are inert to the physiology of the host cell and the undesirable pleiotropic effects are diminished. This chapter reviews two systems based on prokaryotic elements from *Escherichia coli:* the *Lac*-inducible system and the tetracycline-regulated system. Finally, this chapter reviews the development of a novel system based on regulatory elements from *Drosophila melanogaster*, the ecdysone-inducible system.

Lac-Inducible System

The *Lac*-inducible system is based on the well-characterized *E. coli lac* operon (Hu and Davidson, 1987). The *lac* repressor binds to its operator (*lac* op) and prevents formation of an initiation competent complex between RNA polymerase and the promoter that controls transcription of the *E. coli lac* operon. In *E. coli* the *lac* repressor binds as a homotetramer to the *lac* operator and blocks transcription of the β-galactosidase gene (Bourgeois and Pfahl, 1976). Inducers such as alloactose (the natural substrate of β-galactosidase) or isopropyl-β-D-thiogalactopyranoside (IPTG) (a synthetic and non-metabolizable inducer) bind to the repressor, causing a conformational change resulting in a decreased affinity of the repressor for the operator (Kamashev *et al.*, 1995). Once the repressor is removed from the operator, transcription can resume. These properties made the *lac* system a reasonable choice to try and build a system for controlling protein expression in eukaryotic cells. The most straightforward approach would be to place *lac* operator sequences near the TATA box or the start site of a promoter.

The basic format for the *Lac*-inducing system is outlined in Fig. 1 (available commercially from Stratagene as the Lac Switch inducible expression system). The system requires two vectors, a regulatory vector and an expression vector. The regulatory vector encodes the *lac* repressor protein driven by the constitutive F9-1 promoter. The *lac* repressor sequence is present as a fusion with a nuclear localization sequence (NLS) at the carboxyl terminus (Fieck *et al.*, 1992). The vector also contains a hygromycin resistance marker for stable selection in mammalian cells. The expression vector for the *Lac* system contains *lac* operator sequences (*lac* op) upstream of a cloning site for insertion of the gene of interest. The gene of interest is driven by the Rous sarcoma virus (RSV)-LTR promoter. It also contains the neomycin gene to confer G418 resistance in mammalian cells. There are two versions of the expression vector, one has three *lac* operator sequences and the other has two *lac* operator sequences. The vector with two *lac* op sites was reported to yield stable clones consistently whereas the vector with three *lac* op sites produced a clone with the highest fold induction (Lac Switch manual). Cells transfected with both vectors will express the *lac* repressor, which when bound to the *lac* op sites upstream of the sequence encoding the gene of interest should repress expression of this gene. Upon induction with IPTG the *lac* repressor binds IPTG and is removed from the *lac* op sites. This then allows expression of the gene of interest.

Tetracycline-Regulated Expression System

The tetracycline system first described by Gossen and Bujard (1992) utilizes the *tet* repressor from *E. coli.* The antibiotic tetracycline is known to inhibit prokaryotic translation at low concentrations. Operons that produce a resistance response against the antibiotic

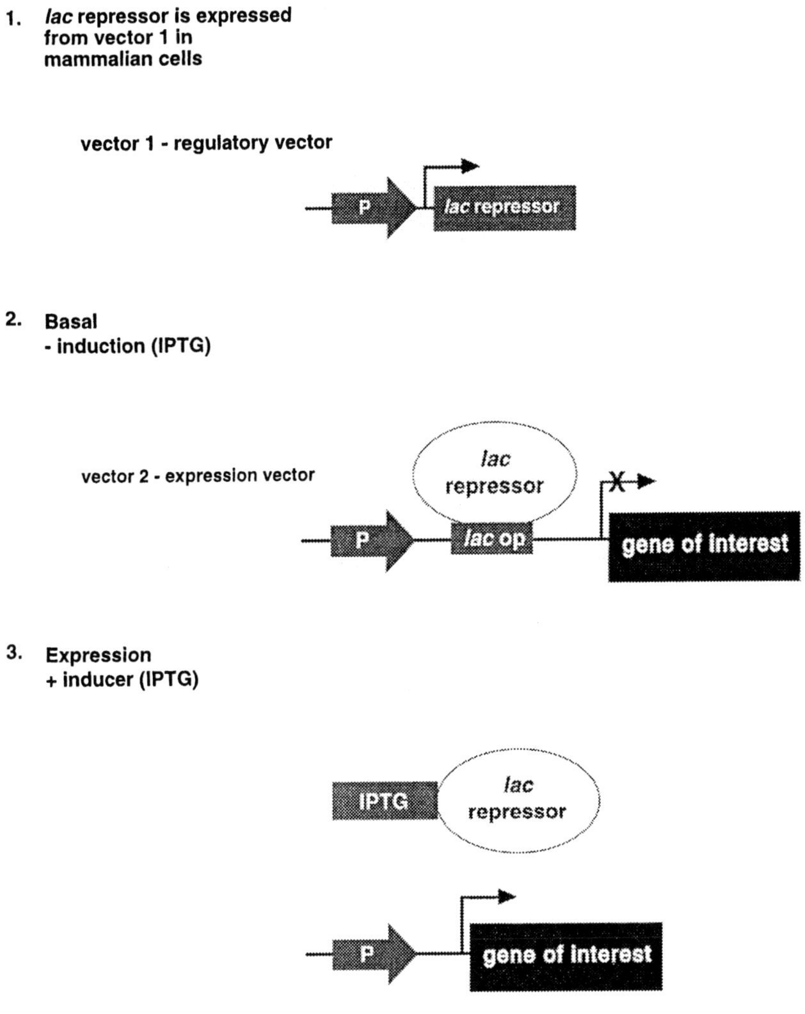

Figure 1 The *Lac*-inducible system.

are optimized to respond to tetracycline. The binding of tetracycline to the repressor of the transposon-10-derived tetracycline resistance operon of *E. coli* is high ($10^9\ M^{-1}$) (Gossen *et al.*, 1993). This, coupled with the fact that the *tet* repressor binds to its operator with high specificity, suggested that these elements may be promising components for a regulatory system in higher eukaryotes. The tetracycline-regulated expression system was built with the *E. coli* tetracycline responsive regulatory element Tn10 tetracycline resistance operon (*tet* op) and the bacterial *tet* repressor protein, which binds this sequence. In the presence of tetracycline the *tet* repressor does not bind to its operators located within the promoter region of the operon and transcription is functional. Combination of the *tet* repressor with the transcriptional activation domain of VP16 produces a hybrid transactivator that is capable of stimulating minimal promoters that are fused to tetracycline operator sequences. The promoters are silent in the presence of tetracycline, which prevents the transactivator from binding to *tet* operator sites.

The tetracycline-regulated mammalian expression system is outlined in Fig. 2. Separate versions of this system are available commercially from GIBCO BRL (Tet-regulated system) and Clontech (Tet-Off gene expression system). The first tetracycline-regulated expression system is a two vector system that includes a regulatory vector and an expression vector. The regulatory vector encodes the transactivator protein. The transactivator is a fusion between the *tet* repressor DNA-binding domain and the transcriptional activation domain of VP16 from herpes simplex virus. The second vector is the expression vector for insertion of the gene of interest. This vector (in both systems) has *tet* op sites upstream of a minimal cytome galovirus (CMV) promoter that drives expression of the gene of interest. In the presence of the repressor, tetracycline, the transactivator protein is unable to bind the *tet* op and the fusion should not transactivate transcription. In the absence of tetracycline, the *tet* repressor/VP16 fusion serves as a potent transactivator and stimulates transcription from a minimal promoter downstream of the *tet* op.

The reverse tetracycline system developed by Gossen *et al.* (1995) utilizes a modified transactivator that is a fusion between a mutant *tet* repressor and the VP16 activation domain (Fig. 3). This system is available commercially from Clontech (Tet-On gene expression system). Just like the original tetracycline-regulated expression system, it is a two vector system. The regulatory vector expresses a modified *tet* repressor fused with the VP16 transactivation domain. The expression vector has *tet* op sites upstream of a minimal CMV promoter to drive transcription of the gene of interest. In this system, in the absence of an inducer the modified transactivator does not bind the *tet* operator sequence (*tet* op), but in the presence of an inducer (the tetracycline derivative doxycycline) the transactivator is able to bind the *tet* op and can drive expression of

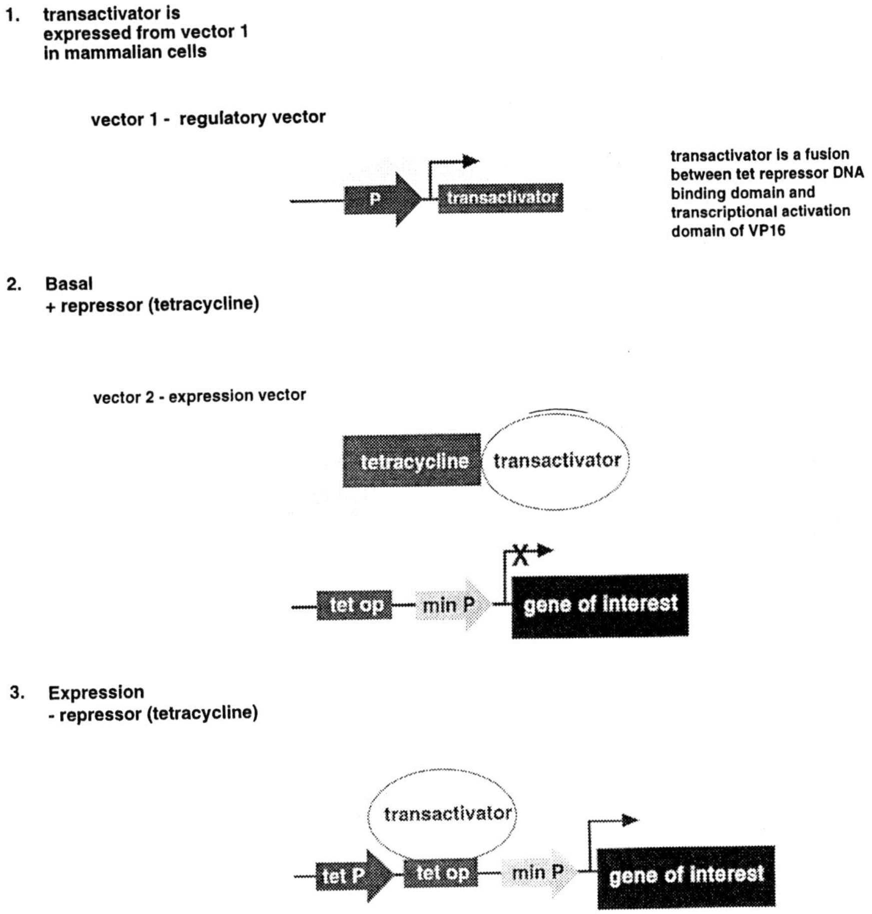

Figure 2 The tetracycline-regulated expression system.

the gene of interest. This system was designed to offer an alternative to the original tetracycline-regulated system, which required the presence of tetracycline for repression. Although tetracycline is nontoxic to eukaryotic cells, for some experiments, like the generation of transgenic mice, the requirement of continuous tetracycline is not advantageous.

Hoffmann and co-workers (1997) created a new expression vector by replacing the original minimal CMV promoter with a modified mouse mammary tumor virus promoter (ΔMTV). Using this promoter in a tetracycline-inducible/repressible system, decreased

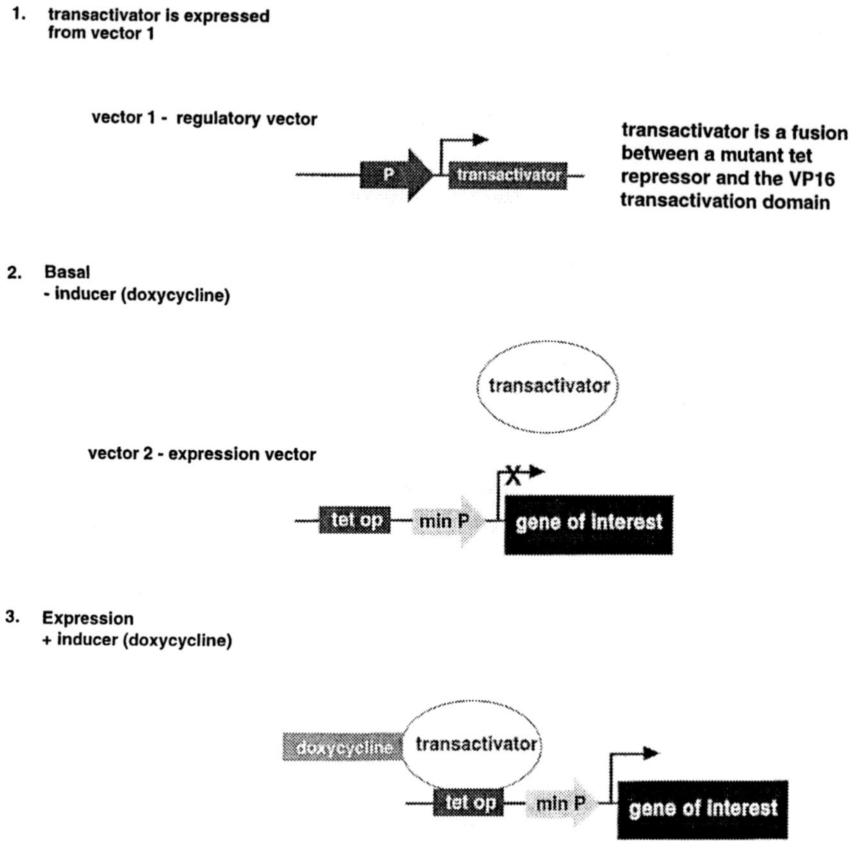

Figure 3 The reverse tetracycline system.

basal levels of the reporter gene luciferase were observed compared to those seen with the CMV promoter. Although the overall transient expression of luciferase from the ΔMTV was less than with the minimal CMV promoter because of the greatly reduced basal levels, the fold induction was much higher when ΔMTV-driven expression was analyzed in several cell lines (Table 1). Stable clones in LLC-PK1 and SaOs2 cells had high fold induction and low basal levels. It appears that just simply changing the promoter used in the expression vector driving the gene of interest is very crucial in exploiting the regulatory properties of tetracycline-controlled gene expression.

One disadvantage of the tetracycline system is the relatively long

Table 1
Comparison of Tetracycline-Sensitive Luciferase Expression Driven by the Mouse Mammary Tumor Virus Promoter (MTV) versus the Minimal Cytomegalovirus Promoter (CMV)[a]

Cell line	Fold induction MTV vs CMV
LLC-PK1 (porcine epithelial cell line)	800 vs 70
SaOs-2 (human osteosarcoma cell line)	1000 vs 20
Sk-N-MC (human neuroblastoma cell line)	3900 vs 1500
C6 (rat glial cell line)	75 vs 30
CV-1 (green monkey kidney cell line)	170 vs 120
HeLa (human epitheloid cervix carcinoma cell line)	1200 vs 800

[a]From Hoffmann *et al.* (1997).

half-lives of tetracycline (4–12 hr) and doxycycline (12–24 hr). This can be a problem if a fast on/off switch is needed in whole animals.

Ecdysone Inducible Mammalian Expression System

The most recent advance in inducible mammalian expression systems has been the development of the ecdysone-inducible expression system. This powerful inducible expression system was initially developed quite elegantly by No *et al.* (1996). Further modifications to make this a two vector system, each with unique selectable markers, have made this system a very user-friendly research tool for the regulated expression of foreign proteins of interest in mammalian cells (Fig. 4). This system is available commercially from Invitrogen (Ecdysone-inducible expression kit).

The ecdysone system is based on the *D. melanogaster* molting induction pathway. During metamorphosis of *Drosophila,* a series of morphological changes are initiated by 20-OH ecdysone. Ecdysone is a steroid hormone that activates a nuclear receptor, the ecdysone receptor. Ecdysone mediates its response through a functional receptor that is actually a heterodimer of the ecdysone receptor and the product of the ultraspiracle gene (Yao *et al.*, 1993). The natural enhancer element is two inverted half-sites of the sequence AGGTCA spaced by one nucleotide. An attractive feature of this inducible process is the fact that the inducer is a steroid hormone, which is lipophilic. This allows for efficient penetration into tis-

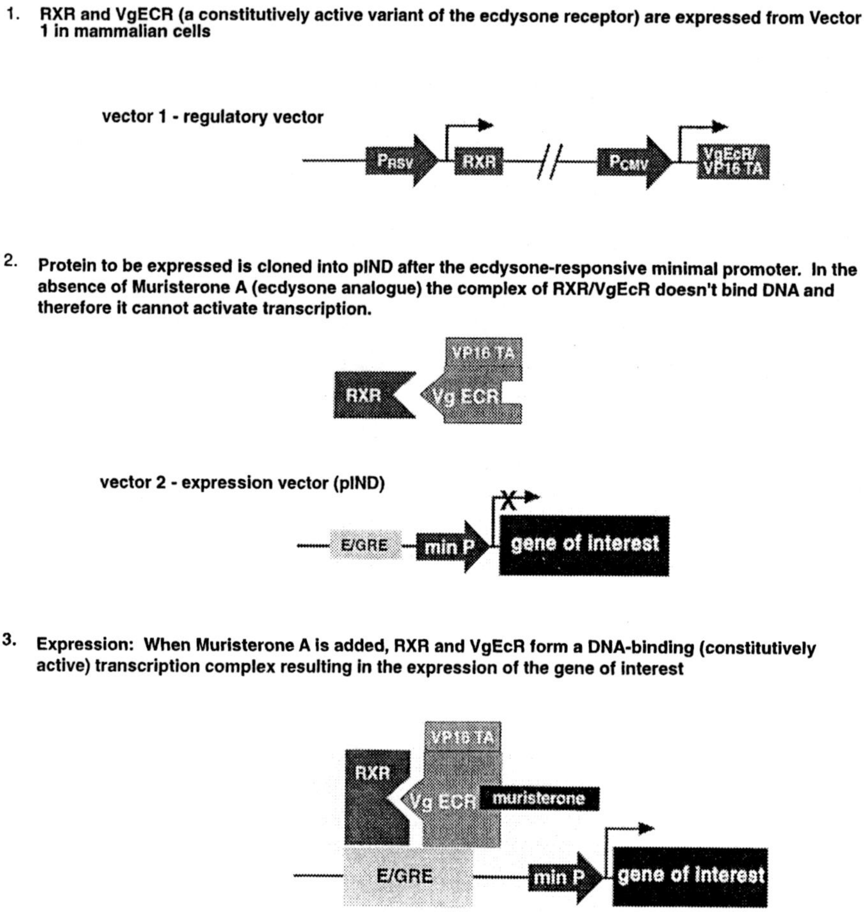
1. RXR and VgECR (a constitutively active variant of the ecdysone receptor) are expressed from Vector 1 in mammalian cells
vector 1 - regulatory vector
PRSV
RXR
PCMV
VgEcR/
VP16 TA
2. Protein to be expressed is cloned into pIND after the ecdysone-responsive minimal promoter. In the absence of Muristerone A (ecdysone analogue) the complex of RXR/VgEcR doesn't bind DNA and therefore it cannot activate transcription.
VP16 TA
RXR
Vg ECR
vector 2 - expression vector (pIND)
E/GRE
min P
gene of interest
3. Expression: When Muristerone A is added, RXR and VgEcR form a DNA-binding (constitutively active) transcription complex resulting in the expression of the gene of interest
VP16 TA
RXR
Vg ECR
muristerone
E/GRE
min P
gene of interest

Figure 4 The ecdysone-inducible system.

sues and a short half-life, which should allow precise and potent induction. Also, because ecdysone (and its analogues) is not toxic and does not affect mammalian physiology, it is an ideal inducer for both cultured cells and transgenic mice.

Key regulatory elements needed for the inducible *Drosophila* molting process have been modified to allow for regulated inducible expression in mammalian cells. The system utilizes a heterodimeric nuclear receptor to activate expression on induction with an ecdysone steroid hormone analogue, muristerone A (No *et al.*, 1996). The system requires two separate vectors; a regulatory vector and a vector for expression of the gene of interest. The regulatory vector expresses both subunits of the heterodimeric nuclear receptor: (1) the retinoid X-receptor (RXR) subunit is driven by the RSV (rous sarcoma virus) promoter; RXR is the mammalian homologue of the endogenous *Drosophila* partner of ecdysone receptor, USP (ultraspiracle) (Yao *et al.*, 1992, 1993), and (2) the ecdysone receptor (VgEcR) expression is driven by the CMV (human cytomegalovirus) promoter. The ecdysone receptor is expressed as a fusion with the VP16 transactivator domain from herpes simplex virus (Cress and Treizenberg, 1991; Sadowski *et al.*, 1988; Treizenberg *et al.*, 1998a, b). The ecdysone receptor VgEcR has also been modified to bind a hybrid DNA response element. The hybrid response element is one-half the natural ecdysone response element 5′-ACCTCA-3′ and one-half the glucocorticoid response element 5′-AGAACA-3′ spaced by one nucleotide (E/GRE) (Umesono and Evans, 1989). This hybrid response element uniquely eliminates background from any endogenous mammalian receptor that may bind the natural ECR element, such as the farsenoid X receptor (Forman *et al.*, 1995). The expression vector contains the hybrid response element in five repeats upstream of the *Drosophila* minimal heat shock promoter (Corces and Pellicer, 1984). In the absence of inducer, the two regulatory subunits do not dimerize and there is no expression of the gene of interest. Upon induction with the synthetic ecdysone analogue muristerone A, there is formation of a heterodimer between RXR and VgEcR and this heterodimer binds the hybrid ecdysone response element. The heterodimer is a transactivator (VgEcR is a fusion with the VP16 transactivation domain) and drives expression of the gene of interest.

All three of the inducible expression systems discussed here have similar basic protocols and time lines for how to utilize each system effectively. The basic steps for using each of these systems are outlined in Table 2. For more detailed protocols, please refer to each of the respective manuals from the commercial suppliers of each system.

Literature Review

Although it is clear researchers have spent a great deal of effort in developing and modifying inducible expression systems, no one

Table 2
Time Line of Experimental Procedures

	Time
1. Clone gene of interest into expression vector	1–2 weeks
2. Test transient-inducible expression	2–3 weeks
Transfect cells with regulatory vectors and expression vectors	
Treat cells with or without inducer for appropriate time	
Lac system	
Basal expression (minus IPTG)	
Expression of protein (plus IPTG)	
Tetracycline systems	
Regulated expression systems	
basal expression (plus tetracycline)	
expression of protein (minus tetracycline)	
Reverse tetracycline system	
basal expression (minus tetracycline)	
expression of protein (plus tetracycline)	
Ecdysone system	
Basal expression (minus muristerone A)	
Expression of protein (plus muristerone A)	
Test for expression of gene of interest[a]	
3. Make stable clones expressing the protein of interest. The *Lac*, tetracycline, and ecdysone systems each have the option of making stable cell lines in one or two steps.	
A. Create stable lines of regulatory and expression vectors	6–8 weeks
B. Make a regulatory vector stable cell line	4–5 months
Make expression stables in that background	
Select and expand clones	
Treat with and without inducer for appropriate time	
Test for expression of gene of interest	

[a]Each system provides a reporter gene that can be tested as well.

system has really been widely accepted yet. A brief list of some of the proteins that have been expressed in each of the three main systems discussed in this chapter is presented in Table 3. Each system has been used to express a variety of proteins, including reporter proteins, kinases, receptors, and regulatory proteins. The tetracycline and ecdysone systems have also been utilized to express and regulate the expression of proteins in transgenic mice. Although each of these systems has clearly been applied successfully, it is very difficult to compare the systems fairly as the failures for each would obviously not appear in the literature.

Table 3
Proteins Expressed in Different Inducible Expression Systems

Protein expressed	Reference
Lac-inducible system	
CAT	Hu and Davidson (1987)
CAT in oocytes	Hu and Davidson (1988)
H-Ras	Liu *et al.* (1992)
H-Ras	Ratnam and Kent (1995)
β-Galactosidase	Jacobsen and Willumsen (1995)
csk tyrosine kinase	Bergman *et al.* (1995)
p53	Zambetti *et al.* (1992)
Tetracycline-based inducible system	
Luciferase	Gossen and Bujard (1992)
ICP47	Fruh *et al.* (1995)
β-Galactosidase	Ackland-Berglund and Leib (1995)
β-Galactosidase	Gossen *et al.* (1995)
RAG1, RAG2	Shockett *et al.* (1995)
G protein of VSG	Iida *et al.* (1996)
CAT	Iida *et al.* (1996)
Dopamine D_3 receptor	Howe *et al.* (1995)
$GluR_6$ glutamate receptor subunit	Howe *et al.* (1995)
rol B in plants	Röder *et al.* (1994)
PG13 in plants	Rieping *et al.* (1994)
BcL-2	Yin and Schimke (1995)
E2F-1	Shan and Lee (1994)
Cyclin E	Wimmel *et al.* (1994)
Cyclin D1 and E	Resnitzky *et al.* (1994)
Dynamin	Damke *et al.* (1994)
cre recombinase in transgenic mice	St. Onge *et al.* (1996)
Luciferase in transgenic mice	Schultze *et al.* (1996)
Luciferase in transgenic mice	Shockett *et al.* (1995)
hGH in transgenic mice	Wang *et al.* (1997)
β-Galactosidase in transgenic mice	Furth *et al.* (1994)
Ecdysone-inducible system	
Luciferase	No *et al.* (1996)
β-Galactosidase	No *et al.* (1996)
β-Galactosidase	Invitrogen (unpublished data)
GFP	Invitrogen (unpublished data)
H-Ras (WT, N17 and V12)	Invitrogen (unpublished data)
αPhOx, human single chain antibody	Invitrogen (unpublished data)
Luciferase in transgenic mice	No *et al.* (1996)

Author's Experience

The author's laboratory has used one version of each of the three systems discussed in this chapter. With the Lac Switch system, a stable cell line in 293 human kidney cells that expressed the *lac* repressor was generated. This was fairly straightforward but time-consuming with the hygromycin selection. Several clones overexpressed the *lac* repressor in the nucleus (an anti-*lac* repressor antibody was utilized to determine this). One of these clones was used to try and generate stable clones expressing chloramphenicol acetyltransferase (CAT). Although the author was able to isolate G418-resistant clones, they did not have detectable expression of CAT by enzymatic assay with IPTG induction in any of the 24 clones tested.

Using the Tet-regulated expression system (GIBCO) to analyze transient expression in 293 cells, there was two- to fourfold induction of luciferase protein (detected by Western blot), but there was also detectable basal levels of luciferase expression. Making stable clones with this system is difficult due to the fact that it requires the transfection of three plasmids and only one has a selectable marker. Analysis of 24 stable 293 clones identified 9 clones that did have detectable luciferase expression. However, none of the 9 clones that expressed luciferase had a fold induction of more than twofold on withdrawal of the repressor tetracycline. The author's laboratory has not worked with the Clontech Tet-Off or Tet-On systems. These systems are designed to be more user friendly; in addition, Clontech offers several cell lines stably expressing the *tet* transactivator (CHO, HeLa, and NIH 3T3).

The ecdysone system was modified in the research laboratory at Invitrogen to generate more user-friendly reagents. Transient expression with this system gives virtually undetectable basal levels, and induction for 20 hr with muristerone gives a strong induction of all the proteins tested to date (ras, β-galactosidase, and GFP). The laboratory has generated several stable cell lines expressing the regulatory components of the system (RXR and VgEcR). These clones were analyzed by Western blot with an anti-RXR antibody. Clones overexpressing RXR were then functionally tested by transient transfection with the ecdysone responsive plasmid pIND/β-galactosidase. Clones for each cell type (293, CHO, and NIH 3T3) were isolated that gave a strong induction of β-galactosidase on induction with the addition of muristerone A.

Table 4
Advantages and Disadvantages of Current Inducible Expression Systems

A. *Lac*-inducible system

Advantages

Based on well-characterized bacterial regulatory operon, the *lac* operon

Synthetic inducer (IPTG) is nontoxic and is transported rapidly into eukaryotic cells allowing for a short induction time (4–12 hr)

Disadvantages

High basal levels of expression are often seen

Because of high basal levels the fold induction of the protein of interest is not very robust

Available vectors have limited cloning options for insertion of the gene of interest

A two vector system

B. Tetracycline-regulated expression systems

Advantages

Based on well-characterized bacterial regulatory element, the *tet* operon

Repression (or induction) requires tetracycline antibiotic (or derivative) at low, nontoxic doses and is transported rapidly into eukaryotic cells

Can generate transgenic mice with this system (tetracycline or derivative can penetrate placenta and blood/brain barrier)

Disadvantages

High basal levels have been reported

Difficult to generate stable clones with current reagents available

One version requires three vectors: regulatory vector, expression vector, and resistance marker vector

Another version requires two vectors: regulatory vector (has marker) and expression vector (no marker)[a]

Vectors available have limited cloning options for insertion of gene of interest

C. Ecdysone-Inducible System

Advantages

Based on the nonmammalian *Drosophila* molting induction pathway

Very tight regulation so basal levels are low to nonexistent

Strong induction with the steroid hormone ecdysone analogue, muristerone A

Can regulate levels of expression of protein of interest by varying the length of induction and concentration of inducer

Utilizes a hybrid response element (half of the ecdysone response element and half of the glucocorticoid response element) to prevent nonspecific transcriptional activation of promoter (minimal promoter driving gene of interest) by endogenous mammalian regulatory molecules

Because basal levels are very low the fold induction of the protein of interest can be quite dramatic

Available vectors have extensive multiple cloning site (16 unique sites) to simplify insertion of gene of interest

Can generate transgenic mice with this system

Disadvantages

Still a two vector system so dual selection is needed when generating stable clones

The system is relatively new so the widespread utility of the system has yet to be proven

[a]This version recommends first making a stable cell line expressing the regulatory vector and then generating a second stable cell line in that background. In order to generate the second stable, two plasmids are needed as the expression vector does not have a resistance marker.

Comparing Inducible Mammalian Expression Systems Currently Available

Each of the three inducible mammalian expression systems discussed in this chapter has important advantages and disadvantages that should be considered before using any of the systems. Table 4 has a summary for each of the systems.

Conclusions and Future Directions

There is a constant desire for systems to evolve and meet the ever-changing needs of researchers. This evolution has persisted in the field of inducible mammalian expression systems. One important use of inducible expression systems has been to generate transgenic mice that have regulated expression and, more specifically, targeted/regulated expression. A unique system has been described by Wang *et al.* (1997) that allows for both inducible and targeted expression in transgenic mice. This system is outlined briefly in Fig. 5. It is a two vector system. The first vector is a regulatory vector that expresses a transactivator from a tissue-specific targeted promoter (liver-specific TTR promoter). The transactivator is a mutated human progesterone receptor ligand-binding domain fused with a hybrid of the Gal4 DNA-binding domain and the VP16 transactivation domain. The vector also encodes insulator sequences that were added to help block suppressive effects of the surrounding chromosome and direct integration site-independent expression of a transgene. The second vector is the expression vector that has Gal4 upstream activation sites upstream of a minimal promoter driving the gene of interest. To create transgenic mice, a mouse expressing the regulatory vector was bred with a mouse expressing the expression vector (with human growth hormone being the gene of interest) to create bigenic mice expressing both transgenes. The bigenic mice were then induced with mifepristone (RU486), and liver-specific expression of human growth hormone was observed. The system has the desired qualities of a low basal expression and a high level of induction of the protein of interest. Because this system requires no other transcription factors or receptor family mem-

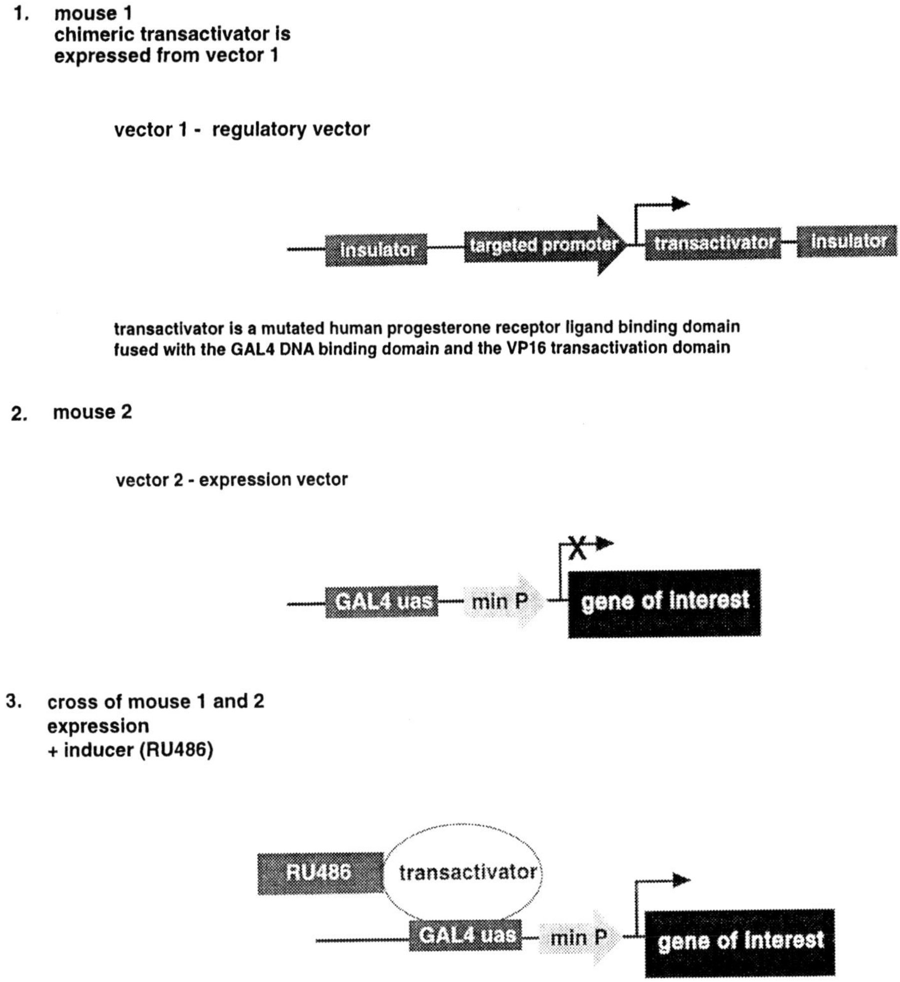

Figure 5 Inducible/targeted expression in transgenic mice.

bers, it should not interfere with any endogenous pathways in the target cell.

Liberles and co-workers (1997) have exploited small chemical inducers of dimerization (CIDs) that bind two proteins simultaneously for inducible gene expression and also protein translocation. This unique system uses the CID rapamycin, which interacts with the FK506-binding protein (FKB12) and FKBP12–rapamycin-associated protein (FRAP) simultaneously. One unique use of these

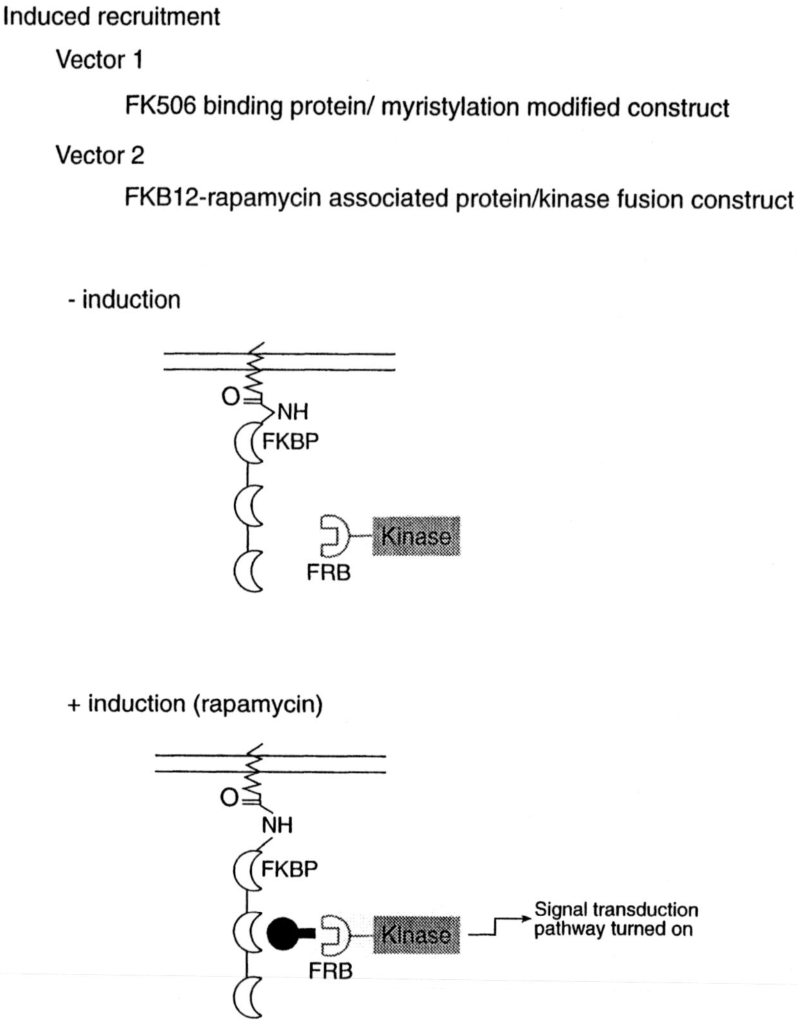

Figure 6 FRAP inducible/recruitment system.

reagents would be to have an induced recruitment pathway outlined in Fig. 6. Vector 1 would encode an FK506-binding protein that is myristylation modified to target it to the membrane. The second vector is FRAP fused with a protein kinase of interest. If these two vectors are expressed and then induced with rapamycin (nontoxic derivative, 9R), this would cause a recruitment of the kinase to the plasma membrane and a signal transduction pathway would be activated. This could be a very powerful tool for researchers studying either mitogenic or apoptosis signal transduction cascades.

Construction of mammalian-inducible expression systems will continue to exploit new regulatory components as they become characterized. The perfect system has yet to be built that has all the desired features. It will be interesting to see how well the ecdysone-inducible expression system works now that more researchers have access to the reagents. As researchers continue to better understand the regulation of gene expression, there will be even more alternatives. There inevitably will be many more clever systems developed as researchers continue to strive for tools that will help them do their experiments in the exact precise manner they want to.

References

Ackland-Berglund, C. E., and Leib, D. A. (1995). Efficacy of tetracycline-controlled gene expression is influenced by cell type. *BioTechniques* **18,** 196–200.

Aubrecht, J., Manivasakam, P., and Schiestl, R. H. (1996). Controlled gene expression in mammalian cells via a regulatory cascade involving the tetracycline transactivator and lac repressor. *Gene* **172,** 227–231.

Baim, S. B., Labor, M. A., Levine, A. J., and Shenk, T. (1991). A chimeric mammalian transactivator based on the lac repressor that is regulated by temperature and isopropyl β-D-thiogalactopyranoside. *Proc. Natl. Acad. Sci. U.S.A.* **88,** 5072–5076.

Baron, U., Freundlieb, S., Gossen, M., and Bujard, H. (1995). Co-regulation of two gene activities by tetracycline via a bi-directional promoter. *Nucleic Acids Res.* **23,** 3605–3606.

Bergman, M., Joukov, V., Virtanen, I., and Alitalo, K. (1995). Overexpressed Csk tyrosine kinase is localized in focal adhesions, causes reorganization of alpha v beta 5 integrin, and interferes with HeLa cell spreading. *Mol. Cell. Biol.* **15,** 711–722.

Bourgeois, S., and Pfahl, M. (1976). Repressors. *Adv. Protein Chem.* **30,** 1–99.

Cayrol, C., and Flemington, E. K. (1995). Identification of cellular target genes of the Epstein-Barr virus transactivator Zta: Activation of transforming growth factor beta igh3 (TGF-beta igh3) and TGF-beta 1. *J. Virol.* **69,** 4206–4212.

Chrast-Balz, J., and van Huijsduijnen, R. H. (1996). Bi-directional gene switching with the tetracycline repressor and a novel tetracycline antagonist. *Nucleic Acids Res.* **24,** 2900–2904.

Copeland, R. A., David, J. P., Dowling, R. L., Lombardo, D., Murphy, K. B., and Patterson, T. A. (1995). Recombinant human dihydrotate dehydrogenase: Expression, purification, and characterization of a catalytically functional truncated enzyme. *Arch. Biochem. Biophys.* **323,** 79–86.

Corces, V., and Pellicer, A. (1984). Identification of sequences involved in the transcriptional control of a *Drosophila* heat-shock gene. *J. Biol. Chem.* **259,** 14812–14817.

Cox, M. E., and Maness, P. F. (1991). Neurite extension and protein tyrosine phosphorylation elicited by inducible expression of the v-src oncogene in a PC12 cell line. *Exp. Cell Res.* **195,** 423–431.

Cress, W. D., and Triezenberg, S. J. (1991). Critical structural elements of the VP16 transcriptional activation domain. *Science* **251,** 87–90.

Damke, H., Baba, T., Warnock, D. E., and Schmid, S. L. (1994). Induction of mutant dynamin specifically blocks endocytic coated vesicle formation. *J. Cell Biol.* **127,** 915–934.

Fieck, A., Wyborski, D. L., and Short, J. M. (1992). Modifications of the *E. coli lac* repressor for expression in eukaryotic cells. Effect of nuclear signal sequences on protein activity and nuclear accumulation. *Nucleic Acids Res.* **90,** 1785–1791.

Forman, B. M., Goode, E., Chen, J., Oro, A. E., Bradley, D. J., Perlmann, T., Noonan, D. J., Burka, L. T., McMorris, T., Lamph, W. W., Evans, R. M., and Weinberger, C. (1995). Identification of a nuclear receptor that is activated by farnesol metabolites. *Cell (Cambridge, Mass.)* **81,** 687–693.

Fruh, K., Gossen, M., Wang, K., Bujard, H., Peterson, P. A., and Yang, Y. (1994). Displacement of housekeeping proteasome subunits by MHC-encoded LMPs: A newly discovered mechanism for modulating the multicatalytic proteinase complex. *EMBO J.* **13,** 3236–3244.

Fruh, K., Ahn, K., Djaballah, H., Sempe, P., van Endert, P. M., Tampe, R., Peterson, P. A., and Yang, Y. (1995). A viral inhibitor of peptide transporters for antigen presentation. *Nature (London)* **3755,** 415–418.

Fuerst, T. R., Fernandez, M. P., and Moss, B. (1989). Transfer of the inducible lac repressor/operator system from *Eschericia coli* to a vaccinia virus expression vector. *Proc. Natl. Acad. Sci. U.S.A.* **86,** 2549–2953.

Furth, P. A., St. Onge, L., Boger, H., Gruss, P., Gossen, M., Kistner, A., Bujard, H., and Hennighausen, L. (1994). Temporal control of gene expression in transgenic mice by a tetracycline-responsive promoter. *Proc. Natl. Acad. Sci. U.S.A.* **91,** 9302–9306.

Gatz, C. (1996). Chemically inducible promoters in transgenic plants. *Bio/Technology* **7,** 168–172.

Gossen, M., and Bujard, H. (1992). Tight control of gene expression in mammalian cells by tetracycline-responsive promoters. *Proc. Natl. Acad. Sci. U.S.A.* **89,** 5547–5551.

Gossen, M., and Bujard, H. (1993). Anhydrotetracycline, a novel effector for tetracycline controlled gene expression systems in eukaryotic cells. *Nucleic Acids Res.* **21,** 4411–4412.

Gossen, M., Bonin, A. L., and Bujard, H. (1993). Control of gene activity in higher eukaryotic cells by prokaryotic regulatory elements. *Trends Biochem. Sci.* **18,** 471–475.

Gossen, M., Freundlieb, S., Bender, G., Müller, G., Hillen, W., and Bujard, H. (1995). Transcriptional activation by tetracyclines in mammalian cells. *Science* **268,** 1766–1769.

Ho, S. N., Biggar, S. R., Spencer, D. M., Schreiber, S. L., and Crabtree, G. R. (1996). Dimeric ligands define a role for transcriptional activation domains. *Nature (London)* **382,** 822–826.

Hoffmann, A., Nolan, G. P., and Blau, H. M. (1996). Rapid retroviral delivery of tetracycline-inducible genes in single autoregulatory cassette. *Proc. Natl. Acad. Sci. U.S.A.* **93,** 5185–5190.

Hoffmann, A., Villalba, M., Journot, L., and Spengler, D. (1997). A novel tetracycline-dependent expression vector with low basal expression and potent regulatory properties in various mammalian cell lines. *Nucleic Acids Res.* **25,** 1078–1079.

Howe, J. R., Skryabin, B. V., Belcher, S. M., Zerillo, C. A., and Schmauss, C. (1995). The responsiveness of a tetracycline-sensitive expression system differs in different cell lines. *J. Biol. Chem.* **270,** 14168–14174.

Hu, M. C., and Davidson, N. (1987). The inducible lac operator-repressor system is functional in mammalian cells. *Cell (Cambridge, Mass.)* **48,** 555–566.

Hu, M. C., and Davidson, N. (1988). The inducible lac operator-repressor system is functional for the control of expression of injected DNA in Xenopus oocytes. *Gene* **62,** 301–313.

Hu, M. C., and Davidson, N. (1990). A combination of derepression of the *lac* operator-repressor system with positive induction by glucocorticoid and metal ions provides a high-level-inducible gene expression system based on the human metallothionein IIA promoter. *Mol. Cell. Biol.* **10,** 6141–6151.

Iida, A., Chen, S. T., Friedmann, T., and Yee, J. K. (1996). Inducible gene expression by retrovirus-mediated transfer of a modified tetracycline-regulated system. *J. Virol.* **70,** 6054–6059.

Israel, D. I., and Kaufman, R. J. (1989). Highly inducible expression from vectors containing multiple GRES in CHO cells overexpressing the glucocorticoid receptor. *Nucleic Acids Res.* **12,** 4589–4604.

Jacobsen, K. D., and Willumsen, B. M. (1995). Kinetics of expression of inducible beta-galactosidase in murine fibroblasts: High initial rate compared to steady-state expression. *J. Mol. Biol.* **252,** 289–295.

Kamashev, D. E., Esipova, N. G., Ebralidse, K. K., and Mirzabekov, A. D. (1995). Mechanism of Lac repressor switch-off: Orientation of the Lac repressor DNA-binding domain is reversed upon inducer binding. *FEBS Lett.* **375,** 27–30.

Ko, M. S. H., Sugiyama, N., and Takano, T. (1989). An auto-inducible vector conferring high glucocorticoid inducibility upon stable transformant cells. *Gene* **84,** 383–389.

Liberles, S. D., Diver, S. D., Austin, D. J., and Schreiber, S. L. (1997). Inducible gene expression and protein translocation using non toxic ligands identified by a mammalian three-hybrid screen. *Proc. Natl. Acad. Sci. U.S.A.* **94,** 7825–7830.

Liu, H. S., Scrable, H., Villaret, D. B., Lieberman, M. A., and Stambrook, P. J. (1992). Control of Ha-ras-mediated mammalian cell transformation by *Escherichia coli* regulatory elements. *Cancer Res.* **52,** 983–989.

Needham, M., Gooding, C., Hudson, K., Antoniou, M., Grosveld, F., and Hollis, M. (1992). LCR/MEL: A versatile system for high-level expression of heterologous proteins in erythroid cells. *Nucleic Acids Res.* **20,** 997–1003.

No, D., Yao, T.-P., and Evans, R. M. (1996). Ecdysone-inducible gene expression in mammalian cells and transgenic mice. *Proc. Natl. Acad. Sci. U.S.A.* **93,** 3346–3351.

Planas-Silva, D., and Means, A. R. (1992). Expression of a constitutive form of calcium/calmodulin dependent protein kinase II leads to arrest of the cell cycle in G2. *EMBO J.* **11,** 507–517.

Ratnam, S., and Kent, C. (1995). Early increase in choline kinase activity upon induction of the H-ras oncogene in mouse fibroblast cell lines. *Arch. Biochem. Biophys.* **323,** 313–322.

Resnitzky, D., Gossen, M., Bujard, H., and Reed, S. I. (1994). Acceleration of the G1/S phase transition by expression of cyclins D1 and E with an inducible system. *Mol. Cell. Biol.* **14,** 1669–1679.

Rieping, M., Fritz, M., Prat, S., and Gatz, C. (1994). A dominant negative mutant of PG13 suppresses transcription from a cauliflower mosaic virus 35S truncated promoter in transgenic tobacco plants. *Plant Cell* **6,** 1087–1098.

Röder, F. T., Schumülling, T., and Gatz, C. (1994). Efficiency of the tetracycline dependent gene expression system—complete suppression and efficient induction of the RolB phenotype in transgenic plants. *Mol. Gen. Genet.* **243,** 32–38.

Sadowski, I., Ma, J., Triezenberg, S., and Ptashne, M. (1988). GA14-VP16 is an unusually potent transcriptional activator. *Nature (London)* **335,** 563–564.

Schultze, N., Burki, Y., Lang, Y., Certa, U., and Bluethmann, H. (1996). Efficient control of gene expression by single step integration of the tetracycline system in transgenic mice. *Nat. Biol.* **14,** 499–503.

Schweinfest, C. W., Jorcyk, C. L., Fujiwara, S., and Papas, T. S. (1988). A heat shock-inducible eukaryotic expression vector. *Gene* **71,** 207–210.

Shan, B., and Lee, W. H. (1994). Deregulated expression of E2F-1 induces S-phase entry and leads to apoptosis. *Mol. Cell. Biol.* **14,** 8166–8173.

Shockett, P., Difilippantonio, M., Hellman, N., and Schatz, D. G. (1995). A modified tetracycline-regulated system provides autoregulatory, inducible gene expression in cultured cells and transgenic mice. *Proc. Natl. Acad. Sci. U.S.A.* **92,** 6522–6526.

Shockett, P. E., and Schatz, D. G. (1996). Diverse strategies for tetracycline-regulated inducible gene expression. *Proc. Natl. Acad. Sci. U.S.A.* **93,** 5173–5176.

St.Onge, L., Furth, P. A., and Gruss, P. (1996). Temporal control of the Cre recombinase in transgenic mice by a tetracycline responsive promoter. *Nucleic Acids Res.* **24,** 3875–3877.

Suzuki, K., and Watanabe, M. (1994). Modulation of cell growth and mutation induction by introduction of the expression vector of human hsp70 gene. *Exp. Cell Res.* **215,** 75–81.

Tasken, K., Andersson, K. B., Erikstein, B. K., Hansson, V., Jahnsen, T., and Blomhoff, H. K. (1994). Regulation of growth in a neoplastic B cell line by transfected subunits of 3′,5′-cyclic adenosine monophosphate-dependent protein kinase. *Endocrinology (Baltimore)* **135,** 2109–2119.

Totzke, F., Marmé, D., and Hug, H. (1992). Inducible expression of human phospholipase C-γ2 and its activation by platelet-derived growth factor A-chain homodimer in transfected NIH3T3 fibroblasts. *Eur. J. Biochem.* **203,** 633–639.

Triezenberg, S. J., Kingsbury, R. C., and McKnight, S. L. (1988a). Functional Dissection of VP16, the trans-activator of herpes simplex virus immediate early gene expression. *Genes Dev.* **2,** 718–729.

Triezenberg, S. J., LaMarco, K. L., and McKnight, S. L. (1988b). Evidence of DNA: Protein interactions that mediate HSV-1 immediate early gene activation by VP16. *Genes Dev.* **2,** 730–742.

Umesono, K., and Evans, R. M. (1989). Determinants of target gene specificity for steroid/thyroid hormone receptors. *Cell (Cambridge, Mass.)* **57,** 1139–1146.

Wang, Y., DeMayo, F. J., Tsai, S. Y., and O'Malley, B. W. (1997). Ligand-inducible and liver-specific target gene expression in transgenic mice. *Nat. Biotechnol.* **15,** 239–243.

Wimmel, A., Lucibello, F. C., Sweing, A., Adolph, S., and Muller, R. (1994). Inducible acceleration of G1 progression through tetracycline-regulated expression of human cyclin E. *Oncogene* **9,** 995–997.

Wyborski, D. L., and Short, J. M. (1991). Analysis of inducers of the *E. coli lac* repressor system in mammalian cells and whole animals. *Nucleic Acids Res.* **19,** 4647–4653.

Yao, T.-P., Segraves, W. Z., Oro, A. E., McKeown, M., and Evans, R. M. (1992). *Drosophila ultraspiracle* modulates ecdysone receptor function via heterodimer formation. *Cell (Cambridge, Mass.)* **71,** 63–72.

Yao, T.-P., Forman, B. M., Jiang, Z., Cherbas, L., Chen, J.-D., McKeown, M., Cherbas, P., and Evans, R. M. (1993). Functional ecdysone receptor is the product of *EcR* and *ultraspiracle* genes. *Nature (London)* **366,** 476–479.

Yarranton, G. T. (1992). Inducible vectors for expression in mammalian cells. *Bio/Technology* **3,** 506–511.

Yin, D. X., and Schimke, R. T. (1995). BCL-2 expression delays drug-induced apoptosis but does not increase clonogenic survival after drug treatment in HeLa cells. *Cancer Res.* **55, 4922–4928.**

Zambetti, G. P., Olson, D., Labow, M., and Levine, A. J. (1992). A mutant p53 protein is required for maintenance of the transformed phenotype in cells transformed with p53 plus ras cDNAs. *Proc. Natl. Acad. Sci. U.S.A.* **89,** 3952–3956.

10

PROTEIN EXPRESSION IN MAMMALIAN CELLS USING SINDBIS VIRUS

Robert P. Bennett

Research and Development, Invitrogen Corporation, Carlsbad, California 92008

Gene Expression Systems: Using Nature for the Art of Expression

Introduction

Viruses have provided many of the tools used by molecular biologists for the expression of proteins in heterologous systems. The relatively small and simple genomes of viruses have led to their obligate requirement of intracellular growth and dependence on host cell components for replication. The viruses of mammalian hosts are useful for heterologous gene expression in higher eukaryotes as viruses and viral components have evolved to function most efficiently in these cell types. Two alphaviruses, Sindbis (SIN) and Semliki Forest virus (SFV), have been described for expression of heterologous proteins in mammalian cells (Schlesinger, 1993; Olkkonen *et al.*, 1994; Piper *et al.*, 1994; Frolov *et al.*, 1996). These viruses have a simple replication pathway and a broad host range that allows very high levels of heterologous protein expression in a variety of eukaryotic cell types.

Virus and Genome Structure

All alphaviruses are composed of a nucleocapsid surrounded by a lipid envelope that is derived from the host plasma membrane. The Sindbis virus nucleocapsid is composed of 240 copies of the capsid protein, which are in close contact with the RNA genome. The lipid envelope contains two glycoproteins, E1 and E2, that are required for binding and entry of the virus into uninfected cells. The E1 and E2 glycoproteins form heterodimers that are arranged as trimers on the virus surface.

The Sindbis nucleocapsid contains a single-stranded RNA genome of positive polarity that is capped with 7-methylguanosine and polyadenylated. The Sindbis genome is 11,703 nucleotides in length and can be divided into two regions (Fig. 1A). The first two-thirds of the genome encode the nonstructural or enzymatic proteins of the virus and the last third encodes the structural proteins of the virus. Located between the nonstructural and the structural open reading frames (ORFs) is a junction sequence or subgenomic promoter that is responsible for expression of the structural proteins. The RNA also contains sequences required for packaging of the RNA genome into newly formed virus and sequences at the 5′ and 3′ ends for replication of the genome.

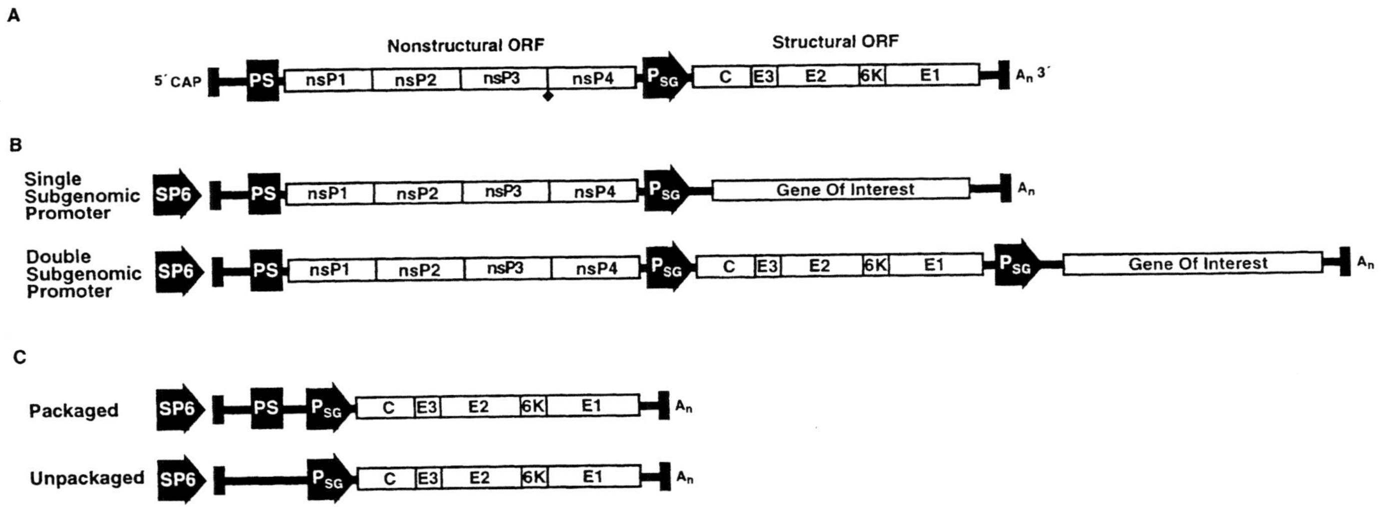

Figure 1 Sindbis genome and expression constructs. (A) Sindbis genome. The Sindbis virus genome is a capped (CAP), polyadenylated (A_n) RNA molecule. The nonstructural ORF encodes four nonstructural proteins of the virus: nsP1, nsP2, nsP3, and nsP4. An opal codon (diamond) separates the nsP3- and nsP4-coding sequences. The structural ORF encodes the structural proteins of the virus: the capsid protein (C) and the glycoproteins (E3, E2, 6K, and E1). The packaging signal (PS) permits encapsidation of the genomic RNA into a newly formed virus. The subgenomic promoter (P_{SG}) directs the synthesis of the subgenomic RNA. (B) Expression constructs. Constructs contain the SP6 promoter (SP6) for the generation of RNA transcripts. In the single subgenomic promoter construct, the structural ORF is replaced with the heterologous gene. The double subgenomic promoter construct encodes nonstructural and structural ORFs and includes an additional P_{SG} for expression of the heterologous gene. (C) Defective helper constructs. Constructs contain the SP6 promoter (SP6) for the generation of RNA transcripts. The defective helper constructs encode only the structural proteins of the virus. These may (packaged) or may not (unpackaged) contain the PS.

Life Cycle

The infection process is initiated when the Sindbis virus binds an uninfected cell via the E1 and E2 glycoproteins and becomes internalized by clathrin-coated vesicles. These vesicles become acidified, promoting fusion of the viral envelope with the vesicle membrane and release of the nucleocapsid into the host cell cytoplasm. Upon introduction into the cell, the RNA genome becomes uncoated and is translated to produce the nonstructural proteins. During the first 3–4 hr of infection, the nonstructural proteins are synthesized as a polyprotein and are processed into two protein species, nsP1–3 and nsP4, that function as the minus-strand replicase (Fig. 2). As the levels of nonstructural proteins and minus-strand

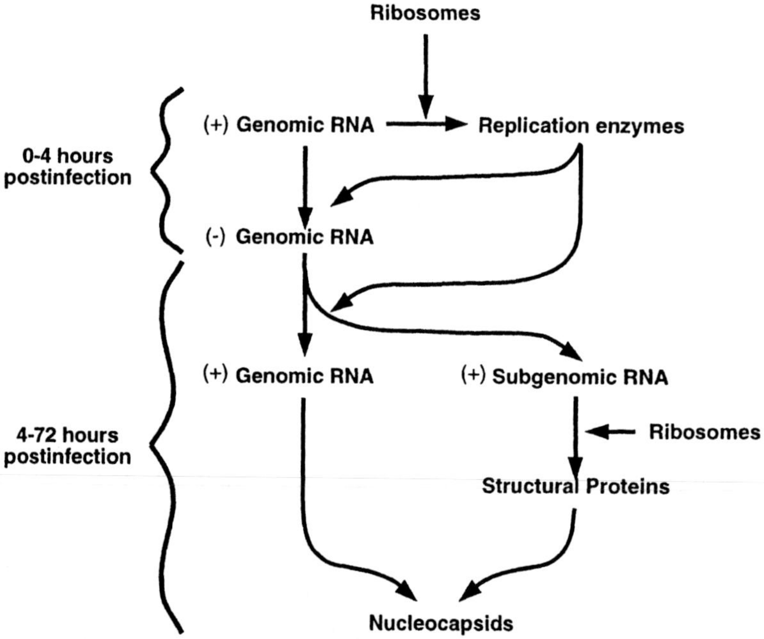

Figure 2 Transcription and replication of sindbis virus. The genomic RNA is a plus-strand (+) RNA and in the cytoplasm is translated to produce replication enzymes. These enzymes direct the synthesis of negative-strand (–) RNA templates for the production of new (+) genomic RNA molecules during the first 4 hr of infection. After the early phase, enzymes direct the synthesis of more (+) genomic RNA and the (+) subgenomic RNA from the (–) genomic RNA. Translation of the subgenomic RNA results in the production of structural proteins that interact with the (+) genomic RNA to form the nucleocapsids.

RNA copies accumulate in the cytoplasm, the nsP1–3 polyprotein is further processed into the mature nonstructural proteins (nsP1, nsP2, nsP3), which together with nsP4 function as the plus-strand RNA replicase. Synthesis of genomic RNA from the minus-strand templates by the plus-strand replicase continues throughout the rest of the infection cycle. The positive-strand replicase also recognizes the subgenomic promoter to produce the shorter RNA transcripts that encode the structural proteins of the virus. The subgenomic RNAs accumulate to very high levels in the cytoplasm [3×10^5 molecules/cell (Frolov *et al.*, 1996)], leading to high-level expression of the structural proteins.

Structural proteins are synthesized initially as a polyprotein and are then processed by a variety of proteases on the way to the plasma membrane, the site of virus assembly. The first cleavage occurs cotranslationally by a protease activity intrinsic to the capsid protein, releasing the capsid protein from the peptide emerging from the ribosome. This cleavage reveals a hydrophobic signal sequence that directs the translating ribosome to the endoplasmic reticulum where the remainder of the peptide is synthesized and anchored in the internal membrane. PE2, a precursor of E3 and E2, and E1 are separated by signal peptidases while in the ER and are then transported as a heterodimer through the Golgi apparatus to the plasma membrane. Processing of PE2 into E3 and E2 occurs in the Golgi apparatus by a furin-like protease. RNA genomes complexed with the capsid protein interact with the tail of the E2 glycoprotein at the plasma membrane to facilitate budding and release of the enveloped virus particle (Lopez *et al.*, 1994). The release of virus particles continues until apoptotic cell death (Levine *et al.*, 1996), approximately 3–4 days postinfection *in vitro*, depending on the type of cell line used.

Host Range

Sindbis virus has a diverse host range both in nature and in the laboratory. In nature, Sindbis generally replicates in vertebrate hosts such as birds or mammals and is transmitted by mosquitoes. In the laboratory, Sindbis has been shown to infect cell lines derived from mammals, birds, reptiles, amphibians, and insects, such as mosquitoes and *Drosophila* (Xiong *et al.*, 1989). In general, infection of insect cells by Sindbis results in a persistent infection with much lower levels of expression than in the lytic infection found in vertebrates. Although the receptors for Sindbis virus in many cell types have not been characterized, it appears that the high-affinity laminin receptor is necessary for the infection of mammalian cells (38). However, infection of each cell line is best determined empirically as infection can be affected by the availability of the cellular receptor or by the ability of the virus to replicate intracellularly. The efficiency of infection may also be influenced by the glycoproteins expressed on the surface of the Sindbis virus. For example, virus containing the glycoproteins of a neurovirulent strain of Sindbis, TE12 (Lustig *et al.*, 1988), infect cells of murine and human cells more efficiently than those derived from the standard laboratory strain, Toto1101 (Bredenbeek *et al.*, 1993).

Sindbis Virus Vector System

Several characteristics of the Sindbis life cycle, including its simple RNA genome, mode of replication, and broad host range, contribute to its effectiveness as an expression system. The first examples of using Sindbis virus as an expression system utilized cDNAs of defective interfering (DI) genomes containing sequences encoding chloromphenicol acetyltransferase (CAT) (Levis *et al.*, 1987) or a complete genome in which the structural genes were replaced with the CAT gene (Xiong *et al.*, 1989). Replacement of the structural genes with the CAT gene resulted in the production of 1×10^8 CAT polypeptides per transfected cell. In each case, the bacteriophage SP6 promoter was used to generate *in vitro* RNA transcripts, which were then transfected into cells competent for Sindbis replication.

In general, two types of vectors have been developed using the cDNAs of an infectious Sindbis genome (Fig. 1B). Each of these vectors utilizes a bacteriophage promoter (e.g., SP6) to generate (+)-stranded RNA molecules that can be introduced into cells for the expression of heterologous genes. A single subgenomic vector substitutes the heterologous gene for the structural gene and places the heterologous gene under control of the subgenomic promoter. Recombinant RNAs generated using this type of vector are self-replicating once introduced into cells, but cannot produce additional virus and spread throughout the culture. A second type of vector containing two subgenomic promoters has also been used to express heterologous proteins. In this vector, one subgenomic promoter is used to express the structural proteins of the virus and the second is used to express the heterologous gene. Introduction of this recombinant RNA into cells results in heterologous expression and the concomitant production of recombinant virus. Because virus infection is such an efficient method to introduce the recombinant genomes, the infection spreads throughout the entire culture, ensuring high levels of protein expression.

The alphavirus vectors containing a single subgenomic promoter can also be packaged into recombinant virus particles using defective helper (DH) constructs (Fig. 1C). These vectors retain elements of the Sindbis genome required for replication of the RNA and expression of the structural proteins. Therefore, cotransfection of a DH RNA with the recombinant RNA results in packaging of the recombinant RNA genome into virus particles. DH RNAs that contain the sequences required for RNA encapsidation allow the DH RNA and the recombinant RNA to be packaged into virus particles. DH constructs that do not contain the encapsidation sequence are not packaged efficiently; therefore, the resulting virus particles are useful for a single round of infection.

Current commercially available expression systems utilizing the alphavirus technology, the Sindbis virus expression system (Invitrogen, Carlsbad, CA) and the Semliki Forest virus expression system (GIBCO BRL, Gaithersburg, MD), rely on the use of a replicon RNA and a defective helper to produce virions that can undergo a single round of infection for expression of a particular gene. A typical protocol is outlined in Fig. 3. The desired gene is cloned into the vector downstream of the subgenomic promoter and the vector linearized by restriction digest. RNA transcripts are then synthesized by *in vitro* transcription of the DNA template using SP6 RNA polymerase. RNA transcripts are capped by including a cap analogue (in addition to the nucleotides) during the synthesis reaction and are polyadenylated due to a stretch of thymidine residues encoded by the vector sequence. These RNA transcripts are then transfected into a cell line that supports alphavirus replication. Typically, BHK cells are used as alphavirus replication is very efficient in this cell line. Heterologous protein expression can then be detected 4–72 hr after infection. To produce recombinant virus, recombinant RNA transcripts can be cotransfected with RNA transcripts synthesized from DH vectors. After 48 hr posttransfection, media containing virus particles can be collected and used directly to infect a variety of cell lines. In general, particles are incubated with the cell line for 1 hr in a small volume and then cell culture media is added without removing the virus. Heterologous protein expression using the

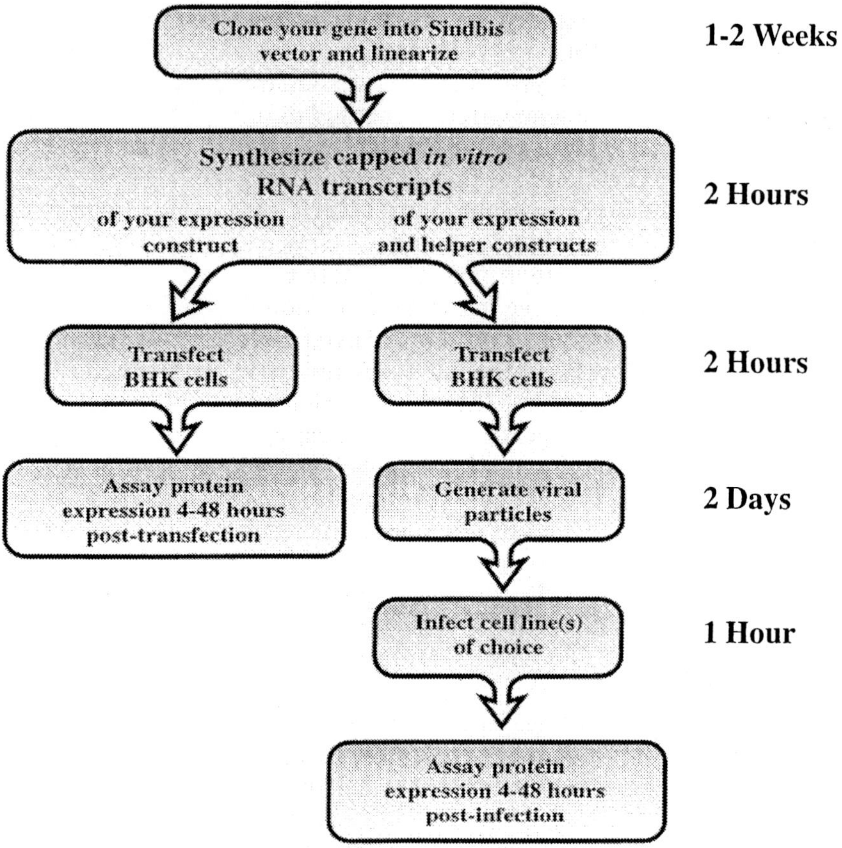

Figure 3 Heterologous expression using the Sindbis virus. Recombinant Sindbis RNAs can be electroporated into cells for expression (left column) or cotransfected with a helper construct for the generation of virus particles (right column). Recombinant particles can then be used to infect a variety of cell lines.

virus can be detected as early as 4 hr postinfection and typically for up to 48–72 hr (until cell death occurs). A detailed protocol for the electroporation of BHK cells and the production of recombinant particles can be found in Liljeström and Garoff (1995).

Sindbis expression vectors have also been developed that use an RNA polymerase (Pol) II promoter (e.g., CMV or MoMLV LTR) to drive expression of the replicative Sindbis genome (Driver *et al.*, 1995; Herweijer *et al.*, 1995; Dubensky *et al.*, 1996). These DNA vectors are introduced into cells that are permissive for Sindbis replication by typical DNA transfection protocols. Synthesis of the Sindbis genomic RNA results in the self-amplification of the RNA and then proceeds according to the Sindbis life cycle. The CMV promoter-based DNA vector produces similar levels of heterologous protein compared to the RNA-based systems on a per cell basis and approximately 10-fold more than conventional DNA plasmids utilizing the same RNA Pol II promoters (Dubensky *et al.*, 1996). However, overall production levels of the protein being expressed are hampered by the lower transfection efficiencies associated with DNA transfections compared to the RNA-based system.

Applications

The Sindbis expression system has a variety of characteristics that make it an extraordinary tool for the expression of heterologous proteins. Introduction of the RNA genome results in amplification of the RNA in the cytoplasm, which leads to high levels of expression. In addition, because replication takes place in the cytoplasm, problems associated with conventional DNA expression systems, such as splicing and mRNA transport, are eliminated. Expression of CAT in chicken embryo fibroblasts reached 3% of the total cell protein (Xiong *et al.*, 1989). High-level expression is useful for structure–function studies of proteins that are normally expressed at very low levels, such as the glucose transporter proteins (GLUT–1 and GLUT–4) and Rab proteins (Piper *et al.*, 1994).

Infection of cells with Sindbis results in the overexpression of the heterologous protein, but is accompanied with changes in the host cells, such as inhibition of host protein synthesis, cytopathic effects, and eventual cell death. Inhibition of *de novo* host protein synthesis during Sindbis virus infection occurs as early as 4–6 hr postinfection and has been postulated to be caused by the high levels of subgenomic RNAs synthesized and/or the synthesis of nonstructural proteins (Frolov and Schlesinger, 1994). Therefore, studies concerned with the cell biology of a particular protein should be performed shortly after infection when expression is not too high and the cytopathic effects on the host cell have not begun.

Cell Biology

Alphavirus expression systems have been used to express a variety of proteins (Table 1). As mentioned earlier, GLUT–1, GLUT–4, and GLUT 1/4 chimeras were expressed in CHO cells using recombinant Sindbis virus (Piper *et al.*, 1992; 1993a,b, 1994). These studies revealed a sequence in GLUT–4 that is responsible for the intracellular sequestration and insulin-stimulated translocation to the plasma membrane. The structure–function relationship of Rab5, a small GTPase involved in endocytosis, has also been studied extensively using the Sindbis expression system (Li and Stahl, 1993a,b; Li *et al.*, 1994). Expression of a variety of mutants identified several dominant-negative mutations and revealed that isoprenylation is required for the proper interaction of Rab5 with the endosomal membranes (Li *et al.*, 1995). Expression of another protein involved in endocytic trafficking, ARF6 (an ADP-ribosylation factor), revealed that myristylation is required for its intracellular targeting and function (D'Souza-Schorey and Stahl, 1995). These examples highlight the usefulness of Sindbis for structure–function studies of proteins in mammalian cells, especially when fatty acid modifications and other targeting signals are important.

Immunology

Sindbis virus has largely been used in immunology for the expression of peptides or proteins to study antigen processing and presentation. Cells expressing a truncated form of influenza hemagglutinin (HA) or two distinct immunodominant cytotoxic T lymphocyte (CTL) HA epitopes were recognized and lysed by HA-specific CTL clones in a major histocompatability complex (MHC) class I-restricted manner *in vitro* (Hahn *et al.*, 1992). Expression of immunodominant peptides using Sindbis virus has also facilitated the identification of a peptide epitope that allows the murine acquired immunodeficiency virus to evade CTLs of certain mice (Coppola *et al.*, 1995). Finally, infection of a recombinant Sindbis virus allowed the identification of the rubella capsid protein as a target of the MHC class I in lymphocytes from rubella virus-infected individuals (Lovett *et al.*, 1993).

Table 1
Heterologous Proteins Expressed Using Alphaviruses

Expressed protein	Cell type(s)	Reference
Proteins expressed using SIN		
β-Galactosidase	BHK	Bredenbeek, 1993
Luciferase	BHK, MDA-MB-435, MC38, SKOV3	Johanning, 1995
Chloramphenicol acetyltransferase	L929, P815, EL4 BHK	Hahn, 1992 Xiong, 1989
ARF6	TRVb-1 (CHO)	D'Souza, 1995
Rab4 and Rab5	BHK, CEF	Li, 1993a; Li, 1993b; Li, 1994
GLUT-4	CHO	Piper, 1993a,b; Piper, 1994
scFv to flavivirus glycoprotein	Porcine kidney (PS) and baby hamster kidney (BSR)	Jiong, 1995
Rubella virus structural proteins	BHK	Chen, 1995
Influenza hemagglutinin and CTL epitopes	L929, P815	Hahn, 1992
Hepadnavirus	BHK	Huang, 1991
HCV glycoproteins	BHK, HepG2, and PK-15	Dubuisson, 1994
Flavirus proteins	BHK, C6/36	Pugachev, 1995
Proteins expressed using SFV		
Dopamine D_3 receptor	CHO	Lundström, 1996
Human neurokinin-1 receptor	BHK, COS, CHO, human osteosarcoma cells	Lundström, 1994; Lundström, 1995
Retinoblastoma (RB) protein	Rat L6 myotubes	Szekely, 1993
Amyloid precursor protein (AAP)	Rat hippocampal neurons	Simons, 1995
Rab8 and VIP21	Rat hippocampal neurons	Olkkonen, 1994
Sso2p (syntaxin)	BHK	Jäntti, 1994
Rab12, Rab22, and Rab24	BHK, MDCK, HeLa, NMuLi (murine liver)	Olkkonen, 1994
VSV G	BHK	Rolls, 1996
HPV capsid proteins	BHK	Heino, 1995

Virology

Recombinant Sindbis virus has also been used for the expression of proteins from a variety of mammalian viruses, including rubella (a Togavirus) (Chen *et al.*, 1995) and Japanese encephalitis virus (JEV; a Flavivirus) (Pugachev *et al.*, 1995). Expression of the JEV structural proteins, prM+E, resulted in the release of subviral particles; however, assembly of these particles was lower than with other expression systems. This was attributed to an effect of the Sindbis nonstructural proteins upon expression. Sindbis virus vectors have also been used to express the hepatitis C virus envelope glycoproteins, E1 and E2 (Dubuisson *et al.*, 1994). The entire pregenome of a hepadnavirus, duck hepatitis B virus, was expressed under control of the subgenomic promoter using a Sindbis vector (Huang and Summers, 1991). Synthesis of the pregenome in the cytoplasm resulted in reverse transcription of the pregenome into the DNA genome and assembly of infectious virions. Finally, the DNA-binding domain of a nonstructural herpesvirus protein, UL9, has been expressed using Sindbis and partially purified from BHK cells (Stabell and Olivo, 1993). In this example, Sindbis was chosen because it provided quantities of the protein in eukaryotic cells and the vector could be easily modified for future structure–function studies.

A double subgenomic Sindbis vector has also been used to express a single chain antibody (scFv) that recognizes a neutralizing epitope on the envelope protein of a flavivirus (Jiang *et al.*, 1995). Expression of the scFv using recombinant Sindbis virus in cells infected with the flavivirus interfered with the release of a new infectious virus. This "intracellular immunization" demonstrates that expression of an antibody can inactivate a target within the cytoplasm of infected cells. Intracellular immunity has also been observed in mosquito cells infected with La Crosse virus (a bunyavirus) when cells were superinfected with a recombinant Sindbis virus expressing an antisense RNA to one of the La Crosse mRNAs (Powers *et al.*, 1994).

In Vivo Expression and Immunization

Sindbis vectors have also been used to direct heterologous protein expression *in vivo*. Introduction of a Sindbis RNA encoding luciferase by direct intramuscular injection resulted in higher expression levels compared to a nonreplicative RNA and expression lasted for at least 14 days, more than twice as long as the nonreplicative RNA (Johanning *et al.*, 1995). Similarly, injection of DNA-based Sindbis vectors has been used for expression of a heterologous gene *in vivo* (Dubiusson *et al.*, 1994; Herweijer *et al.*, 1995; Dubensky *et al.*, 1996). One experiment has also demonstrated that recombinant Sindbis particles can be used for *in vivo* expression (Sallee, 1997). Nonreplicating Sindbis particles encoding the *lacZ* gene were inoculated into regions of mouse brain, resulting in gene expression in differentiated neurons for at least 14 days.

The use of Sindbis to study antigen presentation and for *in vivo* expression first suggested that recombinant Sindbis virus or Sindbis vectors could be used to elicit an immune response. One of the first demonstrations of this utilized a Sindbis virus with a neutralization epitope from Rift Valley fever virus (RVFV) inserted within the E2 glycoprotein (London *et al.*, 1992). Mice immunized with the E2 recombinant virus survived a challenge with a lethal dose of RVFV and survival was attributed to protection by serum antibodies. Introduction of Sindbis vectors has also been shown to induce immune responses. Sindbis DNA vectors encoding the HBe and HBc genes produce a humoral response in mice 10 days following injection with these vectors (Driver *et al.*, 1995). Injection of Semliki forest virus encoding the influenza NP gene resulted in the induction of humoral and CTL responses (Zhuo *et al.*, 1994). Although alphaviruses themselves are not currently vaccine candidates, the use of self-replicating RNA molecules may provide a novel technique for the induction of immune responses for therapeutic purposes.

Semliki Forest Virus

This chapter focuses on the use of the Sindbis virus for heterologous protein expression; however, it should be noted that another member of the alphavirus family, Semliki Forest virus, has also been used for protein expression (Table 1). In general, each of these viral systems provides the same benefits. However, because each virus encodes distinct glycoproteins that are responsible for the interactions between the virus and target cells, certain cell types may be infected more efficiently with SIN or SFV.

Protein Expression

β-Galactosidase

Sindbis virus vectors allow high-level expression in a variety of cell lines in a rapid fashion. To investigate the kinetics and overall levels of protein expression, *in vitro*-synthesized RNA transcripts from a single subgenomic promoter vector containing the β-galactosidase gene were used to transfect BHK cells. Electroporation was used to introduce the RNA transcripts as this method has been reported to result in higher efficiencies than protocols using lipids or DEAE-dextran (Liljeström and Garoff, 1991; Bredenbeek *et al.*, 1993). In fact, electroporation using the SinRep/lacZ RNA results in routine transfection efficiencies greater than 95% (Fig. 4). This is compared to DNA electroporations that generally result in transfection efficiencies between 5 and 40%. One reason for this difference may be that RNA replication occurs in the cytoplasm and does not require transport to the nucleus, which appears to be the limiting step in DNA transfections (Zabner *et al.*, 1995).

Recombinant Sindbis virus encoding the *lacZ* gene was generated by electroporating BHK cells with SinRep/lacZ RNA and the helper RNAs, DH-BB or DH(26S). The glycoproteins of DH-BB and DH(26S) are derived from a laboratory strain (Toto1101) and a neu-

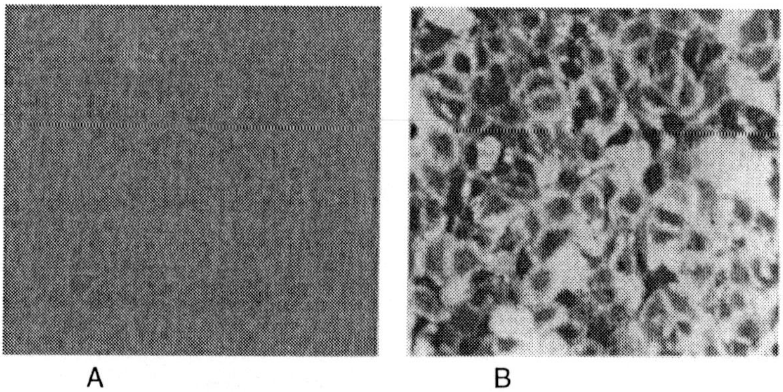

Figure 4 Staining of BHK cells transfected with SinRep/lacZ. Mock transfected cells (A) or cells transfected with 5 μg of SinRep/lacZ RNA using electroporation (B) were stained 48 hr posttransfection for b-galactosidase activity.

rovirulent strain (TE12) of Sindbis virus, respectively. Although both glycoproteins mediate the efficient infection of BHK cells, the DH(26S) helper infects cells of murine and human origin more efficiently. The titer of these virus preparations was determined by serial dilution and infection of BHK cells. The number of infected cells was then determined by staining for β-galactosidase activity and the virus titer was calculated. Typical titers are greater than 10^8 infectious units/ml.

The virus produced using the DH-BB helper was used to infect BHK cells at a multiplicity of infection (MOI) of 30 and the expression levels were monitored by SDS–PAGE and Coomasie blue staining (Fig. 5). Expression of β-galactosidase could be detected as early as 4 hr and continued to increase for 48 hr. Similar results

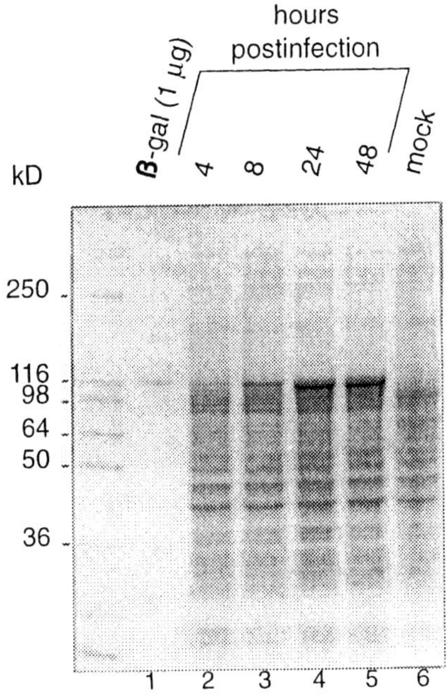

Figure 5 Expression of β-galactosidase in BHK cells. BHK cells (4×10^5) were infected with recombinant virus particles containing the SinRep/lacZ RNA at a MOI of 30. Cells were harvested at 4 (lane 2), 8 (lane 3), 24 (lane 4), and 48 (lane 5) hr postinfection. Mock infected cells were collected at 48 hr (lane 6). Cells were lysed by three freeze–thaw cycles and the postnuclear supernatant was mixed with sample buffer. One-fifth of each sample was separated on a 12% SDS–PAGE gel. Lane 1 contains 1 μg of purified β-galactosidase. Positions of molecular weight markers are indicated on the left.

were observed in BHK cells with virus made with the DH(26S) helper construct. As expected, higher infection efficiencies were observed using the DH(26S) vector in murine and human cell lines, such as NIH 3T3 and HeLa, respectively.

The amount of time that expression can be observed before cell death is affected by both the cell type and the number of virus particles used for the infection (i.e., the MOI). To investigate the effect of MOI on heterologous protein expression, different amounts of SinRep/lacZ virus were used to infect BHK cells. Twenty-four hours posttransfection the monolayers were labeled with [^{35}S]methionine and the proteins separated by SDS–PAGE. An MOI of greater than 0.5 was required for high-level expression and a higher MOI did not appear to increase the levels of protein (Fig. 6). However, infection with a higher MOI did result in the rapid onset of cytopathic effects and cell death, decreasing the overall number of cells expressing protein.

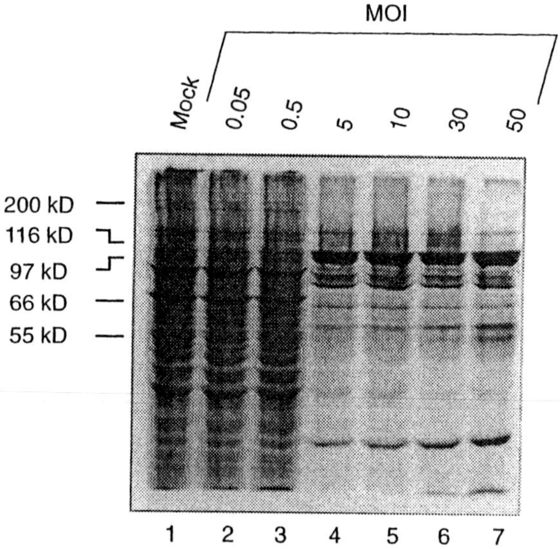

Figure 6 Effect of multiplicity of infection on β-galactosidase expression. BHK cells (4×10^5) were infected with recombinant virus particles containing the SinRep/lacZ RNA at varying MOI. Twenty hours postinfection the cells were labeled metabolically with [^{35}S]methionine for 10 min, followed by a 10-min chase in media without radiolabel. Cells were washed and collected in PBS and then solubilized directly in sample buffer. The proteins were separated on a 12% SDS–PAGE gel and visualized by autoradiography. Positions of the molecular weight markers are indicated on the left.

Expression using the Sindbis virus vectors also results in the inhibition of *de novo* host protein synthesis. This inhibition was not seen in cells infected with virus at a MOI of less than 5, probably due to protein expression from uninfected cells. Metabolic labeling of cells infected with the SinRep/lacZ virus at different times throughout the infection process revealed that inhibition of host protein synthesis begins as early as 4 hr and continues to increase for approximately 12 hr (data not shown). Because the expression of heterologous genes from the Sindbis vectors is not affected by the shutoff of host protein synthesis, metabolic labeling of cells late in the infection process can be used to identify the expressed protein without the use of specific antisera or activity assays.

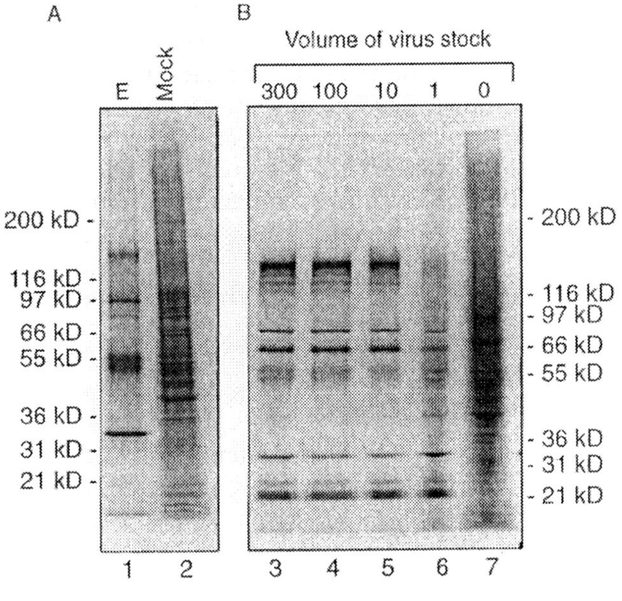

Figure 7 Expression of HIV gp160. (A) Electroporation of BHK cells. BHK cells (5×10^6) were electroporated (E) with approximately 5 μg each of SinRep/gp160 and DH(26S) RNAs. Twenty-four hours post-transfection the cells were labeled metabolically as described in Fig. 6. Thirty-six hours postelectroporation, media covering the cells was collected. Low-speed centrifugation was used to remove any loose cells and debris from the virus stock. A single electroporation produced about 10 ml of virus stock. Positions of the molecular weight markers are indicated on the left. (B) Infection of BHK cells. BHK cells (4×10^5) were infected with varying volumes (μl) of the virus stock. Following infection, cells were labeled metabolically. Positions of the molecular weight markers are indicated on the right.

HIV gp160

The Sindbis virus expression system is useful in expressing transmembrane proteins as well as cytoplasmic proteins. To demonstrate this, the gene encoding the glycoprotein of human immunodeficiency virus, gp160, was cloned into SinRep5. This viral glycoprotein is initially synthesized as a precursor that is cleaved by a Golgi-associated enzyme to produce two components, gp120 and gp41. RNA transcripts of SinRep/gp160 and the DH(26S) helper were used to generate virus particles in BHK cells (Fig. 7). Five micrograms of each RNA transcript was electroporated into 5×10^6 cells and the media (10 ml) harvested 36 hr later. The cell monolayer used to produce the virus particles was then labeled metabolically and the proteins separated by SDS–PAGE. A band migrating at the correct size of the full-length precursor was observed easily due to the inhibition of cellular protein synthesis, as was the Sindbis capsid protein (approximately 34 kDa). Western blot analysis using antiserum to gp120 demonstrated that the precursor protein was being processed into gp120 and gp41 (data not shown). Coexpression of the HIV regulatory protein, Rev, was not required for expression of gp120 (as it is in many systems), as Sindbis replication and expression take place entirely within the cytoplasm.

Different amounts of the virus stock were then used to infect a monolayer of 5×10^5 BHK cells to determine empirically what dilution of virus was best for expression. Virus stock was used straight (300 μl) or diluted into phosphate-buffered saline (PBS) containing 1% fetal bovine serum (FBS) to a final volume of 300 μl. Twenty-four hours postinfection the cells were labeled metabolically and separated by SDS–PAGE. The analysis revealed that 10 μl of the virus stock was sufficient for expression of gp160 and for the inhibition of the cellular protein synthesis. Therefore, a single electroporation produced enough viral stock for 1000 transfections (assuming 5×10^5 cells/transfection).

Interleukin-6

Proteins secreted extracellularly can also be expressed using recombinant Sindbis vectors. The gene encoding interleukin(IL)-6 with its native signal/secretion sequence was expressed using pSinRep5. Cells were electroporated with RNA transcripts synthesized using pSinRep/IL–6 and then pulse-chased with [^{35}S]methionine 20 hr lat-

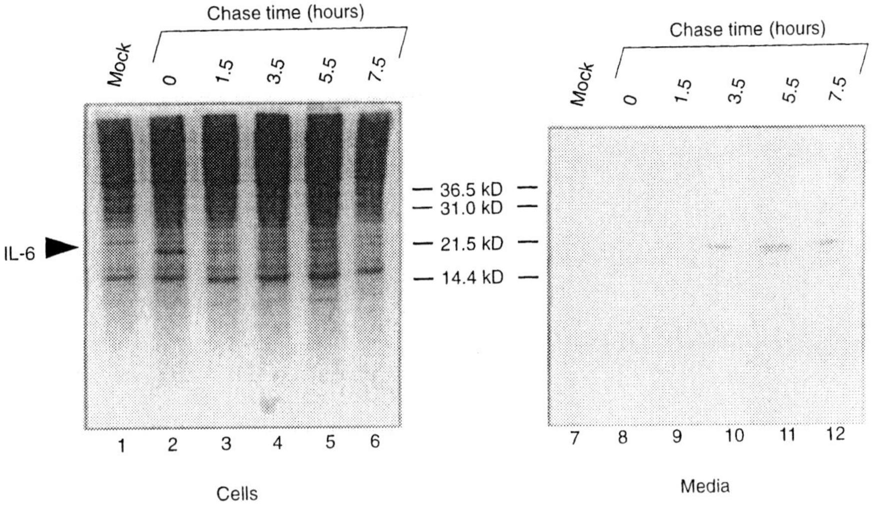

Figure 8 Pulse-chase of BHK cells expressing IL-6. BHK cells (5×10^6) were electroporated with approximately 5 μg of SinRep/IL-6 RNA and cells were plated at a density of 5×10^5 cells/60-mm plate. Twenty-four hours later, each plate was labeled with [^{35}S]methionine for 10 min. After the labeling period, cells were washed with PBS and incubated in regular media lacking FBS. At the indicated times, cells and media were collected. Cell pellets were solubilized directly in sample buffer. IL-6 protein secreted into media was immunoprecipitated with an antiserum reactive to IL-6. Proteins from cells and media were separated on a 7.5% SDS-PAGE gel and visualized by autoradiography.

er (Fig. 8). Cell lysates were examined directly using SDS–PAGE. Proteins in the media were immunoprecipitated with antiserum to IL–6 and then separated by SDS–PAGE. A protein band of the size expected for IL–6 could be observed at time zero in the cell lysates, but was absent at time points after that. The disappearance of this band is attributed to the secretion of IL–6 as this protein could be detected in the media after a 1.5-hr chase. These experiments demonstrate that native secretion signals and the secretion pathway are functioning normally, even 20 hr into the Sindbis replication cycle.

Purification with SinHis

The Sindbis expression system is an ideal method to produce recombinant proteins in mammalian cells for purification due to the high levels of heterologous gene expression and transfection efficiencies approaching 100%. To simplify the purification of proteins expressed using the Sindbis system, sequences encoding the Xpress epitope tag and a six histidine purification sequence were placed upstream of the MCS in pSinRep5. This plasmid, pSinHis, allows the production of fusion proteins that can be purified by Ni^{2+} affinity chromatography and detected with the Xpress antibody. An enterokinase cleavage site was also included to allow the tag to be removed from the expressed protein by proteolysis. To test this purification tool, the gene encoding β-galactosidase was cloned inframe with the purification sequence. The virus was prepared with SinHis/lacZ and the DH(26S) helper and was then used to infect an avian cell line, QT6. The expressed protein was collected 24 hr postinfection and purified using Ni^{2+} chromatography (Fig. 9). Fusion of the histidine tag with β-galactosidase allowed many of the contaminating proteins to be removed in a single purification step. The tag could then be cleaved from the purified protein using en-

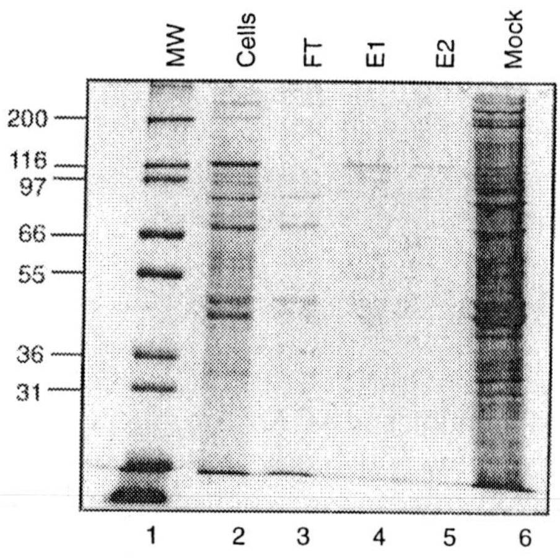

Figure 9 Expression and purification using SinHis/lacZ. Recombinant Sindbis virus containing the SinHis/lacZ RNA was used to infect 10^6 QT-6 cells at a MOI of 30. Twenty-four hours postinfection, cells were lysed by three freeze–thaw cycles in binding buffer (50 m*M* NaH_2PO_4, 300 m*M* NaCl, 20 m*M* imidazole, pH 8.0) and applied to NTA-resin (Probond) using a spin column format. Lysates of cells infected with SinHis/lacZ virus (lane 2) and mock infected (lane 6) are shown. Proteins that did not bind the resin are shown in the flow-through fraction (lane 3). Two successive incubations (lanes 4 and 5) with elution buffer (50 m*M* NaH_2PO_4, 300 m*M* NaCl, 250 m*M* imidazole, pH 8.0) were used to release the bound histidine-tagged β-galactosidase. Equal amounts of each fraction were analyzed by SDS–PAGE and Coomassie staining. Molecular weight markers are shown in lane 1 and their sizes indicated on the left.

terokinase (data not shown). The SinHis plasmid provides a simple way to purify protein expressed using the Sindbis system. The combination of the Sindbis with a simple purification tag facilitates the high-scale production of protein in mammalian cells. In addition, Sindbis virus can be used to infect suspension cells (e.g., CHO cells), increasing the ability to generate and purify large amounts of proteins from mammalian cells.

Comparisons with Other Systems

Many features of Sindbis virus expression can be considered an advantage or a disadvantage depending on the application desired (Table 2). High-level expression, high-transfection efficiencies, shutoff of *de novo* host protein synthesis, and a broad host range are ideal if the desired goal is the synthesis and identification of a gene product in tissue culture cells. The expression levels and high transfection efficiencies exhibited when using Sindbis each contribute to the large amounts of proteins synthesized. The shutoff of

Table 2
Advantages and Disadvantages of Sindbis Virus Expression

Advantages	Disadvantages
Self-replication of the RNA ensures high expression levels	High levels of expression may not be ideal for protein structure–function studies
	Expression is transient due to cell death
Transfection of RNA is more efficient than DNA transfections	RNA can be difficult to work with
RNA replication takes place in cytoplasm, overcomes problems associated with mRNA splicing and transport	Must express cDNAs; no RNA splicing
Construction of recombinant virus is simple	
Broad host range	
Infection of adherent and suspension cells	Large scale virus preparation required for large scale protein expression
Can be used for expression *in vivo*	
Shutoff of host protein synthesis facilitates identification of heterologous gene product	Protein structure–function studies need to be done early in infection to avoid problems associated with CPE
Mammalian posttranslation modifications	

de novo protein synthesis simplifies the identification of the protein product being expressed. However, the high levels of expression and the shutoff of host protein synthesis may not be ideal for protein structure–function studies, although these studies can be performed early in the infection cycle. Nevertheless, the ability to transfect nearly 100% of certain cell lines does provide advantages for the study of certain proteins. For example, the study of transdominant mutants is often hampered by low-transfection efficiencies, where the effects of a particular mutant may be difficult to observe when every cell is not expressing the protein product. Sindbis expression provides a method to ensure high-transfection efficiencies, allowing transdominant effects to be observed, especially if the characterization can be performed within the first few hours of infection.

Other eukaryotic systems are available to direct high levels of protein expression and are useful for the production and identification of gene products. In particular, baculovirus expression has been extremely popular. Indeed, baculovirus and Sindbis virus share a large number of attributes for the expression of heterologous proteins (Table 3). Both employ strong promoters for protein expression (usually the polyhedron promoter in baculovirus and the subgenomic promoter in Sindbis) and infection of cells with each system is very efficient. Furthermore, both systems can be used to infect cells in suspension, which is useful if large amounts of purified protein needs to be expressed and purified. One draw-

Table 3
Comparison of Sindbis Virus and Other Viral Expression Systems

	Sindbis virus	Retrovirus	Adenovirus	Baculovirus
Expression	Transient	Stable	Transient	Transient
Expression level	High	Depends on integration site	High	High
Construction	Plasmid construction	Requires use of cell line	Requires recombination	Requires recombination and plaque purification
Tropism	Wide	Depends on packaging cell line	Variable	Insect cells
Infection efficiency	High	Variable	Variable	High
Host protein shutoff	Yes	No	Yes	Yes

back of each system is that infection results in cell lysis; therefore, protein expression is always transient. Because cell lysis can complicate the purification of proteins due to the activation and release of proteases, protein purification should be performed in advance of cell death.

Sindbis has two main advantages over baculovirus as an expression tool. First, protein expression using the Sindbis virus is performed in mammalian cells, whereas baculovirus expression is confined to leptidopteran insect cell lines. Therefore, posttranslation modifications important for mammalian protein function are more accurate and complete using the Sindbis system than with baculovirus. Second, the construction of recombinant Sindbis virus is much simpler and faster than the baculovirus system. Recombinant Sindbis constructs are produced using simple plasmid-cloning techniques. In addition, high-titer stocks of recombinant virus are produced in a single transfection and, unlike baculovirus, do not require any plaque purification or further amplification steps.

A number of other mammalian viruses, such as retroviruses and adenoviruses, are now being used for heterologous protein expression. Sindbis virus expression has a number of advantages compared to these systems. As mentioned earlier, the construction of recombinant Sindbis virus is simple and fast, requiring just a single transfection to produce the high-titer stock. In addition, this system allows a large number of constructs to be analyzed simply by transfecting the RNA, without the production of virus. In contrast, expression with retroviruses or adenoviruses requires the production of recombinant virus. The construction of recombinant retroviruses requires the use of a packaging cell line, which provides the structural proteins of the virus in *trans* to package the recombinant genome. The generation of adenovirus requires an *in vivo* recombination event to produce virus due to the large size of the genome and the resulting virus typically must also be plaque purified.

Expression levels using Sindbis are typically higher than with the adenovirus or retroviral systems. Problems associated with limiting transcription factors, RNA splicing, and RNA transport are avoided as Sindbis encodes its own replication machinery and replicates entirely within the cytoplasm of infected cells. Finally, Sindbis has a very wide host range and can be used to express proteins in a variety of cell types, including primary and nondividing cells (e.g., neuronal cells). Adenoviruses, too, have a broad host range and can infect nondividing cells. In contrast, the host range of recombinant retroviruses is determined by the packaging cell line used, and currently available systems cannot infect nondividing cells.

A disadvantage of the Sindbis system is that expression is always transient as the outcome of infection is cell death. Similarly, expression with adenovirus vectors is transient as the adenoviral genome generally does not integrate into the host chromosome. In contrast, retoviruses efficiently integrate a cDNA of their genome into the host chromosome and, therefore, is an ideal system if stable protein expression is desired. Characteristics of each expression system are listed in Table 3.

New Directions and Conclusions

Sindbis virus expression would be useful for a wider variety of applications if the time of expression could be lengthened and if the cytopathic effects could be limited. The shut down of host cell protein synthesis and cell death have been attributed to the replication of the genomic RNA and/or the synthesis of nonstructural viral proteins (Frolov and Schlesinger, 1994). Sindbis mutants have been described that are able to maintain a persistent infection (Frolov *et al.*, 1996). Analysis of these mutants has demonstrated that the mutations are localized in the nsP2-coding region. Furthermore, Sindbis replicons containing these changes have no effect on host translation and are able to establish long-term expression. These new replicons should provide a way to combine long-term expression of heterologous genes with the simplicity of the Sindbis virus system.

The Sindbis virus has the potential to be very useful as an expression tool *in vivo*. Sindbis has a very broad host range because it recognizes the high-affinity laminin receptor; however, it may be more desirable to target specific cell types for expression *in vivo*. A chimeric E2 glycoprotein containing the IgG-binding domain of protein A has been shown to bind a monoclonal antibody specific for a cell surface marker that allowed infection of cell lines expressing that surface protein, but not in the absence of antibody (Ohno *et al.*, 1997). In addition, the chimeric glycoprotein had a much lower affinity for the natural receptor. This new development allows a variety of cells to be specifically targeted simply by changing the monoclonal antibody without the need to generate a new recombinant virus. This system could be used to deliver virus to a specific cell for protein expression or could be used to target specific cells for destruction as cells infected with the virus would be eventually eliminated.

DNA expression vectors encoding Sindbis replicons could also be used for protein expression *in vivo* or to target specific cells for death by engineering them with tissue-specific promoters. Introduction of this DNA into an animal would result in cell- or tissue-specific expression of the recombinant Sindbis RNA, resulting in protein expression and/or cell death. Clearly, the ability to target specific cells will further advance the use of Sindbis for vaccine and therapeutic means.

Sindbis virus is growing in popularity for heterologous protein expression in tissue culture cell lines and *in vivo*. The high levels of protein produced in infected cells and the high efficiency of infection contribute to the effectiveness of Sindbis as an expression tool. New advancements such as cell-specific targeting and the use of DNA promoters should further the use of Sindbis as a expression tool *in vivo* and increase the use of the Sindbis system as a candidate for gene therapy protocols.

References

Bredenbeek, P. J., Frolov, I. C. M. Rice and S. Schlesinger. (1993). Sindbis virus expression vectors: Packaging of RNA replicons by using defective helper RNAs. *J. Virol.* **67**, 6439–6446.

Chen, J.-P., Miller, D., Katow S., and Frey, T. K. (1995). Expression of rubella virus structural proteins by an infectious Sindbis virus vector. *Arch. Virol.* **140**, 2075–2084.

Coppola, M. A., Lam, T. M., Strawbridge, R. R., and Green, W. R. (1995). Recognition of endogenous ecotropic murine leukemia viruses by anti-AKR/Gross virus cytotoxic T lymphocytes (CTL): Epitope variation in a CTL-resistant virus. *J. Gen. Virol.* **76**, 635–641.

Driver, D., Latham, E. M., Polo, J. M., Belli, B. A., Banks, T. A., Chada, S., Brumm, D., Chang, S. M. W., Mento, S. J., Jolly, D. J., and Dubensky, Jr., T. W. (1995). Layered amplification of gene expression with a DNA gene delivery system. *Ann. N. Y. Acad. Sciences* **772**, 261–264.

D'Souza-Schorey, C., and Stahl, P. D. (1995). Myristoylation is required for the intracellular localization and endocytic function of ARF6. *Exp. Cell Res.* **221**, 153–159.

Dubensky, Jr., T. W., Driver, D. A., Polo, J. M., Belli, B. A., Latham, E. M., Ibanez, C. E., Chada, S., Brumm, D., Banks, T. A., Mento, S. J., Jolly, D. J., and Chang, S. M. W. (1996). Sindbis virus DNA-based expression vectors: Utility for in vitro and in vivo gene transfer. *J. Virol.* **70**, 508–519.

Dubuisson, J., Hsu, H. H., Cheung, R. C., Greenberg, H. B., Russell, D. G., and Rice, C. M. (1994). Formation and intracellular localization of hepatitis C virus glycoprotein complexes expressed by recombinant vaccinia and Sindbis viruses. *J. Virol.* **68,** 6147–6160.

Frolov, I., and Schlesinger, S. (1994). Comparison of the effects of Sindbis virus and Sindbis virus replicons on host cell protein synthesis and cytopathogenicity in BHK cells. *J. Virol.* **68,** 1721–1727.

Frolov, I., Hoffman, T. A., Prágai, B. M., Dryga, S. A., Huang, H. V., Schlesinger, S., and Rice, C. M. (1996). Alphavirus-based expression vectors: Strategies and applications. *Proc. Natl. Acad. Sci. U.S.A.* **93,** 11371–11377.

Hahn, C. S., Hahn, Y. S., Braciale, T. J., and Rice, C. M. (1992). Infectious Sindbis virus transient expression vectors for studying antigen processing and presentation. *Proc. Natl. Acad. Sci. U.S.A.* **89,** 2679–2683.

Heino, P., Dillner, J., and Schwartz, S. (1995). Human papillomavirus type 16 capsid proteins produced from recombinant Semliki Forest virus assemble into virus-like particles. *Virology* **214,** 349–359.

Herweijer, H., Latendresse, J. S., Williams, P., Zhang, G., Danko, I., Sclesinger, S., and Wolff, J. A. (1995). A plasmid-based self-amplifying Sindbis virus vector. *Hum. Gene Ther.* **6,** 1161–1167.

Huang, M. J., and Summers, J. (1991). Infection initiated by the RNA pregenome of a DNA virus. *J. Virol.* **65,** 5435–5439.

Jäntti, J., Keränen, S., Toikkanen, J., Kuismanen, E., Ehnholm, C., Söderlund, H., and Olkkonen, V. M. (1994). Membrane insertion and intracellular transport of yeast syntaxin Sso2p in mammalian cells. *J. Cell Sci.* **107,** 3623–3633.

Jiang, W., Venugopal, K., and Gould, E. A. (1995). Intracellular interference of tick-borne Flavivirus infection by using a single-chain antibody fragment delivered by recombinant Sindbis virus. *J. Virol.* **69,** 1044–1049.

Johanning, F. W., Conry, R. M., LoBuglio, A. F., Wright, M., Sumerel, L. A., Pike, M. J., and Curiel, D. T. (1995). A Sindbis virus mRNA polynucleotide vector achieves prolonged and high level heterologous gene expression *in vivo*. *Nucleic Acids Res.* **23,** 1495–1501.

Levine, B., Goldman, J. E., Jiang, H. H., Griffin, D. E., and Hardwick, J. M. (1996). Bcl–2 protects mice against fatal alphavirus encephalitis. *Proc. Natl. Acad. Sci. U.S.A.* **93,** 4810–4815.

Levis, R., Huang, H., and Schlesinger, S., (1987). Engineered defective interfering RNAs of Sindbis virus express bacterialchloramphenicol acetyl transferease in avain cells. *Proc. Natl. Acad. Sci. U.S.A.* **84,** 4811–4815.

Li, G., and Stahl, P. D. (1993a). Post-translational processing and membrane association of the two early endosome-associated *rab* GTP-binding proteins (*rab*4 and *rab*5). *Arch. Biochem. Biophys.* **304,** 471–478.

Li, G., and Stahl, P. D. (1993b). Structure-function relationship of the small GTPase *rab*5. *J. Biol. Chem.* **32,** 24475–24480.

Li, G., Barbieri, M. A., Colombo, M. I., and Stahl, P. D. (1994). Structural features of the GTP-binding defective Rab5 mutants required for their inhibitory activity on endocytosis. *J. Biol. Chem.* **269,** 14631–14635.

Li, G., Barbieri, M. A., and Stahl, P. D. (1995). Myristoylation cannot functionally replace the isoprenylation of Rab5. *Arch. Biochem. Biophys.* **316,** 529–534.

Liljeström, P., and Garoff, H. (1991). A new generation of animal cell expression vectors based on the Semliki Forest virus replicon. *BioTechnology* **9,** 1356–1361.

Liljeström, P., and Garoff, H. (1995). Expression of proteins using Semliki Forest Virus. *Curr. Protocols Mol. Biol.* **2,** 16.20.1–16.20.16.

London, S. D., Schmaljohn, A. L., Dalrymple, J. M., and Rice, C. M. (1992). Infectious enveloped RNA virus antigenic chimeras. *Proc. Natl. Acad. Sci. U.S.A.* **89,** 207–211.

Lopez, S., Yao, J.-S. Kuhn, R. J., Strauss, E. G., and Strauss, J. H. (1994). Nucleocapsid-glycoprotein interactions required for assembly of alphaviruses. *J. Virol.* **68,** 1316–1323.

Lovett, A. E., Hahn, C. S., Rice, C. M., Frey, T. K., and Wolinsky, J. S. (1993). Rubella virus-specific cytotoxic T-lymphocyte responses: Identification of the capsid as a target of major histocompatibility complex class I-restricted lysis and the definition of two epitopes. *J. Virol.* **67,** 5849–5858.

Lundström, K., Mills, A., Buell, G., Allet, E., Adami, N., and Liljeström, P. (1994). High-level expression of the human neurokinin-1 receptor in mammalian cell lines using the Semliki Forest virus vector. *Eur. J. Biochem.* **224,** 917–921.

Lundström, K.,Vargas, A., and Allet, B. (1995). Functional activity of a biotnylated human neurokinin-1 receptor fusion expressed in the Semliki Forest virus system. *Biochem. Biophys. Res. Commun.* **208,** 260–266.

Lundström, K., and Turpin, M. P. (1996). Proposed schizophrenia-related gene polymorphism: Expression of the Ser9Gly mutant human dopamine D3 receptor with the Semliki Forest virus system. *Biochem. Biophys. Res. Commun.* **225,** 1068–1072.

Lustig, S., Jackson, A. C., Hahn, C. S., Griffin, D. E., Strauss, E. G., and Strauss, J. H. (1988). Molecular basis of Sindbis virus neurovirulence in mice. *J. Virol.* **62,** 2329–2336.

Ohno, K., Sawai, K., Iijima, Y., Levin, B., and Meruelo, D. (1997). Cell-specific targeting of Sindbis virus vectors displaying IgG-binding domains of protein A. *Nat. Biotechnology* **15,** 763–767.

Olkkonen, V. M., Dupree, P., Simons, K., Liljeström, P., and Garoff, H. (1994). Expression of exogenous proteins in mammalian cells with the Semliki Forest Virus vector. *Methods Cell Biol.* **43,** 43–53.

Piper, R. C., Tai, C., Slot, J. W., Hahn, C. S., Rice, C. M., Huang, H., and James, D. E. (1992). The efficient intracellular sequestration of the insulin-regulatable glucose transporter (GLUT–4) is conferred by the NH2 terminus. *J. Cell Biol.* **117,** 729–743.

Piper, R. C., James, D. E., Slot, J. W., Puri, C., and Lawrence, Jr., J. C. (1993a). GLUT4 phosphorylation and inhibition of glucose transport by dibutyryl cAMP. *J. Biol. Chem.* **268,** 16557–16563.

Piper, R. C., Tai, C., Kulesza, P., Pang, S., Warnock, D., Baenzinger, J., Slot, J. W., Geuze, H. J., Puri, C., and James, D. E. (1993b). GLUT–4 NH2 terminus contains a phenylalanine-based targeting motif that regulates intracellular sequestration. *J. Cell Biol.* **6,** 1221–1232.

Piper, R. C., Slot, J. W., Li, G., Stahl, P D., and James, D. E. (1994). Recombinant Sindbis virus as an expression system for cell biology. *Methods Cell Biol.* **43,** 55–78.

Powers, A. M., Olsen, K. E., Higgs, S., Carlson, J. O., and Beaty, B. J. (1994). Intracellular immunization of mosquito cells to LaCrosse virus using a recombinant Sindbis virus vector. *Virus Res.* **32,** 57–67.

Pugachev, K. V., Mason, P. W., and Frey, T. K. (1995). Sindbis vectors suppress secretion of subviral particles of Japanese Encephalitis Virus from mammalian cells infected with SIN-JEV recombinants. *Virology* **209,** 155–166.

Rolls, M. M., Haglund, K., and Rose, J. K. (1996). Expression of additional genes in a vector derived from a minimal RNA virus. *Virology* **218, 406–411.**

Sallee, F. R. 1997. Utilization of a novel non-replicative Sindbis virus vector as a neuronal gene delivery system. *Invitrogen Express.* **4.5,** 9.

Schlesinger, S. (1993). Alphaviruses-vectors for the expression of heterologous genes. *Trends Biotechnol.* **11,** 18–22.

Simons, M., Tienari, P. J., Dotti, C. G., and Beyenuther, K. (1995). Two-dimensional gel mapping of the processing of the human amyloid precursor protein in rat hippocampal neurons. *FEBS Letters* **368, 363–366.**

Stabell, E. C., and Olivo, P. D. (1993). A truncated herpes simplex virus origin binding protein which contains the carboxy terminal origin binding domain binds to the origin of replication but does not alter its conformation. *Nucleic Acids Res.* **21,** 5203–5211.

Szekely, L., Jin, P., Jiang, W.-Q., Rosén, A., Wiman, K. G., Klein, G., and Ringertz, N. (1993). Position-dependent nuclear accumulation of the Retinoblastoma (RB) protein during *in vitro* myogenesis. *J. Cell. Physio.* **155,** 313–322.

Wang, K. S., Kuhn, R. J., Strauss, E. G., and Strauss, J. H. (1991). High-affinity laminin receptor is a receptor for Sindbis virus in mammalian cells. *J. Virol.* **66,** 4992–5001.

Xiong, C., Levis, R., Shen, P., Schlesinger, S., Rice, C. M., and Huang, H. V. (1989). Sindbis virus: An efficient, broad host range vector for gene expression in animal cells. *Science* **243,** 1188–1191.

Zabner, J., Fasbender, A. J., Moninger, T., Poellinger, K. A., and Welsh, M. J. (1995). Cellular and molecular barriers to gene transfer by cationic lipid. *J. Biol. Chem.* **270,** 18997–9007.

Zhou, X., Berglund P., Rhodes, G., Parker, S. E., Jondal, M., and Liljeström, P. (1994). Self-replicating Semliki Forest virus RNA as recombinant vaccine. *Vaccine* **12,** 1510–1514.

Section III

EXPRESSION IN INSECT SYSTEMS

11

Drosophila S2 SYSTEM FOR HETEROLOGOUS GENE EXPRESSION

Robert B. Kirkpatrick and Allan Shatzman

Department of Gene Expression Sciences, SmithKline Beecham Pharmaceuticals, King of Prussia, Pennsylvania 19406

Introduction

Continuous cell lines of *Drosophila* have now been in use for over a decade as hosts for the expression of heterologous gene products. There are approximately 100 established lines in existence including those derived from wild-type and mutant lines of *Drosophila melanogaster* (Simcox *et al.*, 1985) and a few from other species including *Drosophila immigrans* (Di Nocera and Dawid, 1983), *Drosophila hydei* (Sondermeijer *et al.*, 1980), and *Drosophila virilis* (Braude-Zolotarjova *et al.*, 1986). The most popular are the Schneider lines, S2 and S3 (Schneider, 1972; Schneider and Blumenthal, 1978), and a few derivatives of the Kc line (Echalier and Ohanessian, 1970). These have been widely employed for both stable and transient gene expression.

The Schneider line 2 (S2) in particular is an exceptionally versatile system that has proven to be useful both for the analysis of exogenous gene functions and for high-level protein production. This cell line has been used for the expression and analysis of intracellular, secreted, and membrane-associated proteins. This includes cytokines, oncogenes, antibodies, receptors, and viral antigens, which have all been shown to be authentically processed, biologically active, and produced to high levels. This chapter focuses on the properties, expression methodologies, and key features that make the S2 line amenable to the expression and study of diverse types of proteins. Properties of the many other available *Drosophila* cell lines are discussed elsewhere (for reviews, see Ashburner, 1989a; Cherbas *et al.*, 1994).

Properties of S2 Line

Origin

The S2 line was derived from primary cultures of late stage, 20- to 24-hr old, *D. melanogaster* (Oregan-R) embryos (Schneider, 1972). Cells grew out of this heterogeneous population in approximately 3 months, attaining immortalized nontumorgenic growth. The resultant cell line was originally described as having epithelial-like characteristics, growing in a loose monolayer without piling up into central foci. However, more recent studies on the properties of the cell line indicate that it may actually have been derived from hematopoietic origins (Samakovlis *et al.*, 1990; Abrams *et al.*, 1992; Kirkpatrick *et al.*, 1995b). It has been postulated that specific functions exhibited by the S2 line, including receptor-mediated endocytosis and lipopolysaccharide (LPS) activation, represent conserved components of innate immunity (Hultmark, 1993). This hypothesis is consistent with the discovery of several genes expressed in this cell line that are related to vertebrate genes with known immune function, including NF-κB-like family members (Petersen *et al.*, 1995) and macrophage scavenger receptor dSR-CI 9 (Pearson *et al.*, 1995).

A gene encoding an abundantly secreted protein, DS47, has been cloned from the S2 line (Kirkpatrick *et al.*, 1995b). DS47 is expressed specifically in the fat body and hemolymph, tissues known to play a role in innate immune functions in the fly (Rizki and Rizki, 1984). Other genes cloned from this cell line, including the LPS-inducible cecropins and the macrophage scavenger receptor, are also expressed in this same pattern, consistent with an immunogenic tissue specificity. In addition, DS47 has a human homologue that is expressed in peripheral blood-derived macrophages (Hakala *et al.*, 1993; Kirkpatrick *et al.*, 1997). Thus, the pattern of gene expression in S2 cells appears to reflect conserved hematopoietic origins in the fly.

Growth

S2 cells are maintained in a few types of commercially available media. This includes the original Schneider's medium,[1] as well as the M3 media formulation developed later by Shields and Sang (1977). Both are suitable for growth of S2 cells when supplemented with 10% heat-inactivated serum (65°C for 30 min). A modification of the M3 medium developed in the authors' laboratory enables suspension growth in serum-free conditions. Cells are seeded once per week at 2×10^6 cells/ml and are grown at 25°C. Cell densities between 1 and 2×10^7 cells/ml are achieved in 6–7 days in

[1]The original media formulations used to support *Drosophila* lines *in vitro* were based on the approximate composition of larval hemolymph. Comparable information was not available from embryos. Thus, it may not be surprising that hemocyte growth could be favored in such conditions.

both serum-free and serum-containing conditions. Doubling times of 24 hr are typical. Stably transfected cell lines can be easily scaled to production volumes or frozen for later use.

Cultures can be grown and scaled up in a variety of containers. Cells grown in T flasks grow best when the caps are tightened, indicating that oxygen levels are not significantly limiting in small flasks as it is with similarly grown mammalian cell cultures. Furthermore, sealing flasks maintains a humid environment that is also conducive for growth. In T flasks, the cells grow as a semiadherent monolayer, which is loosened easily by gentle swirling and tapping on the tube without the need for trypsinization. Cultures can also be grown in complete suspension using spinner flasks. Spinner rotation is kept to a minimum to avoid cell shearing but to keep cells floating. The authors also have had excellent success using shake flasks in which case the medium is supplemented with 0.1% pluronic F-68 in order to prevent cell shearing. Optimal culture volumes for T flasks, spinners, and shake flasks are shown in Table A1 (see Appendix).

Stably selected lines can be maintained indefinitely as frozen stocks in liquid nitrogen. Cells are frozen in either serum or serum-free freezing media shortly after selection is complete (see Appendix). Homemade freezing chambers work as well as commercially available containers and can be made easily by placing the 1-ml sample vials in a 6-in. high metal container, interspersed between layers of paper towels, and freezing them in a bucket of dry ice for several hours. Frozen lines are recovered from storage by gently thawing a vial, rolling it between two gloved hands, and transferring it directly into a flask containing growth medium (see Appendix). Vials should never be thawed in a 37°C water bath as this can damage the cells.

Experimental Procedures

Transfection

S2 cells are amenable to both transient and stable transfections. Stable lines are produced by cotransfecting a drug resistance plasmid along with one or more expression plasmids containing the gene of interest under the control of a constitutive or inducible promoter (Fig. 1). Transient expression is achieved without the use of a selection marker and can be analyzed within 1–2 days (Di Nocera and Dawid, 1983; Courey and Tijian, 1988; Chung and Keller, 1990). Stable lines are selected and analyzed in 4–6 weeks. Standard DNA transfection methods can be used, including calcium phosphate DNA precipitation, lipid-mediated transfection, and electroporation. Although each of these methods can achieve high copy expression, the calcium phosphate method is the least expensive and the most likely to yield high expression levels. The time line for generating stable lines is illustrated in Fig. 2.

A hallmark of *Drosophila* stable lines is the occurrence of multicopy insertions of the two vector sequences, which form arrays of more than 1000 gene copies in head-to-tail arrays (van der Straten *et al.*, 1989). The mechanism leading to the insertion of gene copies into the chromosome and the formation of these multicopy insertions is unknown, although homologous recombination has been postulated as a possible mechanism (Cherbas *et al.*, 1994). However, homologous recombination does not seem to be necessary for the initial gene insertion event as this appears to occur at random chromosome locations. Regardless of the mechanism, the formation of these multicopy arrays serves essentially the same purpose as gene amplification of mammalian lines, but is achieved in only one step.

The number of inserted gene copies can be titrated by adjusting the ratio of expression and selection plasmids (van der Straten *et al.*, 1989). This is illustrated in the Southern blot analysis in Fig. 3 showing the result of cotransfections using various ratios of *Drosophila* expression plasmid and hygromycin B selection plasmid. Ratios of expression/hygromycin plasmid were varied from 0.1 to 10 μg/1 μg. The results show a progressive increase in the number of inserted heat shock cognate 3 (HSC3) gene copies (2.3-kb band) as compared to the endogenous genomic copy of HSC3 at 23 kb. A high copy number typically translates into high-level protein expression. Thus, protein expression levels can be varied simply by adjusting the ratios of expression and selection plasmids. An optimal ratio of expression and drug resistance plasmids for high level expression has been generally determined to be 19:1. This can lead to the insertion of more than 1000 copies and protein expression levels approaching 50 mg/liter of secreted material. There are, of course, gene-specific cases where higher or lower ratios have been used to achieve more appropriate expression levels for a particular experiment.

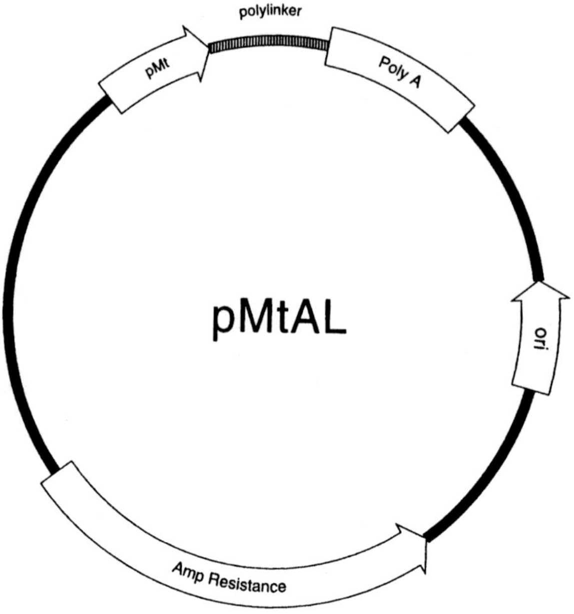

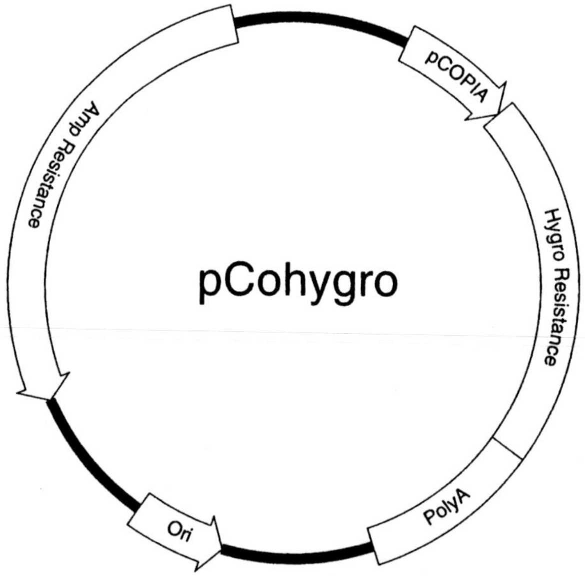

Figure 1 Cotransfection of expression and selection plasmids. Plasmid mtaL, containing the inducible metallothionine promoter, efficient SV40 late poly(A) signal, and polylinker region for subcloning genes for expression is cotransfected with the drug resistance plasmid, pCoHygro, containing the hygromycin acetyltransferase gene under the control of the strong constitutive *Drosophila* copia gene promoter. Selection with hygromycin B leads to the selection of stable lines in 4–6 weeks.

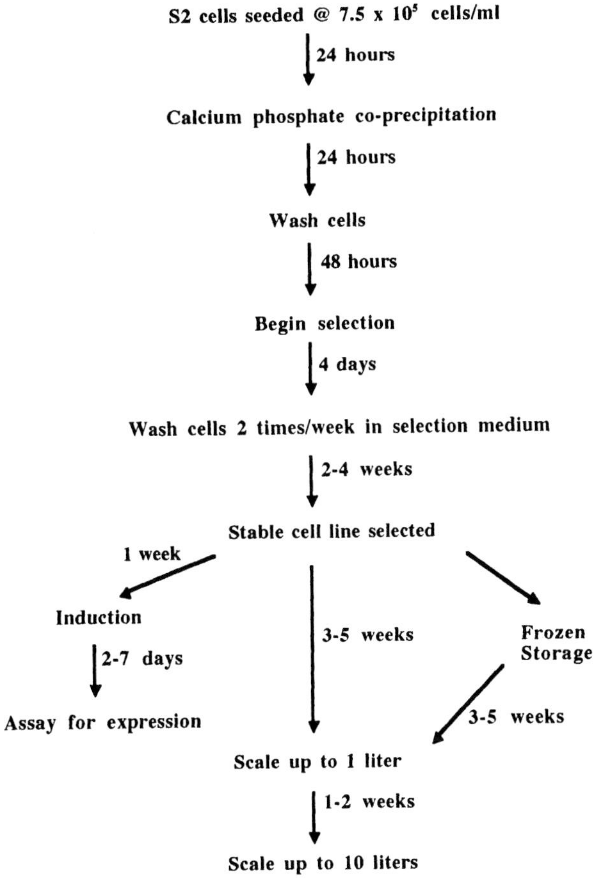

Figure 2 Time line of procedures leading to the selection and scaleup of stable *Drosophila* cell lines for heterologous expression.

Selection

Various resistance markers have been used for the selection of stable *Drosophila* lines (Table 1). Some of the initial attempts were with dihydrofolate reductase (DHFR) and neomycin marker genes using methotrexate and G418 selection, respectively (Bourouis and Jarry, 1983; Moss *et al.*, 1985; Bunch *et al.*, 1988). These methods proved to be problematic in both the time of selection (greater than 6 weeks) and the spontaneous resistance resulting with G418 selection (Rio and Rubin, 1985; van der Straten *et al.*, 1987). A signifi-

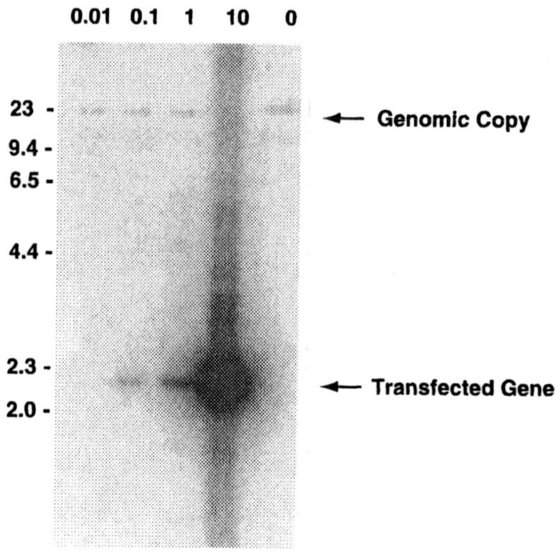

Figure 3 Adjusting gene copy number. The number of HSC3 gene copies inserted into the chromosome was adjusted by varying the ratio of cotransfected expression and resistance plasmids. Various amounts of expression plasmid (0.01, 0.1, 1, and 10 μg) were cotransfected with 1-μg selection plasmid to generate stable lines. DNA was extracted from these lines and the parent S2 line (S2) was digested with *Bam*HI restriction endonuclease for Southern blot analysis. A ^{32}P-labeled *Bam*HI restriction fragment from HSC3 cDNA was used to probe the blot. Multicopy insertions are indicated by the arrow at approximately 2.3 kb. The single genomic copy of HSC3 in S2 cells is shown at 23 kb.

Table 1
Selection Markers

Marker gene[a]	Drug concentration	Comment	Reference
DHFR	0.2 μg ml^{-1} methotrexate	6 to 8-week selection	Bourouis and Jarry (1983); Moss *et al.* (1985); Bunch *et al.* (1988)
Neo	1 mg ml^{-1} G418	3 to 4-week selection, spontaneous resistance	Rio and Rubin (1985); van der Straten *et al.* (1987)
Hygro AT	300 μg ml^{-1} hygromycin B	3 to 4-week selection	van der Straten *et al.* (1987)
α-Ammanatin[r]	5 μg ml^{-1} α ammanatin	3 to 4-week selection	Jokerst *et al.* (1989)

[a]DHFR, dihydrofolate reductase; Neo, neomycin phosphotransferase; Hygro AT, hygromycin acetyltransferase.

cant improvement has been achieved through the use of hygromycin B selection with the hygromycin B–phosphotransferase gene as a selectable marker. This improved the time of selection to approximately 3–4 weeks and did not lead to spontaneous resistance (van der Straten *et al.*, 1989). Another efficient method is α-amanatin selection using an α-amanatin-resistant form of polymerase II as a marker gene. Like hygromycin selection, this method has also been reported to yield stable lines in several weeks (Thomas and Elgin, 1988). However, α-amanatin is approximately 10 times the cost of hygromycin B (using 5 μg/ml α-amanatin and 300 μg/ml hygromycin B, respectively). Furthermore, extra care must be taken when handling α-amanatin, which is extremely toxic.

Clonal cell lines can be selected either directly from transfected cells or from preexisting polyclonal populations. The major difficulty in selecting clones from any source is the inability of *Drosophila* cells to grow at low density. They apparently require one or more secreted factors that are produced in sufficient quantities only when the cells are grown at high density. Methods for cloning using conditioned medium to support growth have been reported by Lindquist (Ashburner, 1989b). More efficient methods involve the use of irradiated feeder layer cells (Cherbas *et al.*, 1994). Briefly, S2 cells are irradiated with 24 kR X rays or γ rays, mixed with transfected cells in soft agar containing selective media, and plated onto 100-mm petri dishes (1×10^6 feeder cells/ml: 200 viable transfectants per plate). Clonal growth is supported by feeder layer cells, which are themselves unable to proliferate. Clones are then picked from the plates as agar plugs and transferred to appropriate growth medium containing selective agents.

Cloning with the use of a feeder layer is also amenable to a 96-well plate format. Cells can be cloned efficiently in this format using a feeder layer of untransfected S2 cells in place of X-irradiated cells (see Appendix). In this method, transfected cells are plated at one cell per well onto 1×10^5 feeder layer cells in 0.2 ml. Resultant clones appear within 2–3 weeks.

This method relies on the capacity of the feeder layer to remain viable just long enough to support the high density growth of clonal populations and works best for cloning previously selected polyclonal populations, which tend to grow more quickly anyway. In addition, the authors have also used this method for the direct cloning of transfected cells.

Although the selection of clonal lines may sometimes be advantageous in increasing expression and reproducibility, the selection of polyclonal lines is generally suitable for high level protein expression. Chromosomal position effects do not seem to pose a significant detriment to high level expression because they are averaged within the population. With the exception of toxic gene products, high level expression does not correlate with low growth. Therefore, it is unlikely that very high level expressing cells within a population will be overgrown by faster-growing, low-expressing clones. Furthermore, even toxic gene products can be maintained in a transcriptionally inactive state under the control of the inducible metallothionine (Mtn) promoter (Johansen *et al.*, 1989). Therefore, in most cases, the use of stable clonal lines offers little advantage over the use of stable polyclonal lines, which are produced with considerably less effort.

Expression Vectors

Expression vectors containing strong constitutive or inducible promoters are available for heterologous gene expression in the S2 line (Table 2). The inducible Mtn promoter has proven to be extremely efficient and tightly regulated even at high copy number. The authors' original vector, MTaL (Fig. 1), combines the use of the Mtn promoter with the efficient SV40 late polyadenylation signal for high level mRNA expression, inducible by adding 500 μM copper sulfate directly to the cell culture. This permits the overexpression of even lethal gene products to high levels (Johansen *et al.*, 1989). An inducible promoter system is also convenient for studies on heterologous gene regulation and protein function where the uninduced condition provides a useful negative control.

Of the constitutive *Drosophila* promoters that have been tested, those from the copia transposable element long terminal repeat (Bourouis and Jarry, 1983), α tubulin gene, and actin 5C gene (Angelichio *et al.*, 1991) have proven to be the most effective for high level constitutive mRNA expression (Table 2). In addition, a number of minimal constitutive promoters have been used, including those from the *Drosophila* Adh and men genes (Krasnow *et al.*, 1989; Yoshinaga and Yamamoto, 1991). The actin promoter has been the most widely used and works more efficiently in the S2 system than the others for high level expression. A strong constitutive promoter has been isolated from the DS47 gene that drives high level expression of the DS47 product specifically in S2 cells

Table 2
Promoter Vectors for High-Level Expression in *Drosophila* S2 Cells

Vector	Promoter	Note[a]	Reference
pMta	Metallothionine	Tightly regulated inducible *Drosophila* promoter, SV40 early poly(A)	van der Straten *et al.* (1987); Ivey-Hoyle (1991); Johansen *et al.* (1989)
pMTaL	Metallothionine	Contains SV40 late poly(A), which is three times more efficient than SV40 early poly(A)	Angelichio *et al.* (1991)
pRmHa	Metallothionine	Contains alcohol dehydrogenase poly(A)	Bunch *et al.* (1988)
pMttBNS	Metallothionine	Contains human tPA signal peptide sequence for directing secretion	Ivey-Hoyle *et al.* (1991a); Brighty and Rosenberg (1994)
pMtBL2	Metallothionine	Contains *Drosophila* BiP signal peptide sequencing for directing secretion	R. B. Kirkpatrick (unpublished)
pHt4	hsp70	Inducible expression under heat shock	Snow *et al.* (1989); Elkins *et al.* (1990)
pDS47SV40	DS47	Strong constitutive promoter from an abundantly expressed gene in *Drosophila* S2 cells, SV40 late poly(A)	R. B. Kirkpatrick (unpublished)
pPac	Actin 5C	Strong constitutive *Drosophila* promoter, actin poly(A)	Krasnow *et al.* (1989)
pA5CSV40	Actin 5C	Strong constitutive *Drosophila* promoter, SV40 late poly(A)	Angelichio *et al.* (1991)
pCopia	Copia	Strong constitutive *Drosophila* promoter	Bourouis and Jarry (1983)

[a]poly(A), polyadenylation site; tPA, tissue plasminogen activator; BiP, immunoglobulin-binding protein.

(Kirkpatrick *et al.*, 1995b). This promoter has been found to drive transcription levels comparable to those achieved with the actin 5C gene promoter (R. Kirkpatrick, unpublished).

The S2 system also supports very efficient secretion of heterologous proteins. Secreted protein levels of 10 mg/liter are commonly achieved after 1 week in culture (Ivey-Hoyle, 1991), although levels approaching 50 mg/liter have also been reported (Ivey-Hoyle *et al.*, 1991b). Efficient secretion can be directed using either native signal peptide sequences or signal sequences derived from other genes. In most cases the native signal peptide sequence associated with the heterologously expressed gene of interest is suitable for efficient secretion. However, for genes that do not contain signal sequences, two vectors containing optimized signal sequences derived from the human tissue plasminogen activator gene (tPA) and *Drosophila* immunoglobulin-binding protein (BiP) signal peptide sequences have been developed. The efficiency of the tpa signal was discovered fortuitously through studies with tPA itself. The signal sequence of *Drosophila* BiP was specifically chosen because of the high levels of BiP secreted into the endoplasmic reticulum in the S2 line (Kirkpatrick *et al.*, 1995a). Both vectors, mttBNSS and mtBL2, contain restriction sites for cloning in frame with the signal peptide-coding sequence (Fig. 4) (Table 2). Thus, secretion of virtually any protein to high levels is possible, regardless of whether a natural secretion sequence is present.

Literature Review

Large-Scale Protein Expression

The *Drosophila* S2 system is exceptionally well suited for heterologous protein expression. Being of eukaryotic origin, S2 cells contain processing signals for subcellular localization, secretion, and glycosylation,[2] enabling the expression and secretion of authentic proteins of mammalian or viral origin. Compared to the lytic expression produced by recombinant baculovirus infection of *Spodoptera* cells, the generation of stable *Drosophila* lines allows continuous

[2]*Drosophila* glycosylation differs from mammalian glycosylation only in its ability to make complex-type oligosaccharides and add sialic acid.

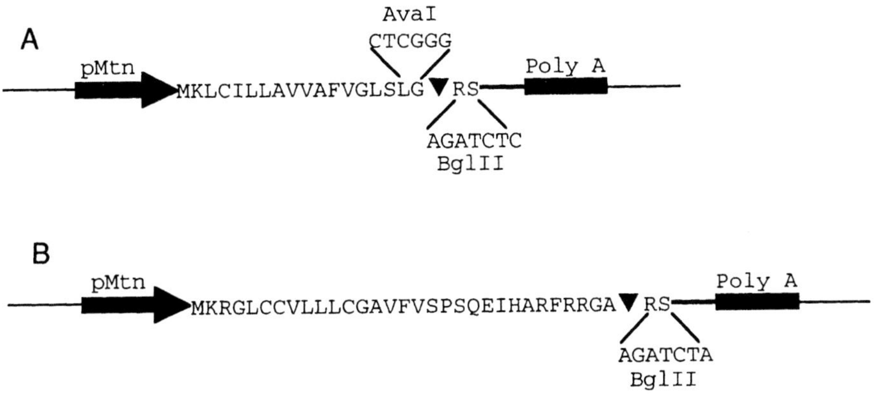

Figure 4 Secretion vectors. (A) MtBL2 contains the metallothionine promoter (Mtn) for regulated expression at a high gene copy number, an 18 amino acid signal peptide sequence derived from the *Drosophila* HSC3 gene, and the SV40 late poly(A) signal. Heterologous gene expression and secretion are achieved by subcloning in frame at the *Ava*I or *Bgl*II restriction sites. Because the *Ava*I site comprises the last two codons of the signal peptide sequence, secreted proteins are produced without any extraneous residues. Cloning into the *Bgl*II site leaves two residues (RS) on the N terminus of the expressed protein. ▼, signal peptide cleavage site. (B) MttBNS contains the Mtn promoter, a 31 amino acid signal peptide sequence derived from the human tissue plasminogen activator gene, and the SV40 early poly(A) signal. Secretion is achieved by cloning in frame at the *Bgl*II restriction site. Proteins secreted from MttBNS contain two extra residues (RS) at the N terminus.

protein production, which enhances reproducibility and facilitates large-scale production in bioreactors. Furthermore, the high protein levels produced without amplification (~50 mg/liter) after a short selection period (3–4 weeks) are compatible with most expression needs.

One important advantage of the S2 system for large-scale production is the availability of both tightly regulated inducible expression vectors and strong constitutive vectors. Inducible expression under the tightly regulated Mtn promoter makes possible the coordinated expression of genes at a very high copy number. Remarkably, induced expression under the Mtn promoter even allows for the high level production of lethal gene products. This was clearly demonstrated by the ability to overexpress the toxic H-ras gene product under metallothionine induction, which under continuous induction was inhibitory to cell growth and led to the eventual loss of H-ras expression (Johansen *et al.*, 1989). Thus, maintaining the cells in the promoter-off state in the absence of a heavy metal inducer permitted normal cellular growth and allowed for high level H-ras expression following induction.

The high level production of biologically active proteins has become routine through the generation of stable S2 lines as illustrated by the examples in Table 3. Although protein yields are reproducibly high, variations in expression levels sometime reflect gene-specific differences. For example, the expression of various HIV isolates of recombinant envelope glycoprotein (gp120) yielded protein levels ranging from 5 to 35 mg/liter (Ivey-Hoyle, 1991; Ivey-Hoyle *et al.*, 1991b). These differences did not appear to represent experimental variation as they were reproducible in independent transfections. This effect is apparently related to the inherent properties of *Drosophila* transcription, translation, and/or secretion that favor expression of certain genes over others. perhaps it will be possible in the near future to identify specific components of the *Drosophila* expression machinery (e.g., rare tRNAs) that can be supplemented or modified in order to eliminate these gene-specific expression differences.

An important advantage of the S2 system is the ability to secrete

Table 3
Expression Yields in Stable S2 Lines

Product[a]	Vector	Expression level (mg/liter)	Reference
Secreted hIL-5	pMttBNSS	22	Johanson *et al.* (1995)
hIL-5Rα	pMtaL	1 million sites/cell	Johanson *et al.* (1995)
Secreted shIL-5Rα	pMtaL	17	Johanson *et al.* (1995)
Secreted shIL-5RαFc	pMtaL	10	Johanson *et al.* (1995)
Secreted gp120	pMttBNSS	5–35	Ivey-Hoyle *et al.* (1991a)
Secreted shIL-1RFc	pMtaL	5	Kumar *et al.* (1995)
Galactokinase	pMta	3	van der Straten *et al.* (1987)
Human glucagon receptor	pRmHa	250 pmol mg^{-1}	Tota *et al.* (1995)
Drosophila ionotropic GABA receptor	pRmHa	35,000 sites/cell	Millar *et al.* (1994)

[a]hIL-5, human interleukin-5; shIL-5Rα, soluble IL-5 receptor α; shIL-5RαFc, soluble IL-5 receptor α–human immunoglobulin Fc fusion; shIL-1RFc, soluble human interleukin-1–human immunoglobulin Fc fusion; GABA, γ-aminogutyric acid; gp120, HIV envelope glycoprotein.

proteins over a long period of time. In contrast to baculovirus-infected insect cultures that secrete proteins over a narrow window of time before the lytic stage ensues, proteins secreted from S2 cells can accumulate over many days or even weeks in the culture medium. Therefore, proteins secreted from S2 cells are generally more stable than those secreted from baculovirus-infected cultures in which intracellular proteases are released following cell lysis. Furthermore, purification from S2 culture supernatants is facilitated greatly by growth in serum-free media conditions. In most cases, the protein of interest represents one of only a few major protein-secreted species present in the medium. Fractionation via ion-exchange capture followed by a sizing column is often all that is required in order to purify an expressed protein.

Receptor Expression

Stable receptor expression on the *Drosophila* S2 cell membrane or as secreted soluble proteins has been exceptionally useful in the study of receptor–ligand kinetics (Johanson *et al.*, 1995; Kumar *et al.*, 1995; Einstein *et al.*, 1996). One notable example is the human interleukin (IL)-5 receptor (Johanson *et al.*, 1995). In this study, a full-length α chain receptor was secreted onto the cell surface as directed by the native 20 amino acid signal sequence. There it was assembled into a single transmembrane receptor that was functionally identical to the mammalian-derived IL-5 receptor in terms of its ligand-binding kinetics. The iodinated IL-5 ligand (also produced in S2 cells) bound to these membranes with high affinity and was saturatable with cold ligand (Fig. 5). A soluble form of IL-5 lacking the transmembrane and cytoplasmic domains was also expressed in this study. This soluble form was secreted efficiently into the culture fluid as directed by the human tissue plasminogen activator signal peptide sequence. The high protein yields obtained from this expression greatly facilitated cocrystallization studies with the human IL-5 ligand.

The four transmembrane human α-aminobutyric acid (GABA-A) receptor and the seven transmembrane endothelin receptor (ET) have been expressed in S2 cells (E. Appelbaum, S. Ganguly, and A. Swift, unpublished). Both were expressed to high density as authentic membrane receptors on the cell surface. In order to achieve GABA-A receptor expression, each of the three receptor subunits (α, β, and γ) was cotransfected as an individual plasmid construct in combination with each of the other receptor subunit genes. Each

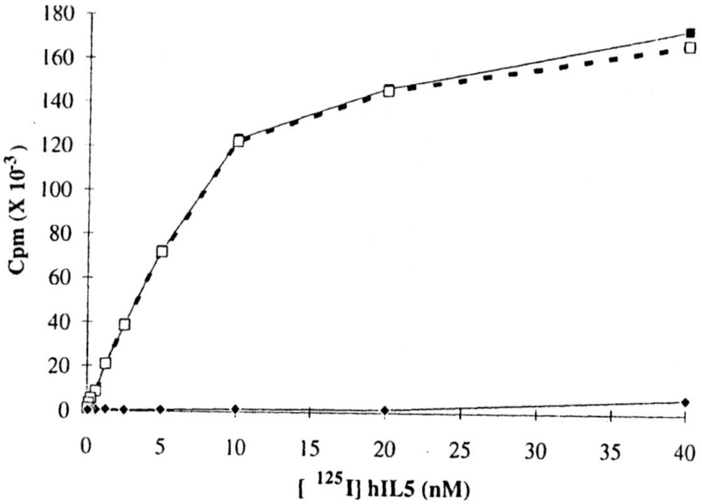

Figure 5 Binding of iodinated interleukin-5 (^{125}hIL-5) to full-length IL-5 receptor (hIL-5Rα) expressed on *Drosophila* S2 cells. Specific binding (□) is defined as total binding (■) minus binding in the presence of 300-fold excess hIL-5 (◆).

of the expressed subunits was authentically processed, folded together, and secreted onto the cell membrane. The resultant receptor bound authentic ligand with specificity comparable to that of the endogenous receptor found in rat cerebellum (Fig. 6). Similarly, ET was expressed to high density (40,000 sites/cell) on S2 membranes and showed specific binding to ET ligand (Fig. 7). High density expression combined with low background binding greatly facilitated receptor–ligand analysis for both ET and GABA-A.

The stable expression of receptors to high density for pharmacological analyses represents another important advance in the use of S2 cells (Millar *et al.*, 1994, 1995; Tota *et al.*, 1995; Buckingham *et al.*, 1996). In the first of these reports, the *Drosophila* ionotropic GABA receptor subunit was expressed stably as a functional homoligomeric ion channel (Millar *et al.*, 1994). Functionality was assessed by electrophysiological recording using patch-clamp techniques similar to those developed for use with *Xenopus* oocytes. This technique proved to be useful for the pharmacological characterization of both wild-type and mutant GABA receptors expressed in stable S2 lines (Buckingham *et al.*, 1996). The smaller size of S2 cells compared to *Xenopus* oocytes facilitated rapid agonist mea-

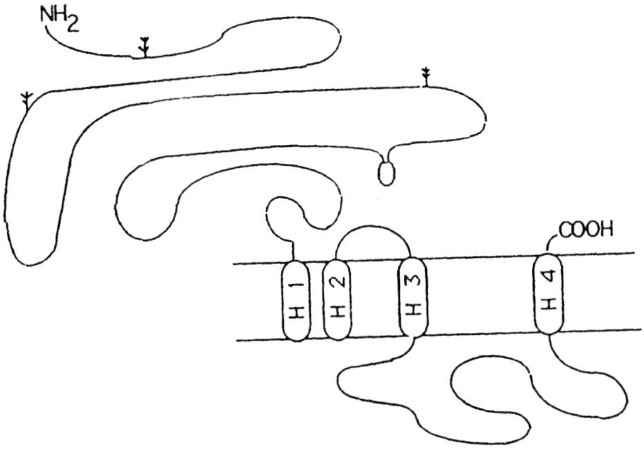

S2 cells transfected with	*GABA analog		
	Background Counts	Binding Counts	Ratio -/+
Control	6.2	6.0	0.97
α3β1γ2 Induced (7 day)	6.8	107.8	15.85
Rat Cerebellum	7.9	206.1	33.81

Figure 6 Binding of the iodinated human γ-aminobutyric acid (GABA) analogue to GABA receptor expressed on *Drosophila* S2 cells. (Top) Four membrane-spanning domains of GABA. (Bottom) Results of binding to cells uninduced (control) or induced for GABA receptor expression or to rat cerebellum membranes containing the endogenous GABA receptor. The ratio is the total counts minus background binding in the absence of cells.

surements in these studies. Furthermore, S2 cells uninduced for GABA expression were found to have few, if any, voltage-activated ion channels.

In a second study, the G-coupled *Drosophila* muscarinic acetylcholine receptor was expressed and shown to couple through calcium (Millar *et al.*, 1995). This was done using quantitative fluorescence imaging techniques as is commonly done with mammalian cells to measure receptor activation. Ligand-specific coupling was demonstrated clearly and could be blocked by anticholinergic com-

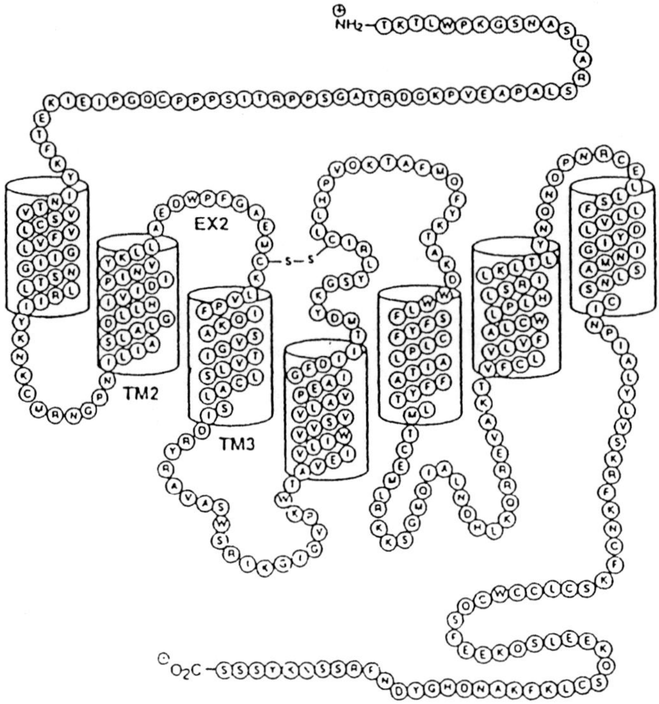

Sample	^{125}I - ET Binding
S2 cells (untransfected)	------
S2 cells - ET receptor (uninduced)	----
S2 cells - ET receptor (induced)	++++

Figure 7 Binding of iodinated human enothelin (ET) to ET receptor expressed in *Drosophila* S2 cells. (Top) Seven transmembrane-spanning domains of ET. (Bottom) Results of binding assay showing high binding to cells induced for ET expression but not to uninduced or untransfected cells.

pounds. Furthermore, this coupling correlated with the specific binding of tritiated ligand to cell membranes, which was readily dissociable with cold ligand (K_d 0.75 nM).

In a third study, the human glucagon receptor was expressed stably in S2 cells so that whole cell-binding assays could be performed. Ligand-induced coupling to cyclic AMP was demonstrated in these assays (Tota *et al.*, 1995). Biophysical and structural studies on this receptor were facilitated greatly by the high level expression that was attained (250 pmol/mg membrane protein). This represents an improvement over mammalian expression systems in which significant G-coupled protein receptor expression has been notoriously difficult to achieve.

These studies illustrate the utility of the S2 system for the measurement of functional coupling through expressed receptors and channels. The functional characterization of these receptors and channels on *Drosophila* cells offers advantages for stability, inducibility, and expression levels. Another important advantage of using stable S2 lines for receptor expression, as illustrated in the earlier examples, is their compatibility with whole cell functional assays. In contrast, the constitutive lytic expression produced by recombinant baculovirus does not lend itself readily to this purpose. In addition, because S2 cells lack endogenous mammalian receptors, they provide a suitably low background in which to assay these receptors. Finally, the tightly regulated expression made possible by metallothionine induction offers a unique experimental control that is not possible in other expression systems for the study of signal transduction mechanisms.

G-Proteins

The fact that human G-coupled protein receptors function in S2 cells demonstrates that in at least some cases *Drosophila* G-proteins can substitute for their mammalian homologues. Still, little is known about specific G-proteins in *Drosophila* or the receptors to which they are linked. To date, only a few *Drosophila* G-protein α subunit homologues have been studied (Provost *et al.*, 1988; de Sousa *et al.*, 1989; Thambi *et al.*, 1989; Yoon *et al.*, 1989; Lee *et al.*, 1990, 1994; Quan *et al.*, 1991, 1993; Scott *et al.*, 1995; Talluri *et al.*, 1995; Wolfgang and Forte, 1995). Perhaps the most well characterized is Dgq, which is thought to play an important function in the phototransduction pathway analogous to that of transducin in mammalian cells (Lee *et al.*, 1990, 1994). Other less studied *Drosophila* G-proteins include DGs, DGo, and DGi. Of these, DGs and DGo are expressed in the brain (de Sousa *et al.*, 1989; Quan *et al.*, 1989) whereas DGi is expressed in the embryo and pupae (Provost *et al.*, 1988; Wolfgang and Forge, 1995). Despite the similarity to mammalian G-proteins, *Drosophila* G-proteins are unique from their mammalian counterparts. For example, DGi has no pertussis toxin recognition site, and Gs is larger than its mammalian homologue. Consequently, it is unclear if these represent true homologues at all. Therefore, functional coupling of vertebrate G-coupled receptors in *Drosophila* cells will need to be determined empirically for each receptor. However, in cases where G-proteins are not represented, coupling may still be made possible by coexpression of appropriate vertebrate G-protein subunits.

In order to screen for possible G-protein homologues in S2 cells, cell lysates were probed on Western blots with antisera raised to human Gα subunits. Preliminary results from this analysis are consistent with the presence of Gs, Gz, and Gq subunits, but not Gi or Go (Table 4). Stable lines expressing human β-adernergic receptor in which stimulation is functionally coupled to increases in intracellular cyclic AMP levels have been constructed (S. Ganguly and S. Konchar, personal communication). This suggests that Gs, which is known to couple to cyclic AMP in vertebrates, has an analogous function in *Drosophila.* This result is also consistent with the demonstrated ability of DGs to complement Gsα function in Gsα-deficient mammalian cell lines (Quan *et al.*, 1989).

Cell Adhesion Molecules

A novel application of the S2 system is in the study of cell adhesion. S2 cells apparently lack intrinsic self-adhesive properties, making them exceptionally amenable to adhesion assays. This has

Table 4
Expression of G-α Subunits in S2 Cells

				Western data	
Drosophila G-protein	Specific expression	M_r	Apparent size (kDa)	Human antibody probe	Size band (kDa) in S2
dGsL	Brain	45,003	51	Anti GS	51
dGsS	Brain	44,704	48		48
dGi	Embryo/pupae	40,612	41	Anti Gi-1, Gi-2, or Gi-3	None detected
dG0-1	Brain	40,430	40	Anti Go	None detected
dG0-2	Brain	40,491	40		
dG1	Chemosensory and taste			Anti Gq and G11	43
				Anti Gz	46
				Anti-Gα-16	46

been shown for a number of cell adhesion molecules that have been expressed stably in S2 cells (Hortsch and Bieber, 1991). Two notable examples are fasciclin I and III (Snow *et al.*, 1989; Wang *et al.*, 1993). These glycosylphosphatidylinositol-linked surface glycoproteins belong to the immunoglobulin superfamily and have been implicated to play a role in nervous system development. Both fasciclin I and III can independently promote cellular aggregation when their expression is induced under metallothionine or heat shock promoter control. Remarkably, when fasciclin I-expressing S2 cells are mixed with fasciclin III-expressing cells, a cell-type-specific sorting occurs, which leads to the formation of two homogeneous populations of fasciclin I- and fasciclin III-associated cells (Elkins *et al.*, 1990).

It is tempting to draw parallels between S2 cell sorting *in vitro* and the sorting of cells into tissues during development. Such comparisons have been the impetus for additional experiments aimed at defining specific cell adhesion interactions associated with various aspects of development. Other examples include chaoptin polypeptide, which is expressed specifically in photoreceptor cells in the developing eye (Krantz and Zipursky, 1990), the maternal effect Toll protein, which is expressed in early embryonic development (Keith and Gay, 1990), neurotactin, which is expressed in central nervous system development (Barthalay *et al.*, 1990), and gliolectin, which is expressed during the development of the embryonic nervous system (Tiemeyer and Goodman, 1996). As with faciclin, each of these receptors can promote homophilic cellular aggregation when expressed in S2 cells.

Heterophilic cell–cell interactions have also been shown for the products of Notch and Delta. These transmembrane glycoproteins are purported to play a role in the differentiation of neuronal and epidermal cell types. When expressed in S2 cells, mixed aggregates of notch- and delta-expressing cells are produced, consistent with a possible mechanism by which notch and delta control cell fate during development (Fehon *et al.*, 1990).

A human cell adhesion molecule has been expressed in *Drosophila* S2 cells. The receptor tyrosine kinase ARK (adhesion-related kinase) was shown to mediate cellular aggregation by homophilic binding, independent of the kinase activation domain (Bellosta *et al.*, 1995). Similar results were obtained when ARK was expressed in Chinese hamster ovary (CHO) cells, indicating a highly conserved adhesion mechanism in insects and vertebrates. The expression of vertebrate adhesion molecules in S2 cells may therefore be useful in the characterization of these mechanisms.

Conserved Functions

Many transcription, translation, and secretion mechanisms are conserved between *Drosophila* and vertebrates. Nevertheless, it was surprising to find that some functions thought to be unique to vertebrate systems were also conserved in S2 cells. One example is the HIV-1 rev protein, which is known to transactivate expression of viral gp-160 envelope protein, enabling export from the nucleus to the cytoplasm in mammalian cells. Remarkably, S2 cells expressing gp160 mRNA in the absence of rev are unable to produce gp120 protein (Fehon *et al.*, 1990; Brighty *et al.*, 1991; Brighty and Rosenberg, 1994). Furthermore, when gp160 and rev are coexpressed, message escapes the nucleus and is translated efficiently into the gp160 product, which is subsequently processed into gp120 and gp41 envelope glycoproteins. These later become noncovalently associated at the cell surface, similar to the way in which gp120 and gp41 are expressed on HIV-infected mammalian cell membranes. Apparently, the mechanisms enabling this to occur, including nuclear transport, proteolytic processing, protein folding, and secretion, are well conserved in S2 cells. The S2 system has proven to be an exceptional model system in the study of these mechanisms. It has enabled the delineation of a region of gp160 mRNA responsible for nuclear retention as well as the identification of a region of gp120 protein responsible for its association with gp41 (Ivey-Hoyle *et al.*, 1991a).

Another example of an unexpectedly conserved mechanism present in S2 cells is the mechanism leading to the expression and secretion of immunoglobulins (Kirkpatrick *et al.*, 1995a). Remarkably, coexpression in S2 cells of humanized heavy and light chain immunoglobulins resulted in the secretion of fully folded immunoglobulins. The resultant *Drosophila*-produced antibody had an apparent molecular weight identical to that of immunoglobulin expressed in CHO cells as detected by sodium dodecyl sulfate (SDS)–polyacrylamide gel electrophoresis and Coomassie blue staining (Fig. 8, left), which resolved to heavy and light chain monomers under reducing conditions (Fig. 8, right). Furthermore, the secreted antibody was fully functional in antigen-binding assays, exhibiting binding properties identical to those of CHO-expressed antibody (Fig. 9). Thus, the mechanisms of antibody secretion and folding appear to be conserved in S2 cells, even though *Drosophila* do not naturally make antibodies.

In mammalian cells, antibody folding and secretion are regulated precisely through the interactions of chaperone proteins with heavy and light chains. It is through these interactions that heavy and light chain dimers are thought to pair and fold into complete antibodies. One important component in this process is the endoplasmic reticulum chaperone called immunoglobulin-binding protein, which is considered to be a critical component of the antibody-folding mechanism. In addition, BiP is also believed to prevent the secretion of heavy chain dimers in the absence of light chain by retaining them in the endoplasmic reticulum.

Even though *Drosophila* cells do not make antibodies naturally, they do contain conserved chaperone proteins, including a close homologue of immunoglobulin-binding protein called heat shock cognate 72 (hsc 72) (Rubin *et al.*, 1993). Given the close homology (80% identity), it was of interest to determine if *Drosophila* BiP (hsc72) could be involved in the antibody secretion in S2 cells. Indeed, when immunoglobulins were precipitated from cell lysates with protein G, hsc72 was found to be associated, indicating that it could substitute for vertebrate BiP function in folding antibodies (Fig. 10, lane 3). Furthermore, dissociation of the heavy chain from hsc72 in these cell lysates was achieved by the addition of ATP, which is consistent with the mechanism of HC–BiP association and dissociation in vertebrates (Fig. 10, lane 6). However, when the BiP retention function was assessed by expressing heavy chain in the absence of light chain, heavy chain dimers were found to be secreted efficiently in the absence of light chain, despite the specific hsc72 interaction. This result suggests an important mechanistic

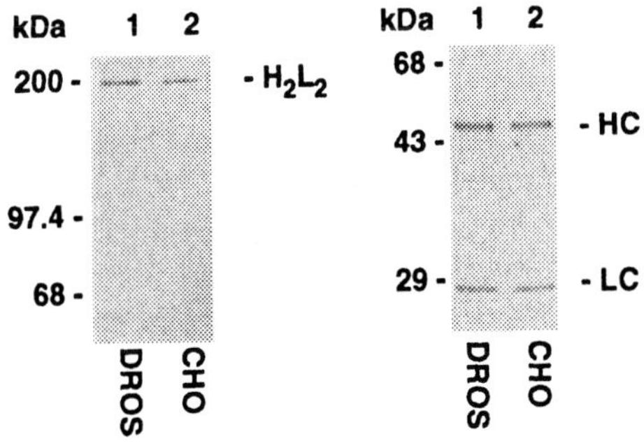

Figure 8 Expression of a fully folded humanized RSHZ19 monoclonal antibody. Analysis of purified antibody by nonreducing (left) or reducing (right) SDS–polyacrylamide gel electrophoresis and Coomassie blue staining. Lane 1, *Drosophila* S2-expressed protein; lane 2, CHO-expressed protein. Reprinted with permission from Kirkpatrick *et al.* (1995a). *The Journal of Biological Chemistry* **270,** 19800–19805.

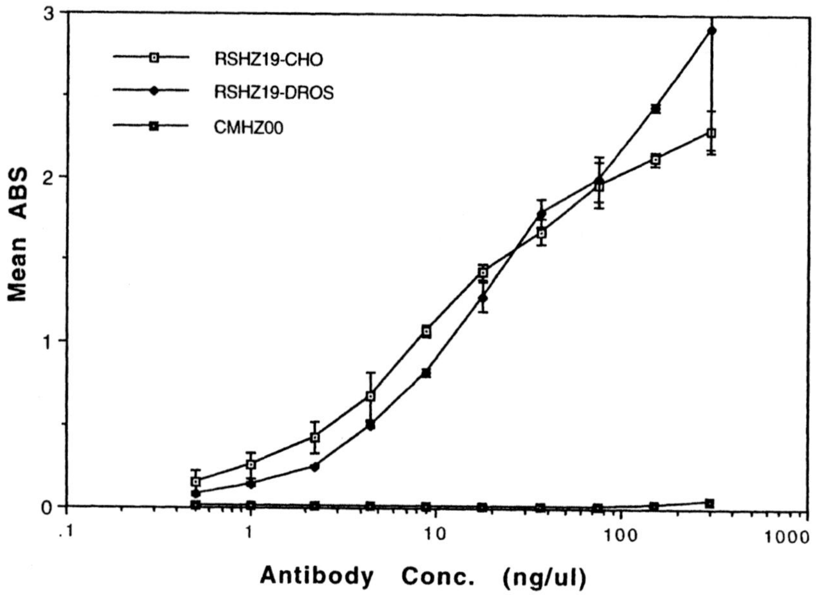

Figure 9 Antigen-binding ELISA assay. RSHZ19 mAb expressed in CHO cells, RSHZ19 mAb produced in *Drosophila* S2 cells, and a mutant form of RSHZ19 mAb, CMHZ00, expressed in CHO cells were compared for binding to RSV F protein. The mAb RSHZ19 served as a positive control (S. Ganguly, unpublished). The mAb CMHZOO lacks binding affinity for F protein and was used as a negative control (S. Ganguly, unpublished). RSHZ19 and CMHZ00 antibodies are equivalent to the reshaped and nonreshaped mAbs HuRSV19VHFNS/VK and HuRSV19VH/VK, respectively (Tempest *et al.*, 1991). ED_{50} = 19 and 13 ng/ml for *Drosophila* CHO-expressed mAbs, respectively. ED_{50} > 300 ng/ml for CMHZ00. Reprinted with permission from Kirkpatrick *et al.* (1995a). *The Journal of Biological Chemistry* **270,** 19800–19805.

difference that may relate to the evolved need for retaining immature immunoglobulins in vertebrate cells.

Another conserved mechanism that has been examined in S2 cells is glycoinositol phosphate (GPI) anchoring. In one study, components required for the GPI anchoring of *Drosophila* acetylcholinesterase were defined (Incardona and Rosenberry, 1996). Mammalian GPI anchors also appear to be recognized efficiently in *Drosophila* S2 cells, as was shown by the expression of Syrian hamster prion protein (Raeber *et al.*, 1995). Membrane-associated prion protein could be dissociated from the membrane by treatment with the GPI-specific lipase, phosphotidylinositol-specific phospholipase C, indicating a specific membrane attachment through a conserved GPI-anchoring mechanism. The S2 system may therefore prove useful in dissecting components of this conserved anchoring system.

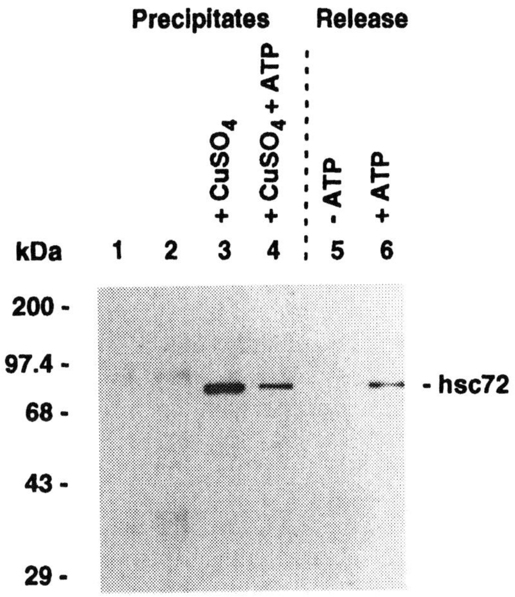

Figure 10 Association of hsc72 with HC immunoglobulins in *Drosophila* cells. Western detection of hsc72 from mtHZ19 cell lysates precipitated with protein G. Lanes: 1, nonrecombinant S2 cells; 2, uninduced mtRZ19 cells; 3 and 4, induced mtHZ19 cells washed in the absence (lane 3) or in the presence of 1 m*M* ATP (lane 4); 5 and 6, protein released from the final wash of protein G precipitations in the absence (lane 5) or in the presence of 1 m*M* ATP (lane 6). Reprinted with permission from Kirkpatrick *et al.* (1995a). *The Journal of Biological Chemistry* **270,** 19800–19805.

Studies on Gene Regulation

In addition to its more recent use as a stable expression system, the *Drosophila* S2 line has also been used extensively as a transient expression system in which to study transcriptional regulatory mechanisms. It has proven to be particularly useful for the study of homeotic genes as well as the homeodomain-containing proteins that bind and activate them. One reason for this is the absence of endogeneously expressed homeotic proteins, which makes this an ideal "null" background in which to do these studies. Another is the ease with which cells can be transiently cotransfected with both target and activator constructs. *Drosophila* S2 cultures apparently contain a subset of cells with an enhanced capacity for DNA uptake (Winslow *et al.*, 1989). Thus, there is a high probability of taking up plasmid DNA in an equimolar ratio in this subset, enabling the study of subcellular interactions between the two constructs. In many cases, the result appears to be an excellent representation of subcellular events occurring in the fly.

The basic experimental design used for the study of homeotic gene transcription in S2 cells involves coexpressing a homeotic gene with a reporter gene construct, containing a homeobox target domain fused to a minimal promoter. Homeobox proteins expressed from one construct bind to the homeobox target-binding site of another construct, thereby enhancing or suppressing the level of reporter gene expression (Fig. 11A). For example, the binding of homeobox proteins fushi taruzu, paired, and zen to the homeobox-binding region of the engrailed was found to enhance reporter gene transcription synergistically, whereas products of the even-skipped and engrailed genes suppressed transcription through binding to the same site (Han *et al.*, 1989). Using this method of analysis, the actions of homeobox proteins, including fushi tarazu, zen, paired, even-skipped, engrailed, ultrabithorax, and antennapedia, were studied on their respective homeotic gene targets, helping to create a better picture of the transcriptional cascade controlling early *Drosophila* development (Jaynes and O'Farrell, 1988; Krasnow *et al.*, 1989; Winslow *et al.*, 1989).

Entire promoters have been dissected in S2 cells. Examples include the actin 5C gene promoter (Chung and Keller, 1990), the retrotransposon mdgl gene promoter (Arkhipova and Ilyin, 1991), and Doc promoter elements (Contursi *et al.*, 1995). In these experiments, a single plasmid was transiently transfected containing a promoter fused to the bacterial chloramphenicol acetyltransferase (CAT) gene as a reporter. Interactions with endogenous S2 transcription factors at the promoter caused enhanced CAT transcription (Fig. 11B). In this way, deletion constructs were used to define important control regions in the promoter regions. Heat shock control elements in the untranslated leader of the hsp70 promoter were also defined in this way. However, these experiments were performed in stably transformed lines (McGarry and Lindquist, 1985). In all of these studies, both positive and negative regulatory components were identified, which are required for transcriptional activation in S2 cells.

The S2 system has also been useful in the study of certain human transcription factors for which there are no homologues in S2 cells. For example, cotransfection assays were used to identify functional domains in the human transcription factor SP1 (Courey and Tijian, 1988). This was done by measuring the effect of wild-type and mutant SP1 expression on the activity of a target promoter fused to the CAT gene. Examples of mammalian transcription

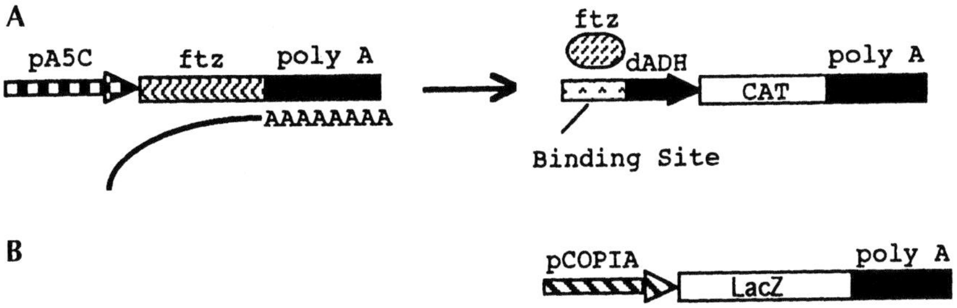

Figure 11 Promoter activation analysis. (A) The study of homeotic gene interactions in a transient assay system. Cotransfections are performed with effector, reporter, and control plasmids. In this example, the effector construct consists of the fushi taruzu (*ftz*) gene, which is transcribed constitutively under control of the actin 5C promoter. The translated ftz protein binds subsequently to a putative activation-binding domain fused to a minimal promoter sequence (dADH, distal alcohol dehydrogenase gene promoter), which on activation through binding of ftz drives the expression of chloramphenicol acetyltransferase (CAT). Expression of β-galactosidase from the control plasmid (*LacZ* gene under constitutive copia promoter activation) is used to normalize expression levels. (B) Promoter delineation in a transient assay system. Deletion constructs from a putative promoter region are prepared and transfected transiently into S2 cells. Endogenous transcription factors in S2 cells bind to regions in the promoter sequence contained in some of the deletions, activating CAT gene expression, thereby delineating positive and negative promoter elements. Expression levels are normalized by comparison to levels of β-galactosidase from the control plasmid (*LacZ* gene under constitutive glyceraldehyde phosphate dehydrogenase promoter activation).

interactions characterized in S2 cells include c-Krox binding at the α 1 collagen promoter (Galera *et al.*, 1994), Kruppel-like factor/SP1 interaction at erythroid promoters (Gregory *et al.*, 1996), and SP1 interaction at both nexin-1 and insulin-like growth factor receptor gene promoters (Werner *et al.*, 1992). Because *Drosophila* S2 cells lack homologous transcription factors, they provide an ideal background in which to study such interactions.

Choosing an Expression System

The S2 system offers a number of advantages as a protein expression system. These are illustrated in Table 5. Compared to other available systems, the S2 system is arguably the simplest to use. In contrast to mammalian cell culture systems (e.g., CHO and COS), which require maintenance twice per week, including trypsinization of adherent cultures, S2 cells are split only once per week simply by inoculating an aliquot of suspension culture into fresh medium. Washing cells is unnecessary. Transfection protocols are also simple and involve very little "hands on" time for the selection of polyclonal lines.

S2 cells are also inexpensive to grow. Media composition does not include expensive supplements such as insulin and nucleotides, which are required for mammalian cell cultures. Serum-free medium conditions can further reduce the cost of S2 cell culture. S2 cells are maintained essentially at room temperature without CO_2 supplementation. One of the drawbacks of this lower temperature growth, of course, is slower growth compared to 37°C expression systems. However, this slower growth does not interfere with production levels.

Compared to transient baculovirus infection of *Spodoptera* cells,

Table 5
Advantages and Disadvantages of *Drosophila* S2 Expression

Advantages	Disadvantages
High density growth at 25°C without CO_2 supplementation	Unable to grow at low density
Growth in suspension/no need for trypsinization	Growth is slower than mammalian cultures
Inexpensive media/serum free or supplemented with 10% serum	
Stable or transient expression	
Stable lines in 3–4 weeks	
One-step gene amplification	
Stable lines amenable to growth in bioreactors for large-scale expression	
Stable lines can be stored indefinitely at −70°C	
Regulated expression at high copy number	Expression levels fall short of those achieved in amplified mammalian cultures or in *E. coli*
High level constitutive or inducible expression	Mammalian promoters function poorly in S2 cells
Higher eukaryotic signals for protein folding, processing, secretion, subcellular localization, and glycosylation	Insect glycosylation is simpler in structure than vertebrates, containing high mannose content and lacking sialic acid
Membrane receptor expression in stable lines	Signaling pathways are uncharacterized
Null background for vertebrate protein expression	

the S2 system is amenable to both transient and stable expression. The ability to produce stably expressing lines is perhaps the most important feature distinguishing these two insect systems. Because baculovirus titers must be optimized each time an expression run is performed, there can be considerable variability in expression levels between experiments. This variability is compounded because viral stock titers tend to drop over time in storage. Furthermore, the growth of *Spodoptera* cultures infected with recombinant baculovirus is lytic.

Harvests of secreted proteins must therefore be timed precisely prior to cell lysis, and intracellular proteins collected after cell lysis sometimes contain proteases that can alter the resultant product. Upon scaleup, these issues can create enormous difficulties. Significant time and effort must be devoted toward optimizing a particular viral stock prior to large-scale culture in bioreactors.

In contrast to baculovirus-infected *Spodoptera* cells, expression in stable S2 cultures is exceptionally reproducible. The authors have carried stable lines for more than 60 passages without any change in expression levels. In addition, when the lines are not in use, they can be frozen indefinitely at –70°C. Cell recovery from these frozen stocks is very efficient. Lines stored for more than 5 years have been recovered without any adverse effects. This reproducibility facilitates large-scale growth greatly in bioreactors. Because growth is nonlytic, there is tremendous flexibility in harvest times. Furthermore, regulated expression under metallothionine induction enables protein production over a defined period.

Tightly regulated inducible expression under Mtn promoter induction in S2 cells is a unique feature of the S2 system (inducible promoters are not as tightly regulated in other eukaryotic systems). Inducible expression is desirable for functional studies in S2 cells in which the comparison of induced and uninduced states provides an important internal control. In addition, maintaining cell lines in the promoter-off state appears to favor cell growth in some cases, which is also likely to contribute to the stability of *Drosophila* lines. Even genes that encode toxic gene products can be maintained under the Mtn promoter. Following induction, the production of these proteins to high levels over a short period is tolerated without adversely affecting cell growth (Johansen *et al.*, 1989).

A high copy number is achieved in mammalian cultures only after months of selection and amplification. In *Drosophila* S2 cells, amplification is a one-step process involving direct insertion of multicopy gene arrays of more than 1000 copies each (Johanson *et al.*, 1989). Each gene copy within the array is apparently functional as the number of gene copies is proportional to the level of transcription. However, even at the highest copy numbers, transcription can still be controlled tightly under the metallothionine promoter. This regulated high copy number expression translates into high level protein production. Unfortunately, the levels do not approach the grams per liter amounts produced in some mammalian expression systems following amplification; however, the levels are still considerably higher than unamplified expression in other systems.

As is the case with other eukaryotic systems, factors required for the proper processing, folding, secretion, subcellular localization, and glycosylation are all present in S2 cells. This enables the high level production of authentic proteins. Except for differences in the pattern of glycosylation added to proteins expressed in S2 cells, protein processing is indistinguishable from that of mammalian expression systems. Furthermore, S2 cells are very effective at secreting proteins into the extracellular fluid as directed either by native signal peptide sequences or by one of two efficient signal peptide sequences engineered into expression vectors (Table 2).

As with mammalian expression systems, S2 cells can express receptors and channels onto the cell surface that can then be studied in whole cell binding or functional assays. In contrast, the lytic nature of baculovirus-infected cultures precludes whole cell studies. The S2 cell expression of receptors also offers advantages over mammalian cell expression because of the null background of endogenous receptors. This enables binding and functional assays to be carried out with low background. However, S2 cells may also lack some of the components required for the signal transduction from vertebrate receptors. Endogenous signal transduction pathways are not well characterized in *Drosophila* S2 cells.

Future Directions

The exceptional versatility of the *Drosophila* S2 system lends itself beautifully to a variety of potential uses. For example, the ability to express vertebrate receptors to high density on the cell surface can be utilized to configure screens based on the high-affinity interaction with radiolabeled ligand. This high density expression combined with a null background of vertebrate receptors should be particularly amenable to high throughput screen development. Indeed, this has already been proven to be a useful approach in screening for antagonists that block the binding of IL-5 to the IL-5 receptor expressed in S2 cells (Fig. 12). This screen led to the identification of a number of new exciting IL-5 antagonists that are currently under study (E. Appelbaum, unpublished). A stable cell line provided a plentiful source of membrane-associated receptors for these studies.

Another feature of the S2 system that can be exploited toward screen development is the efficient transcriptional control achieved under Mtn promoter induction. Combined with a suitable reporter system, the inducible expression of an effector molecule can be used to produce a quantifiable readout in whole S2 cells. Specifically, this can be achieved by cotransfecting a target construct, consisting of a transcriptional-binding site fused to minimal promoter–reporter gene cassette, in combination with an effector construct, consisting of a transcription factor gene under the control of the Mtn promoter (Fig. 13). The reporter gene is activated in *trans* by the production of transcription factor under copper sulfate induction. A transactivation assay to screen for inhibitors of herpes immediate early gene promoter activation has been configured (Ho and Rosenberg, 1998). The ability to titrate the number of reporter genes inserts has proven to be critical for these experiments in order to reduce background expression to acceptable levels.

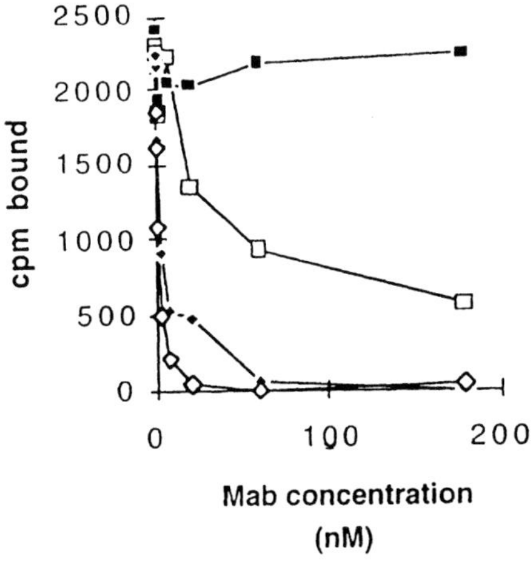

Figure 12 Screen for antagonists of IL-5/IL-5R interaction. Iodinated interleukin-5 was bound to the full-length IL-5 receptor (hIL-5Rα) expressed on *Drosophila* S2 cells in the presence of antibody antagonists 2B6R (■), 3H10 (□), and 2B6 (◆), or with excess IL-5 as the competitor (◇).

Potentially, screens for virtually any subcellular interaction could be configured using the basic inducible promoter design. For example, this design has been adapted to the configuration of an assay screen for inhibitors of a viral protease (Amegadzie, 1998). This was accomplished by cotransfecting the protease gene under the control of the Mtn promoter along with a substrate gene under a constitutive promoter. The induction of protease in this system led to the proteolysis of a constitutively expressed substrate. This assay gives a very effective measure of proteolytic activity as monitored by the direct detection of protein products by SDS–polyacrylamide gel electrophoresis. In the future, this assay could also be adapted for screening by monitoring the increased production of an active product in a secondary activity assay linked to a colorimetric readout.

The future application of S2 cells toward pharmacological assay development, including measures of signal transduction through G-coupled protein receptors expressed in S2 cells, is an exciting prospect. Although the demonstrated expression of functional receptors and channels in S2 cells is an important step in this direction, the limited understanding of signal transduction pathways in *Drosophila* in general will present a significant challenge to the development of the S2 system for this purpose. Nevertheless, the null background of vertebrate receptors in S2 cells should facilitate the

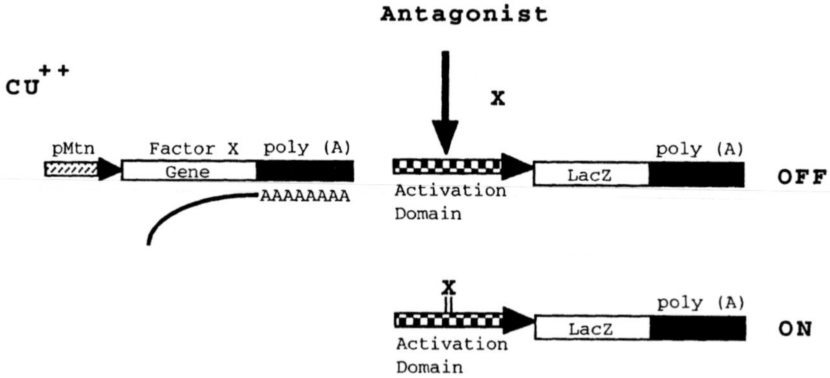

Figure 13 Screen for antagonists of transcriptional activation. Stable lines are constructed by cotransfecting effector and reporter constructs. In this example, the effector gene (factor X) is under inducible metallothionine promoter control. Upon activation, the factor X gene is transcribed and translated into the factor X protein, which binds to the activation domain of a minimal promoter construct, thereby simulating the expression of β-galactosidase from the *LacZ* reporter gene (bottom right). In the presence of an antagonist, factor X cannot bind to the activation domain and *LacZ* remains off (upper right).

analysis of vertebrate receptors in S2 cells greatly, provided that they are functional in this system.

Conclusions

Continuous *Drosophila* cell lines, including the S2 line, are being used increasingly as hosts for recombinant gene expression. This is mainly attributable to the regulated high copy number expression that can be achieved in just a few weeks. Of equal importance is the fact that proteins expressed in this system routinely exhibit authentic properties and functions. This includes membrane-linked, transmembrane, intracellular, and secreted proteins from insect and vertebrate species alike. Furthermore, the system has proven to be an ideal background in which to study mechanisms of DNA transcription, RNA transport, protein folding, protein secretion, cell adhesion, and receptor function. The S2 system provides an attractive alternative to mammalian expression systems for the inducible expression of recombinant proteins in a null background. In addition, stably expressing lines are preferable to transient baculovirus expression for many applications, including large-scale protein production in bioreactors. In the future, applications of the *Drosophila* S2 system toward the development screening assays will add to its many uses as an expression tool.

Appendix

Maintenance of *Drosophila* S2 Lines

- Maintain cells between 23 and 25°C (Table A1).
- The seeding density is 2×10^6 cells/ml. *Note:* Optimal cell growth requires that conditioned medium be carried along with the cells
- Flasks should be shaken gently to loosen semiadherent cells in order to obtain an accurate cell count before splitting

Stable Expression: Selection of Polyclonal Lines Using Hygromycin

1. One day prior to transfection, seed a T25 flask with 3×10^6 cells

in 4 ml (i.e., 0.75×10^6 cells/ml) M3 medium supplemented with 10% heat-inactivated fetal bovine serum albumin (heat activated for 30 min at 65°C to inactivate complement). Incubate flask lying down overnight at 25°C. Seed one flask for each transfection to be performed.

2. Add 40 μl penicillin/streptomycin to each flask prior to adding DNA.

3. Set up calcium phosphate coprecipitation for hygromycin B selection as described in the following table in Falcon No. 2058 polystyrene tubes. *Note:* The copy number can be adjusted by varying the amount of DNA used.

Tube No.	DNA	pCoHygro	1 *M* $CaCl_2$	dH_2O
1	19 μg	1 μg	125 μl	Up to 500 μl
2 (positive)	19 μg (MtaL)	1 μg	125 μl	Up to 500 μl
3 (negative)	19 μg (salmon DNA)	0	125 μl	Up to 500 μl

4. Add each mixture dropwise to another tube containing 0.5 ml 2× Hepes-buffered saline (HBS).

5. Incubate for 40 min at room temperature.

6. Slowly add precipitate mixture to the flasks prepared the day before containing 3×10^6 cells in 4 ml. Mix gently and incubate for

Table A1

Optimized Cell Volumes for Growth of Different Vessels

Vessel	Optimum volume for ideal surface area	Maximum volume for sustained growth	Optimum rpm	Caps
96-well plate	50 μl/well	200 μl/well		Sealed with Parafilm
24-well plate	300 μl/well	600 μl/well		Sealed with Parafilm
12-well plate	600 μl/well	1 ml/well		Sealed with Parafilm
6-well plate	1.5 ml/well	2.5 ml/well		Sealed with Parafilm
T25	5 ml			Closed tightly
T75	15 ml			Closed tightly
T150	30 ml			Closed tightly
125-ml spinner	70 ml	100 ml	100	Loosened
250-ml spinner	120 ml	150 ml	100	Loosened
500-ml spinner	250 ml	500 ml	80–90	Loosened
1000-ml spinner	400 ml	600 ml	70–80	Loosened
3000-ml spinner	800 ml	1000 ml	70–80	Loosened
250-ml shake flash	100 ml	150 ml	115	Loosened
500-ml shake flash	200 ml	300 ml	115	Loosened
1000-ml shake flash	500 ml	700 ml	115	Loosened
3000-ml shake flash	1000 ml	1200 ml	115	Loosened

24 hr at 25°C.

7. The next day, wash cells twice in fresh M3 media containing 10% heat-inactivated fetal bovine serum and 1% penicillin/streptomycin. Collect cells at 1000 × *g* for 30 sec (setting 5 on IEC clinical centrifuge) in a 15-ml conical tube. Use fresh media for the first wash (5 ml) to rinse out the T flask in which the cells were incubated before adding it to the cell pellet, which is first homogenized by tapping the bottom of the tube vigorously. Add fresh medium for the second wash (5 ml) directly to the homogenized cell pellet.

8. Following the second wash, resuspend cells in 5 ml fresh media, transfer back to their original flask, and incubate without drug selection for 48 hr at 25°C.

9. After 48 hr, spin down cells, homogenize the cell pellet as described in step 7, and resuspend in fresh M3 medium containing 10% heat-inactivated fetal bovine serum, 1% penicillin/streptomycin, and 300 μg/ml hygromycin B to begin selection.

10. Replace with fresh selection media every 4–5 days as desdribed in step 9. *Note:* Cells tend to stick to the wall of the flask at lower densities toward the beginning of selection. These will loosen as the densities increase. Therefore, it is essential that the same flask be used throughout the selection procedure.

11. Resistant cell lines should develop in 3–4 weeks. At this time, viable cell counts should be taken on a hemacytometer to determine that positive and experimental cells are growing and that the negative control is dead.

12. After the cells reach a density greater than 4×10^6 cells/ml, they should be split into fresh selection medium to a density of 2×10^6 cells/ml. *Note:* The innoculum should contain cells and conditioned medium.

13. The cells should adapt rapidly to high density growth in serum medium. At this time the cells can be transferred to serum-free medium for protein production. Polyclonal lines should be maintained under selection except in large-scale cultures. *Note:* Antibiotics cannot be used in serum-free media conditions as they cause the cells to die.

Limit Dilution Cloning of S2 Cells in a 96-Well Format

1. Set up calcium phosphate cotransfections as described for stable selection of polyclonal lines (steps 1–9).

1 μg pCoHygro
19 μg expression plasmid

125 μl 1 *M* $CaCl_2$
to 500 μl with DH_2O
Add dropwise to 500 μl HBS.

2. Spin down and resuspend recovered cells in M3 plus 10% heat-inactivated fetal bovine serum, 1% penicillin/streptomycin, and 300 μg/ml hygromycin B.
3. Set up a 96-well plate containing feeder S2 cells: 100 μl/well at 2.5×10^6 cells/ml.
4. Determine cell count of transfection and dilute between 10 and 1000 cells/ml in selection medium. For example, serially dilute a culture at 1×10^7 cells/ml as follows:

a. 100 μl:10 ml = 1 × 105 cells/ml
b. 250 μl of dilution 1 in 25 ml = 1000 cells/ml
c. 2.5 ml of dilution 2 in 25 ml = 100 cells/ml
d. 2.5 ml of dilution 3 in 25 ml = 10 cells/ml

5. Perform dilutions b–d in reagent reservoirs and transfer each directly to 96-well plates using a multichannel pipette (100 μl/well or between 1 and 100 cell/well).
6. Seal in Parafilm and a zip-lock bag.
7. Select 3–5 weeks until transformants outgrow dying feeder layer.
8. Transfer selected transfections to 24-well plates and then to 6-well plates.

Alternatively, clonal lines can be selected from a preexisting polyclonal line starting from step 3.

Reagents

1 M $CaCl_2$: 14.7 g/100 ml
2× HBS: 8.0 g NaCl, 0.135 g Na_2HPO_4, 6.0 g Hepes, pH to 7.1 and raise volume to 500 ml
Penicillin (5000 U), streptomycin (5000 μg) GIBCO-BRL

Induction of Cell Lines

- Cells should be induced with 500 μ*M* $CuSO_4$ when they have reached log phase: approximately 4–6 × 10^6 cells/ml. *Note:* This may need to be determined empirically for different stable cell lines exhibiting different growth properties.

- Typical induction time is 6–7 days but can be varied from 2 to 10 days depending on specific requirements.

Freezing and Reviving Cell Lines

- Pellet cells at 1000×g and then resuspend to a density between 1 and 2×10^7 cells per/ml in freezing medium consisting of M3 medium supplemented with 20% heat-inactivated serum and 10% dimethyl sulfoxide (DMSO).
- Alternatively, cells can be resuspended to a density of 5.0×10^7 cells per/ml in freezing medium consisting of conditioned serum-free medium plus 10% DMSO.
- Aliquot a 1-ml cell suspension into a 1.8-ml NUNC cryo vial.
- Freeze vial gradually to –70°C using a control rate freezer or equivalent.
- Store vials in liquid nitrogen.
- To revive a vial of cells, quick thaw by rubbing the tube between two hands and sterilize with 70% alcohol.
- As soon as cells are thawed, resuspend in 5 ml fresh medium: M3 plus 10% heat-inactivated fetal bovine serum and 1% penicillin/streptomycin for cultures frozen in serum conditions or in serum-free medium for cultures frozen in serum-free conditions.
- Incubate at 25°C overnight.
- The next day, pellet and resuspend the revived culture into fresh medium to remove DMSO, which can interfere with cell growth.

References

Abrams, J. M., Lux, A., Steller, H., and Krieger, M. (1992). Macrophages in *Drosophila* embryos and L2 cells exhibit scavenger receptor-mediated endocytosis. *Proc. Natl. Acad. Sci. U.S.A.* **89**(21), 10375–10379.

Amegadzie, B. (1998). In preparation.

Angelichio, M. L., Beck, J. A., Johansen, H., and Ivey-Hoyle, M. (1991). Comparison of several promoters and polyadenylation signals for use in heterologous gene expression in cultured *Drosophila* cells. *Nucleic Acids Res.* **19,** 5037–5043.

Arkhipova, I. R., and Ilyin, Y. V. (1991). Properties of promoter regions of mdg1 *Drosophila* retrotransposon indicate that it belongs to a specific class of promoters. *Embo J.* **10**(5), 1169–1177.

Ashburner, M. (1989a). "*Drosophila*: A Laboratory Handbook." Cold Spring Harbor Laboratory Press, Cold Spring Harbor, NY.

Ashburner, M. (1989b). "*Drosophila*: A Laboratory Manual." Cold Spring Harbor Laboratory Press, Cold Spring Harbor, NY.

Barthalay, Y., Hipeau-Jacquotte, R., de la Escalera, S., Jimenez, F., and Piovant, M. (1990). *Drosophila* neurotactin mediates heterophilic cell adhesion. *EMBO J.* **9**(11), 3603–3609.

Bellosta, P., Costa, M., Lin, D. A., and Basilico, C. (1995). The receptor tyrosine kinase ARK mediates cell aggregation by homophilic binding. *Mol. Cell. Biol.* **15**(2), 614–625.

Bourouis, M., and Jarry, B. (1983). Vectors containing a prokaryotic dihydrofolate reductase gene transform *Drosophila* cells to methotrexate-resistance. *EMBO J.* **2**(7), 1099–1104.

Braude-Zolotarjova, T., Kakpakov, V., and Schuppe, N. G. (1986). Male diploid embryonic cell line of *Drosophila virilis. In Vitro Cell Dev. Biol.* **22**(8), 481–484.

Brighty, D. W., and Rosenberg, M. (1994). A cis-acting repressive sequence that overlaps the Rev-responsive element of human immunodeficiency virus type 1 regulates nuclear retention of env mRNAs independently of known splice signals. *Proc. Natl. Acad. Sci. U.S.A.* **91**(18), 8314–8318.

Brighty, D. W., Rosenberg, M., Chen, I. S., and Ivey-Hoyle, M. (1991). Envelope proteins from clinical isolates of human immunodeficiency virus type 1 that are refractory to neutralization by soluble CD4 possess high affinity for the CD4 receptor. *Proc. Natl. Acad. Sci. U.S.A.* **88**(17), 7802–7805.

Buckingham, S., Matsuda, K., Hosie, A. M., Baylis, H. A., Squire, M. D., Lansdell, S. J., Miller, N. S., and Sattelle, B. (1996). Wild-type and insecticide-resistant homo-oligomeric GABA receptors of *Drosophila melanogaster* stably expressed in a *Drosophila* cell line. *Neuropharmacology* **35**(9/10), 1393–1401.

Bunch, T. A., Grinblat, Y., and Goldstein, L. S. (1988). Characterization and use of the *Drosophila* metallothionine promoter in cultured *Drosophila melanogaster* cells. *Nucleic Acids Res.* **16**(3), 1043–1061.

Cherbas, L., Moss, R., and Cherbas, P. (1994). Transformation techniques for *Drosophila* cell lines. *Methods Cell Biol.* **44,** 161–179.

Chung, Y. T., and Keller, E. B. (1990). Positive and negative regulatory elements mediating transcription from the *Drosophila melanogaster* actin 5C distal promoter. *Mol. Cell. Biol.* **10**(12), 6172–6180.

Contursi, C., Minchiotti, G., and Di Nocera, P. P. (1995). Identification of sequences which regulate the expression of *Drosophila melanogaster* Doc elements. *J. Biol. Chem.* **270**(44), 26570–26576.

Courey, A., and Tijian, R. (1988). Analysis of SP1 *in vivo* reveals multiple transcriptional domains, including a novel glutamine-rich activation motif. *Cell (Cambridge, Mass.* **55,** 887–898.

de Sousa, S. M., Hoveland, L. L., Yarfitz, S., and Hurley, J. B. (1989). The *Drosophila* Go alpha-like G protein gene produces multiple transcripts and is expressed in the nervous system and in ovaries. *J. Biol. Chem.* **264**(31), 18544–18551.

Di Nocera, P., and Dawid, I. (1983). Transient expression of genes introduced into cultured *Drosophila* cells. *Proc. Natl. Acad. Sci. U.S.A.* **80,** 7095–7098.

Echalier, G., and Ohanessian, A. (1970). In vitro culture of *Drosophila* melanogaster embryonic cells. *In Vitro* **6**(3), 162–172.

Einstein, R., Jackson, J., D'Alessio, K., Lillquist, J. S., Sathe, G., Porter, T., and Young, P. R. (1996). Type I IL-1 receptor: Ligand specific conformation differences and the role of glycosylation in ligand binding. *Cytokine* **8**(3), 206–213.

Elkins, T., Hortsch, M., Bieber, A.J., Snow, P. M., and Goodman, C. S. (1990).

Drosophila fasciclin I is a novel homophilic adhesion molecule that along with fasciclin III can mediate cell sorting. *J. Cell Biol.* **110**(5), 1825–1832.

Fehon, R. G., Kooh, P. J., Rebay, I., Regan, C. L., Xu, T., Muskavitch, M. A. T., and Artavanis-Tsakonas, S. (1990). Molecular interactions between the protein products of the neurogenic loci Notch and Delta, two EGF-homologous genes in *Drosophila*. *Cell (Cambridge, Mass)* **61**(3), 523–534.

Galera, P., Musso, M., Ducy, P., and Karsenty, G. (1994). c-Krox, a transcriptional regulator of type I collagen gene expression, is preferentially expressed in skin. *Proc. Natl. Acad. Sci. U.S.A.* **91**(20), 9372–9376.

Gregory, R. C., Taxman, D. J., Seshasayec, D., Kensinger, M. H., Bieker, J. J., and Wojchowski, D. (1996). Functional interaction of GATA1 with erythroid Kruppel-like factor and Sp1 at defined erythroid promoters. *Blood* **87**(5), 1793–1801.

Hakala, B. E., White, C., and Recklies, A. D. (1993). Human cartilage gp-39, a major secretory product of articular chondrocytes and synovial cells, is a mammalian member of a chitinase protein family. *J. Biol. Chem.* **268**(34), 25803–25810.

Han, K., Levine, M. S., and Manley, J. L. (1989). Synergistic activation and repression of transcription by *Drosophila* homeobox proteins. *Cell (Cambridge, Mass.)* **56**(4), 573–583.

Ho, Y.-S., and Rosenberg, M. (1998). In preparation.

Hortsch, M., and Bieber, A. J. (1991). Sticky molecules in not-so-sticky cells. *Trends Biochem. Sci.* **16**(8), 283–287.

Hultmark, D. (1993). Immune reactions in *Drosophila* and other insects: A model for innate immunity. *Trends Genet.* **9.**

Incardona, J., and Rosenberry, T. (1996). Construction and characterization of secreted and chimeric transmembrane forms of *Drosophila* acetylcholinesterase: A large truncation of the C-terminal signal peptide does not eliminate glycoinositol phospholipid anchoring. *Mol. Biol. Cell* **7**, 595–611.

Ivey-Hoyle, M. (1991). Recombinant gene expression in cultured *Drosophila melanogaster* cells. *Curr. Opin. Biotechnol.* **2**(5), 704–707.

Ivey-Hoyle, M., Clark, R. K., and Rosenberg, M. (1991a). The N-terminal 31 amino acids of human immunodeficiency virus type 1 envelope protein gp120 contain a potential gp41 contact site. *J. Virol.* **65**(5), 2682–2685.

Ivey-Hoyle, M., Culp, J. S., Chaikin, M. A., Hellmig, B. D., Matthews, T. J., Sweet, R. W., and Rosenberg, M. (1991b). Envelope glycoproteins from biologically diverse isolates of immunodeficiency viruses have widely different affinities for CD4. *Proc. Natl. Acad. Sci. U.S.A.* **88**(2), 512–516.

Jaynes, J. B., and O'Farrell, P. H. (1988). Activation and repression of transcription by homeodomain-containing proteins that bind a common site. *Nature (London)* **336**, 744–749.

Johansen, H., van der Straten, A., Sweet, R., Otto, E., Maroni, G., and Rosenberg, M. (1989). Regulated expression at high copy number allows production of a growth-inhibitory oncogene product in *Drosophila* Schneider cells. *Genes Dev.* **3**(6), 882–889.

Johanson, K., Appelbaum, E., Doyle, M., Hensley, P., Zhao, B., Abdel-Mequid, S. S., Young, P., Cook, R., Carr, S., Matico, R., Cusimano, D., Dul, E., Angelichio, M., Brooks, I., Winborne, E., McDonnell, P., Morton, T., Bennett, D., Sokolski, T., McNulty, D., Rosenberg, M., and Chaiken, I. (1995). Binding interactions of human interleukin 5 with its receptor alpha subunit. Large scale production, structural, and functional studies of *Drosophila*-expressed recombinant proteins. *J. Biol. Chem.* **270**(16), 9459–9471.

Jokertst, R. S., Weeks, J. R., Zehring, W. A., and Greenleaf, A. L. (1989). Analysis of the gene encoding the largest subunit of RNA polymerase II in Drosophila. *Mol. Gen. Genet.* **215**(2), 266–275.

Keith, F. J., and Gay, N. J. (1990). The *Drosophila* membrane receptor Toll can function to promote cellular adhesion. *EMBO J.* **9**(13), 4299–4306.

Kirkpatrick, R. B., Ganguly, S., Angelichio, M., Griego, S., Shatzman, A., Silverman, C., and Rosenberg, M. (1995a). Heavy chain dimers as well as complete antibodies are efficiently formed and secreted from *Drosophila* via a BiP-mediated pathway. *J. Biol. Chem.* **270**(34), 19800–19805.

Kirkpatrick, R. B., Matico, R. E., McNulty, D. E., Strickler, J. E., and Rosenberg, M. (1995b). An abundantly secreted glycoprotein from *Drosophila melanogaster* is related to mammalian secretory proteins produced in rheumatoid tissues and by activated macrophages. *Gene* **153**(2), 147–154.

Kirkpatrick, R. B., Emery, J. G., Connor, J. R., Dodds, R., Lysko, P. G., and Rosenberg, M. (1997). Induction and expression of human cartilage glycoprotein-39 in rheumatoid inflammatory and peripheral blood monocyte-derived macrophages. *Exper. Cell Res.* **237,** 46–54.

Krantz, D. E., and Zipursky, S. L. (1990). *Drosophila* chaoptin, a member of the leucine-rich repeat family, is a photoreceptor cell-specific adhesion molecule. *EMBO J.* **9**(6), 1969–1977.

Krasnow, M. A., Saffman, E. E., Kornfeld, K., and Hogness, D. S. (1989). Transcriptional activation and repression by Ultrabithorax proteins in cultured *Drosophila* cells. *Cell (Cambridge, Mass.)* **57**(6), 1031–1043.

Kumar, S., Minnich, M. D., and Young, P. (1995). ST2/T1 protein functionally binds to two secreted proteins from Balb/c 3T3 and human umbilical vein endothelial cells but does not bind interleukin 1. *J. Biol. Chem.* **270**(46), 27905–27913.

Lee, Y. J., Dobbs, M. B., Verardi, M. L., and Hyde, D. R. (1990). dgq: A *Drosophila* gene encoding a visual system-specific G alpha molecule. *Neuron* **5**(6), 889–898.

Lee, Y. J., Shah, S., Suzuki, E., Zars, T., O'Day, P. M., and Hyde, D. R. (1994). The *Drosophila* dgq gene encodes a G alpha protein that mediates phototransduction. *Neuron* **13**(5), 1143–1157.

McGarry, T. J., and Lindquist, S. (1985). The preferential translation of *Drosophila* hsp70 mRNA requires sequences in the untranslated leader. *Cell (Cambridge, Mass.)* **42**(3), 903–911.

Millar, N., Buckingham, S. (1994). Stable expression of a functional homo-oligomeric *Drosophila* GABA receptor in a Drosophila cell line. *Proc. R. Soc. London, Ser. B* **258,** 307–314.

Millar, N. S., Baylis, H. A., Reaper, C., Bunting, R., Mason, W. T., and Sattelle. (1995). Functional expression of a cloned *Drosophila* muscarinic acetylcholine receptor in a stable Drosophila cell line. *J. Exp. Biol.* **198,** 1843–1850.

Moss, R., Cherbas, L., and Cherbas, P. (1985). Transformation of *Drosophila* cells in culture: plasmid recombination and expression of nonselectable markers. *In Vitro Cell Dev. Biol.* **21**(3), 42A.

Pearson, A., Lux, A., and Krieger, M. (1995). Expression cloning of dSR-Cl, a class C macrophage-specific scavenger receptor from *Drosophila melanogaster. Proc. Natl. Acad. Sci. U.S.A.* **92**(9), 4056–4060.

Petersen, U. M., Bjorklund, G., , Ip, Y. T., and Engström, Y. (1995). The dorsal-related immunity factor, Dif, is a sequence-specific trans-activator of *Drosophila* Cecropin gene expression. *EMBO J.* **14**(13), 3146–3158.

Provost, N. M., Somers, D. E., and Hurley, J. B. (1988). A *Drosophila melanogaster* G protein alpha subunit gene is expressed primarily in embryos and pupae. *J. Biol. Chem.* **263**(24), 12070–12076.

Quan, F., Wolfgang, W. J., and Forte, M. A. (1989). The *Drosophila* gene coding for the alpha subunit of a stimulatory G protein is preferentially expressed in the nervous system. *Proc. Natl. Acad. Sci. U.S.A.* **86**(11), 4321–4325.

Quan, F., Thomas, L., and Forte, M. A. (1991). *Drosophila* stimulatory G protein alpha subunit activates mammalian adenylyl cyclase but interacts poorly with mammalian receptors: Implications for receptor-G protein interaction. *Proc. Natl. Acad. Sci. U.S.A.* **88**(5), 1898–1902.

Quan, F., Wolfgang, W. J., and Forte, M. A. (1993). A *Drosophila* G-protein alpha subunit, Gf alpha, expressed in a spatially and temporally restricted pattern during Drosophila development. *Proc. Natl. Acad. Sci. U.S.A.* **90**(9), 4236–4240.

Raeber, A. J., Muramoto, T., Kornberg, T. B., and Prusiner, S. B. (1995). Expression and targeting of Syrian hamster prion protein induced by heat shock in transgenic *Drosophila melanogaster. Mech. Dev.* **51**(2-3), 317–327.

Rio, D., and Rubin, G. (1985). Transformation of cultured *Drosophila melanogaster* cells with a dominant selectable marker. *Mol. Cell. Biol.* **5,** 1833–1838.

Rizki, T. M., and Rizki, R. M. (1984). The cellular defense system of *Drosophila melanogaster. In* "Insect Ultrastructure" (R. C. King and H. Akail, eds.), Vol. 2, pp. 579–604. Plenum, New York.

Rubin, D. M., Mehta, A. D., Zhu, J., Shohan, S., Chen, X., Wells, Q. R., and Palter, K. B. (1993). Genomic structure and sequence analysis of *Drosophila melanogaster* HSC70 genes. *Gene* **128**(2), 155–163.

Samakovlis, C., Kimbrell, D. A., Kylsten, D. A., Engstrom, P., and Hultmark, D. (1990). The immune response in *Drosophila*: Pattern of cecropin expression and biological activity. *EMBO J.* **9**(9), 2967–2976.

Schneider, I. (1972). Cell lines derived from late embryonic stages of *Drosophila melanogaster. J. Embryol. Exp. Morph.* **27,** 353–365.

Schneider, I., and Blumenthal, A. B. (1978). Drosophila Cell and tissue culture. *In* "The Genetics and Biology of *Drosophila*: (M. Ashburner and T. R. F. Wright, eds.), Vol. 2a, pp. 266–305. Academic Press, New York.

Scott, K., Becker, A., Sun, Y., Hardy, R., and Zuker, C. (1995). Gq alpha protein function in vivo: Genetic dissection of its role in photoreceptor cell physiology. *Neuron* **15**(4), 919–927.

Shields, G., and Sang, J. H. (1977). Improved medium for culture of *Drosophila* embryonic cells. *Drosophila Inf. Serv.* **52,** 161.

Simcox, A. A., Sobeih, M. M., and Shearn, A. (1985). Establishment and characterization of continuous cell lines derived from temperature-sensitive mutants of *Drosophila melanogaster. Somatic Cell Mol. Genet.* **11**(1), 63–70.

Snow, P. M., Bieber, A. J., and Goodman, C. S. (1989). Fasciclin III: A novel homophilic adhesion molecule in *Drosophila. Cell (Cambridge, Mass.)* **59**(2), 313–323.

Sondermeijer, P. J., Derksen, J. W., and Lubsen, N. H. (1980). New cell line: Established cell lines of *Drosophila hydei. In Vitro* **16**(11), 913–914.

Talluri, S., Bhatt, A., and Smith, D. P. (1995). Identification of a *Drosophila* G protein alpha subunit (dGq alpha-3) expressed in chemosensory cells and central neurons. *Proc. Natl. Acad. Sci. U.S.A.* **92**(25), 11475–11479.

Tempest, P. R., Bremner, P., Lambert, M., Taylor, G., Furze, J. M., Carr, F. J., and Harris, W. J. (1991). Reshaping a human monoclonal antibody to inhibit human respiratory syncytial virus infection in vivo. *Biotechnology* (NY) **9**(3), 266–271.

Thambi, N. C., Quan, F., Wolfgang, W. J., Spiegel, A., and Forte, M. (1989). Immunological and molecular characterization of Go alpha-like proteins in the *Drosophila* central nervous system. *J. Biol. Chem.* **264**(31), 18552–18560.

Thomas, G., and Elgin, S. (1988). The use of the gene encoding α-amanitin-resistant subunit of RNA polymerase II as a selectable marker in cell transformation. *Drosophila Inf. Serv.* **67**, 85.

Tiemeyer, M., and Goodman, C. S. (1996). Gliolectin is a novel carbohydrate-binding protein expressed by a subset of glia in the embryonic *Drosophila* nervous system. *Development (Cambridge, UK)* **122**(3), 925–936.

Tota, M. R., Xu, L., Sirotina, A., Strader, C. D., and Graziano, M. P. (1995). Interaction of [fluorescein-Trp25]glucagon with the human glucagon receptor expressed in *Drosophila* Schneider 2 cells. *J. Biol. Chem.* **270**(44), 26466–26472.

van der Straten, A., Johansen, H., and Rosenberg, M. (1987). Efficient expression of foreign genes in cultured *Drosophila melanogaster* cells using hygromycin B selection. *In* "Invertebrate Cell Systems Applications" (J. Mitsuhashi, ed.), Vol. 1, pp. 131–134. CRC Press, Boca Raton, FL.

van der Straten, A., Johansen, H., Rosenberg, M., and Sweet, R. W. (1989). Introduction and constitutive expression of gene products in cultured *Drosophila* cells using hygromycin B selection. *Curr. Methods Mol. Cell Biol.* **1**, 1–8.

Wang, W. C., Zinn, K., and Bjorkman, P. (1993). Expression and structural studies of fasciclin I, an insect cell adhesion molecule. *J. Biol. Chem.* **268**(2), 1448–1455.

Werner, H., Bach, M. A., Bach, M. A., Stannard, B., Roberts, C. T., and LeRoith, D. (1992). Structural and functional analysis of the insulin-like growth factor I receptor gene promoter. *Mol. Endocrinol.* **6**(10), 1545–1558.

Winslow, G. M., Hayashi, S., Krasnow, M., Hogness, D. S., and Scott, M. P. (1989). Transcriptional activation by the Antennapedia and fushi tarazu proteins in cultured *Drosophila* cells. *Cell (Cambridge, Mass.)* **57**(6), 1017–1030.

Wolfgang, W. J., and Forte, M. (1995). Posterior localization of the *Drosophila* Gi alpha protein during early embryogenesis requires a subset of the posterior group genes. *Int. J. Dev. Biol.* **39**(4), 581–586.

Yoon, J., Shortridge, R. D., Bloomquist, B. T., Schneuwly, S., Perdew, M. H., and Pak, W. L. (1989). Molecular characterization of *Drosophila* gene encoding G0 alpha subunit homolog. *J. Biol. Chem.* **264**(31), 18536–18543.

Yoshinaga, S. K., and Yamamoto, K. R. (1991). Signaling and regulation by a mammalian glucocorticoid receptor in *Drosophila* cells. *Mol. Endocrinol.* **5**(6), 844–853.

12

BACULOVIRUS EXPRESSION VECTOR SYSTEM

Michael Galleno and August J. Sick
Invitrogen Corporation, Carlsbad, California 92008

Introduction

The baculovirus expression system (BEVS) has become a vital gene expression "tool" in the production and characterization of numerous recombinant proteins. Traditional baculovirus expression systems are based on an *in vivo* (cultured insect cells) low frequency recombination event between a shuttle vector carrying the gene of interest and the circular *Autographa californica* nuclear polyhedrosis virus (AcMNPV) double-stranded (ds) DNA viral genome. Recombinant viruses were subsequently identified by the absence of polyhedra formation on microscopic visualization of infected cells. Polyhedrin is a protein produced by wild-type baculovirus to protect the virus from environmental pressures such as temperature and osmolarity. However, polyhedrin is not required for the formation of viral particles in tissue culture. The recombination efficiency of the original BEVS was very low (about 0.1–2% foreground). With increased use of the system, numerous improvements were made to facilitate the generation and screening of recombinant baculoviruses. These improvements included linearizing the dsDNA AcMNPV genome to increase foreground efficiencies to >50%, liposome-mediated transfection of insect cells to increase the transfected viral titers to 2–4 × 10^4 plaque forming units (PFU)/mg DNA, and "Blue" screening vectors facilitating the visualization of recombinant viruses. Other improvements focused on generating a replication-deficient baculovirus by removing essential sequences from the AcMNPV genome. These sequences were then subsequently replaced by homologous recombination with a baculovirus-bacterial shuttle vector. The use of the "deleted" baculovirus allowed recombination efficiencies to approach 99% foreground and represented a significant advancement in improving the ease of use of the BEVS. Additional advancements focused on using bacterial artificial chromosomes (BACs) (Luckow *et al.*, 1993) or yeast artificial chromosomes (YACs) (Patel *et al.*, 1992), harboring the entire baculovirus genome (117 kb) as an independent episome. The use of BACs or YACs reduced the time necessary to generate a recombinant baculovirus significantly; however, the manipulation of the genomic viral DNA is tedious and yields inconsistent results. Among all of these innovations in the BEVS, the method most utilized because of its reproducibility and ease of use is linearization of AcMNPV DNA with specific restriction endonucleases, thereby deleting viral sequences essential to virus formation (see Fig. 1). Consequently,

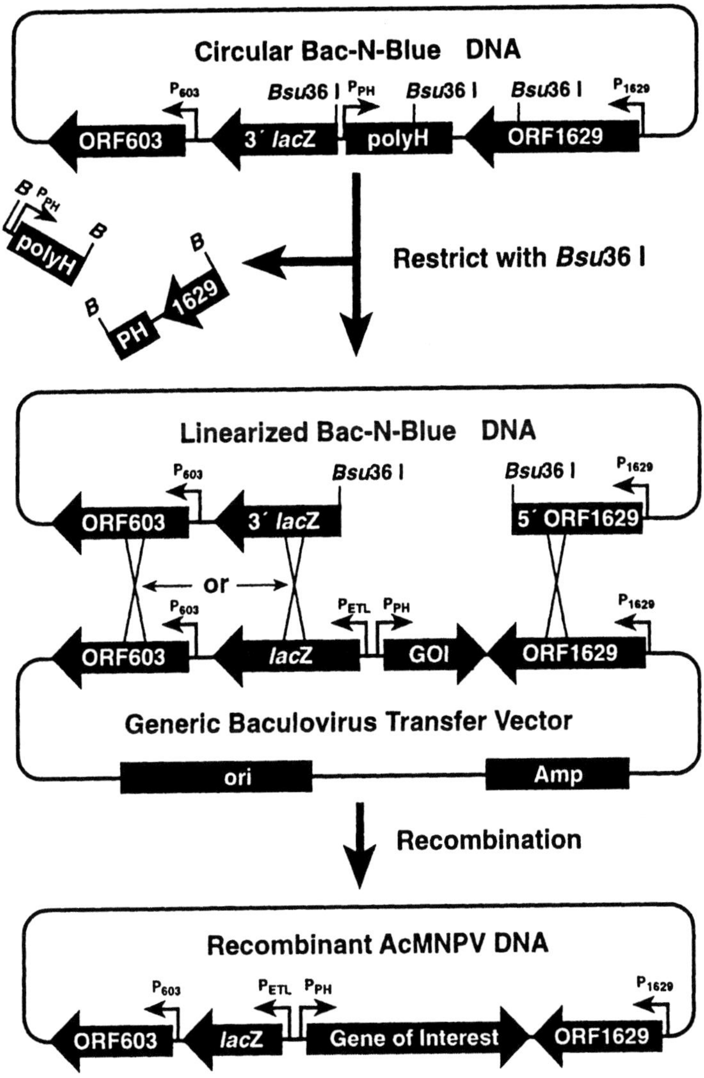

Figure 1 Homologous recombination between a baculovirus transfer vector and AcMNPV Bac-N-Blue DNA. Digestion of a modified AcMNPV DNA with the restriction endonuclease *Bsu*36 I results in removal of the essential C-terminal ORF1629 sequence, the polyhedrin promoter, and the polyhedrin ORF. During transfection, recombination between homologous sequences in the viral DNA and the transfer vector supply the essential sequence needed for replication of recombinant virus. In addition, cutting the DNA at three sites increases the chances of obtaining completely linearized DNA and removing the essential part of ORF1629 (Possee *et al.*, 1991). The result is an increase in the percentage of recombinant virus produced following cotransfection. GOI, gene of interest.

using linear AcMNPV DNA, the number of progeny viruses that are derived from unrecombined or uncut viral DNA is reduced greatly. This process allows for the generation of recombinant baculovirus in a simple two-step process: the construction of a baculovirus shuttle vector containing the cloned gene of interest and the production of recombinant virus by cotransfection of insect cells with linear AcMNPV DNA and the shuttle vector.

The baculovirus expression system has many desirable features that are undoubtedly responsible for its success. The BEVS is a potent eukaryotic expression system that can be utilized to express a variety of recombinant proteins at levels as high as 500 mg per liter (O'Reilly *et al.*, 1994). Another advantage of BEVS is that insect cells possess similar posttranslational modifications (processing pathways) to mammalian cells (O'Reilly *et al.*, 1994), a feature that prokaryotic expression systems lack. These posttranslational modifications include phosphorylation, glycosylation, myristolation, and palmitolation. In addition, proteins expressed in BEVS are almost always soluble, correctly folded, and biologically active. Other attributes to the system include simplified cell growth of several insect cell lines that grow at room temperature (22–27°C), cell growth does not require CO_2 (humidified environment), and most cells are adaptable to growth in serum-free media and suspension culture. Because the baculovirus expression system is a transient (lytic) system, the multiplicity of infection (MOI), choice of cell line, and time of harvest postinfection provide elements of control that allow for the optimal expression of cytoplasmic, secreted, and membrane-bound proteins. Figure 2 outlines the process of using the BEVS from cloning into s shuttle vector to scaleup expression of the recombinant protein.

Background

Baculoviruses are a large family of dsDNA viruses that can infect over 600 different types of invertebrates. The genome size can be between 80 and 220 kb in length, and their host range is limited to mainly insects. The potential of baculoviruses as mediators for heterologous gene expression was investigated when it was noticed that several proteins accumulated to very high amounts (>30% total cell protein) in cells late in the infection cycle. AcMNPV is the

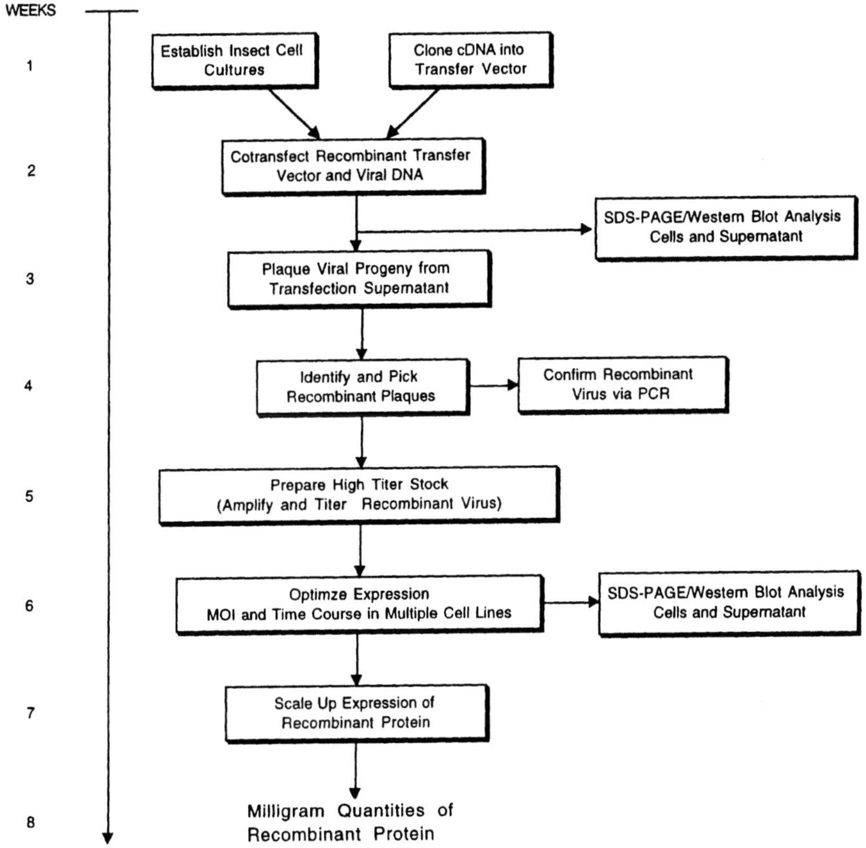

Figure 2 Flow chart of BEVS outlining the generation and characterization of recombinant baculoviruses.

most common baculovirus used as a gene expression system. In AcMNPV-infected cells, two proteins accumulated to large amounts, polyhedrin and p10. For all baculoviruses, replication occurs in the nucleus of the infected cell. Early in the infection cycle, virions are transported from the nucleus into the cytoplasm, where they bud from the cell and infect other cells. In addition, some progeny virions are surrounded by a polyhedrin protein that aggregates and encases the viral particles, forming a characteristic nuclear-localized occlusion body (occ^+). After the insect dies, virions encapsulated in polyhedra are deposited on the surface of vegetation to be ingested by other insects, thereby reinitiating the cycle for viral replication. Propagation of the virus in cell culture renders mechanisms for insect-to-insect dissemination dispensable (i.e., polyhedrin), and replacement with a foreign gene results in virions that can replicate and express the foreign gene product as a recombinant protein (Luckow *et al.*, 1988; Webb *et al.*, 1990).

Originally, the manipulation of baculoviruses was tedious and time-consuming. The heterologous gene was cloned into a shuttle plasmid and introduced into the circular viral genome via low-frequency homologous recombination. Recombinant viruses were subsequently identified via visual inspection of viral plaque with a microscope. Once a possible recombinant virus was identified, it was purified by three independent viral amplifications and plaque assays. As the baculovirus expression system gained widespread success in expressing a variety of genes, the system was modified to make it easier to use. In 1993, Kitts and Possee engineered a baculovirus cloning and expression system that yielded very high (>90%) recombination efficiencies with reproducible results. This system utilized a recombinant AcMNPV derivative (BacPAK6) that included target sites for the restriction endonuclease *Bsu*36I upstream of the polyhedrin gene and within the essential open reading frame (ORF) 1629. In baculovirus, the ORF1629 gene encodes a protein implicated in viral replication and viral capsid formation. Because *Bsu*36I sites are absent in the AcMNPV backbone, digestion with *Bsu*36I deletes part of the ORF1629 and renders the linear virus nonviable. However, the "lethal" deletion could be rescued by recombination with a plasmid that included a wild-type copy of ORF1629. In addition, to make visualizing the recombinant plaques easier, a β-galactosidase gene was incorporated into the baculovirus shuttle vector. Adding X-Gal during the plaques assay (discussed later) allows the recombinant plaques to appear blue.

The polyhedrin gene is not the only dispensable gene in the baculovirus genome. There are a wide variety of proteins that are nonessential for replication and infection in tissue culture conditions (p10, p35, p6.9, and p39) and are expressed at very high levels. This feature allows the construction of a wide variety of baculovirus shuttle vectors, each with a different promoter, to drive expression of a foreign gene. In addition, each promoter in baculovirus is regulated differently and at different time points in the infection cycle. Examples are the polyhedrin and p10 promoters. Both of these promoters are activated very late in the viral life cycle. Unfortunately, this might not be advantageous for certain heterologous proteins as most of the cellular processing machinery is turned over to the virus. Heterologous proteins that are targeted for secretion or the cell surface might be inhibited by the decrease in host protein production. The flexibility of the BEVS can assist in the optimization of gene expression by choosing a viral promoter that is turned on earlier in the life cycle prior to the cessation of host protein synthesis. This option may provide a better environment for posttranslational processing and targeting to specific organelles or extracellular environment. Examples of alternative promoters are the IE1(immediate-early 1), which is an early promoter, and vp39 and p6.9, which are late promoters. The versatility of the BEVS is probably best emphasized by the ability to integrate at more than one loci and express more than one gene from several different viral and cellular promoters.

Certainly this chapter is a simplified review of the basic system that has evolved in the baculovirus expression system used today. Several elements critical to the overall efficiency of the baculovirus expression vector system include (1) insect cell culture, (2) transfer vectors, (3) viral DNA, and (4) transfection reagents. These are all available commercially in many different variations. However, they are used in basically the same manner as outlined in the flow chart in Fig. 2. The many variations, including a wide range of transfer vectors, allow easy identification of recombinant virus, simplified cloning, different promoters, simultaneous expression of more than one gene, protein secretion, and the simple detection and purification of recombinant proteins. A more detailed discussion of how the system works and what should be thought about prior to commencing any baculovirus expression project follows.

Insect Cell Culture

Several insect (*Lepidopteran*) cell lines are available for the manipulation of the virus and for the production of recombinant protein. Differences between the various cell lines should be considered prior to selecting an insect cell line. Table 1 illustrates some of the commonly used insect cell lines with the baculovirus expression system. Some of the criteria for selecting an optimal insect line are growth rate, doubling time, ability to support viral replication and plaque formation, ability to grow in suspension (facilitates scaleup for expression), and ability to support protein synthesis, posttranslational modifications, and secretion. Three of the most commonly used cell lines are Sf9, Sf21, and BTI-TN-5B1-4 (Wickham, 1993;

Table 1
General Characteristics of Commonly Used Cell Lines

Cells	Doubling time	Cell appearance	Preferred medium
Sf9	18–24 hr	Spherical with some granular appearance; regular in size Loose attachment to surfaces	Complete TNM-FH
Sf21	18–24 hr	Spherical with some granular appearance; different sizes Loose attachment to surfaces	Complete TNM-FH
MG-1 (*Trichoplusia ni* midgut cells)	≤ 24 hr	Adherent cells are fibroblast like with multiple nuclei Very difficult to transfer	Complete TNM-FH or serum-free medium (e.g., EXCELL-400)
High Five	≤ 24 hr	Adherent cells are fibroblast like with multiple nuclei and appear as a mix of round and spindle shapes Firm attachment to surfaces	Serum-free medium (e.g., EXCELL-400)

Davis *et al.*, 1992, 1993), known as High Five from Invitrogen. All three cell lines have different morphologies under the microscope but have similar growth characteristics. Sf9 and Sf21 cell lines are the traditional cell lines used with BEVS. These two cell lines originated from the IPLBSF-21 cell line, derived from the pupal ovarian tissue of the fall army worm, *Spodoptera frugiperda* (Vaughn *et al.*, 1977). The Sf9 cell line is a clonal isolate of Sf21 (O'Reilly *et al.*, 1992). The small, regular shape of Sf9 cells makes them exceptional for the formation of monolayers and plaques. Sf21 cells are more disparate in size and form monolayers and plaques that are more irregular. Both cell lines are suitable for transfection, plaque purification, generating high-titer viral stocks, and expressing recombinant proteins. The High Five cell line was developed by the Boyce Thompson Institute for Plant Research and originated from the ovarian cells of the cabbage looper, *Trichoplusia ni* (Davis *et al.*, 1992). High Five cells grow well in monolayer but do not form as regular a monolayer as Sf9 cells, making plaque formation difficult to identify. However, they are easily adaptable to suspension culture using standard spinner or shake flasks.

Sf9, Sf21, and High Five cells are usually maintained at 27°C in Graces insect media supplemented with 10% fetal bovine serum in a nonhumidified environment (Vaughn *et al.*, 1977). However, all are suitable for maintenance under serum-free conditions in commercially available media such as Excel 400 (JRH Biosciences), Insect Express (Biowhittaker), and TC-100. It is recommended that all manipulations of the virus be carried out in Sf9 or Sf21 cells and that High Five cells be reserved for expression purposes. High Five cells have been shown to express secreted proteins at up to 25-fold higher levels than Sf9 cells (Davis *et al.*, 1992) and should always be included in an expression time course optimization study. Optimal culture volumes for spinner flasks and T flasks are shown in the Appendix.

Viral DNA for Cotransfection

In the early years of the BEVS, generating a recombinant baculovirus was very time-consuming and required the experience of a trained eye. Homologous recombination between circular wild-type baculovirus DNA and a shuttle vector *in vivo* occurred at a very low efficiency (less than 1%). Hence a large number of viral plaques needed to be screened by microscopy in order to find a few recombinant viruses. Low recombination efficiency, the difficulty of plating confluent insect cell monolayers, and the difficulty of distinguishing between occ^- from occ^+ plaques made heterologous gene expression in baculovirus an art form that was not mastered easily. However, this problem was overcome by the finding that linearizing baculovirus DNA could render the virus incapable of replicating (Kitts *et al.*, 1990), increasing the recombination efficiency to 25–30%. The use of linearized viral DNA is based on a theory that baculovirus requires circular DNA for efficient replication. Linear viral DNA would then be incapable of replication in insect cells. Homologous recombination of linear viral DNA with a suitable transfer vector restores the replication capabilities to the virus. In addition, the visualization of recombinant viruses is simplified by the incorporation of a β-galactosidase gene (*lacZ*) on the shuttle vector. Therefore, recombinant virus would express the *lacZ* gene, producing blue viral plaques, whereas nonrecombinant virus would not express β-galactosidase, making the identification of recombinant virus less time-consuming.

The linearization of viral DNA is accomplished by placing specific restriction endonuclease sites in viral genes that are common to both transfer vectors and the baculovirus genome (polyhedrin and p10 loci). Unfortunately, the number of possible candidates for restriction sites is hindered by the large size of the viral genome (117 kb). In addition, most restriction enzymes could not cleave the viral DNA very efficiently (<80%). This was compounded by the fact that linear baculovirus DNA (at one site) could recircularize once transfected. The next generation of the BEVS was marked by the development of a modified baculovirus genome with specific restriction sites, *Bsu*36I, positioned at several places in the baculovirus genome, particularly within an essential viral gene, ORF1629. The ORF1629 is a protein implicated in both baculovirus replication and virion assembly. Several restriction sites were placed within the baculoviral genome to both increase the potential of obtaining linear viral DNA and eliminate an essential replication gene. Using the "modified" genome increased the recombination efficiency between the viral DNA and the shuttle vector to greater than 95% (Kitts and Possee, 1993). It must be stressed that because the deletion of the essential gene is created by digestion with a restriction enzyme and that it is impossible to obtain 100% digestion, there is always a low level of background.

Many different varieties of linearized DNA are available from commercial suppliers. Most, if not all, carry the lethal deletion and, depending on the efficiency of digestion and to some extent purification of the DNA, the level of recombination efficiency will vary from 90 to 99%. Other features to be concerned with in addition to the lethal deletion when choosing a viral DNA are the absence of the polyhedrin gene (Occ^- vs Occ^+) and blue/white visualization and the identification of recombinant plaques using β-galactosidase. A more recent advancement (Wittwer and A. J. Sick, unpublished) was the development of a virus that contained all the desired characteristics: a linearized genome with a lethal deletion and Occ^+ parental morphology, with the added benefit of being able to use smaller shuttle vectors.

Baculovirus Transfer Vectors

In addition to the host insect cells and the viral DNA, the shuttle vector is probably the most important aspect of the system. This is the element that introduces the gene of interest into the baculovirus genome and where the expression engineering takes place. Because the BEVS is eukaryotic, a suitable shuttle vector would require specific transcriptional and translational elements that direct gene expression in cultured insect cells as well as whole organisms (insects). One example is a baculovirus-specific promoter to drive expression of the foreign gene. Other examples include an origin of replication, antibiotic selection markers that facilitate maintenance of the shuttle vector in bacteria, a suitable multiple cloning region to facilitate cloning of the foreign gene into the transfer vector, and viral sequences for homologous recombination with the baculovirus genome. Prior to generation of a recombinant virus, the gene of interest must be cloned into a shuttle vector. Although there have been some reports of direct cloning (Peakman *et al.*, 1992) in the viral DNA, the least cumbersome method uses a transfer vector with the foreign cloned. The baculovirus genome is too large to permit easy manipulation for direct cloning. The early baculovirus shuttle vectors only allowed replacement of the polyhedrin gene with a gene of interest. These vectors were very large, ranging between 10 and 14 kb and had only a single cloning site. They were based on the polyhedrin gene locus where an *Eco*RI fragment (7.3 kb) containing the polyhedrin promoter was excised from the viral DNA and subcloned into a bacterial shuttle vector. Many advancements have been made to baculovirus transfer vectors since then to make a versatile paired system between the vector and the linear viral DNA for simplified manipulation.

In addition to carrying the gene of interest with flanking homologous sequences into proximity with the parental viral DNA, the transfer vector is the vehicle where most of the strategy and manipulation occur in order to further optimize gene expression. The transfer vector contains a viral promoter, most often the polyhedrin promoter, followed by various features. These can include a multiple cloning site with and without an ATG start codon, leader sequences with identification and purification tags, leader sequences for signaling secretion, and fusions with reporter proteins.

Cotransfection of Viral DNA and Transfer Vectors

Once the insect cell line has been established, viral DNA prepared, and the gene of interest cloned into a shuttle plasmid, the next step is to cotransfect insect cells with the transfer plasmid and viral DNA. There are several methods for efficiently transferring the DNA into insect cells, including $CaPO_4$, electroporation, and lipid-mediated transfection. The use of lipids can produce greater than 1×10^6 plaque-forming units (PFU) per microgram of circular baculovirus DNA (Hartig *et al.*, 1991). The cotransfection typically involves adding 2×10^6 cells to a 60-mm plate per transfection and allowing them to adhere to the plate. A DNA lipid emulsion is made combining the shuttle plasmid, linear viral DNA, and lipid, with brief vortexing. The cell monolayer is washed several times with serum-free media and finally replaced with the lipid/DNA cocktail by dropwise addition over the cells. The plates are rocked for several hours to allow even application of DNA followed by incubation at 27°C. After 24 hr, the cells begin to show signs of infection. These morphological changes can be visualized under the microscope as progressive events characterized by an increased cell diameter, a bulging nucleus, and a granular appearance (the presence of viral occlusions in wild-type occ^+ virus) culminating in cell lysis after 4–5 days. Table 2 summarizes some morphological changes that occur during baculovirus infection of insect cells. Approximately

Table 2
Morphological Changes of Insect Cells during Infection with Baculovirus

Early phase	
	Increased cell diameter: A 25–50% increase in the diameter of the cells may be observed
	Increased size of cell nuclei: Nuclei may appear to occupy the cell
Late phase	
	Cessation of cell growth: Cells appear to stop growing when compared to cell-only control
	Granular appearance: Signs of viral budding; vesicular appearance to cells
	Viral occlusions: A few occlusion bodies (wild-type virus) may be observed that appear as refractive crystals in the nucleus of the cell
	Detachment: Cells release from the dish or flask
Very late	
	Cell lysis: Cells will burst, leaving signs of clearing in the monolayer and releasing virus to the medium

72 hr posttransfection, the budded recombinant virus will be released into the medium. Using an inverted phase microscope at 250–400× magnification, cells can be inspected for signs of infection. Signs of wild-type viral infection are classified as early (within the first 24 hr), late (24–72 hr), and very late (>72 hr). However, the kinetics of infection may differ between recombinant virus and wild-type virus. The supernatant should contain active viable virus, >90% of which should be recombinant. It is very important to purify the recombinant virus away from any uncut viral DNA (occlusion body plus, occ$^+$) and/or illegitimate (nonhomologous) recombinants that do not contain the gene of interest. Contamination of recombinant DNA with uncut (occlusion$^+$) DNA will lead to dilution of the recombinant virus over time. Wile-type (uncut) virus infects and replicates at higher efficiency than recombinant virus (Huang *et al.*, 1991). In addition, long-term expression studies with a pure, single virus population will ensure reproducible results. The transfection supernatant is harvested and assayed for recombinant plaques via plaque assay (isolating focal points of infection from an agarose overlay). However, if the question is simply whether the baculovirus system can express a certain recombinant protein, a "Fast-Track" approach could be used. This involves using the transfection viral stock directly without plaque purification to infect insect cells and assaying for recombinant protein production via immunological or enzymatic methods.

Plaque Assay

Until recent advancements made with the BEVS, recombination efficiency was so low that the resulting virus from the transfection was mostly wild-type virus. With today's technology, recombinant virus can be isolated away from wild-type virus in the supernatant through use of a visual marker such as β-galactosidase. The plaque assay allows for independent isolation of recombinant virus from wild-type virus. Serial dilutions of the transfection supernatant are added to a monolayer of cultured insect cells and subsequently overlaid with a soft agarose (Appendix). After 7 days, plaques begin to appear in the monolayer. Figure 3 illustrates how a viral plaque is formed in soft agarose. In visualization with an inverted microscope, a recombinant occ$^-$ (occlusion minus) plaque is a dull white color and has a distinctively different morphology as compared to wild-type occ$^+$ (occlusion plus) plaques, which are very refractive

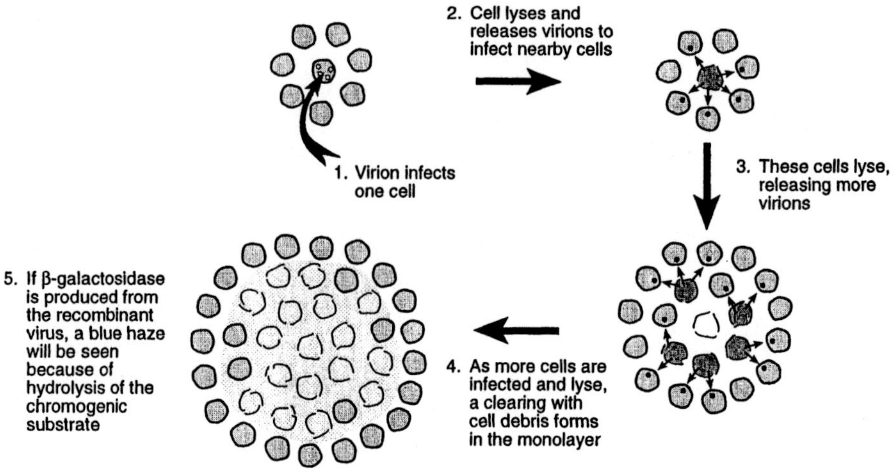

Figure 3 Graphic representation of plaque formation from AcMNPV-infected insect cells.

and shiny. If the recombinant virus harbors a β-galactosidase gene, positive plaques are visual via hydrolysis of a chromogenic substrate (5-bromo-4-chloro-3-indolyl-β-D-galactoside, X-Gal) in the agarose overlay. Recombinant plaques are scored and then selected from the agarose. Most often the density of plaques on the plates is very high, making it almost impossible to pick a well-isolated plaque. Several rounds of plaque purification may be required to obtain a well-isolated pure recombinant.

Once a pure recombinant plaque is chosen, there are several methods to verify that it contains the gene of interest. The most obvious test is to reinfect a small number of cultured cells and harvest media at time intervals of 0, 24, 48, and 72 hr postinfection. These samples are then analyzed by SDS–PAGE, followed by Western blot or ELISA if antibody is available. If recombinant protein is detected, other methods such as Northern hybridization analysis can be carried out to detect the presence of the gene or if a RNA transcript was produced. Most frequently, polymerase chain reaction is used to analyze the putative positive plaques quickly to verify if the gene is present and only those positively identified recombinants carried forward (Appendix).

Advancements in viral DNA have increased the efficiency of recombination from <1% to almost 100%; however, it must be stressed again that even a low percentage of wild-type virus must be removed if long-term use of the recombinant virus is to be anticipated. There is no substitute to a plaque assay for isolating a single viral recombinant. It is completely analogous to plating an *Escherichia coli* transformation of a supercoiled plasmid. One would still pick a well-isolated colony, prepare and characterize the plasmid DNA, and then make a master glycerol stock.

Generation of High-Titer Stock

Once a positive plaque has been identified, it is necessary to amplify or expand the recombinant virus through the generation of a high-titer viral stock (HTS). Isolation and expansion of a pure recombinant virus are absolutely critical to the future generation of viral progeny and to accurate and consistent expression results. The generation of a HTS usually involves infection of insect cells at volumes greater than 100 ml and incubation to near complete cell lysis. Infection volumes depend on how much is needed. Therefore, enough HTS should be produced to carry out several gene expression analyses and the titer determination by plaque assay with several orders of magnitude dilution to determine the number of viable virions (titer). Usual titers range between 5×10^7 and 5×10^8 PFU/ml. Once the titer is determined, some primary HTS should be aliquoted further into 5-ml quantities and frozen at –70°C. This serves as a master stock of recombinant virus for future expansion. On defrosting, the viral titer will have dropped significantly due to freezing, but there will be enough of the original virus intact to infect 100 ml to 1 liter of cells. The same viral can be reamplified several times. However, no more than three generations should be passaged from the original HTS due to degeneration of the virus and to the formation of interfering particles (Wickham *et al.*, 1992) that will compete with active virus.

Expression Optimization Study

Once the HTS has been generated and titered, expression levels need to be optimized. A typical optimization study includes testing variables such as the multiplicity of infection (MOI), time of harvesting, and several different cell lines. MOI can range from 0.5 to 10 (virions per cells) and will assist in achieving a synchronous infection where cells are infected at the same time point. Harvesting at different periods postinfection can allow for the recovery of intact recombinant protein without significant protein degradation. Each cell line (Sf9, Sf21, MG-1, and High Five) has slightly different growth characteristics, and both posttranslational modifications and the kinetics of infection may vary. The study can be performed in T flasks or in suspension cultures. It should be noted that expression levels are not comparable between infected monolayers and suspension cultures. If large amounts of recombinant protein are to be required, then it is recommended to proceed with cells in suspension as they are more easily scaled up. In either case, cells and supernatants harvested are analyzed by SDS–PAGE and or Western blot at several time points postinfection (i.e., 0, 24, 48, 72, and 96 hr) to determine optimal conditions. An example of expression optimization is the expression of respiratory syncytial virus fusion (F) protein in both Sf9 and *T. ni* (High Five) cells (Parrington *et al.*, 1997). The yield of secreted RSV-Fa protein was over sevenfold higher in High Five cells than Sf9 cells. In addition, N-terminal sequencing of the recombinant protein reveals that the secretory signal was processed completely in High Five cells but not in Sf9 cells.

Scaled-Up Expression

Once expression levels are optimized successfully, the next step is to proceed to the large-scale expression of recombinant protein. For some researchers, this may mean moving to suspension culture for the first time or moving up to larger vessels (1 liter or more). Others may wish to go still larger and utilize airlift bioreactors and/or fermentors. Table 3 summarizes the requirements and options that exist for large-scale expression of protein using the baculovirus expression system.

Scaling up expression means increasing the volume of media, cells, and virus used. The increasing ratio of vessel volume to surface area decreases the availability for oxygen transfer to the growing cells. Sparging oxygen becomes necessary to reach high cell densities. However, with sparging oxygen there is a tendency for foaming, resulting in a loss of cell viability. The availability of serum-free media and the use of antifoam decrease foaming and have made large-scale production feasible in airlift fermentors and stirred bioreactors. Table 3 summarizes some of the requirements and benefits of scaleup production of recombinant protein in the BEVS.

Table 3
Methods, Requirements, and Benefits for Scale-Up Production of Recombinant Protein

Method	Requirement	Benefit
Stirred bioreactor	For a 5 liter bioreactor 5-liter bioreactor Sterilized tubing Microbial air filters High purity nitrogen, oxygen, and air pH, dissolved oxygen and temperature probes External dissolved oxygen controller External pH controller Peristaltic pump for acid/base lines Linear recorder to monitor dissolved oxygen and pH control Laminar flow hood in close proximity to the bioreactor	Addresses increased oxygen needs of large-scale culture Controlled growth and optimization of variables in the culture Increased cell densities Elevated protein production Reproducible results for batch production of protein
Airlift fermentor	For a 5 liter airlift fermentor 5-liter airlift fermentor system Dissolved oxygen control module Dissolved oxygen electrode Microbial air filters 1/4 in. stainless-steel tubing Silicone tubing Circulating water bath	Addresses increased oxygen needs of large-scale culture Control over environmental variables in the culture Increased cell densities Elevated protein production

Basic Research of Baculoviruses as a Tool for Gene Expression

Several unique features have made the baculovirus (AcMNPV) the expression system of choice for many applications. The widespread use of the baculovirus as an expression system and as a natural biological insecticide is reflected by the progress that has been made in the study of baculovirus genetics and molecular biology. Research efforts aimed at studying and characterizing elements involved in the infection and replication functions have increased the utility of baculoviruses. Some of these advances are outlined briefly in the following paragraphs.

Examination of the 5′ and 3′ regions of various baculoviral genes has led to some interesting observations relative to promoter structure and essential sequences for overexpression of the polyhedrin and p10 proteins. Although certain consensus eukaryotic sequence motifs are present (i.e., TATA and CAAT boxes), sequences surrounding the transcriptional start site are essential for high-level expression of proteins. The promoter sequence encompasses a TAAG motif and is conserved in other baculoviruses. Mutational analysis suggests that even single point mutations in this region can reduce expression levels by as much as 1000-fold (Rankin *et al.*, 1988; Gross and Rohrmann, 1993). Similarly, deletions in the 3′ regions of the polyhedrin gene, in some cases encompassing the polyadenylation site, also reduced expression levels. Additional transcriptional studies have focused on analyzing alternative polyadenylation sites that can be used in place of native sequences. One example utilized synthetic sequences based on the 3′ end of rabbit β-globin compared to 3′ ends of an SV40-based, baculovirus early gene poly(A) signal and the AcMNPV ORF 1629 poly(A) signal (Westwood *et al.*, 1993). Data indicated that the rabbit β-globin 3′ region was more proficient than the other poly(A) regions when transcription was driven by the polyhedrin promoter. Results from the aforementioned studies have facilitated the construction of baculovirus multigene expression cassettes. Along with numerous other studies, these investigations expanded the utility of baculoviruses as expression hosts. Similar investigations to increase the yield of the desired protein are currently being undertaken. Examples include identifying sequences responsible for viral replication, characterization of proteins involved in the infection process, and the development of alternative insect cell lines.

AcMNPV possesses a complex genome structure encoding over 70 proteins. Only a limited number of these genes are well characterized. The characterization of several baculovirus genes and their respective ORFs has led to advances in identifying their role in viral multiplication, infection, heterologous protein production, and pathogenesis in insects. Two such advancements involve the identification of genes required for DNA replication as well as viral genes that influence cellular apoptosis. AcMNPV replication involves over six genes and eight origins of replication. Two gene products, LEF-1 and LEF-2, have been shown to interact via protein–protein interaction assay and through their interaction support transient DNA replication (Evans *et al.*, 1997). How the genes implicated in viral replication will affect the use of the baculovirus system as a heterologous gene expression tool remains to be seen.

There is a distinct area of research that examines the roles baculoviral gene products have in cellular apoptosis. The p35 gene is required to block apoptosis during baculovirus infection in some insect cell lines. Apoptotic cell death results in the fragmentation of nuclear DNA and subsequent compartmentalization into membrane vesicles. Baculoviruses lacking the p35 gene grow poorly because the cells die before efficient budding or occlusion body formation takes place (Clem *et al.*, 1991). LaCount and Friessen (1997) suggested that baculovirus-induced apoptosis in insect cells can be initiated by early and late viral replication events after virus entry. In addition, expression of the AcMNPV p35 in the developing embryo and eye of *Drosophila melanogaster* eliminates the vast majority of programmed cell death (Hay *et al.*, 1994). The identification of molecules that interact with AcMNPV p35 provides a model system for studying how cellular death is regulated and represents an alternative use for baculovirus and its gene products.

Expression Examples

Heterologous gene expression systems, such as the BEVS, are employed for the production of recombinant proteins because they are able to express various genes to higher levels than they are normally expressed in their native systems. What makes the BEVS so widespread is their ability to generate high amounts of recombinant protein without compromising authenticity. It provides a eukaryotic environment that is conducive to disulfide bridge formation, proper folding, oligomerization, and other posttranslational modifications required for biological activity. Expression examples include proteins that have been N-glycosylated (Kuroda *et al.*, 1990), O-glycosylated, myristilated (Luo *et al.*, 1990), palmitoylated, carboxy methylated (Lowe *et al.*, 1990), phosphorylated, and acylated. In some cases, the sites of the posttranslational modifications of proteins produced in the BEVS are identical to mammalian systems. Table 4 summarizes some of the proteins that have been expressed in the baculovirus expression system.

The flexibility of the capsid structure allows the insertion of

Table 4
Heterologous Protein Expression

Protein	Origin	Reference
β-Hexoaminidase B	Human	Pennybacker *et al.* (1997)
Myosin light chain kinase	Human	Lin *et al.* (1997)
DNA helicase II	Human	Zhang and Grosse (1997)
Bullous pemphigoid antigen 2	Human	Masunaga *et al.* (1997)
Lymphotoxin-α1β2 complex	Human	Williams-Abbott *et al.* (1997)
Werner's syndrome (WRN) helicase	Human	Suzuki *et al.* (1997)
Angiostatin	Murine	Wu *et al.* (1997)
Serotonin 5-HT7 receptor	Murine	Obosi *et al.* (1997)
Replication factor C (hRFC)	Human	Cai *et al.* (1997)
Endothelial nitric oxide (eNOS)	Bovine	Ju *et al.* (1997)
Growth hormone receptor	Human	Bieth *et al.* (1997)
MAL	—	Puertollano *et al.* (1997)
Leptin receptor	—	Devos *et al.* (1997)
Insulin receptor substrate-1	Human	Algenstaedt *et al.* (1997)
Manganese-dependent superoxide dismutase	Human	Wright *et al.* (1997)
Parvovirus VP2 capsid protein	Porcine	Sedlik *et al.* (1997)
Varicella-zoster virus (VZV) glycoproteins E and I (heterodimer)	—	Kimura *et al.* (1997)
Hawaii calicivirus capsid protein	—	Green *et al.* (1997)
Cardiac Ca^{2+} pump and phospholamban	Canine	Autry and Jones (1997)
Type IV collagenase/gelatinase	Human	George *et al.* (1997)
Eukaryotic initiation factor-2B (eIF-2B)	Human	Fabian *et al.* (1997)
Cellobiohydrolase	Fungal	von Ossowski *et al.* (1997)
Respiratory syncytial virus fusion protein	—	Parrington *et al.* (1997)

very large genes in the AcMNPV genome. Genes as large as 10 kb have been efficiently recombined and expressed in baculovirus (O'Reilly, 1994). The theoretical upper size limit for the insertion of foreign genetic elements into the AcMNPV genome is not known.

Multigene Expression in Baculoviruses

One attractive feature of the BEVS is the ability to coexpress multiple genes. Hazama *et al.* (1995) demonstrated the coexpression of both subunits of *Xenopus* bone morphogenetic protein in Sf9 cells cotransfected with two recombinant viruses, each harboring an individual subunit. This approach has been utilized extensively to express heterodimeric proteins such as various receptors, growth factors, transcription factors, trimeric enzymes (St. Angelo *et al.*, 1987), and functional immunoglobulins (Hasemann and Capra, 1990). As an alternative approach to infection with two or more recombinant baculoviruses, a single baculovirus transfer vector could be constructed to facilitate the coexpression of five or more genes. Belyaev *et al.* (1995) engineered a single expression vector with tandem copies of the polyhedrin and p10 promoters to drive expression of five different genes (blue tongue virus VP4, VP6, VP7, NS1, NS2 proteins). This type of flexibility offered by the BEVS could prove useful in the future structure/function characterization of various complex enzymes, and in operons, such as the *Helicobacter pylori* urease operon (9 genes) (Cussac, 1992). In addition, coexpression of two or more related proteins in the baculovirus expression system can provide an ideal environment for the assembly of heterologous viral particles.

Use of Baculovirus to Produce Potentially Therapeutic Agents

The ability to express multiple genes and modify them correctly holds immense potential in the development of the BEVS as the system of choice for producing vaccines and other therapeutic agents. Significant contributions for the further development of baculovirus as a system to assess therapeutic agents were made by the expression of both FIV and HIV *gag* and *pol* genes (Overton *et al.*, 1990; Royer, 1991) and their use in examining potential HIV proteinase inhibitors. Intact HSV-1 (herpes simplex virus type 1) capsid particles have been formed using the BEVS as well as proteins involved in virion maturation studies (Robertson *et al.*, 1997). Other examples where heterologous virions were expressed and assembled include bluetongue virus (Belyaev *et al.*, 1995), bovine rotavirus (Labbé *et al.*, 1991), parvovirus B19, and Newcastle disease virus. The use of baculoviruses to express potential mammalian vaccines has been cited throughout the literature. One such example utilized the BEVS to produce both soluble and secreted forms of human parainfluenza virus type 3 glycoproteins (Lehmann *et al.*, 1993). The proteins were coexpressed in both CHO cells and baculovirus and were found to be comparable in immunogenicity and efficacy when injected into cotton rats. CHO cells and baculovirus-synthesized human parainfluenza virus type 3 glycoproteins were shown to elicit similar levels of neutralizing antibodies. Other therapeutic agents synthesized include a vaccine against swine transmissible gastroenteritis virus spike glycoprotein (Shoup *et al.*, 1997).

Baculovirus Expression System as an Alternative to Bacterial and Other Eukaryotic Systems

Baculovirus can provide environments where overexpressed proteins can fold properly. Although most bacterial expression systems can generate amounts of protein in excess of eukaryotic systems, they are unable to perform complex posttranslational modifications to the recombinant protein, such as disulfide bond formation and glycosylation. Baculovirus, yeast, and mammalian expression systems all possess the ability to produce heterologous proteins that resemble their native counterpart closely. The baculovirus system compared to other higher eukaryotic systems (i.e., mammalian) can produce proteins in the hundreds of milligrams per liter range or 50% of total cellular protein (O'Reilly *et al.*, 1992). Mammalian systems can modify expressed genes efficiently, but often yield a low amount of recombinant protein. Therefore, the BEVS combines the ability to generate large amounts of recombinant protein with "higher" eukaryotic posttranslational modifications.

Guidelines to Optimize Heterologous Gene Expression in Baculovirus

As in most gene expression systems, the efficiency of heterologous gene expression in baculovirus can vary from gene to gene, depending on the nature of the protein. There is little that can be done to influence the intrinsic nature of the gene without changing the overall genetic content. However, there are several guidelines to optimizing expression. The most important ones are good laboratory practices and an understanding of basic virological techniques. This is especially relevant during the plaque assay to estimate viral titer and/or purify recombinant baculovirus. Even though recent advances have allowed researchers to generate recombinant virus at high efficiencies (>90% positives), small traces of wild-type virus can potentially reduce the heterologous gene expression level at later stages (Huang *et al.*, 1991). Other aspects of heterologous gene expression that influence expression levels are related to mRNA transcription/processing and protein translation/processing. As in most eukaryotes, the length of the 5′-untranslated region can influence expression levels by interfering with translational initiation (Kozak, 1989). Eliminating certain AU-rich regions in the 3′-untranslated end of gene of mRNAs can improve RNA stability. AU-rich regions have been implicated in mRNA stability in most eukaryotes, including insects (*Drosophila*) (Yost *et al.*, 1990). Translational initiation in insects can be influenced by the context of the initiation codon AUG (Kozak, 1989). However, several reports to the contrary demonstrate a minimal effect on translation initiation when the context of the initiation codon was altered (Hills and Crane-Robinson, 1995). Similarly, codon usage in baculovirus does not seem to influence gene expression levels significantly. This is supported by the expression of *E. coli lacZ,* which has many rare codons but can be produced to levels comparable to native polyhedrin (500 mg/liter) (Pennock *et al.*, 1984). However, in-depth codon usage tables have been compiled from the codons of highly expressed AcMNPV genes (Ranjan and Hasnain, 1995) and can be used as a guide in assembling synthetic genes for expression in the BEVS.

Several publications are available that cover the baculovirus expression system and baculoviral genetics in great detail (O'Reilly *et al.*, 1992; Richardson, 1995; Summers and Smith, 1988). Because the factors that determine how well a foreign gene is expressed in the BEVS have not yet been well characterized, it is difficult to predict how efficiently different genes will be expressed. However, this feature is common among all gene expression systems (bacterial, yeast, and mammalian). Fortunately, several methods can be used to facilitate the expression of foreign genes in the baculovirus expression system. These include high transfection efficiency through the cotransfection of linear AcMNPV DNA with a shuttle plasmid into insect cells, use of the plaque assay to assess recombination efficiency and viral titer, the time course study/expression analysis, and the relative ease in the growth/maintenance of insect cell lines.

Summary and Future

Selecting the optimal expression system for heterologous genes can be critical in obtaining biologically active recombinant protein. Deciding on what expression system to use can be influenced by the overall complexity of the system, the proposed use of the recombinant protein, and the quantities of recombinant protein required. The BEVS has several unique features that make it the system of choice for many applications. These features include the ability to express functional recombinant protein at high levels with "higher eukaryotic" posttranslational modifications. It has the capacity to express large genes (>8 kb), can express multiple genes simultaneously, and is easy to use. Advancements in baculovirus technology have facilitated the use of the baculovirus expression system both in agricultural settings as an effective biological insecticide (see O'Reilly *et al.*, 1992, for detail) and in academia and industry as a potent producer of recombinant proteins.

Future developments of the baculovirus expression system will focus on expanding its use in pharmaceutical and academic environments. A major area studied will be the generation of nonlytic, stable baculoviruses that can produce high levels of recombinant proteins. Advances in this area are already underway using the IE promoters of baculoviruses (Jarvis *et al.*, 1990) and gaining an understanding on the mechanism by which baculovirus mediates host protein production. Other research areas will focus on the generation and screening of cDNA expression libraries in the baculovirus expression system. Ward *et al.* (1995) have demonstrated that expression libraries could be generated and screened in the baculovirus. The possibility of using a eukaryotic virus to clone, screen, and express cDNA libraries could offer significant advantages over traditional bacteriophage methods. Nevertheless, it is clear that the system has yet to reach its full potential. The prospect of expressing 10 or more genes simultaneously may offer new insights into unraveling complex multiprotein or multicomponent vaccine interactions with high structural complexity.

Appendix

Seeding Densities and Volumes for Infections

The table gives approximate seeding densities and volumes for typical vessel sizes. Infection at these densities in the minimal volumes listed will yield optimal infection.

Minimal volumes: The total volumes used are lower than those used in general cell culture and maintenance. This is done so that the virus added is kept concentrated and can infect cells more readily.

Cell density: Cell density in adherent culture is approximately 50% confluent to allow maximal cell surface area for contact with virus and subsequent infection.

MOI: Use an MOI of 5.0–10.0 for a time course of protein expression or a large-scale protein preparation.

Amount of virus to add: The amount of virus to add depends on MOI.

Type of vessel	Cell density	Final volume (culture medium + added virus)
96-well plate	2.0×10^4 cells/well	100 μl
24-well plate	5.0×10^5 cells/well	500 μl
12-well plate	7.0×10^5 cells/well	750 ml
6-well plate	1.0×10^6 cells/well	1 ml
60-mm² plate	2.5×10^6 cells/plate	3 ml
25-cm² flask	3.0×10^6 cells/flask	5 ml
75-cm² flask	9.0×10^6 cells/flask	10 ml
150-cm² flask	1.8×10^7 cells/flask	15–20 ml
Spinners (all)	$2.0\text{–}2.5 \times 10^6$ cells/ml	No more than half of the total volume of the flask

PCR Analysis of Recombinant Baculovirus

Polymerase chain reaction technology allows a quick, safe, nonradioactive method to determine the presence of an insert in a putative recombinant virus and confirms the isolation of a pure,

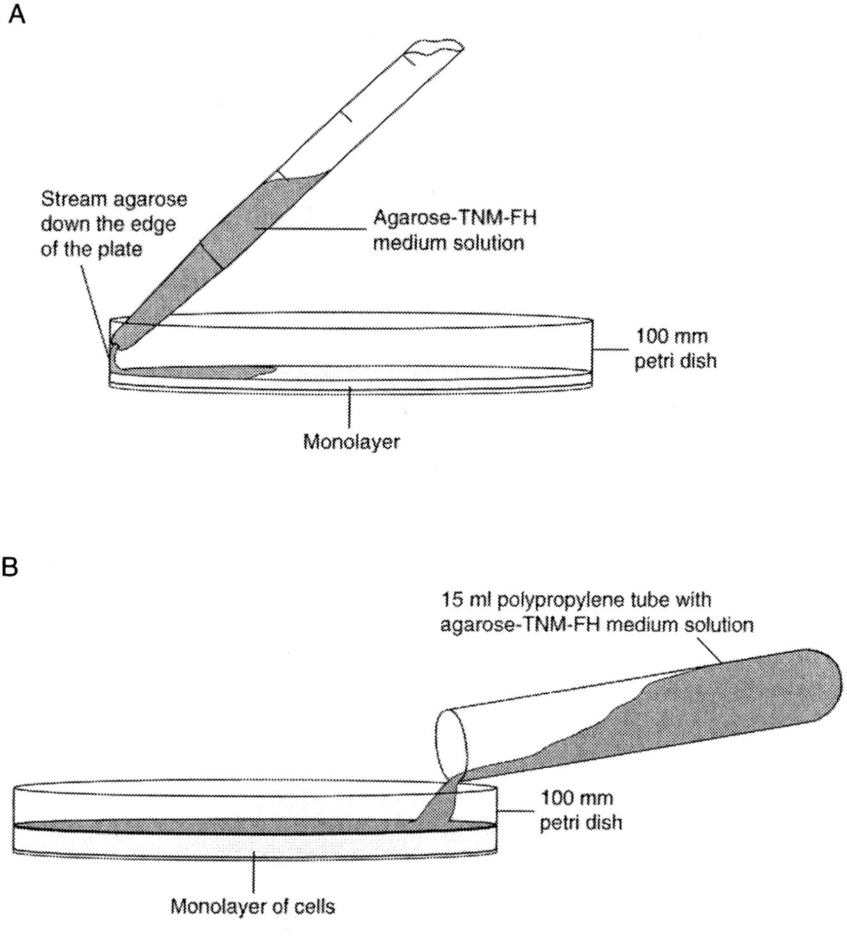

Figure A1 Graphic representation of pouring an agarose overlay on baculovirus-infected insect cells. The agarose medium mixture can either be aliquoted using a pipette (A) or poured gently across the plate from one edge (B); however, be sure not to disturb the monolayer.

recombinant plaque. PCR analysis is a fast and efficient way to rule out false positives at an early stage. Because the size of the foreign gene insert is known, the size of the PCR product expected can be determined easily and wild-type contamination detected.

Outline of PCR Analysis This table outlines the steps needed to perform PCR analysis to confirm isolation of recombinant virus.

Stage	Description
1	Calculate the expected size of the PCR fragment based on the primers that are being used
2	Grow putative recombinants in 12-well microtiter plates
3	Inspect wells and select only those that are occ⁻
4	Resuspend cells into the medium and harvest 0.75 ml of the suspension for PCR analysis
5	Continue to incubate cells in microtiter plates until they lyse
6	Isolate viral DNA from the 0.75-ml sample and perform PCR on the viral DNA
7	Analyze the results on a 1% agarose gel and select only recombinant virus that contains the expected fragment (there should be no contamination with wild-type virus)
8	When cells have lysed, return to the microtiter plates and harvest all the medium from wells containing only pure recombinant virus for the P-1 viral stock

Primers Recombinant baculovirus forward and reverse PCR primers are available commercially. The following forward and re-

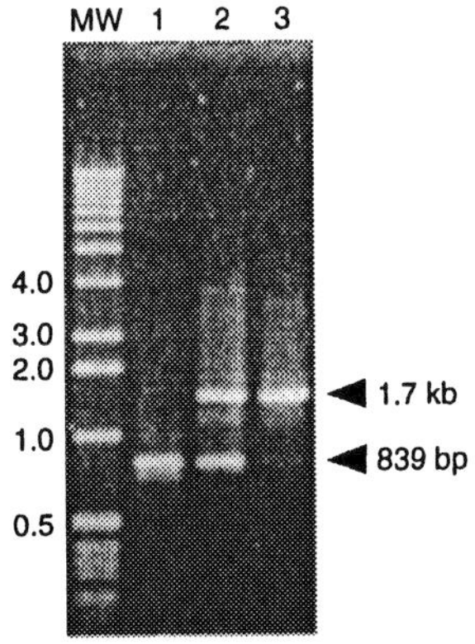

Figure A2 A typical PCR analysis of recombinant viral clones is shown. Two samples were analyzed for the recombinant PCR fragment (1.7 kb) and the wild-type Bac-N-Blue DNA PCR fragment (839 bp). DNA was isolated from 0.75 ml of the P-1 viral stock and 5 μl was used as a template in a PCR reaction with the recombinant baculovirus forward and reverse primers. Ten microliters from each sample was analyzed on a 1% gel. DNA was also isolated from a wild-type (occ$^+$) plaque as a negative control. Lane 1 is the wild-type (occ$^+$) control; lane 2 demonstrates that a putative recombinant plaque is actually a mixture of recombinant and wild-type virus; and lane 3 is a pure, recombinant plaque.

verse primers have been designed to flank the polyhedrin region. These primers are compatible with all polyhedrin promoter-based baculovirus transfer vectors. The forward PCR primer binds from –44 (nt 4049) to –21 (nt 4072) in front of the start of the polyhedrin gene, using the nomenclature of O'Reilly *et al.* (1992). The reverse PCR primer binds at 794 (nt 4886) to 774 (nt 4866) 3′ to the polyhedrin gene.

Forward primer (-44): 5′-TTTACTGTTTTCGTAACAGTTTTG-3′ T_m = 62°C

Reverse primer (794): 5′-CAACAACGCACAGAATCTAGC-3′ T_m = 58°C

The following table provides solutions to common problems with the plaque assay.

Problem	Reason	Solution
No plaques	The kinetics of infection are variable for each recombinant virus	If it is less than Day 6 postplating, be patient. If no plaques appear after Day 8 or 9, proceed to investigate other possibilities
	There was not a confluent monolayer on Day 2 or Day 3 postinfection	Seed 5 × 10^6 cells in a 100-mm plate (50% confluence). Cells should double at least once before infection stops growth
		Agar was too hot. More cells growing on one side of the plate where the agar cooled, than on the other.
		Cell viability was not >95% or cells were not in log phase. Check viability of cells before seeding plates
	Incubation in a humidified incubator can cause condensation that runs down the side of the agar and onto the monolayer, interfering with plaque formation	Incubate in a sealed container or bag with damp paper towels. Remove towels or open container if condensation appears

Problem	Reason	Solution
	Virus titer too low	Do not dilute virus as much. Use undiluted, 10^{-1}, or 10^{-2} dilutions. Transfection may have to be repeated to increase viral titer
Small plaques difficult to visualize	Too many cells seeded over area used	Seed fewer cells and redo the plaque assay
Too many plaques or complete cell lysis	Virus titer too high	Dilute virus to 10^{-5} or 10^{-6} and repeat plaque assay

References

Algenstaedt, P., Antonetti, D. A., Yaffe, M. B., and Kahn, C. R. (1997). Insulin receptor substrate proteins create a link between the tyrosine phosphorylation cascade and the Ca^{2+}-ATPases in muscle and heart. *J. Biol. Chem.* **272,** 23696–23702.

Autry, J. M., and Jones, L. R. (1997). Functional co-expression of the canine cardiac Ca2+ pump and phospholamban in *Spodoptera frugiperda* (Sf21) cells reveals new insights on ATPase regulation. *J. Biol. Chem.* **272,** 15872–15880.

Belyaev, A. S., Hails, R. S., and Roy, P. (1995). High-level expression of five foreign genes by a single recombinant baculovirus. *Gene* 229–233.

Bieth, E., Cahoreau, C., Cholin, S., Molinas, C., Cerutti, Rochiccioli, P., Devauchelle, G., and Tauber, M. (1997). Human growth hormone receptor: Cloning and expression of the full-length complementary DNA after site-directed inactivation of a cryptic bacterial promoter. *Gene* **194**(1), 97–105.

Cai, J., Gibbs, E., Uhlmann, F., Philips, B., Yao, N., O'Donnell, M., and Hurwitz, J. (1997). A complex consisting of human replication factor C p40, p37, and p36 subunits is a DNA-dependent ATPase and an intermediate in the assembly of the holoenzyme. *J. Biol. Chem.* **272**(30), 18974–18981.

Clem, R., Fechheimer, M., and Miller, L. K. (1991). Prevention of apoptosis by a baculovirus gene during infection of insect cells. *Science* **254,** 1388–1390.

Davis, T. R., Trotter, K. M., Granados, R. R., and Wood, H. A. (1992). Baculovirus expression of alkaline phosphatase as a reporter gene for evaluation of production, glycosylation, and secretion. *Bio/Technology* **10,** 1148–1150.

Davis, T. R., Wickham, T. J., McKenna, K. A., Granados, R. R., Shuler, M. L., and Wood, H. A. (1993). Comparative recombinant protein production of eight insect cell lines. *In Vitro Cell Dev. Biol.* **29A,** 388–390.

Devos, R., Guisez, Y., Van der Heyden, J., White, D. W., Kalai, M., Fountoulakis, M., and Plaetinck, G. (1997). Ligand-independent dimerization of the extracellular domain of the leptin receptor and determination of the stoichiometry of leptin binding. *J. Biol. Chem.* **272**(29), 18304–18310.

Evans, J. T., Leisy, D. J., and Rohrmann, G. F. (1997). Characterization of the interaction between the baculovirus replication factors LEF-1 and LEF-2. *Virology* **71,** 3114–3119.

Fabian, J. R., Kimball, S. R., Heinzinger, N. K., and Jefferson, L. S. (1997). Subunit assembly and guanine nucleotide exchange activity of eukaryotic initiation factor-2B expressed in Sf9 cells. *J. Biol. Chem.* **272,** 12359–12365.

George, H. J., Marchand, P., Murphy, K., Wiswall, B. H., Dowling, R., Giannaras, J., Hollis, G. F., Trzaskos, J. M., and Copeland, R. A. (1997). Recombinant human 92-kDa type IV collagenase/gelatinase from baculovirus insect cells: Expression, purification, and characterization. *Protein Express. Purif.* **10,** 154–161.

Green, K. Y., Kapikian, A. Z., Valdesuso, J., Sosnovtsev, S., Treanor, J. J., and Lew, J. F. (1997). Expression and self-assembly of recombinant capsid protein from the antigenically distinct Hawaii human calicivirus. *J. Clin. Microbiol.* **35,** 1909–1914.

Gross, C. H., and Rohrmann, G. F. (1993). Analysis of the role of 5′ promoter elements and 3′ flanking sequences on the expression of a baculovirus polyhedron envelope protein gene. *Virology* **192,** 273–281.

Hartig, P. C., Cardon, M. C., and Kawanishi, C. Y. (1991). Generation of recombinant baculovirus via liposome-mediated transfection. *BioTechniques* **11,** 310.

Hasemann, C. A., and Capra, J. D. (1990). High-level production of a functional immunoglobulin heerodimer in a baculovirus expression system. *Proc. Natl. Acad. Sci. U.S.A.* **87,** 3942–3946.

Hay, B. A., Wolff, T., and Rubin, G. M. (1994). Expression of baculovirus P35 prevents cell death in Drosophila. *Development (Cambridge, UK)* **120,** 2121–2219.

Hazama, M., Aono, A., Ueno, N., and Fujisawa, Y. (1995). Efficient expression of a heterodimer of bone morphogenetic protein subunits using a baculovirus expression system. *Biochem. Biophys. Res. Commun.* **209,** 859–866.

Hills, D., and Crane-Robinson, C. (1995). Baculovirus expression of human basic fibroblast growth factor from a synthetic gene: Role of the Kozak consensus and comparison with bacterial expression. *Biochim. Biophys. Acta* **1260,** 14–20.

Hu, L. T., Foxall, P. A., Russell, R., and Mobley, H. L. (1992). Purification of recombinant heliobacter pylori urease apoenzyme encoded by ureA and ureB. *Infect. Immun.* **60,** 2657–2666.

Huang, Y. S., Bobseine, K. L., Setzer, R. W., and Kawanishi, C. Y. (1991). Selection kinetics during serial cell culture passage of mixture of wild-type *Autographa californica* nuclear polyhedrosis virus and its recombinant Ac360-β-gal. *J. Gen. Virol.* **72,** 2653–2660.

Jarvis, D. L., Fleming, J. A., Kovacs, G. R., Summers, M. D., and Guarino, L. A. (1990). Use of early baculovirus promoters for continuous expression and efficient processing of foreign gene products in stably transformed lepidopteran cells. *Bio/Technology* **8,** 950–955.

Ju, H., Zou, R., Venema, V. J., and Venema, R. C. (1997). Direct interaction of endothelial nitric-oxide synthase and caveolin-1 inhibits synthase activity. *J. Biol. Chem.* **272**(30), 18522–18525.

Kimura, H., Strauss, S. E., and Williams, R. K. (1997). Varicella-zoster virus glycoproteins E and I expressed in insect cells form a heterodimer that requires the N-terminal domain of glycoprotein I. *Virology* **233,** 382–391.

Kitts, P. A., and Possee, R. D. (1993). A method for producing recombinant baculovirus expression vectors at high frequency. *BioTechniques* **14,** 810–817.

Kitts, P. A., Ayers, M. D., and Possee, R. D. (1990). Linearization of baculovirus

DNA enhances the recovery of recombinant virus expression vectors. *Nucleic Acids Res.* **18,** 5667–5672.

Kozak, M. (1989). The scanning model for translation: An update. *J. Cell Biol.* **108,** 229–241.

Kuroda, K., Geyes, H., Gayer, R., Doerfler, W., and Klenk, H. D. (1990). The oligosaccharides of influenza virus haemagglutinin expressed in insect cells by a baculovirus vector. *Virology* **174,** 418–429.

Labbé, M., Charpilienne, A., Crawford, S. E., Estes, M. K., and Cohen, J. (1991). *J. Virol.* **65,** 2946–2952.

LaCount, D. J., and Friesen, P. D. (1997). Role of early and late replication events in induction of apoptosis by baculovirus. *J. Virol.* **71,** 1530–1573.

Lehmann, D. J., Roof, L. L., Brideau, R. J., Aeed, P. A., Thomsen, D. R., Elhammer, A. P., Wathen, M. W., and Homa, F. L. (1993). Comparison of soluble and secreted forms of human parainfluenza virus type 3 glysoproteins expressed from mammalian and insect cells as subunit vaccines. *J. Gen. Virol.* **74,** 459–469.

Lin, G., Wang, X., Long, Q., Pang, Y., Wong, A. O., and Yu, K. (1997). Production of recombinant goldfish hormone I in a baculovirus expression system. *Chin. J. Biotechnol.* **13,** 91–97.

Lowe, P. N., Sydenham, M., and Page, M. J. (1990). The Ha-ras protein, p21, is modified by a derivative of mevalonate and methyl-esterified when expressed in the insect baculovirus system. *Oncogene* **5,** 1045–1048.

Luckow, V. A., and Summers, M. D. (1988). Trends in the development of baculovirus expression vectors. *Bio/Technology* **6,** 47–55.

Luckow, V. A., Lee, S. C., Barry, G. F., and Olins, P. O. (1993). Efficient generation of infectious recombinant baculoviruses by site-specific transposon-mediated insertion of foreign genes into a baculovirus genome propagated in *Escherichia coli. J. Virol.* **67,** 4566–4579.

Luo, L., Li, Y., and Kang, C. Y. (1990). Expression of gag precursor protein and secretion of virus-like gag particles of HIV-2 from recombinant baculovirus infected insect cells. *Virology* **179,** 874–890.

Masunaga, T., Shimizu, H., Yee, C., Borradori, L., lazarova, Z., Nishikawa, T., and Yancey, K. B. (1997). The extracellular domain of BPAG2 localizes to anchoring filaments and its carboxyl terminus extends to the lamina densa of normal human epidermal basement membrane. *J. Invest. Dermatol.* **109**(2), 200–206.

Obosi, L. A., Hen, R., Beadle, D. J., Bermudez, I., and King, L. A. (1997). Mutational analysis of the mouse 5-HT7 receptor: Importance of the third intracellular loop for receptor-G-protein interaction. *FEBS Lett.* **412**(2), 321–324.

O'Reilly, D. R., Miller, L. K., and Luckow, V. A. (1994). "*Baculovirus* Expression Vectors: A Laboratory Manual." Freeman, New York.

Overton, H. A., McMillan, D. J., Gridley, S. J., Brenner, J., Redshaw, S., and Mills, J. S. (1990). *Virology* **179,** 508–511.

Parrington, M., Cockle, S., Wyde, P., Du, R. P., Snell, E., Yan, W. Y., Wang, Q., Gisonni, L., Sanhueza, S., Ewasyshyn, M., and Klein, M. (1997). Baculovirus expression of the respiratory syncytial virus fusion protein using Trichoplusia in insect cells. *Virus Genes* **14,** 63–72.

Patel, G., Nasmyth, K., and Jones, N. C. (1992). A new method for the isolation of recombinant baculovirus. *Nucleic Acids Res.* **20,** 97–104.

Peakman, T. C., Harris, R., and Gewert, D. R. (1992). Highly efficient generation of recombinant baculovirus by enzymatically mediated site-specific in vitro recombination. *Nucleic Acids Res.* **20,** 495–500.

Pennock, G. D., Shoemaker, C., and Miller, L. K. (1984). Strong and regulated expression of Escherichia coli β-galactosidase in insect cells with a baculovirus vector. *Mol. Cell. Biol.* **4,** 399–406.

Possee, R. D., Sun, T. P., Howard, S., Ayres, M., Hill-Perkins, M., and Gearing, K. (1991). Nucleotide sequence of the Autographa californica nuclear polydedosis 9.4 kbp EcoRI-I and -R (polyhedrin gene) region. *Virology* **185,** 229–241.

Puertollano, R., Li, S., Lisanti, M. P., and Alonso, M. A. (1997). Recombinant expression of the MAL proteolipid, a component of glycolipid-enriched membrane microdomains, induces the formation of vesicular structures in insect cells. *J. Biol. Chem.* **272**(29), 18311–18315.

Ranjan, A., and Hasnain, S. E. (1995). Codon usage in the prototype baculovirus-*Autographa californica* nuclear polyhedrosis virus. *Indian J. Biochem. Biophys.* **32,** 424–428.

Rankin, C., Ooi, B. G., and Miller, L. K. (1988). Eight base pairs encompassing the transcriptional start point are the major determinant for baculovirus polyhedrin gene expression. *Gene* **70,** 39–50.

Richardson, C. D. (1995). Baculovirus expression protocols. *Methods Mol. Biol.* **39.**

Robertson, B. J., McCann, P. J., III, Matusick-Kumar, L., Preston, V. G., and Gao, M. (1997). Na, an autoproteolytic product of the herpes simplex virus type 1 proteinase can functionally substitute for the assembly protein ICP35. *J. Virol.* **71,** 1683–1687.

Royer, M., Cerutti, M., Gay, B., Hong, S. S., Devauchelle, G., and Boulanger, P. (1991). Functional domains of HIV-1 gag-polyprotein expressed in baculovirus-infected cells. *Virology* **184,** 417–422.

Sedlik, C., Saron, M., Sarraseca, J., Casal, I., and Leclerc, C. (1997). Recombinant parvovirus-like particles as an antigen carrier: A novel nonreplicative exogenous antigen to elicit protective antiviral cytotoxic T cells. *Proc. Natl. Acad. Sci., U.S.A.* **94,** 7503–7508.

Shoup, D. I., Jackwood, D. J., and Saif, L. J. (1997). Active and passive immune responses to transmissible gastroenteritis virus (TGEV) in swine inoculated with recombinant baculovirus-expressed TGEV spike glycoprotein vaccines. *Am. J. Vet. Res.* **58,** 242–250.

St. Angelo, C., Smith, G. E., Summers, M. D., and Krug, R. M. (1987). Two of three influenza viral polymerase proteins expressed by using baculovirus vectors form a complex in insect cells. *J. Virol.* **61,** 361–365.

Summers, M. D., and Smith, G. E. (1988). A manual of methods for baculovirus vectors and insect cell culture procedures. *Tex. Agric. Exp. Stn. [Bull.]* **B-1555.**

Suzuki, N., Shimamoto, A., Imamura, O., Kuromitsu, J., Kitao, S., Goto, M., and Furuichi, Y. (1997). DNA helicase activity in Werner's syndrome gene product synthesized in a baculovirus system. *Nucleic Acids Res.* **25**(15), 2973–2978.

Vaughn, J. L., Goodwin, R. H., Tompkins, G. J., and McCawley, P. (1977). The establishment of two cell lines from the insect *Spodoptera frugiperda* (Noctuidae). *In Vitro* **13,** 213–217.

von Ossowski, I., Teeri, T., Kalkkinen, N., and Oker-Blom, C. (1997). Expression of a fungal cellobiohydrolase in insect cells. *Biochem. Biophys. Res. Commun.* **233,** 25–29.

Ward, V. K., Kreissing, S. B., Hammock, B. D., and Choudary, P. V. (1995). Generation of an expression library in the baculovirus expression vector system. *J. Virol. Methods* **53,** 263–277.

Webb, N. R., and Summers, M. D. (1990). Expression of proteins using recombinant baculoviruses. *Technique* **2,** 173–188.

Westwood, J. A., Jones, I. M., and Bishop, D. H. (1993). Analysis of alternative poly(A) signals for use in baculovirus expression vectors. *Virology* **195,** 90–99.

Wickham, T. J., Davis, T., Granados, R. R., Shuler, M. L., and Wood, H. A. (1992). Screening of insect cell lines for the production of recombinant proteins and infectious virus in the baculovirus expression system. *Biotechnol. Prog.* **8,** 391–396.

Williams-Abbott, L., Walter, B. N., Cheung, T. C., Goh, C. R., Porter, A. G., and Ware, C. F. (1997). Th elymphotoxin-alpha (LT-alpha) subunit is essential for the assembly, but not for the receptor specificity, of the membrane-anchored LTalpha1beta2 heterotrimeric ligand. *J. Biol. Chem.* **272**(31), 19451–19456.

Wright, G., Reichenbecher, V., Green, T., Wright, G. L., and Wang, S. (1997). Paraquat inhibits the processing of human manganese-dependent superoxide dismutase by Sf-9 insect cell mitochondria. *Exp. Cell Res.* **234,** 78–84.

Wu, Z., O'Reilly, M. S., Folkman, J., and Shing, Y. (1997). Suppression of tumor growth with recombinant murine angiostatin. *Biochem. Biophys. Res. Commun.* **236**(3), 651–654.

Yost, H., Peterson, R., and Lindquist, S. (1990). RNA metabolism: Strategies for regulation in heat shock response. *Trends Genet.* **6,** 223–227.

Zhang, S., and Grosse, F. (1997). Domain structure of human nuclear DNA helicase II (RNA helcase A). *J. Biol. Chem.* **272,** 11487–11494.

Section IV

TRANSGENIC EXPRESSION

13

RECOMBINANT PROTEIN EXPRESSION IN TRANSGENIC MICE

Rula Abbud and John H. Nilson

Department of Pharmacology, Case Western Reserve University, Cleveland, Ohio 44106

Gene Expression Systems: Using Nature for the Art of Expression

Overview and Background

As the century closes, the entire scientific community enthusiastically awaits completion of both human and mouse genome projects (Lander, 1996). As evident from the sequencing of the entire genome of the yeast *Saccharomyces cerevisiae* (Goffeau *et al.*, 1996), a plethora of genes of unknown function will emerge with the sequencing of each new eukaryotic genome. This information, in turn, will spawn new revolutionary approaches for further understanding gene function in health and disease as well as increase our understanding of animal and human biology. Although this chapter focuses on reviewing advances made in using transgenic mice as experimental tools for understanding gene function, these approaches draw heavily on the progress made in human and mouse genomic sequencing projects.

Transgenic mice offer a variety of paradigms for probing gene function. This chapter discusses several transgenic strategies that lead to overexpression, altered expression, or underexpression of specific genes, with emphasis on using these strategies to probe complex physiological pathways. This chapter also reviews how transgenic mice can be used to model human disease and discusses the authors' experience in targeting overexpression of a chimeric gonadotropin to establish transgenic mice that model a human reproductive disorder. This chapter describes the use of transgenic mice as bioreactors to generate large amounts of therapeutically useful proteins. Finally, the future of this field is considered by emphasizing the need for inducible control of the temporal and spatial pattern of transgene expression.

Transgenic Mice: General Overview

Palmiter *et al.* (1982) first demonstrated that introducing a foreign gene into the germ line can alter the phenotype of an animal. Microinjection and subsequent stable integration of a metallothionine (MT)-growth hormone fusion gene resulted in a giant mouse phenotype. With the ensuing reiteration of this approach by several investigators (Palmiter and Brinster, 1986), the mouse soon became the preferred animal for transgenic studies. Advances in mouse genetics and embryology allow for the routine manipulation of both the mouse and its genome (Bieberich *et al.*, 1993; Polites and Pinkert, 1993). In fact, facilities for generating transgenic mice are now available around the globe, and the scientific community via the Internet is compiling several important databases for use. A list of these databases can be found at the following address: http://www.gdb.org/Dan/tbase/docs/dblist.html. The existence of these Internet resources is a strong indication of the widespread use and applicability of transgenic mice.

Although several methods have been devised for introducing foreign DNA into the germ line of mice, only two are widely used. Direct injection of cloned DNA into the male pronucleus of a fertilized mouse egg (illustrated in Fig. 1) remains the method of choice for adding new genes into the mouse genome (Polites and Pinkert, 1993), especially where the site of integration is not a major factor. Cloned DNA can also be introduced into primary cultures of embryonic stem cells prepared from the inner cell mass of blastocysts derived from developing embryos (Doetschman, 1993; see also Fig. 1). This approach is used for assessing the functional significance of a specific gene via targeted disruption ("knockout") through homologous recombination. Because this chapter is concerned primarily with the overexpression of added genes, the authors will concentrate on transgenic paradigms that utilize DNA microinjection.

Technical aspects of DNA microinjection and important parameters regarding the preparation of cloned DNA for optimal integration have been described in detail (Nilson *et al.*, 1995; Polites and Pinkert, 1993) and therefore are only reviewed here briefly. Among the more important parameters are linearization of the cloned DNA and the removal of plasmid vector sequences. The presence of the latter often suppresses promoter activity in cloned genes injected into the male pronucleus for reasons that still remain unexplained. Typically, several hundred copies of a cloned gene are injected into the male pronucleus of the fertilized egg. Injected eggs are then implanted into the oviduct of a pseudopregnant foster mother mouse. Although the efficiency of integration varies with each construct, approximately 25% of the mice born from the

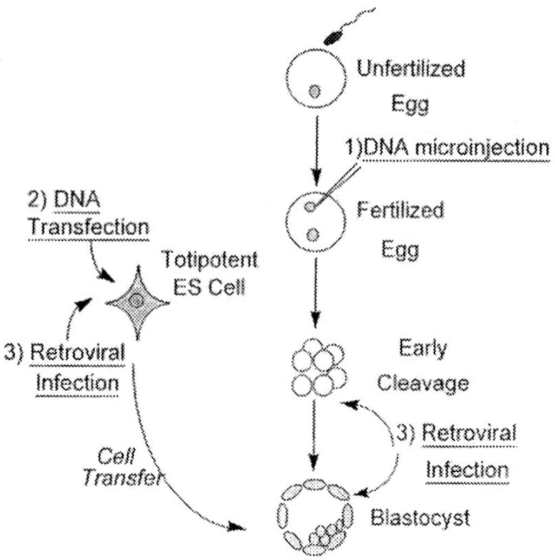

Figure 1 Different strategies for the generation of transgenic mice: (1) DNA microinjection into the pronucleus of a fertilized egg, (2) transfection of DNA into embryonic stem (ES) cells, or (3) retroviral infection of ES cells or embryos. Fertilized eggs are transferred into pseudopregnant mice. ES cells are injected into blastocysts prior to transfer into the animal.

pseudopregnant animals will harbor one to several hundred copies of the injected DNA. These founder animals are termed "transgenic," and the injected DNA is referred to as the "transgene."

With microinjection, integration of the transgene occurs randomly and generally before the first cell division. Approximately 70% of the founder animals will carry the transgene in all their cells, including the germ line. The remaining 30% of the founder animals are mosaic and occasionally fail to harbor the transgene in germ line tissue. Genetically uniform animals can be distinguished from mosaic animals by following the frequency of transgene transmission during selective outbreeding of the original founder animals. Detection of the transgene in founder animals and outbred stocks can be accomplished by analyzing DNA prepared from small segments of tail by "slot blot" hybridization or by polymerase chain reaction analysis (Bieberich *et al.*, 1993).

Stable integration of a transgene does not always guarantee expression. To optimize expression, the transgene should contain a strong promoter. If it is desirable to have the transgene expressed in the proper cell type at the appropriate time, the promoter region must contain DNA sequences required for correct temporal and spatial expression. Even with these features in place, expression of the transgene may be influenced by the neighboring chromatin as integration occurs randomly. Chromatin effects can either silence or stimulate transgene expression (Dorer and Henikoff, 1994; Wilson *et al.*, 1990). Therefore, an ideal transgene contains additional elements that block integration site effects, thereby allowing copy number-dependent expression. The following sections of this chapter describe several transgenic approaches that illustrate most of these features.

Examples of Recombinant Protein Expression in Transgenic Mice

There are numerous reasons for expressing recombinant proteins in transgenic mice. As mentioned earlier, transgenic mice provide a means of exploring the interaction among genes in complex physiological pathways without pharmacological intervention or surgical manipulation. The following section discusses how ectopic and eutopic overexpression of a transgene could be used to dissect the interplay between genes within a physiological context. In addition, it addresses how transgenic mice could be used to disrupt the function of a particular protein or ablate a particular cell type.

Transgene Overexpression: A Means for Dissecting the Interplay between Genes in a Complex Hierarchical Pathway

Ectopic Overexpression Expression of many tissue-specific genes is often regulated by the products of other genes expressed in different tissues. Achieving ectopic transgene expression with a heterologous promoter can release a specific protein from regulation by its cognate axis. This optimizes the prospect of overexpression and provides a mechanism for evaluating how dysregulation of one gene affects expression of the other genes in the axis.

Genes involved in the regulation of somatic growth were early targets of the ectopic expression paradigm and provide a convenient illustration of a complex physiological axis. These genes encode the following proteins: growth hormone (GH), produced in the somatotropes of the pituitary; two hypothalamic releasing hormones—growth hormone-releasing hormone (GHRH) and somatostatin (SS); insulin-like growth factor (IGFI and IGFII) synthesized in hepatocytes, and insulin-like growth factor-binding proteins (IGF-BP), also synthesized in the liver. In this axis GH regulates somatic growth indirectly by controlling the synthesis of IGF. Levels of GH are controlled positively by GHRH and negatively by somatostatin. Serum GH can also exert its own negative feedback. Finally, IGF-BP neutralizes the activity of IGF, providing yet another level of control to this axis.

The impact of ectopic overexpression of GH was initially explored by Palmiter *et al.* (1982). As a first approach, they achieved overexpression of human GH by using a mouse metallothionine-1 (MT) promoter. This promoter targeted expression of GH to several organs that normally synthesize metallothionine, including liver, kidney, and spleen. Expression of GH was also observed in the intestines, skin, and gonads and in gonadotropes of the pituitary. Although activity of the MT promoter is usually dependent on stimulation by heavy metal ions, expression of this chimeric transgene in mice occurred early in fetal life and persisted throughout adulthood without metal supplementation. In the end, the chronic production of GH from multiple tissues led to excessive overproduction of the hormone. Moreover, although ectopically expressed, GH encoded by the transgene remained biologically active as evidenced by the giant phenotype of the mice. Ectopic expression of the GH transgene resulted in elevated levels of negative regulators of GH secretion that in turn suppressed endogenous GH synthesis and secretion. The input from the hypothalamus was also altered. Somatostatin levels increased and GHRH levels dropped in attempt to control the elevation in serum GH levels (Bartke *et al.*, 1994). All of these changes would have diminished the synthesis of GH were it not under the regulation of a heterologous promoter.

Many additional transgenic mice have been produced to date to dissect the interplay between members of the GH axis. Mice that overexpress GHRH under the control of the mouse MT promoter also demonstrated an increase in body size accompanied by massive hyperplasia of somatotropes (Mayo *et al.*, 1988). These studies revealed that GHRH not only regulates GH synthesis and secretion, but also has an effect on somatotrope cell proliferation. There has also been some debate as to whether the growth-promoting effects of GH are solely mediated by IGF-1. To address this question, mice that overexpress IGF-1 under control of the MT promoter were generated (D'Ercole, 1993). Although the mice were giant, they did not exhibit all the phenotypes observed in GH-overexpressing mice. These studies suggested that the actions of GH and IGF-1 are not always coupled. Ectopic overexpression has also been used to study IGFBPs that bind IGF-I and IGF-II and limit their bioavailability (Rajkumar *et al.*, 1996). Overexpression of IGFBP–1 was obtained using the phosphoglycerate kinase promoter that directs expression to the brain, uterus, lung, kidney, and heart. Homozygous mice exhibited significant growth retardation accompanied with fasting hyperglycemia, impaired glucose tolerance, and reduced fertility. Collectively, these studies underscore the importance of GHRH and IGFBP in the regulation of GH and IGF-1 action.

Cell-Specific (Eutopic) Overexpression The studies described earlier illustrate the principle that overexpression of genes normally regulated by a complex physiological axis involving several tissues can be achieved through the use of an ectopic promoter that releases them from control by the axis. In contrast, dissecting the functional significance of proteins in a complex intracellular pathway within a specific cell type requires a eutopic approach. For example, consider the role of cAMP-dependent protein kinase A (PKA) in mediating the effects of GHRH at the level of the somatotrope. Presumably, GHRH regulates GH synthesis and secretion by activating G_s that in turn activates PKA. To test the physiological significance of this interaction, the GH promoter was used to target the expression of cholera toxin to somatotropes (Burton *et al.*, 1991). Cholera toxin is known to activate G_s in a nonreversible fashion, leading to the stimulation of protein kinase A (PKA). Significantly, cholera toxin-expressing mice were giants and developed pituitary hyperplasia in the same manner as transgenic mice that ectopically overexpressed GHRH. Thus, targeted activation of the PKA pathway in somatotropes mimicked the effects of GHRH. This identifies the kinase as a physiologically significant component of the GHRH signaling pathway.

In summary, ectopic and eutopic overexpression strategies have been useful tools for dissecting the role of hypothalamic, pituitary, and hepatic factors in the regulation of growth. Similar approaches have been used to study other complex hierarchical pathways.

Transgenic Strategies for Disrupting Function of Proteins within a Specific Cell Type: Avenues for Dissecting Functionally Significant Components of Complex Intracellular Pathways

The functional significance of proteins within a multicomponent pathway can also be examined through targeted disruption of the candidate protein. This can be achieved through one of three avenues: gene disruption, mRNA ablation, or dominant interference of the targeted protein. All require use of a cell-specific promoter. The repertoire of such promoters is growing exponentially and makes a comprehensive review problematic. Examples of such promoters are described throughout this chapter.

Cell-specific elimination of a targeted gene in transgenic mice can be achieved through the use of *cre* recombinase (Barinaga, 1994; Rajewsky *et al.*, 1996), a bacteriophage enzyme that mediates site-specific recombination between *loxP* sites (Hoess *et al.*, 1982; Sternberg and Hamilton, 1981). The approach is bigenetic, requiring two strains of mice with different transgenes (Fig. 2) (Byrne and Ruddle, 1989). An activator strain of mice is constructed by direct pronuclear injection and contains a transgene encoding *cre* recombinase. This transgene also contains a cell-specific promoter to direct expression of the *cre* recombinase to the desired location. The other line of mice (target strain) is generated through a homologous recombination approach that yields homozygous mice where loxP sequences flank the gene targeted for disruption. When the two lines are mated, 25% of the resulting progeny should contain a cell-specific deletion of the candidate gene. Thus far, however, application of this approach has been limited to the deletion of the DNA β polymerase gene in T lymphocytes (Gu *et al.*, 1994) and liver (Kuhn *et al.*, 1995). Thus, its feasibility for other cell types, especially those with low rates of division, remains unknown.

Another approach that has been developed for targeted disruption of gene expression is the use of ribozymes. Ribozymes are RNA molecules that inhibit gene expression by catalytic cleavage of mRNA (Marschall *et al.*, 1994). The hammerhead ribozyme, when flanked with specific antisense sequences, cleaves the mRNA of interest in a sequence-specific manner (Fig. 3). The catalytic site is often dependent on specific secondary structure. Furthermore, although many ribozymes exhibit catalytic activity in *in vitro* cleavage assays, this property is often compromised in transgenic mice (Lieber and Strauss, 1995). Ribozyme inefficiency has

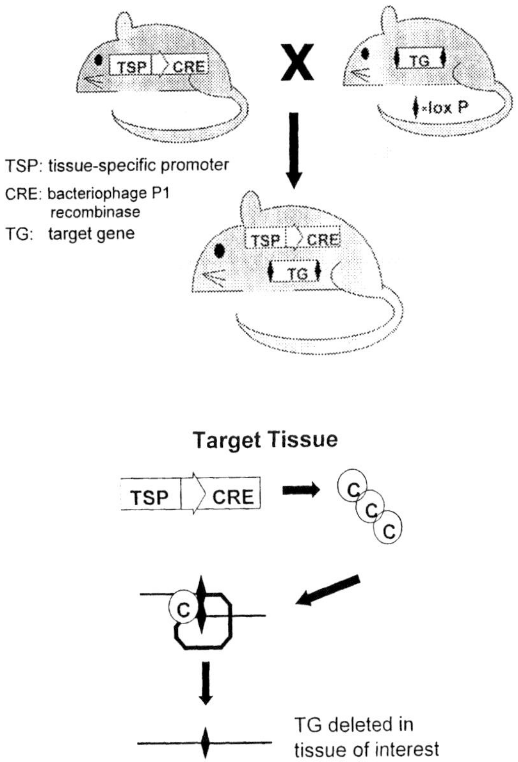

Figure 2 Directing site-specific recombination in transgenic mice. An activator strain of mice that expresses the cre recombinase (CRE) under the regulation of a tissue specific promoter (TSP) is generated by pronuclear injection, and a target strain of mice that have lox P sequences flanking the target gene (TG) is obtained by homologous recombination. Upon mating of animals from the two strains, animals expressing CRE and carrying loxP-flanked TG are obtained. Expression of CRE in the target tissue results in the deletion of the TG.

been attributed to (1) a short half-life and low abundance of the ribozyme, which fails to reach significant levels in the cytoplasm, (2) inaccessibility of the target mRNA, or (3) a ribozyme catalytic domain that becomes embedded within the carrier mRNA. Despite these limitations, however, this approach was used successfully to alter glucokinase mRNA levels selectively in pancreatic β cells of transgenic mice (Efrat *et al.*, 1994). The transgene contained a ribozyme hammerhead flanked by antisense sequences for the putative ATP-binding site of glucokinase. Expression was targeted to

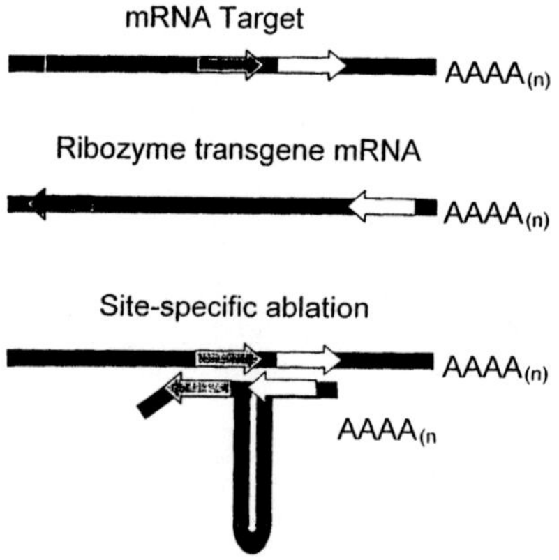

Figure 3 Targeting cleavage of specific mRNAs with transgenic ribozymes. Ribozymes are catalytic RNA molecules that cleave mRNA. They act in *trans* when flanked with antisense sequences. To obtain cell-specific ablation of the target mRNA, transgenic mice carrying the ribozyme transgene are generated. This transgene consists of a tissue-specific promoter (TSP) that drives the expression of the ribozyme flanked by antisense sequences specific for the target mRNA. The antisense sequences (arrows) dock the ribozyme to the mRNA of interest, which results in the catalytic cleavage of the intervening sequences.

pancreatic β cells using the rat insulin promoter. Transgenic mice carrying this construct exhibited a 70% reduction in glucokinase activity and protein levels, demonstrating the potential usefulness of this approach.

Transgenes containing a cell type specific promoter and encoding a protein with a dominant negative effect can also be used to disrupt protein function. This approach works particularly well with multimeric proteins where activity depends on the integrity of all the subunits in the complex. For example, a dominant negative strategy was used to examine the role of the dimeric cAMP-binding protein (CREB) in mediating the effects of GHRH in somatotropes. The transgene in this study encoded a mutant form of CREB placed under control of the rat GH promoter (Struthers *et al.*, 1991). The mutant CREB was capable of binding to DNA but lacked a critical PKA phosphorylation site required for transcriptional activity. When expressed in transgenic mice, this mutant CREB resulted in dominant interference, leading to a dwarf phenotype accompanied by hypoplasia of the somatotropes. These studies indicated that activation of CREB is not only important for the stimulation of

GH synthesis, but also in the developmental appearance of somatotropes.

Transgenic Paradigms for Achieving Cell-Specific Ablation

Tissue- and cell-specific promoters can also be used transgenically to assess cell lineage. At least two different approaches have been described. The first approach is direct and utilizes a cell-specific promoter to target expression of the Diphtheria toxin A chain (DT-A) (Palmiter *et al.*, 1987). This strategy was used to demonstrate that a 315-bp promoter fragment of the bovine α subunit of glycoprotein hormones gene contains all the information required for directing expression specifically to pituitary gonadotropes. Transcription of DT-A occurred only in gonadotropes and resulted in their ablation and subsequent development of a hypogonadal phenotype (Kendall *et al.*, 1991).

Cell lineage has also been investigated through use of a vector encoding thymidine kinase of herpes simplex virus 1 (HSVTK). In contrast to the mammalian enzyme, HSVTK has the capacity to phosphorylate nucleoside analogues such as acyclovir, gancyclovir, or FIAU [1-(2-deoxy–2-fluoro-β-δ-arabino-furanosyl)–5-iodouracil]. Incorporation of any one of these analogues into replicating DNA causes chain termination and ultimately blocks cell replication (Borrelli *et al.*, 1988, 1989; Heyman *et al.*, 1989). The advantage of this approach is that the viral enzyme is usually not harmful until the animals are treated pharmacologically with a nucleoside analogue. This permits temporal control over cell-specific ablation. One disadvantage, however, is that nonreplicating cells may not be responsive, implying the need to start treatment early in fetal life, although there have been instances where ablation was obtained in adult animals [thyrotropes (Wallace *et al.*, 1991) and gonadotropes (Markkula *et al.*, 1995)]. There may also be situations where transgenic expression in the absence of pharmacological activation may be lethal. For example, in the authors' hands, targeted expression of HSVTK in β cells of the endocrine pancreas resulted in the onset of diabetes without the need for treatment with a nucleoside analogue (data not shown). The testis is another tissue where pharmacologically independent expression of HSVTK has proven toxic (Braun *et al.*, 1990). Therefore, the usefulness of this approach depends on the cell type targeted for destruction and requires a case-by-case assessment.

Modeling Human Disease through Targeted Overexpression

Animal models of human diseases can enhance our understanding of the etiology and progression of a particular disorder and enhance the design of therapeutic strategies. With technical advances in the manipulation of the mouse genome, several transgenic models of human diseases have emerged (see Bedell *et al.*, 1997, for a recent review). These include animal models of cancer (Adams and Cory, 1991; Kappel *et al.*, 1994; Lovejoy *et al.*, 1997), neurodegenerative disorders (Aguzzi *et al.*, 1996; Brown, 1995), atherosclerosis (Breslow, 1996; Chien, 1996), diabetes (Tisch and McDevitt, 1996), cardiovascular disease (Chien, 1996), Down syndrome (Cabin *et al.*, 1995; Groner, 1995; Mirochnitchenko and Inouye, 1996; Sumarsono *et al.*, 1996), and others.

Transgenic animal models also help determine the relationship between underlying genetic, epigenetic, and environmental factors and the onset of a disorder. These same factors may also explain the predisposition of some individuals to a particular disease. Thus, transgenic approaches can be particularly informative for complex multifactorial disorders with epigenetic contributions (Erickson, 1996).

Several of the transgenic models described earlier were established by mutation or deletion of a specific gene. Other disease models have resulted from the overexpression of a candidate gene, either in wild type or mutant form. This approach is particularly powerful when targeted expression of a candidate gene is used to rescue a particular phenotype.

When contemplating an overexpression paradigm, much emphasis is placed on selection of a promoter with sufficient strength and cell-type specificity. Although integration site effects seldom compromise promoter specificity, they often interfere with the extent of expression. Furthermore, many candidate promoters are regulated by complex feedback pathways, making their overexpression even more problematic. As described next, most of these limitations can be overcome by careful promoter selection, by consideration of the half-life of the protein encoded by the transgene, and by using the biology of the system to limit feedback effects.

The authors' laboratory has developed transgenic mice that chronically hypersecrete luteinizing hormone (LH) (Risma *et al.*,

1995). This pituitary hormone belongs to a family of glycoprotein hormones that include thyroid-stimulating hormone (TSH), follicle-stimulating hormone (FSH), and human chorionic gonadotropin (hCG). Members of this family are heterodimers containing a common α subunit and a unique β subunit. LH and FSH are secreted into the circulation as distinct pulses that ultimately regulate gametogenesis and gonadal steroidogenesis. The secreted pulses of LH and FSH derive from identical pulses of GnRH secreted from the hypothalamus. Although LH and FSH stimulated production of estrogen and androgens from the gonads, these sex steroids also feed back at the level of both the hypothalamus and the pituitary to limit secretion of additional LH.

Several clinical studies have shown a correlation between hypersecretion of LH and functional ovarian hyperandrogenism (FOH), infertility, and miscarriage in women (Barnes *et al.*, 1994; Ehrmann *et al.*, 1995; Franks, 1995; Regan *et al.*, 1990; Shoham *et al.*, 1993), suggesting that chronically elevated LH impairs fertility. Unfortunately, no studies show a direct relationship between hypersecreted LH and reproductive abnormalities. Because LH is secreted in regulated pulses (Gibson *et al.*, 1991) and its serum half-life is short (20–30 min; Niswender *et al.*, 1974), it is difficult to devise protocols for chronic administration of exogenous LH that mimic endogenous pulse patterns of LH. To circumvent this limitation, the authors devised a transgenic approach where elevated hormone levels are maintained chronically, without requiring multiple injections, supraphysiologic dosing, or appreciable dampening of the hypothalamic–pituitary–gonadal axis.

A two-pronged approach was utilized to achieve elevated levels of serum LH. First, to increase delivery of hormone from the pituitary, expression of a transgene encoding an additional LHβ subunit was directed to gonadotropes using a previously characterized bovine α subunit promoter (Hamernik *et al.*, 1992; Kendall *et al.*, 1991). This promoter effectively directs expression of a varicty of reporter genes to gonadotropes and renders them responsive to GnRH, estrogen, and androgen (Clay *et al.*, 1993; Keri *et al.*, 1991). Second, the LHβ subunit encoded by the transgene contained a peptide extension that the authors proposed would slow the elimination of LH from the serum. This peptide is normally found at the carboxyl terminus of the β subunit of human chorionic gonadotropin (hCG) and is referred to as the carboxyl terminal peptide (CTP). This peptide is thought to be a major determinant of the serum half-life of hCG (Matzuk *et al.*, 1990) and has been shown to increase the half-life of FSH two- to threefold when fused to its β subunit. Accordingly, a transgene was constructed that links the α subunit promoter with the coding region of bovine LHβ fused in frame to the coding region of CTP (αLHβ-CTP).

Female transgenic mice carrying the αLHβ-CTP transgene chronically hypersecreted LH. Concomitant with elevated LH, serum levels of androgens and estrogens were elevated (Risma *et al.*, 1995). In contrast, FSH and prolactin levels were normal, suggesting that the hypothalamic–pituitary–gonadal axis remains functionally intact. Hypersecretion of LH and androgens occurred early during neonatal development, causing premature vaginal opening and ovarian follicular development (hallmarks of precocious puberty) (Risma *et al.*, 1997). Although follicles continued to develop, ovulation failed to occur. This led initially to the formation of follicular cysts with pronounced hemorrhagia and ultimately to a large granulosa tumor mass. In addition, some of these mice had enlarged bladders and developed hydronephrosis. In other instances the bladder became herniated. These renal phenotypes probably reflect elevated levels of serum steroids. Hydronephrosis has been reported in rats chronically treated with high levels of estradiol as well as in pregnant women and nonhuman primates (Au *et al.*, 1985; Buhl *et al.*, 1985; Roberts, 1976). Furthermore, bladder hernia has been previously reported in mice that overexpress the wild-type estrogen receptor under the control of the metallothionine promoter (Davis *et al.*, 1994).

Although elevated levels of LH can be attributed to the combined effects of a strong gonadotrope-specific promoter and increased stability of the recombinant subunit, an additional biological component that underlies overexpression was discovered. Normally, high levels of sex steroids reduce the concentration of serum LH. In the case of the αLHβ-CTP mimic, elevated LH in the face of elevated estrogen and androgen suggested that resistance to steroid negative feedback must have occurred. Preliminary data indicate that the αLHβ-CTP transgene retains responsiveness to negative feedback from androgen but has developed a selective resistance to estrogen negative feedback (Abbud *et al.*, 1997). This resistance is transgene specific and therefore contributes to the overall elevation in serum LH.

In summary, a transgenic strategy has been successfully developed for demonstrating that chronic hypersecretion of LH causes functional ovarian hyperandrogenism, a leading cause of infertility in women. Although the strategy required targeted overexpression of LH, this outcome relied on designing a composite transgene that capitalized on unique features of a complex physiological pathway. This model of chronic LH hypersecretion joins a growing list of transgenic animal models of human disease. Because most diseases are multifactorial, increasingly novel approaches will be required to provide a full array of disease models. The ability to manipulate the mouse genome and test alterations in gene expression in different genetic background offers further assurance that transgenic mice will remain as invaluable tools in studying complex human diseases and designing efficient strategies for gene therapy.

Biomass: Transgenic Mice as Model Bioreactors

Although the mouse is smaller than many mammals, it is an ideal vehicle for laying the groundwork for recombinant protein expression in large animals. It is essential that the desired protein be made in large quantities and be readily accessible. For this reason, a large body of work has focused on the targeted expression of recombinant proteins to the mammary gland or blood (Echelard, 1996).

Several promoter regions from genes that encode milk proteins, such as β-lactoglobulin (Hurwitz *et al.*, 1992), β-casein (Persuy *et al.*, 1995), αs1-casein (Meade *et al.*, 1990), and whey acidic protein (Devinoy *et al.*, 1994), have been used to target the expression of recombinant proteins to the mammary gland. Achieving high levels of protein yields, however, is difficult when cDNA constructs are placed downstream of these promoters. Although higher levels of expression are often associated with the use of larger genomic fragments (<20 kb), these constructs remain subject to position effects (Wilson *et al.*, 1990). Coinjection strategies with cDNA constructs and genomic DNA can increase yield, presumably because the integration site contains a larger transgene array that attenuates position effects (Clark *et al.*, 1992; McKnight *et al.*, 1995). In the end, however, unless the transgene can be completely shielded from its genetic neighborhood, the site of integration limits yield.

Studies on the expression of the β-globin cluster of genes have provided important insights for designing constructs that insulate transgenes from integration site effects (Grosveld *et al.*, 1989; Needham *et al.*, 1992; Strouboulis *et al.*, 1992). High-level and position-independent transgene expression has been obtained when the transgene is linked to the β-globin locus control region (LCR) (Hanscombe *et al.*, 1989). This region contains structural elements that establish a boundary domain to prevent influence by the integration site (Strouboulis *et al.*, 1992). Similar regions from other genes have been isolated and shown to confer position-independent expression of transgenes (McKnight *et al.*, 1992; Phi-Van *et al.*, 1990; Wilson *et al.*, 1990).

The recombinant hemoglobin expression system has been used to generate large amounts of peptides that have therapeutic value. Both magainin (26-mer) and β-endorphin (16-mer) were produced in transgenic mice by fusion of either peptide to the C terminus of human α-globin (Sharma *et al.*, 1994). A conveniently positioned protein cleavage site allows the separation of the peptide of interest from α-globin. The production of both recombinant proteins in blood was obtained and more than 1 g/liter of peptide was obtained. The expression of these peptides in mice did not appear to cause any detrimental effect, suggesting that peptides lack biological activity when joined to the carrier protein (α-globin) inside the erythrocyte. This approach has great implications as chemical synthesis of peptides is extremely laborious and expensive.

Although the globin LCR transgenesis system offers an important avenue for producing larger amounts of a desired peptide, the requirement for production of a fusion protein and subsequent enzymatic recovery limits its overall efficiency. In addition, there may be instances where it is necessary to direct transgene expression to tissue sites other than blood. Work with yeast artificial chromosomes (YAC) and bacterial artificial chromosomes (BAC) suggests that concern over integration site effects and site of expression may soon become nonexistent. Yang and colleagues (1997) have reported a simple method for modifying BAC directly in a recombination deficient *Escherichia coli* strain by homologous recombination. To demonstrate feasibility, the authors introduced a *LacZ* reporter gene by homologous recombination into a 131-kb BAC containing the murine zinc finger gene, *RU49*. The site of recombination placed the *LacZ* reporter under control of the *RU49* promoter. Transgenic mice were then generated by pronuclear injection of the modified BAC. Brain-specific expression of the *LacZ* transgene occurred when transgenic mice were made with the modified BAC. When transgenic mice were made with a conventional plasmid-based *LacZ* transgene, expression failed to occur. These results strongly suggest that placing a transgene within the context of a BAC provides complete protection from integration site effects, thereby offering a new and improved platform for achieving maximum transgene expression.

Although transgenesis with BACs is in its infancy, there are several noteworthy features. First, eukaryotic genes often exceed 100-kb pairs and require large cloning vectors. Although YACs permit the cloning of extremely large DNA fragments (Lamb and Gearhart, 1995; Schedl *et al.*, 1993, 1996), rearrangements resulting from aberrant recombination are frequent, mitigating their routine use. Because BACs are manipulated in a recombinant deficient strain of *E. coli*, rearrangement is not a problem. Although BACs are smaller than YACs (100 kb vs 250 kb), they are still large enough to harbor eukaryotic genes greater than 100 kb. This increases the likelihood of constructing transgenes that contain a complete array of promoter-regulatory elements. The smaller size of BACs also makes them much more amenable to pronuclear injection. Finally, efforts are underway to annotate BAC libraries by identifying those linked with all of the expressed sequence tags (ESTs) (Simon, 1997). The potential significance of this outcome is remarkable as BAC-EST libraries will provide a vast template of reporter genes for transgenesis that await only the insertion of desired eukaryotic promoters by homolgous recombination.

Future of Transgenic Mouse Approaches: Need for Inducible Expression

Most promoter sequences used in transgenic mice confer the same temporal and spatial patterns of expression as their respective endogenous gene. Consequently, the onset of transgene transcription reflects a unique stage of embryonic or postnatal development. Furthermore, transgene expression usually persists over the life of the animal. Although this type of pattern is essential for validating the presence of all the appropriate promoter-regulatory elements, there are many scenarios where tighter control over transgene expression would be beneficial. For example, it remains unclear whether the ovarian phenotype caused by chronic hypersecretion of LH in the αLHβ-CTP transgenic mice described earlier is reversible. This could be readily addressed by supplying the transgene with a pharmacological on–off switch that permits restriction of LH hypersecretion to defined periods of time. Similarly, many proteins with potential therapeutic activity may be toxic if expressed at high levels over long periods of time. On/off control of transgene expression would undoubtedly enhance the bioreactor capability of transgenic mice.

Many eukaryotic promoters are under natural inducible control, providing a simple means for controlling the extent and duration of expression. For example, in the absence of heavy metals, the MT promoter has low activity. Treatment with cadmium or zinc, however, increases promoter activity several fold. The effectiveness of this induction scheme was demonstrated in transgenic mice harboring a mutated sheep MT promoter linked to an ovine GH-coding sequence. When maintained on water supplemented with zinc, these mice secreted excessive levels of GH and displayed a giant phenotype (Shanahan *et al.*, 1989). Mice overexpressing GH have also been made with a transgene that utilizes the PEPCK promoter (McGrane *et al.*, 1988). Transcription of the chimeric gene occurred after birth in the kidney, liver, and adipose tissue. This promoter is regulated by composition of the diet: a diet high in carbohydrates reduces expression whereas a diet high in protein and low in carbohydrate stimulates expression. Thus, the PEPCK promoter permits even tighter control over transgene activity, with diet representing the essential trigger. In other instances, heat shock promoters and steroid hormone response elements have been used for inducible gene expression (Yarranton, 1992).

Although each of the examples just described provides inducible control over transgene expression, each has at least one limitation that restricts usefulness. These include expression in multiple tissues, high basal activity, low levels of induction, and generalized effects of the inducer on gene expression in a variety of target tissues. Several strategies have been described that permit tighter control over transgene expression. The ultimate promoter should have a low level of basal activity and a rapid and substantial response to the inducer. In addition, the inducer should be easy to administer and lack side effects. Most of these strategies rely on a "binary" approach that requires two transgenes (Barinaga, 1994; Byrne and Ruddle, 1989). One transgene encodes the protein of interest. The product of the second transgene controls expression of the target transgene. This product can act as either an activator or a repressor. Three examples of such binary systems are described below.

Tetracycline Repressor System

This incorporates features of the tetracycline resistance operon from the transposon Tn10 of *E. coli*. In bacteria, the tetracycline repressor (*tet*R) binds regulatory sequences (*tet*O) located upstream of the tetracycline resistance gene to suppress gene expression. Promoter activity resumes when tetracycline binds *tet*R, rendering it incapable of binding to *tet*O (Gossen *et al.*, 1993). This property could be modified by converting *tet*R to a transcriptional activator protein (tetracycline-controlled transactivator, *t*TA) through fusion of the *tetR* DNA-binding domain (DBD) to the transactivation domain (TAD) of VP16 (Gossen and Bujard, 1992; Gossen *et al.*, 1995). Thus, in the absence of tetracycline, *t*TA activates an otherwise dormant promoter. The addition of tetracycline renders the complex incapable of binding *tet*O, thereby reverting the promoter to dormancy.

Two modified versions of the *t*TA system have also been reported. One approach uses a mutant *t*TA (r*t*TA) with altered DNA-binding properties that reverse its mode of action (Gossen *et al.*, 1995). r*t*TA binds to DNA only in the presence of tetracycline, making the protein ideal for inducible gene expression. The other approach involves the fusion of *tet*R to the KRAB silencer domain of human Kox-1(Deuschle *et al.*, 1995). This chimeric protein acts as a dominant transcriptional repressor when bound to *tet*O sequences. Restoration of gene expression occurs on treatment with tetracycline.

The tetracycline *t*TA and r*t*TA systems have been tested in transgenic mice (Efrat *et al.*, 1995; Ewald *et al.*, 1996; Furth *et al.*, 1994; Kistner *et al.*, 1996; Passman and Fishman, 1994; Schultze *et al.*, 1996; Shockett *et al.*, 1995). In most instances, the approach utilized two lines of mice: one contained the target transgene and the other one harbored either the transactivator or the transrepressor (Fig. 4). The target transgene encoded the recombinant protein of interest with expression controlled by a *tet*O-containing eukaryotic promoter with ubiquitous activity. Activator or repressor transgenes were placed under control of a strong viral promoter or a tissue-specific promoter. These encoded either the *t*TA or the r*t*TA fusion proteins. Transgenes were brought together through breeding of the appropriate lines of mice. Transcription of the target transgene occurred only in cells that expressed the transactivator or transrepressor. For *t*TA, suppression of the target transgene

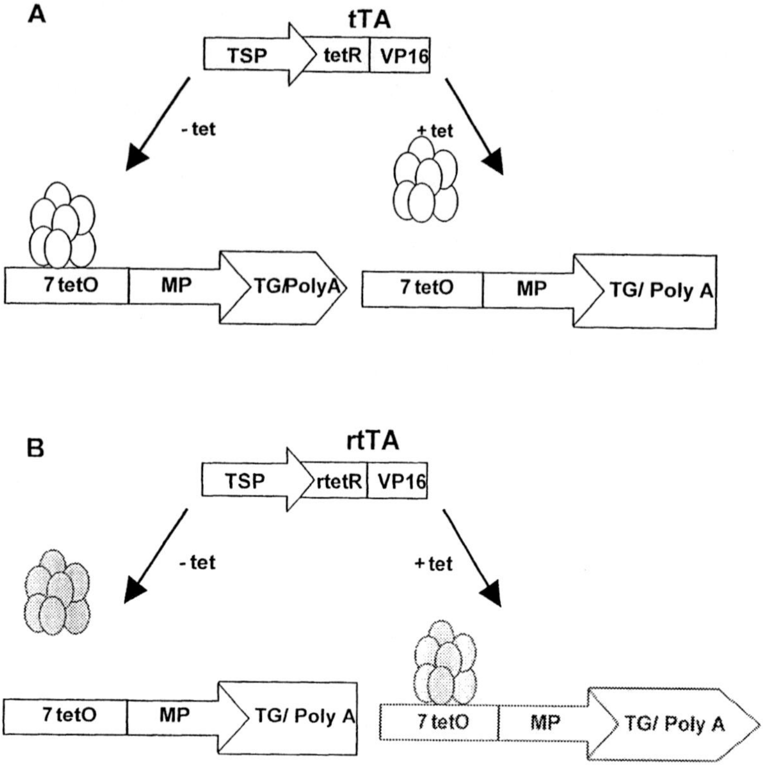

Figure 4 Inducible control of transgene expression: the tetracycline repressor. A tissue-specific promoter (TSP) targets expression of either the tetracycline transactivator protein (A; tTA) or the mutant (B; rtTA) to the tissue of interest in transgenic mice that carry the gene of interest (TG/PolyA) under the control of a minimal promoter (MP) linked to seven tetracycline operator sequences (tetO). (A) The tTA system where the administration of tetracycline represses transgene expression. (B) The rtTA system where gene expression occurs only in the presence of tetracycline. [Adapted, with permission, from Kistner, A., Gossen, M., Zimmermann, F., Jerecic, J., Ullmer, C., Lubbert, H., and Bujard, H. (1996). Doxycycline-mediated quantitative and tissue-specific control of gene expression in transgenic mice. *Proc. Natl. Acad. Sci.* **93,** 10933–10938. Copyright (1996) National Academy of Sciences, U.S.A.]

required the continuous presence of tetracycline. In contrast, continuous tetracycline activated the target transgene only in cells with r*t*TA. This binary approach was subsequently refined and simplified by combining the transactivator and the target transgenes into one construct (Schultze *et al.*, 1996). The utility of the *t*TA system has also been demonstrated in studies where tetracycline-dependent suppression of SV40 large T antigen reversed cellular hyperplasia in both the pancreas (Efrat *et al.*, 1995) and the submandibular gland of transgenic mice (Ewald *et al.*, 1996).

Ecdysone Receptor System

Although the tetracycline system has proven useful, the pharmacokinetics of the antibiotic make the on/off switch relatively slow. For example, tetracycline clears slowly from bone, thereby preventing a quick and precise response. In addition, robust and prolonged responsiveness requires continuous treatment with the drug. To circumvent these limitations, No *et al.* (1996) took advantage of the properties of the insect molting hormone ecdysone to design a system for conditional expression in transgenic mice. Ecdysone is a steroid hormone that regulates transcription by binding to the ecdysone nuclear receptor. In *Drosophila,* the ecdysone receptor forms a heterodimer with ultraspiracle (USP) and binds to specific consensus sequences flanking ecdysone-responsive genes. In contrast to tetracycline, ecdysone is highly lipophilic and efficiently distributes to all tissues in the body, including the brain. In addition, it has a short half-life, allowing for rapid and potent induction. Furthermore, ecdysone lacks any toxic or teratogenic effects in mammals. To optimize the use of this system in transgenic mice, the potency of ecdysone was enhanced by altering the properties of the ecdysone response elements. Two lines of mice were subsequently established (see Fig. 5). The "reporter" line contained a transgene with a novel ecdysone-inducible promoter upstream of the recombinant protein of interest. The "receptor" line harbored two transgenes that use a T-cell-specific promoter to obtain tissue-specific expression of a modified ecdysone receptor (VpEcR or VgEcR) and RXR. Expression of these receptors in T cells had no effect on the health of the animal. Cell-specific induction of the reporter gene with very low basal expression occurred upon breeding of the two lines of mice followed by treatment with the ecdysone receptor agonist (muristerone).

Rapamycin-FRAP-FKBP System

This strategy has been tested in mammalian cells in culture and in mice implanted with cells that stably express the requisite compo-

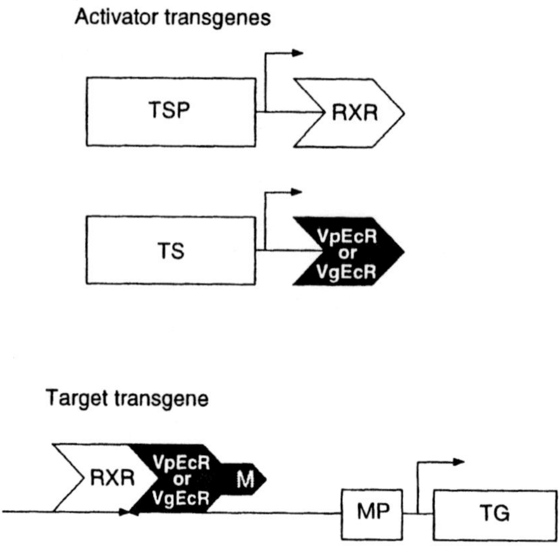

Figure 5 Inducible control of transgene expression: the ecdysone receptor. Two lines of transgenic mice are generated for ecdysone-inducible gene expression. One line harbors the activator transgenes whereas the other carries the target transgene. The activator transgenes encode for RXR and a modified ecdysone receptor (VpEcR or VgEcR) under the control of a tissue-specific promoter (TSP). The target transgene (TG) is regulated by several ecdysone response elements upstream of a minimal promoter (MP). Upon treatment with muristerone the ecdysone receptor forms heterodimers with RXR and binds to DNA to activate gene expression. [Adapted, with permission, from No, D., Yao, T., and Evans, R. M. (1996) Ecdysone-inducible gene expression in mammalian cells and transgenic mice. *Proc. Natl. Acad. Sci. U.S.A.* **93,** 3346–3351. Copyright (1996) National Academy of Sciences, U.S.A.]

nents. Although not yet tested in transgenic mice, results obtained thus far are promising (Ho *et al.*, 1996; Rivera *et al.*, 1996). The immunosuppressant rapamycin plays a central role in this binary scheme. It causes dimerization of two cellular proteins [FK506-binding protein (FKBP12) and FKBP12-rapamycin-associated protein (FRAP)] involved in a eukaryotic signal transduction pathway. As diagrammed in Fig. 6, use in transgenic mice requires that one of the two proteins (e.g., FKBP) be fused to the DNA-binding domain of a known transcription factor (e.g., Gal4 DBD). The target transgene would necessarily contain a promoter with the cognate *cis*-acting elements for this transcription factor. The other transgene would encode a fusion protein containing FRAP and a transactivation domain, such as the VP16 TAD. This transgene would be

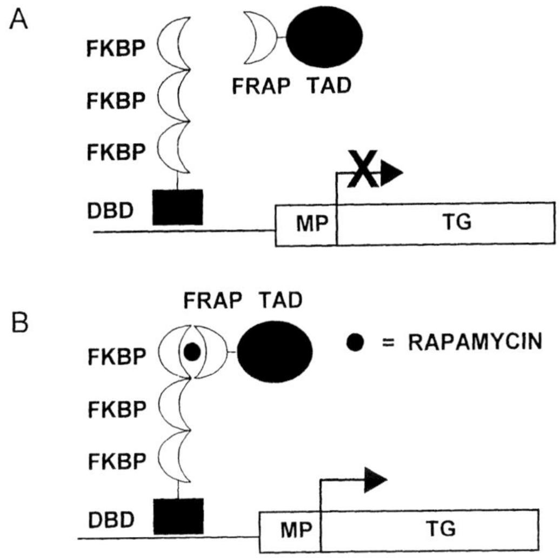

Figure 6 inducible control of transgene expression: a hypothetical scheme employing rapamycin, FKBP, and FRAP. (A) Recombinant proteins and the transgene used in this system. The DNA-binding domain of a specific transcription factor (DBD) is fused to FKBP whereas a cognate transactivation domain (TAD) is fused to FRAP. The target gene (TG) is under the control of a minimal promoter with several tandem-binding sites for the DBD (TBS). (B) How rapamycin causes the formation of FKBP and FRAP heterodimers resulting in the reconstitution of the transcriptional activator and activation of gene expression. [Adapted, with permission, from Liberles, S. D., Diver, S. T., Austin, D. J., and Schreiber, S. L. (1997). Inducible gene expression and protein translocation using nontoxic ligands identified by a mammalian three-hybrid screen. *Proc. Natl. Acad. Sci. U.S.A.* **94,** 7825–7830. Copyright (1997) National Academy of Sciences, U.S.A.]

placed under control of a cell-specific promoter. Treatment with rapamycin should result in the dimerization of the chimeric proteins and reconstitution of the transcriptional activator and induction of transcriptional activity. Advantages of this system include low levels of basal activity, high induction ratios, the lack of dependence on any cell-type-specific transcription factor, and the bioavailability of the inducer drug. In short, it appears to have a high likelihood for triggering precise and controlled induction of gene expression. Although promising, there is one major drawback: rapamycin is a potent immunosuppressor that affects cell cycle progression. A nontoxic analogue of rapamycin (rap*) that binds a mutant form of the rapamycin-binding domain of FRAP (FRB*) has been developed and used to target gene expression in Jurkat T cells (Liberles *et al.*,

1997). The usefulness of this approach remains to be tested in transgenic mice.

Conclusions

This chapter highlighted multiple approaches that are currently used or contemplated for expressing recombinant proteins in transgenic mice. Whenever possible, both strengths and limitations associated with the different strategies were discussed (Table 1). Clearly, these examples represent only a limited sampling of how this technology has and will have an impact on the production and evaluation of recombinant proteins in a whole animal setting. There is little doubt that transgenic mice have and will continue to provide an extremely useful tool for bridging molecular genetics and mammalian physiology.

Table 1
Advantages and Disadvantages of Overexpression in Transgenic Mice

Advantages	Disadvantages
Study regulation of gene expression in a physiological context	Generation of transgenic mice is time-consuming and expensive
Establish function of a particular protein by overexpressing either the protein in question or its inhibitors	Integration site effects that could result in the lack of copy number-dependent expression
Study cell lineage	Strain-specific effects
Generation of cell lines	
Obtain cell-specific ablation or mutation of a particular gene	
Model human disease and gene therapy	
Study polygenic disorders and identify extragenic modifiers of diseases	
Generation of large quantities of any gene product	
Using inducible expression approaches, it is possible to precisely target expression to the appropriate cell type at the right time for the desired duration	

References

Abbud, R., Nett, T. M., and Nilson, J. H. (1997). Resistance of the alpha subunit of glycoprotein hormones to estradiol negative feedback contributes to the hypersecretion of LH in female transgenic mice that carry a chimeric LH beta subunit. *Soc. Study Reprod.* **56,** 119.

Adams, J. M. and Cory, S. (1991). Transgenic models of tumor development. *Science* **254,** 1161–1167.

Aguzzi, A., Brandner, S., Marino, S., and Steinbach, J. P. (1996). Transgenic and knockout mice in the study of neurodegenerative diseases. *J. Mol. Med.* **74,** 111–126.

Au, K. K., Woo, J. S., Tang, L. C., and Liang, S. T. (1985). Aetiological factors in the genesis of pregnancy hydronephrosis. *Aust. N. Z. J. Obstet. Gynaecol.* **25,** 248–251.

Barinaga, M. (1994). Knockout mice: Round two [news; comment]. *Science* **265,** 26–28.

Barnes, R. B., Rosenfield, R. L., Ehrmann, D. A., Cara, J. F., Cuttler, L., Levitsky, L. L., and Rosenthal, I. M. (1994). Ovarian hyperandrogynism as a result of congenital adrenal virilizing disorders: Evidence for perinatal masculinization of neuroendocrine function in women. *J. Clin. Endocrinol. Metab.* **79,** 1328–1333.

Bartke, A., Cecim, M., Tang, K., Steger, R. W., Chandrashekar, V., and Turyn, D. (1994). Neuroendocrine and reproductive consequences of overexpression of growth hormone in transgenic mice. *Proc. Soc. Exp. Bio. Med.* **206,** 345–359.

Bedell, M. A., Largaespada, D. A., Jenkins, N. A., and Copeland, N. G. (1997). Mouse models of human disease. Part II: Recent progress and future directions. *Genes Dev.* **11,** 11–43.

Bieberich, C. J., Ngo, L., and Jay, G. (1993). Molecular approaches involved in mammalian gene transfer: Evaluation of transgene expression. *In* "Transgenic Animal Technology: A Laboratory Handbook" (C. A. Pinkert, ed.), pp. 235–260. Academic Press, San Diego, CA.

Borrelli, E., Heyman, R., Hsi, M., and Evans, R. M. (1988). Targeting of an inducible toxic phenotype in animal cells. *Proc. Natl. Acad. Sci. U.S.A.* **85,** 7572–7576.

Borrelli, E., Heyman, R. A., Arias, D., Sawchenko, P. E., and Evans, R. M. (1989). Transgenic mice with inducible dwarfism. *Nature (London)* **339,** 538–541.

Braun, R. E., Lo, D., Pinkert, C. A., Widera, G., Flavell, R. A., Palmiter, R. D., and Brinster, R. L. (1990). Infertility in male transgenic mice: Disruption of sperm development by HSV-TK expression in postmeiotic germ cells. *Biol. Reprod.* **43,** 684–693.

Breslow, J. L. (1996). Mouse models of atherosclerosis. *Science* **272,** 685–688.

Brown, R. H., Jr. (1995). Amyotrophic lateral sclerosis: Recent insights from genetics and transgenic mice. *Cell (Cambridge, Mass.)* **80,** 687–692.

Buhl, A. E., Yuan, Y. D., Cornette, J. C., Frielink, R. D., Knight, K. A., Ruppel, P. L., and Kimball, F. A. (1985). Steroid-induced urogenital tract changes and urine retention in laboratory rodents. *J. Urol.* **134,** 1262–1267.

Burton, F. H., Hasel, K. W., Bloom, F. E., and Sutcliffe, J. G. (1991). Pituitary hyperplasia and giantism in mice caused by a cholera toxin transgene. *Nature (London)* **350,** 74–77.

Byrne, G. W. and Ruddle, F. H. (1989). Multiplex gene regulation: A two-tiered approach to transgene regulation in transgenic mice. *Proc. Natl. Acad. Sci. U.S.A.* **86,** 5473–5477.

Cabin, D. E., Hawkins, A., Griffin, C., and Reeves, R. H. (1995). YAC transgenic mice in the study of the genetic basis of Down syndrome. *Prog. Clini. Biol. Res.* **393,** 213–226.

Chien, K. R. (1996). Perspective series: Molecular medicine in genetically engineered animals. Genes and physiology: Molecular physiology in genetically engineered animals. *J. Clin. Invest.* **97,** 901–909.

Clark, A. J., Cowper, A., Wallace, R., Wright, G., and Simons, J. P. (1992). Rescuing transgene expression by co-integration. *Bio/Technology* **10,** 1450–1454.

Clay, C. M., Keri, R. A., Finicle, A. B., Heckert, L. L., Hamernik, D. L., Marschke, K. M., Wilson, E. M., French, F. S., and Nilson, J. H. (1993). Transcriptional repression of the glycoprotein hormone alpha subunit gene by androgen may involve direct binding of androgen receptor to the proximal promoter. *J. Biol. Chem.* **268,** 13556–13564.

Davis, V. L., Couse, J. F., Goulding, E. H., Power, S. G. A., Eddy, E. M., and Korach, K. S. (1994). Aberrant reproductive phenotypes evident in transgenic mice expressing the wild-type mouse estrogen receptor. *Endocrinology (Baltimore)* **135,** 379–386.

D'Ercole, A. J. (1993). Expression of Insulin-like growth factor–1 in transgenic mice. *Annals New York Academy of Sciences* **692,** 149–160.

Deuschle, U., Meyer, W. K.-H., and Thiesen, H.-J. (1995). Tetracycline-reversible silencing of eukaryotic promoters. *Mol. Cell. Biol.* **15,** 1907–1914.

Devinoy, E., Thepot, D., Stinnakre, M. G., Fontaine, M. L., Grabowski, H., Puissant, C., Pavirani, A., and Houdebine, L. M. (1994). High level production of human growth hormone in the milk of transgenic mice—The upstream regulatory region of the rabbit whey acidic protein (Wap) gene targets transgene expression to the mammary gland. *Transgenic Res.* **3,** 79–89.

Doetschman, T. (1993). Gene transfer in embryonic stem cells. *In* "Transgenic Animal Technology: A Laboratory Handbook" (C. A. Pinkert, ed.), pp. 115–140. Academic Press, San Diego, CA.

Dorer, D. R. and Henikoff, S. (1994). Expansions of transgene repeats cause heterochromatin formation and gene silencing in drosophila. *Cell (Cambridge, Mass.)* **77,** 993–1002.

Echelard, Y. (1996). Recombinant protein production in transgenic animals. *Curr. Opin. Biotechnol.* **7,** 536–540.

Efrat, S., Leiser, M., Wu, Y.-J., Dusco-DeMane, D., Emran, O., Surana, M., Jetton, T. L., Magnuson, M. A., Weir, G., and Fleischer, N. (1994). Ribozyme-mediated attenuation of pancreatic b-cell glucokinase expression in transgenic mice results in impaired glucose-induced insulin secretion. *Proc. Natl. Acad. Sci. U.S.A.* **91,** 2051–2055.

Efrat, S., Fusco-DeMane, D., Lemberg, H., Al Emran, O., and Wang, X. (1995). Conditional transformation of a pancreatic b-cell line derived from transgenic mice expressing a tetracycline-regulated oncogene. *Proc. Natl. Acad. Sci. U.S.A.* **92,** 3576–3580.

Ehrmann, D. A., Barnes, R. B., and Rosenfield, R. L. (1995). Polycystic ovary syndrome as a form of functional ovarian hyperandrogenism due to dysregulation of androgen secretion. *Endocr. Rev.* **16,** 322–353.

Erickson, R. P. (1996). Mouse models of human genetic disease: Which mouse is more like a man. *BioEssays* **18,** 993–998.

Ewald, D., Minglin, L., Efrat, S., Auer, G., Wall, R., Furth, P., and Hennighausen, L. (1996). Time-sensitive reversal of hyperplasia in transgenic mice expressing SV40 T antigen. *Science* **273,** 1384–1386.

Franks, S. (1995). Polycystic Ovary Syndrome. *N. Eng. J. Med.* **333,** 853–861.

Furth, P., Onge, L. C., Boger, H., Gruss, P., Gossen, M., Kistner, A., Bujard, H., and Hennigausen, L. (1994). Temporal control of gene expression in transgenic mice by a tetracycline-responsive promoter. *Proc. Natl. Acad. Sci. U.S.A.* **91,** 9302–9306.

Gibson, M. J., Miller, G. M., and Silverman, A.-J. (1991). Pulsatile luteinizing hormone secretion in normal female mice and in hypogonadal female mice with preoptic area implants. *Endocrinology (Baltimore)* **128,** 965–971.

Goffeau, B., Barrell, G., Bussey, H., Davis, W., Dujon, B., Feldman, H., Galibert, F., Hoheisel, J. D., Jacq, C., Johnston, M., Louis, E. J., Mewes, H. W., Murakami, Y., Philippsen, P., Tettelin, H., and Oliver, S. G. (1996). Life with 6000 genes. *Science* **274,** 546–567.

Gossen, M. and Bujard, H. (1992). Tight control of gene expression in mammalian cells by tetracycline-responsive promoters. *Proc. Natl. Acad. Sci. U.S.A.* **89,** 471–475.

Gossen, M., Bonin, A., and Bujard, H. (1993). Control of gene activity in higher eukaryotic cells by prokaryotic regulatory elements. *Trends Biol. Sci.* **18,** 471–475.

Gossen, M., Freundlieb, S., Bender, G., Muller, G., Hillen, W., and Bujard, H. (1995). Transcriptional activation by tetracyclines in mammalian cells. *Science* **268,** 1766–1769.

Groner, Y. (1995). Transgenic models for chromosome 21 gene dosage effects. *Prog. Clini. Biol. Res.* **393,** 193–212.

Grosveld, F., Assendelft, G. B., Greaves, D. R., and Kollias, G. (1989). Position-independent, high-level expression of the human beta-globin gene in transgenic mice. *Cell (Cambridge, Mass.)* **51,** 975–985.

Gu, H., Marth, J. D., Orban, P. C., Mossmann, H., and Rajewsky, K. (1994). Deletion of a DNA polymerase b gene segment in T cells using cell type-specific gene targeting. *Science* **265,** 103–106.

Hamernik, D. L., Keri, R. A., Clay, C. M., Clay, J. N., Sherman, G. B., Sawyer, H. R., Jr., Nett, T. M., and Nilson, J. H. (1992). Gonadotrope- and thyrotrope-specific expression of the human and bovine glycoprotein hormone alpha-subunit genes is regulated by distinct Cis-Acting elements. *Mol. Endocrinol.* **6,** 1745–1755.

Hanscombe, O., Vidal, M., Kaeda, J., Luzzatto, L., Greaves, D. R., and Grosveld, F. (1989). High-level, erythroid-specific expression of the human alpha-globin gene in transgenic mice and the production of human hemoglobin in murine erythrocytes. *Genes Dev.* **3,** 1572–1581.

Heyman, R. A., Borrelli, E., Lesley, J., Anderson, D., Richman, D. D., Baird, S., Hyman, R., and Evans, R. M. (1989). Thymidine kinase obliteration: creation of transgenic mice with controlled immune deficiency. *Proc. Natl. Acad. Sci. U.S.A.* **86,** 2698–2702.

Ho, S. N., Biggar, S. R., Spencer, D. M., Schreiber, S. L., and Crabtree, G. R. (1996). Dimeric ligands define a role for transcriptional activation domains in reinitiation. *Nature (London)* **382,** 822–826.

Hoess, R. H., Ziese, M., and Sternberg, N. (1982). P1 site-specific recombination: Nucleotide sequence of the recombination sites. *Proc. Natl. Acad. Sci. U.S.A.* **79,** 3398–3402.

Hurwitz, D. R., Nathan, M., Barash, I., Ilan, N., and Shani, M. (1992). Specific combinations of human serum albumin introns direct high level expression of albumin in transfected COS cells and in the milk of transgenic mice. *Transgenic Res.* **1,** 195–208.

Kappel, C. A., Bieberich, C. J., and Jay, G. (1994). Evolving concepts in molecular pathology. *FASEB J.* **8,** 583–592.

Kendall, S. K., Saunders, T. L., Jin, L., Lloyd, R. V., Glode, L. M., Nett, T. M., Keri, R. A., Nilson, J. H., and Camper, S. A. (1991). Targeted ablation of pituitary gonadotropes in transgenic mice. *Mol. Endocrinol.* **5,** 2025–2036.

Keri, R. A., Andersen, B., Kennedy, G. C., Hamernik, D. L., Clay, C. M., Brace, A. D., Nett, T. M., Notides, A. C., and Nilson, J. H. (1991). Estradiol inhibits transcription of the human glycoprotein hormone a-subunit gene despite the absence of a high affinity binding site for estrogen receptor. *Mol. Endocrinol.* **5,** 725–733.

Kistner, A., Gossen, M., Zimmermann, F., Jerecic, J., Ullmer, C., Lubbert, H., and Bujard, H. (1996). Doxycycline-mediated quantitative and tissue-specific control of gene expression in transgenic mice. *Proc. Natl. Acad. Sci. U.S.A.* **93,** 10933–10938.

Kuhn, R., Schwenk, F., Aguet, M., and Rajewsky, K. (1995). Inducible gene targeting in mice. *Science* **269,** 1427–1429.

Lamb, B. T. and Gearhart, J. D. (1995). YAC transgenics and the study of genetics and human disease. *Curr. Opin. Genet. Dev.* **5,** 342–348.

Lander, E. S. (1996). The new genomics: Global views of biology. *Science* **274,** 536–539.

Liberles, S. D., Diver, S. T., Austin, D. J., and Schreiber, S. L. (1997). Inducible gene expression and protein translocation using nontoxic ligands identified by a mammalian three-hybrid screen. *Proc. Natl. Acad. Sci. U.S.A.* **94,** 7825–7830.

Lieber, A. and Strauss, M. (1995). Selection of efficent cleavage sites in target RNAs by using a ribozyme expression library. *Mol. Cell. Biol.* **15,** 540–551.

Lovejoy, E. A., Clarke, A. R., and Harrison, D. J. (1997). Animal models and the molecular pathology of cancer. *J. Pathol.* **181,** 130–135.

Markkula, M., kananen, K., Paukku, T., Mannisto, A., Loune, E., Frojdman, K., Pelliniemi, L. J., and Huhtaniemi, I. (1995). Induced ablation of goadotropins in transgenic mice expressing herpes simplex virus thymidine kinase under the FSH beta-subunit promoter. *Mol. Cell. Endocrinol.* **108,** 1–9.

Marschall, P., Thomson, J. B., and Eckstein, F. (1994). Inhibition of gene expression with ribozymes. *Cell. Mol. Neurobiol.* **14,** 523–538.

Matzuk, M. M., Hsueh, A. J. W., Lapolt, P., Tsafriri, A., Keene, J. L., and Boime, I. (1990). The biological role of the carboxyl-terminal extension of human chorionic gonadotropin beta subunit. *Endocrinology (Baltimore)* **126,** 376–383.

Mayo, K. E., Hammer, R. E., Swanson, L. W., Brinster, R. L., Rosenfeld, M. G., and Evans, R. M. (1988). Dramatic pituitary hyperplasia in transgenic mice expressing a human growth hormone-releasing factor gene. *Mol. Endocrinol.* **2,** 606–612.

McGrane, M. M., deVente, J., Yun, J. S., Bloom, J., Park, E. A., Wynshaw-Boris, A., Wagner, T., Rottman, F. M., and Hanson, R. W. (1988). Tissue specific expression and hormonal regulation of a chimeric PEPCK/bGH gene in transgenic mice. *J. Biol. Chem.* **263,** 11443–11451.

McKnight, R. A., Shamay, A., Sankaran, L., Wall, R. J., and Hennighausen, L. (1992). Matrix-attachment regions can impart position-independent regulation of a tissue-specific gene in transgenic mice. *Proc. Natl. Acad. Sci. U.S.A.* **89,** 6943–6947.

McKnight, R. A., Wall, R. J., and Hennighausen, L. (1995). Expression of genomic and cDNA after co-integration in transgenic mice. *Transgenic Res.* **4,** 39–45.

Meade, H., Gates, L., Lacy, E., and Lonberg, N. (1990). Bovine alpha S1-casein gene sequences direct expression of active human urokinase in mouse milk. *Bio/Technology* **8**, 443–446.

Mirochnitchenko, O. and Inouye, M. (1996). Effect of overexpression of human Cu,Zn superoxide dismutase in transgenic mice on macrophage functions. *J. Immunol.* **156**, 1578–1586.

Needham, Gooding, C., Hudson, K., Antoniou, M., Grosveld, F., and Hollis, M. (1992). LCR/MEL: A versatile system for high-level expression of heterologous proteins in erythroid cells. *Nucleic Acids Res.* **20**, 997–1003.

Nilson, J. H., Keri, R. A., Reed, D. K. (1995). Transgenic mice provide multiple paradigms for studies in molecular endocrinology. *In* "Molecular Endocrinology: Basic Concepts and Clincial Correlations" (B. Weintraub, ed.), pp. 77–94. Raven Press, New York.

Niswender, G. D., Nett, T. M., Akbar, A. M. (1974). *In* "Reproduction in Farm Animals" (E. S. E. Hafez, ed.), pp. 57–81. Lea & Febiger, Philadelphia.

No, D., Yao, T., and Evans, R. M. (1996). Ecdysone-inducible gene expression in mammalian cells and transgenic mice. *Proc. Natl. Acad. Sci. U.S.A.* **93**, 3346–3351.

Palmiter, R. D. and Brinster, R. L. (1986). Germ-line transformation of mice. *Annu. Rev. Genet.* **20**, 465–499.

Palmiter, R. D., Brinster, R. L., Hammer, R. E., Trumbauer, M. E., Rosenfeld, M. G., Birnberg, N. C., and Evans, R. M. (1982). Dramatic growth of mice that develop from eggs microinjected with metallothionein-growth hormone fusion genes. *Nature (London)* **300**, 611–615.

Palmiter, R. D., Behringer, R. R., Quaife, C. J., Maxwell, F., Maxwell, I. H., and Brinster, R. L. (1987). Cell lineage ablation in transgenic mice by cell-specific expression of a toxin gene. *Cell (Cambridge, Mass.)* **50**, 435–443.

Passman, R. S. and Fishman, G. I. (1994). Regulated expression of foreign genes in vivo after germline transfer. *J. Clin. Invest.* **94**, 2421–2425.

Persuy, M. A., Legrain, S., Printz, C., Stinnakre, M. G., Lepourry, L., Brignon, G., and Mercier, J. C. (1995). High-level, stage- and mammary tissue specific expression of a caprine kappa-casein-encoding minigene driven by a beta-casein promoter in transgenic mice. *Gene* **165**, 291–296.

Phi-Van, L., Dries, J. P. ., Ostertag, W., and Stratling, W. H. (1990). The chicken lysozyme 5′ matrix attachment region increases transcription from a heterologous promoter in heterologous cells and dampens position effects on the expression of transfected genes. *Mol. Cell. Biol.* **10**, 2302–2307.

Polites, H. G., and Pinkert, C. A. (1993). DNA microinjection and transgenic animal production. *In* "Transgenic Animal Technology: A Laboratory Handbook" (C. A. Pinkert, ed.), pp. 15–67. Academic Press, San Diego, CA.

Rajewsky, K., Gu, H., Kuhn, R., Betz, U. A. K., Muller, W., Roes, J., and Schwenk, F. (1996). Perspective series: Molecular medicine in genetically engineered animals. Conditional gene targeting. *J. Clin. Invest.* **98**, 600–603.

Rajkumar, K., Dheen, S. T., and Murphy, L. J. (1996). Hyperglycemia and impaired glucose tolerance in IGF binding protein–1 transgenic mice. *Am. J. Physiol.* **270**, E565-E571.

Regan, L., Owen, E. J., and Jacobs, H. S. (1990). Hypersecretion of luteinizing hormone, infertility, and miscarriage. *Lancet* **336**, 1141–1144.

Risma, K. A., Clay, C. M., Nett, T. M., Wagner, T., Yun, J., and Nilson, J. H. (1995). Targeted overexpression of luteinizing hormone in transgenic mice leads to infertility, polycystic ovaries, and ovarian tumors. *Proc. Natl. Acad. Sci. U.S.A.* **92,** 1322–1326.

Risma, K. A., Hirshfield, A. H., and Nilson, J. H. (1997). Elevated LH in prepubertal transgenic mice causes hyperandrogenemia, precocious puberty, and substantial ovarian pathology. *Endocrinology (Baltimore)* **138,** 3540–3547.

Rivera, V. M., Clackson, T., Natesan, S., Pollock, R., Amara, J. F., Keenan, T., Magari, S. R., Phillips, T., Courage, N. L., Cerasoli, F., Holt, D. A., and Gilman, M. (1996). A humanized system for pharmacological control of gene expression. *Nat. Med.* **2,** 1028–1032.

Roberts, J. A. (1976). Hydronephrosis of pregnancy. *Urology* **8,** 1–4.

Schedl, A., Montoliu, L., Kelsey, G., and Schutz, G. (1993). A yeast artifical chromosome covering the tyrosinase gene confers copy number dependent expression in transgenic mice. *Nature (London)* **362,** 258–261.

Schedl, A., Grimes, B., and Montoliu, L. (1996). YAC transfer by microinjection. *Methods Mol. Biol.* **54,** 293–306.

Schultze, N., Burki, Y., Lang, Y., Certa, U., and Bluethmann, H. (1996). Efficient control of gene expression by single step integration of the tetracycline system in transgenic mice. *Nat. Biotechnol.* **14,** 499–503.

Shanahan, C. M., Rigby, N. W., Murray, J. D., Marshall, J. T., Townrow, C. A., Nancarrow, C. D., and Ward, K. A. (1989). Regulation of expression of a sheep metallothionein 1a-sheep growth hormone fusion gene in transgenic mice. *Mol. Cell. Biol.* **9,** 5473–5479.

Sharma, A., Khoury-Christianson, A. M., White, S. P., Dhanjal, N. K., Huang, W., Paulhiac, C., Friedman, E. J., Manjula, B. N., and Kumar, R. (1994). High-efficiency synthesis of human alpha-endorphin and magainin in the erythrocytes of transgenic mice: A production system for therapeutic peptides. *Proc. Natl. Acad. Sci. U.S.A.* **91,** 9337–9341.

Shockett, P., Difilippantonio, M., Hellman, N., and Schatz, D. G. (1995). A modified tetracycline-regulated system provides autoregulatory, inducible gene expression in cultured cells and transgenic mice. *Proc. Natl. Acad. Sci. U.S.A.* **92,** 6522–6526.

Shoham, Z., Jacobs, H. S., and Insler, V. (1993). Luteinizing hormone: its role, mechanism of action, and detrimental effects when hypersecreted during the follicular phase. *Fertil. Steril.* **59,** 1153–1161.

Simon, M. I. (1997). Dysfuntional genomics: BACs to the rescue. *Nat. Biotechnol.* **15,** 839

Sternberg, N. and Hamilton, D. (1981). Bacteriophage P1 site-specific recombination. I. Recombination between loxP sites. *J. Mol. Biol.* **150,** 467–486.

Strouboulis, J., Dillon, N., and Grosveld, F. (1992). Developmental regulation of a complete 70-kb human b-globin locus in transgenic mice. *Genes Dev.* **6,** 1857–1864.

Struthers, R. S., Vale, W. W., Arias, C., Sawchenko, P. E., and Montminy, M. R. (1991). Somatotroph hypoplasia and dwarfism in transgenic mice expressing a non-phosphorylatable CREB mutant. *Nature (London)* **350,** 622–624.

Sumarsono, S. H., Wilson, T. J., Tymms, M. J., Venter, D. J., Corrick, C. M., Kola, R., Lahoud, M. H., *et al.* (1996). Down's syndrome-like skeletal abnormalities in Ets2 transgenic mice. *Nature (London)* **379,** 534–537.

Tisch, R. and McDevitt, H. Insulin-dependent diabetes mellitus. *Cell (Cambridge, Mass.)* (1996). **85,** 291–297.

Wallace, H., Ledent, C., Vassart, G., Bishop, J. O., and Al-Shawi, R. (1991). Specific ablation of thyroid follicle cells in adult transgenic mice. *Endocrinology (Baltimore)* **129,** 3217–3226.

Wilson, C., Bellen, H. G., and Gehring, W. J. (1990). Position effects on eukaryotic gene expression. *Annu. Rev. Cell Biol.* **6,** 679–714.

Yang, X. W., Model, P., and Heintz, N. (1997). Homologous recombination based modification in *Esherichia coli* and germline transmission transgenic mice of a bacterial artificial chromosome. *Nat. Biotechnol.* **15,** 859–865.

Yarranton, G. T. (1992). Inducible vectors for expression in mammalian cells. *Curr. Opin. Biotechnol.* **3,** 506–511.

14

EXPRESSION OF RECOMBINANT PROTEINS IN THE MILK OF TRANSGENIC ANIMALS

H. M. Meade,* Y. Echelard,* C. A. Ziomek,* M. W. Young,* M. Harvey,* E. S. Cole,+ S. Groet,* T. E. Smith,* and J. M. Curling†

**Genzyme Transgenics Corp. and +Genzyme Corp., Framingham, Massachusetts 01701 and †John Curling Consulting AB, S-753 29 Uppsala, Sweden*

Gene Expression Systems: Using Nature for the Art of Expression

Introduction

Introduction of foreign DNA into the murine genome by microinjection and the generation of transgenic offspring were first reported by Gordon *et al.* (1980) and Gordon and Ruddle (1981). This technique has been applied to the production of transgenic livestock. Using mammary gland-specific promoters, a wide range of proteins of biopharmaceutical interest have been expressed in rodents, pigs, and dairy animals. An expression vector, comprising a gene encoding the target protein of interest fused to a milk promoter gene, is introduced by microinjection into the pronucleus of a one-cell embryo. Upon germ line integration and expression, the transgene becomes a dominant Mendelian genetic characteristic that is inherited by the progeny of the founder animal. The transgenic offspring may express the target protein in gram per liter quantities, most frequently as a soluble whey protein. Mammalian mammary epithelial cells have the capacity to carry out complex protein synthesis with a variety of posttranslational modifications and proper folding. Coexpression of modifying enzymes in the epithelial cell Golgi apparatus may allow the heterologous protein to be engineered to confer specific or desired pharmacokinetic characteristics. The milk of transgenic livestock presents an excellent starting material from which human diagnostic or pharmaceutically active, therapeutic proteins may be purified using established technologies of the dairy and biopharmaceutical industries.

To date (1997), probably more than 50 proteins have been expressed in the milk of transgenic mice, rats, rabbits, goats, sheep, pigs, and dairy cows. Phase I clinical trials with antithrombin III produced in transgenic goats were completed successfully in 1996, and phase II clinical trials are currently being performed in the United States. α_1-Antitrypsin produced in transgenic sheep is currently in phase I clinical trials in the United Kingdom. In addition, human lactoferrin and fibrinogen expressed in the milk of transgenic cows and sheep, respectively, are in the late stages of development or preclinical evaluation.

The field of transgenic research has been reviewed periodically since 1990. The reader will find the following articles, which trace the development of transgenic technology applied to dairy animals, of interest: Henninghausen (1990), Henninghausen *et al.* (1990), Bialy (1991), Wilmut *et al.* (1991), Wall *et al.* (1992), Jänne *et al.* (1992, 1994), Logan (1993), Ebert and Schindler (1993), Lee and de Boer (1994), Houdebine (1994, 1995), and Echelard (1996).

This chapter discusses the expression of therapeutically useful proteins in mammalian milk with a focus on dairy animals as the production systems of choice and reviews expression constructs, milk-specific transgenes, transgene insertion, transgenic animal production, protein biosynthesis and secretion, lactation, milk, protein purification, and quality and regulatory issues.

Expression of Heterologous Proteins in Milk

Laboratory mice have served as a model expression system for foreign proteins since the inception of animal transgenesis and are used frequently for feasibility studies concomitant with or prior to the generation of larger, founder transgenic animals. Many recombinant proteins of interest in human therapy are, by their very nature, biologically active in most mammals. The murine model is useful in determining, at an early stage of research, the transgene expression characteristics and potential effects on animal health. Transgenic mice are generally predictive of what will be observed in larger animals. Many transgenic proteins have been expressed in the milk of transgenic livestock. Table 1 summarizes the published results of expression in transgenic farm animals during the period from 1990 to 1996. Several of these proteins have been expressed at high levels, demonstrating the usefulness of the production system. Examination of a great number of transgenic lines has shown that mammary gland-specific expression of a target protein is associated with increased plasma levels of this protein, even in the absence of ectopic expression. The most likely explanation is leakage of the protein from the milk into the circulation through the junctional complexes of the mammary epithelial cells. Therefore, certain proteins and peptides, such as highly active hormones and cytokines, cannot be expressed in the mammary system as their secretion into the blood may have severe detrimental effects on the host.

Table 1
Biologically Active Proteins Expressed in Caprine, Porcine, Ovine, and Bovine Milk

Expressed protein	Animal	Promoter	Expression level	Reference
Human growth hormone	Goat	Retroviral promoter	12 ng/ml[a]	Archer *et al.* (1994)
Human long-acting tPA	Goat	Murine WAP	3 μg/ml	Ebert *et al.* (1991)
Human long-acting tPA	Goat	Caprine β-casein	3.5–8.0 mg/ml	Ebert *et al.* (1994)/GTC[b]
Human protein C	Pig	Murine WAP	1 mg/ml	Velander *et al.* (1992)
Human α-antitrypsin	Sheep	Ovine β-lactoglobulin	35 mg/ml	Wright *et al.* (1991)
Human α_1-proteinase inhibitor	Goat	Caprine β-casein	20 mg/ml	GTC[b]
Human antithrombin III	Goat	Caprine β-casein	20 mg/ml	GTC[b]
Human factor VIII	Sheep	Ovine β-lactoglobulin	Not given	Halter *et al.* (1993)
Human factor IX	Sheep	Ovine β-lactoglobulin	5 μg/ml	Clark *et al.* (1989)
Human fibrinogen	Sheep	Ovine β-lactoglobulin	5 mg/ml	Carver (1996)
Colon cancer MAb	Goat	Caprine β-casein	10 mg/ml	GTC[b]
Human lactoferrin	Cow	Bovine αs1 casein	Not given	Krimpenfort *et al.* (1991) Lonnerdal and Iyer (1995)

[a]Gene insertion via teat canal using retroviral vectors GaLV and MoMLV.
[b]Data from Genzyme Transgenics Corp. (1996).

Milk-Specific Transgenes

Transgenes containing sequences of several milk protein genes, reviewed by Maga and Murray (1995) and Echelard (1996), have been used to direct the expression of exogenous proteins to the lactating mammary gland. These transgenes are usually chimeric, being derived from the fusion of a target protein gene and mammary-specific regulatory sequences. Although both genomic DNAs and cDNAs coding for target proteins have been used for expression, higher levels are normally obtained with genomic DNAs. The incorporation of untranslated exons and introns may contribute to increased expression of the transgene (Whitelaw *et al.*, 1991). Addition of a signal sequence is necessary if the exogenous protein is not normally secreted. This will cause the protein to be secreted out of the mammary tissue into the milk.

Regulatory sequences from several milk-specific genes have been isolated and tested in transgenic animals: ovine β-lactoglobulin; murine, rat, and rabbit whey acidic protein (WAP); bovine α-s1 casein; rat, rabbit, and goat β-casein; and guinea pig, ovine, caprine, and bovine α-lactalbumin. Of these promoters, several have permitted grams per liter expression of target proteins in the milk of transgenic offspring, sometimes in large dairy animals; some of this work is summarized next.

The ovine β-lactoglobulin gene contains seven exons and six introns spanning a 4.2-kb region (Harris *et al.*, 1988). The first reported ovine β-lactoglobulin chimeric transgenes (Archibald *et al.*, 1990) were composed of 4 kb of 5′ flanking fused to the α_1-antitrypsin genomic sequences. Other configurations using variable amounts of 5′- and 3′-flanking sequences have also been used (Whitelaw *et al.*, 1992) and reviewed by Maga and Murray (1995). With the ovine β-lactoglobulin gene, high-level expression (g/liter) of α_1-antitrypsin, fibrinogen, and HSA have been reported (Wright *et al.*, 1991, Shani *et* al., 1992; Prunkard *et al.*, 1996). However, similar results have not been observed with transgenes containing cDNA sequences (Clark *et al.*, 1989, Shani *et al.*, 1992, Hansson *et al.*, 1994, Yull *et al.*, 1995).

Rodent WAP genes consist of four exons and three introns: the middle two exons encode the two cysteine-rich regions, which probably form separate protein domains (Campbell *et al.*, 1984). WAP is present in the milk of mouse, rat, rabbit, and camel, but is absent from the milk of cow, sheep, pig, goat, and human. No recognizable WAP gene homologues have been isolated from these species. Rat (Wei *et al.*, 1995, Yarus *et al.*, 1997), mouse (Gordon *et al.*, 1987; Ebert *et al.*, 1991, Reddy *et al.*, 1991; Velander *et al.*, 1992; Drohan *et al.*, 1994; Hansson *et al.*, 1994; Limonta *et al.*, 1995), and rabbit (Bischoff *et al.*, 1992; Devinoy *et al.*, 1994; Thépot *et al.*, 1995) WAP regulatory sequences have been used to direct expression of exogenous proteins to the mammary gland. High-level expression of the target protein was observed with mouse and rabbit WAP transgenes. Surprisingly, relatively high expression levels (up to 1 g/liter) were observed in transgenic pigs with a construct containing 2.6 kb of 5′ and 1.3 kb of 3′ mouse WAP sequences linked to a human protein C cDNA (Velander *et al.*, 1992). This result seems to indicate that mWAP regulatory sequences can function efficiently in species that do not have an endogenous WAP gene.

The bovine αs1-casein gene contains 9 exons and spans 17.5 kb (Koczan *et al.*, 1991). Originally (Meade *et al.*, 1990), a transgene containing 21 kb of 5′ and 2 kb of 3′ flanking sequence fused to the genomic sequences of human urokinase was shown to direct high milk expression levels (1–2 mg/ml) in mice. Promising results were also obtained in transgenic rabbits with a construct containing the human IGF-1 cDNA (Brem *et al.*, 1994), in mice with the human lysozyme cDNA (Maga *et al.*, 1994), and in human lactoferrin and human granulocyte–macrophage colony-stimulating factor genomic constructs (Nuijens *et al.*, 1995; Uusi-Oukari *et al.*, 1997). Conversely, a bovine αs1-casein–human lactoferrin cDNA construct only permitted low-level expression in milk of transgenic mice (Platenburg *et al.*, 1994), as was the case with a human tPA cDNA construct fused to 1.6 kb of bovine α-s1 casein 5′-flanking sequences (Riego *et al.*, 1993).

The bovine α-lactalbumin gene contains four exons and three introns. Early reports (Vilotte *et al.*, 1989; Stinnakre *et al.*, 1991) indicated that a construct containing 750 bp of 5′ and 336 bp of 3′ flanking region was sufficient to direct intermediate expression levels in transgenic mouse milk when fused to bovine α-lactalbumin or ovine trophoblastin cDNAs. By using a construct containing the same amount of 5′-flanking sequence linked to hGH genomic sequences, higher levels of the target protein (up to 4.3 mg/ml) were obtained in the milk of transgenic rats (Ninomiya *et al.*, 1994). However, at this point, results obtained with α-lactalbumin-driven transgenes in large animals have not been reported.

The caprine β-casein gene (CSN2) has been cloned and sequenced (Roberts *et al.*, 1992; Persuy *et al.*, 1992). The intron/exon organization of the 9-kb goat gene is similar to that of other CSN2 genes and its expression is limited principally to the mammary gland during lactation. High-level expression was observed in goats transgenic for a construct containing 6.2 kb of 5′ and 7.1 kb of 3′ goat β-casein flanking noncoding sequence fused to a variant of the human tPA cDNA (Ebert *et al.*, 1991). High-level expression with caprine β-casein-containing transgenes has also been observed, in mouse milk, with bovine κ-casein (Persuy *et al.*, 1995; Gutierrez *et al.*, 1996), antithrombin III (cDNA and genomic, mice and goats), HSA (cDNA and genomic), α_1-antitrypsin (genomic, mice and goats), and both heavy and light chains of several humanized antibodies (H. Meade *et al.*, unpublished data).

Insertion of the Transgene into the Germ Line

Transgenic animals may be generated by direct microinjection of the foreign gene into the pronuclei of one-cell stage embryos. Microinjection techniques have been reviewed exhaustively by Hogan *et al.* (1986) and Pinkert (1994), among others. Techniques initially developed for gene insertion into murine pronuclei (Gordon *et al.*, 1980; Gordon and Ruddle, 1981; Brinster *et al.*, 1985; Palmiter and Brinster, 1986) have been adapted to gene transfer into the pronuclei of ruminants and pigs (Hammer *et al.*, 1985; Pinkert, 1994). If the microinjected DNA integrates into the genome of the recipient before the first cell division occurs, a heterozygous founder can be created. Later integration leads to genetic mosaics consisting of normal cells with a normal genome and cells with transgenomes.

Generally, fertilized eggs are flushed from the oviduct of a superovulated female donor, microinjected with a few hundred copies of the transgene, transferred to the oviduct or uterus of a pseudopregnant recipient animal, and developed to term. The first transgenic offspring, or founder animals, are at best hemizygous as the transgene is not integrated into both copies of a pair of homologous chromosomes. In the case of mosaic founders, germ line transmission is not always observed. In addition, multiple transgene integration sites have been detected in 10–20% of transgenic founders.

To optimize the collection and transfer of microinjectable goat embryos, Selgrath *et al.* (1990) established regimens for superovulation/synchronization and timing of pronuclear embryo collection. Does were synchronized with Norgestomet ear implants and superovulation was induced with pregnant mare serum gonadotropin (PMSG) or follicle-stimulating hormone (FSH-P). Does were hand mated with the average female being mated six to eight times by two different males over a 24- to 36-hr period. Embryos were recovered surgically and pronuclei could be visualized without centrifugation. After microinjection the embryos were transferred to the reproductive tracts of recipient females using a surgical procedure similar to that used for the embryo collection.

Ewes may be similarly synchronized and superovulated during the breeding season using Progestin (30 mg fluorogestone acetate) pessaries and FSH injection (Rexroad and Wall, 1987). Fertilized sheep oocytes are semiopaque and not readily visible. However, differential interference contrast microscopy allows visualization, and successful microinjection may be determined by the swelling of the pronucleus, which occurs on injection of DNA (Simons *et al.*, 1988). Microinjected embryos are then transferred to recipient ewes.

Bovine oocytes are generally collected from slaughtered heifers or obtained from cows superovulated with PMSG or prostaglandin. To circumvent surgical procedures and *in vivo* fertilization, Krimpenfort *et al.* (1991) have used an *in vitro* fertilization and embryo production procedure. The technique for DNA microinjection is similar to those described earlier: centrifugation, e.g., at 12,000 × *g* for 10 min, is necessary to visualize pronuclei. Microinjected cow embryos are usually cultured *in vitro* to the morula-blastocyst age at which time they are transferred nonsurgically to suitable recipients. The techniques and efficiency of gene transfer in cows have been described by Roschlau *et al.* (1989) and McEvoy and Sreenan (1990).

Synchronization of sows is accomplished with hormonal treatment and superovulation induced with PMSG or human chorionic gonadotropin. Synchronization, however, has been shown to affect the farrowing rate (Pursel *et al.*, 1990). Pig ova are opaque and no nuclear structures can be seen, even using interference contrast microscopy. Centrifugation at 15,000 × *g* for 5 min leaves pronuclei visible in the equatorial segment of the cytoplasm (Brem *et al.*, 1985; Hammer *et al.*, 1985). Microinjected embryos are then transferred to recipient pigs.

Factors affecting the success of microinjection as a gene insertion technique have been reviewed by Rexroad *et al.* (1990). Experience in microinjection is an important factor as well as DNA concentration and the gene construct. The stage of egg development and the quality of eggs may affect the efficiency of producing transgenic animals. Three factors contribute to problems of microinjection of livestock embryos compared to mice. Cytoplasmic vesicles may obscure the view of pronuclei, fewer embryos are available for microinjection, and there is considerable variability in the stage of embryo development at the time of embryo collection. Despite these challenges, there has been considerable success in developing transgenic goats, sheep, pigs, and cows. Key reproduction parameters and the success rate of transgenesis are summarized in Table 2.

In summary, oocytes may be fertilized *in vivo* or *in vitro*. *In vivo* fertilization may be controlled by artificial insemination of a superovulated animal at the stage where the oocyte has matured in the ovary. An alternative pathway is to isolate follicular oocytes from the ovary of the donor animal and proceed to *in vitro* maturation and fertilization prior to microinjection and implantation in the recipient female. In goats, sheep, and cows the rate of transgenic births is 5–10%.

Embryonic stem (ES) cells may offer an alternative to pronuclear microinjection for achieving transgenesis. However, pluripotent ES cells able to contribute to the germ line have only been described in mice. There have been descriptions of chimeric animals generated with rat, pig, and cow ES cells (Iannaconne *et al.*, 1994; Wheeler, 1994; Stice *et al.*, 1996), but in the case of rats and pigs no evidence of germ line transmission from these cells has been reported. Cow

Table 2
Comparison of Reproduction Data for Dairy Animals Used to Produce Biopharmaceutical Products by Transgenesis[a]

	Goat	Sheep	Cow
Seasonal breeding	Yes	Yes	No
Number of one-cell embryos	1–3	1–3	1
Number of one-cell embryos (superovulated)	4–8	4–10	3–6
Gestation time (months)	5	5	9
Rate of transgenic births (%)	5–10	5–10	5–10
Litter size	1–2	1–2	1
Time to sexual maturity (months)	6–8	6–8	15

[a]Data in part from Clark *et al.* (1987).

fetuses obtained following nuclear transfer of ES cell-derived nuclei in recipient oocytes died *in utero,* exhibiting major defects in placental development (Stice *et al.*, 1996). In sheep, the first large animals derived from cultured cells were described in 1996 by Campbell *et al.* Two healthy phenotypicaly female lambs were born from embryos generated by transferring nuclei isolated from embryo-derived cells into enucleated oocytes. DNA analysis demonstrated that all the nuclear transfer lambs and fetuses were derived from the cell line. It is not yet clear whether foreign DNA can be introduced in this type of cell line, and there are questions about the health and reproductive fitness of the recovered offspring. Nevertheless, this experiment is certainly a step toward the possibility of replacing pronuclear microinjection as the method of choice for the generation of transgenic large animals.

Another potential alternative to the pronuclear microinjection of the transgene is the use of replication-defective retrovirus vectors. Archer *et al.* (1994) have described a procedure in which the construct is infused directly into the mammary gland, via the teat canal, during a period of hormone-induced mammogenesis. A gibbon ape leukemia virus was used to deliver the structural gene encoding for human growth hormone resulting in expression of the hormone in goat mammary epithelial cells. The advantage of this method is that expression of the recombinant protein is obtained quickly, without the delay caused by the generation interval required to generate a producing transgenic animal (18 months for goats). However, reported production levels (Archer *et al.*, 1994) are very low (see Table 1) and at this point in time can only be used for analytical purposes.

Transgenic Animal Production

Transgenic animals for the production of therapeutic and diagnostic proteins are produced by transferring fertilized, transgene-carrying embryos to recipient animals. Following natural gestation and birth, offspring are subject to tissue biopsy and blood sampling: usually ear tissue and blood are screened by polymerase chain reaction and/or Southern analysis for the presence of the transgene. Animals identified as being transgenic are mated with nontransgenic animals: transgenic founder females will produce the protein for which the DNA codes in their blood or milk depending on the tissue-specific, regulatory promoter sequence of the transgene. Subsequently, transgenic female progeny derived from breeding founder females and males will also express the transgenic protein.

In goats, 16–18 months are required before the first milk is obtained from a natural lactation of a female transgenic animal. However, milk samples can be obtained from founder transgenic females, as well as approximately 30% of male transgenic animals by hormonally induced lactation at about 13 months after microinjection. Induced lactation is useful in checking the expression level and integrity of the heterologous protein.

Biosynthesis of Milk Proteins

The original precursors of most of the milk constituents are cellulose, starch, protein, fat, minerals, and vitamins of the plant materials of the ruminant diet. Water is also a prerequisite, and dairy cows require 3–4 liters of water per liter of milk produced. Rumen microorganisms synthesize amino acids, which are adsorbed into the bloodstream. The milk protein precursor amino acids are adsorbed from the bloodstream via the extracellular fluid between the capillaries and the epithelial cells, across the basement membrane of the mammary epithelial cells. Protein biosynthesis is carried out in epithelial cells, and milk proteins are discharged into the lumen of the alveolus by exocytosis and then into the ducts. In ruminants, the ducts empty into a single, primary duct or cistern, which provides extra milk storage capacity. Up to 30% of the milk in the udder is held in the cistern.

As measured by the concentration difference in arteriovenous blood, the mammary gland is particularly efficient at extracting amino acids: 80% of the arterial methionine; 70% of the phenylalanine, leucine, and threonine; 60% of the lysine, arginine, and isoleucine; 55% of the histidine; and 50% of valine are adsorbed. More than 25% of the arterial blood glucose is removed during passage through the mammary gland and is used to power protein synthesis, fat, and lactose production. Intermediates required for protein synthesis are produced in the cytosol and mitochondria. Protein is synthesized at polyribosomes in the rough endoplasmic reticulum where removal of signal peptides occurs. Glycosylation is carried out in the endoplasmic reticulum and the Golgi apparatus where phosphorylation, other posttranslational modifications, and the assembly of casein micelles are carried out. Studies with radiolabeled amino acids indicate that proteins are synthesized in 3–15 min. Radioactivity is seen in the Golgi apparatus after 15–30 min, and the label concentration in the lumen increases after a further 30–60 min. Immunoglobulins and albumin present in milk are not synthesized in the mammary gland but by plasma cells and hepatocytes, respectively, and enter the milk by active transport or filtration. During lactation, B lymphocytes migrate to the mammary gland where they become plasma cells. Plasma cells lodged in the interstitial space may contribute to the high IgG concentration present in the colostrum of early lactation.

In the cow, approximately 500 liters of blood is required to provide the precursors for 1 liter of bovine milk: the blood flow in the udder is ca. 280 ml per second. The goat uses about 400 liters of blood to produce 1 liter of milk: the blood flow in the udder is ca. 1200 liters per day. The mammary gland is able to secrete about 2 g of milk, containing approximately 18 mg of whey protein, per gram of tissue per day. A gram of tissue contains about 2×10^8 cells and the milk output is therefore of the order of 10^{-8} g per cell per day. Lymph drains from the udder of the cow at the rate of 1300 ml per hour and in the goat at the rate of 6.5–35 ml per hour. Leukocytes, which account for the major part of the somatic cells found in milk, are derived from lymph.

Morcöl *et al.* (1994) have calculated the synthesis rate of human recombinant protein C in transgenic swine expressing the protein at 0.1–1 mg/ml milk to be approximately 14 mg per gram of mammary cell per day or about 14 pg/cell/day. In contrast, the rate of normal synthesis in hepatocytes was calculated to be 0.02 mg protein C per gram cells per day.

It has been suggested that mammalian species phylogenetically close to humans may be expected to have more elements of the glycosylation machinery in common (Jenkins *et al.*, 1996). Initial reports indicate that human glycoproteins expressed in the mammary gland of transgenic animals contain glycosylation patterns that differ from those found on human plasma-derived proteins. In general, the glycosylation found on human proteins secreted into transgenic animal milk has been generally similar to plasma protein glycosylation. Sites are mainly biantennary complex oligosaccharides with some variations consistent with the tissue and species of origin. Cole *et al.* (1994) have shown that human antithrombin III (AT III) and a long-acting form of htPA expressed in goat milk had some GalNAc replacing galactose on complex N-linked oligosaccharides. Both LAtPA and AT III were shown to be more fucosylated than their recombinant or plasma counterparts. Goat plasma AT III contains *N*-glycolylneuraminic acid and *N*-acetylneuraminic acid, as do transgenically expressed AT III and LAtPA. An additional difference observed between plasma and transgenic goat AT III is in the degree of sialylation (Edmunds *et al.*, 1994), with the transgenic protein less sialylated than plasma AT III. Denman *et al.*, (1991) have also noted that significantly lower levels of galactose, *N*-acetylglucosamine, and sialic acid are present in goat transgenic LAtPA compared to the murine C 127 cell line and Chinese hamster ovary (CHO) cell-derived LAtPA.

The glycosylation of interferon-γ at asparagine 25 and 97 is influenced dramatically by CHO cell culture conditions (Curling *et al.*, 1990); considerable variations in site occupancy are seen. Transgenic mouse-derived interferon-γ has predominantly complex sialated biantennary N-glycans at asparagine 25, and oligosaccharides are α 1-6 core fucosylated similar to the Asn_{25} of CHO cell-derived interferon-γ (James *et al.*, 1995). There is an increased incidence of oligomannose at Asn_{97} compared to the CHO-derived counterpart, suggesting that murine mammary epithelial cells may be deficient in the α 1-2 mannosidase I and GlcNAc transferase I activities in the endoplasmic reticulum.

Although an ultimate goal may be to produce proteins with authentic human glycosylation patterns, a more realistic and possibly more desirable objective is the production of proteins with defined, engineered glycosylation characteristics and therefore predictable pharmacokinetics. Glycoproteins can be remodeled *in situ* by the transgenic coexpression of human glycosyltransferase. Prieto *et al.* (1995) have demonstrated that the heterologous, transgenic expression of human α1,2-fucosyltransferase results in expression of both transgene and secondary gene products. Their work also suggests that the mammary gland may be a unique bioreactor for the production of biologically active oligosaccharides and glycoconjugates.

In studies of the expression of rh protein C in transgenic pigs, Subramanian *et al.* (1996) have shown that there are rate limitations of γ-carboxylation in mice and pigs, partly dependent on the transgene. Their study indicates that a rate limitation of γ-carboxylation in mammary epithelial cells occurs at expression levels of >20 μg/ml in mice and at >500 μg/ml milk in pigs.

Milk Secretion from the Mammary Gland

Mammary glands are skin glands that have no counterpart in nonmammals and are located in the inguinal region of cows, sheep, and goats. In the sow they are located along the thoracic, abdominal, and inguinal walls. Streak canals link the internal milk secretory system with the external environment. The number of teats, teat orientation, and length of the let-down reflex affect the ease and periodicity of milking. Milk is released as a result of a neuroendocrine reflex of the nervous system to tactile, auditory, and visual stimuli. Negative psychological and environmental disturbances have a detrimental effect on milk production. Milk let-down is a response to the release of oxytocin, synthesized in the hypothalamus, by the posterior pituitary gland. Oxytocin is transported to the mammary gland in the arterial blood and binds to myoepithelial cells that contract, causing a release or let-down of the milk. Sheep and goats have let-down periods of 1–2 and 2–4 min, respectively. The let-down reflex in the cow lasts for 5–8 min. In the sow, however, the reflex is extremely short, of the order of 10–20 sec, with a frequency of 1 hr or less.

From the point of view of primary milk production, goats and cows are preferred animals because of the relatively long let-down reflexes, vertical teat orientation, milk volume, and duration of lactation. However, other factors, such as protein concentration, may favor a choice of sheep. Goats, sheep, and cows respond to familiar signals, such as the sound of a milking machine, whereas lactating, nonsuckling sows require injections of oxytocin to elicit the let-down response.

Lactation and Milk Output

At parturition, a series of programmed hormonal changes take place that transform the mammary cells to the fully secretory state. Stage 2 lactogenesis, or the copious production of milk, is brought about by a synchronous drop in progesterone, an increase in estrogen, and the release of prolactin from the anterior pituitary and follows the immediate postpartum production of colostrum.

In the cow, milk production increases in the first 3–6 weeks of lactation and then slowly declines. A similar pattern is seen in goats, sheep, and pigs. Milk secretion continues as long as milk is regularly withdrawn, although production declines during lactation. Dairy animals are capable of undergoing estrus and pregnancy while maintaining their lactation. This enables dairy management to breed the animal such that it will give birth and begin lactation on a yearly basis. In order to maximize production, the animals are allowed to lactate until 2 months before they are due to give birth. They are then "dried off" by cessation of milking. The 2-month rest before the restart of lactation allows the animal to rebuild her energy reserves for the coming birth and lactation. Following birth of the progeny, the animal once again begins the yearly lactation cycle. In all species, yield and energy content are related to body size.

The expression level of exogenous protein tends to follow the normal milk output as can be seen from the plots in Fig. 1 which shows (a) the production of a recombinant monoclonal antibody over a 220-day lactation of a transgenic goat and (b) the levels of antithrombin III in the milk of a transgenic goat during a 300-day lactation.

It has been noted that "considering the yearly milk output of dairy cattle (6000–8000 liters) and the milk content of αs1-casein (10 g/liter), one cow carrying a transgene under the control of αs1-casein promoter would theoretically produce 60–80 kg/year of the transgene-derived protein" (Jänne *et al.*, 1992). At an expression level of 5 g/liter a transgenic goat is capable of producing about 4 kg of target protein per year; at 20 g/liter the output is 16 kg. It is, therefore, quite conceivable to produce 100-kg quantities of target protein in small goat herds or sheep flocks.

Milk Composition and Purification of the Target Protein

Milk is a multiphasic fluid composed of a fat emulsion, a micellular casein dispersion, a colloidal suspension of lipoproteins, and a solution of proteins, mineral salts, vitamins, organic acids, and minor components. When collected, milk is not sterile and contains

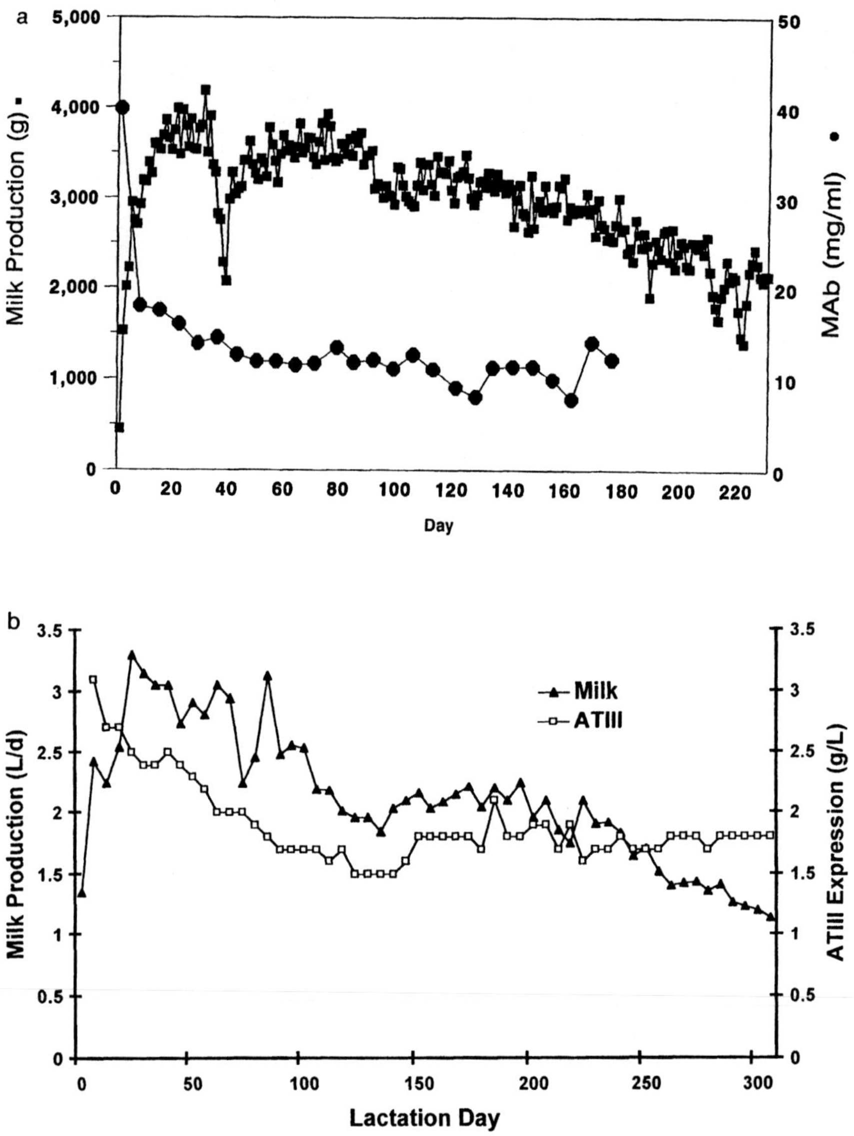

Figure 1 (a) Milk and recombinant antibody production during the first natural lactation of a transgenic goat during the 1996 season. (b) Milk and Antithrombin III production during the second natural lactation of a transgenic goat during the 1995 season. Work from Genzyme Transgenics Corp. (1996).

bacteria derived from the teat as well as somatic cells derived mostly from the lymphatic ducts of the udder. According to the Pasteurized Milk Ordinance (U.S. Department of Health and Human Services, 1993), the bacterial plate count should be less than 100,000 per milliliter.

Milk composition is species specific; major differences are seen between human and ruminant milk. Within a species, the composition varies with breed, diet, and other factors. Volume and composition also vary during lactation. Table 3 gives the average percentage compositions of livestock milk. Milk is approximately 85–90% water; the pH is 6.5–6.7 and as high as pH 6.8 in ewe's milk.

It is important from a purification point of view that the fat is present in globular form in the size range of 0.1–10 μm and with a density higher than the other constituents. The fat globule is enclosed in a membrane of polar lipids and proteins. Triglycerides make up 97% of the fat.

The three subgroups of casein, αs-, β-, and κ-casein, display genetic polymorphism and consist of two to eight variants. Casein is present as 10 to 300-nm micelles formed of submicelles held together by phosphate and hydrophobic bonding. Each submicelle has a polar core and a heterogeneous distribution of αs- and β-casein with surface κ-casein, αs- and β-caseins are almost insoluble, whereas the glycoprotein, β-casein is highly soluble in water. Casein may be precipitated somewhat below the isoelectric point range (pH 5.1–5.3) at about pH 4.6–4.7 by acidification or by the addition of chymosin, which attacks the 105 (phenylalanine) and 106 (methionine) peptide bond of the κ-casein. The hydrophilic amino

Table 3
Percentage Composition of Major Milk Constituents[a]

Animal	Total protein (%)	Casein (%)	Whey protein (%)	Fat (%)	Lactose (%)
Goat	3.6	2.7	0.9	4.1	4.7
Sheep	5.8	4.9	0.9	7.9	4.5
Pig	5.8	4.9	0.9	8.2	4.8
Cow[b]	3.5	2.8	0.7	3.7	4.8

[a]Data from Bylund (1995).
[b]Holstein.

acid terminal peptide (106–169) of κ-casein solubilizes in the whey fraction and all other caseins precipitate.

The soluble whey proteins consist primarily of α-lactalbumin and β-lactoglobulin with serum albumin and immunoglobulins being derived from the bloodstream. β-Lactoglobulin is the major protein, accounting for 50% or more of the total whey protein in ruminants and pigs. α-Lactalbumin accounts for approximately 25% of the whey protein fraction and is essential for lactose synthesis and the control of milk secretion: α-lactalbumin binds calcium and zinc. Both α-lactalbumin and β-lactoglobulin have an amino acid composition close to the nutritional optimum and provide amino acids that are essential for the neonate. Whey acid protein is present only in rodent milk.

The temperatures of raw milk at collection is about 37°C. To prevent bacterial growth, oxidation, and proteolysis, milk should be chilled immediately to 4°C and processed within 48 hr unless it is frozen and stored. In the dairy, milk is standardized with regard to the fat content by the centrifugal separation of cream and skim milk: 100 kg of 4% (fat) bovine milk yields 90.35 kg of 0.05% fat skim milk and 9.65 kg of 40% fat cream. The cream fraction is remixed with the skim milk to the required fat content. In processing for a target protein in the whey it is clear that the bulk of the fat can be removed using this standard procedure. However, other standard dairy procedures, particularly microfiltration, may be used to remove fat, casein, and cellular components in a single step. The high lactose and salt content may also be reduced by ultrafiltration. Membrane techniques are generally the initial recovery methods of choice and, when correctly used, may yield a 60% pure protein in a single or a tandem microultrafiltration step, thus providing a clarified whey concentrate for further processing. The use of such techniques also provides a barrier to the entry of adventitious viruses, bacteria, and other microorganisms into the final product.

An alternative pathway is to apply expanded bed technology. After initial processing to remove fat, skim milk may be passed through an adsorbent for the target, allowing the casein, lactose, and the bulk of the whey proteins to pass through the bed. Partially purified target protein may be recovered by desorbtion from the matrix in a fixed bed mode. This type of separation or direct feed capture may be used in a totally fixed bed mode applied to the whey fraction after tangential flow filtration. Subsequent processing by various chromatographic techniques, including affinity, hydrophobic interaction, ion exchange, and metal chelate chromatography, is applicable to achieve a protein of required purity. The number of steps should be kept to a minimum as even small step losses can lead to a low yield over an extensive process.

A procedure for the purification of human recombinant protein C expressed in porcine milk has been described (Degener *et al.*, 1996). Milk fat was removed by centrifugation, and casein was precipitated in the presence of zinc. Direct feed capture was performed using an expanded bed, and the protein was purified by hydrophobic interaction and anion-exchange chromatography.

Quality Issues in Transgenic Production

Farm activities associated with transgenic production include founder development, progeny testing, and dairying. To minimize production risks and achieve validatable procedures, "good agricultural practices" (GAPs), which are in the spirit of good manufacturing practices and good laboratory practice, should be used. GAPs are based on high standards of animal husbandry, adherence to standard operating procedures, on-site veterinary care, rigorous animal health monitoring programs, and state-of-the-art milking practices to maximize product quality. Standard operating procedures should cover areas such as generating founder animals, herd maintenance, herd health, breeding, milk production, and other special procedures. Master and working transgenic banks should be kept under strictly controlled conditions.

Animal feeds should be specially blended free of animal fat and protein. Hay for transgenic production animals should be screened for residual chemicals that may have been used in the growing process. Both written and computerized records should be kept to track animal lineages, health, and performance records.

Careful attention should be paid to milk facilities and personnel sanitation. Production animals should be observed on a daily basis, and animal side testing of milk can be used to detect early indications of mastitis and other animal illnesses. Milk collection and processing areas should be physically separated, and state-of-the-art pharmaceutical grade milking equipment and practices should be used in dedicated milking parlors.

Regulatory Considerations

"Points to Consider in the Manufacture and Testing of Therapeutic Products for Human Use Derived from Transgenic Animals" [Food and Drug Administration (FDA), 1995] is the result of an iterative process between the FDA and the biopharmaceutical industry, and the FDA is supportive of transgenic production. The European Union's Committee on Proprietary Medicinal Products (CPMP) is similarly positive to the use of transgenic animals, stating in its guidance document: "Transgenic animals may produce higher quantities of material in more concentrated form than existing culture methods, and therefore have considerable advantages in both the cost of producing the starting material and in its downstream processing. In some instances where very large amounts of material are required for therapy the use of transgenic animals may be one of the few viable production strategies" (CPMP, 1995).

The FDA has addressed topics such as the structure of the transgene, creation, characterization, and maintenance of the transgenic herd, fidelity of transgenic inheritance, consistency of transgene expression, analysis of product identity and purity, and the avoidance of contamination by drugs, chemicals, and adventitious agents. Many issues of purity, consistency, safety, and potency of transgenically produced therapeutic proteins overlap with various CBER points to consider for monoclonal antibodies and other biologicals produced by recombinant DNA technology. To quote the FDA: "The considerations that apply are therefore a blend of those relevant to recombinant DNA derived materials and materials from less defined sources" (FDA, 1995).

Despite the fact that goats and other farm animals are susceptible to species-specific viral and prion infections, viruses may be more of a concern for animal health than they are a human risk. A CBER official has pointed out that prions, such as those responsible for bovine spongiform encephalitis and scrapie, have "less than a theoretical" risk of transmission via milk because they do not occur in the mammary gland and, if introduced, they do not persist (Rudolph, 1995). No transmission of scrapie has ever been reported in humans. Milk and semen are noninfectious for scrapie according to the World Health Organization (1996). Although scrapie has been detected in significant numbers of sheep in the United States, only five goat scrapie cases have occurred in goats, all comingled with scrapie-infected sheep. The use of animals from closed, scrapie-free herds or flocks has been adopted by the leading transgenic production companies to obviate any such concerns. Goat viruses relevant in North America are shown in Table 4.

The risk of viral and bacterial contamination of human biopharmaceutical products expressed in the milk of transgenic goats can be minimized by a multistage or combinatorial approach at three levels: the goat, the milk, and the final product (Ziomek, 1996). Minimization of the risk of contamination of the goat production herd can be accomplished by strict selection, animal husbandry, and adherence to GAPs. Bacterial contamination can be minimized by careful, state-of-the-art milking procedures and rapid GMP processing. Purification processes from raw milk are designed in a manner similar to the manufacturing schemes for recombinant therapeutics from fermentation and cell culture with the inclusion of steps to remove and/or inactivate adventitious agents, microbial contaminants, and pyrogens.

For advantages and disadvantages of transgenic expression in the milk of dairy animals, see Table 5.

Table 4
Goat Viruses Relevant to Production of Proteins in Transgenic Goat Milk[a]

	Family	Nucleic Acid[b]	Shape	Size (nm)
Enveloped				
Rabies (z)	Rhaboviridae	RNA (ss)	Bullet	75 × 180
Coronavirus	Coronovi ridae	RNA (ss)	Unsymmetric	70–120
Caprine arthritis encephalitis	Retroviridae	RNA (ss)	Spherical	80 –110
Caprine herpes virus	Herpesviridae	DNA (ds)	Spherical	120–200
Pseudorabies	Herpesviridae	DNA (ds)	Icosahedral	180–200
Nonenveloped				
Bluetongue	Reoviridae	RNA (ds)	Spherical	60–80
Rotavirus	Reoviridae	RNA (ds)	Spherical	60–80
Adenovirus	Adenoviridae	DNA (ds)	Icosahedral	70–90
Complex				
Capripox (z)	Poxviridae	DNA (ds)	Complex	200–300
Contageous ecthyma (z)	Poxviridae	DNA (ds)	Complex	200–350

[a]From Ziomek (1996).
[b]z, zoonotic: ss, single stranded; ds, double stranded.

Table 5
Advantages and Disadvantages of Transgenic Expression in the Milk of Dairy Animals

Advantages
High level expression at multigram/liter level
Expression directed to and located in mammary gland
Expression can be tested in rodents
Genetic and lactation-to-lactation stability
Animal-to-animal consistency
Mammary epithelial cells carry out post-translational modification
Product recovered at high concentration in milk
Aseptic milk processing technology well proven
Bulk impurities removed easily
Low cost of pre-purification product
Low investment costs
Capability of combining transgenics with nuclear transfer (cloning)
Disadvantages
Possible adventitious virus and prion issues in dairy animals
Control of animal environment and feed
Microinjection required as a technology for species where cloning is not yet available
Time to clinical product slower than cell culture (without process development)

Current Status and Future Directions

Expression of therapeutically beneficial proteins in the milk of transgenic dairy animals present an unparalleled opportunity for the large-scale production of monoclonal antibodies, plasma, and other proteins.

Mammary gland epithelia have a cell density that is 100- to 1000-fold greater than the cell densities used, for example, in mammalian cell culture using CHO cells. The cells are some of the most productive protein synthesis sources designed by nature to produce large amounts of correctly processed protein and which are switched on and off by hormonal changes. Thirty-five transgenic goats expressing a monoclonal antibody at 8 g/liter in their milk are equivalent to a 8500-liter batch CHO cell culture running 200 days/year with a 1-g/liter final expression level (Young *et al.*, 1997). In production terms, 170,000 liters of culture is equivalent to 21,000 liters of milk; with the expression levels noted earlier and process yields of 60%, both of these systems would produce 100 kg of purified monoclonal antibody.

It has also been discussed that the mammary bioreactor is capable of most posttranslational modifications and protein folding and can, therefore, be used to produce complex proteins.

A current limitation of the technology is the low rate (5–10%) of transgenesis. Nuclear transfer techniques, in which embryonic or adult cells are engineered and cultured to produce a master cell bank from which nuclei are transferred to recipient oocytes, are under development; these methods hold great promise for improved rates of transgenesis.

Because of concerns of transmissible spongiform encephalopathies, animal sourced proteins and amino acids are currently under scrutiny. However, it should be kept in mind that the expression system described in this chapter is the mammary gland. Given the safety of milk and milk products in combination with the quality and regulatory precautions described, there is every reason to consider that proteins derived from transgenic animals should be safe with respect to possible virus and prion transmission. Many of the proteins that are expression targets today are currently available as human plasma derivatives, which have a measurable degree of risk for the transmission of HIV, hepatitis, parvo-, and other viruses, and Creutzfeld-Jakob disease.

As indicated, several proteins are in phase I and II clinical trials. This number can be expected to increase dramatically in the near future. Transgenic dairy animals provide a bulk protein production system that is capable of making proteins available that are either not available today or are only recoverable from other sources at very low yields. Transgenic production may thus enable a move from "on-demand" patient treatment to prophylaxis and far wider indications and use of many proteins. In addition, the production of proteins in the milk of transgenic dairy animals is highly cost effective (Young *et al.*, 1997), opening up real possibilities for nutraceutical product development.

References

Archer, J. S., Kennan, W. S., Gould, M. N., and Bremel, R. D. (1994). Human growth hormone (hGH) secretion in milk of goats after direct transfer of the hgH gene into the mammary gland by using replication-defective retrovirus vectors. *Proc. Natl. Acad. Sci. U.S.A.* **91,** 6840–6844.

Archibald, A. L., McClenaghan, M., Hornsey, V., Simons, J. P., and Clark, A. J. (1990). High level expression of biologically active human a1-antitrypsin in the milk of transgenic mice. *Proc. Natl. Acad. Sci. U.S.A.* **87,** 5178–5182.

Bialy, H. (1991). Barnyard biotechnology: "Transgenic pharming comes of age." *Bio/Technology* **9,** 786–788.

Bischoff, R., Degryse, E., Perraud, F., Dalemens, W., Ali-Hadji, D., Thépot, D., Devinoy, E., Houdebine, L. M., and Pavirani, A. (1992). A 17.2 kbp region located upstream of the rabbit WAP gene directs high level expression of a functional human protein variant in transgenic mouse milk. *FEBS Lett.* **305,** 265–268.

Brem, G., Brenig, B., Goodman, H. M., Selden, R. C., Graf, F., Kruff, B., Springman, K., Hondele, J., Meyer, J., Winnacker, E.-L., and Krässlich, H. (1985). Production of transgenic mice, rabbits and pigs by microinjection into pronuclei. *Zuechthygiene* **20,** 251–252.

Brem, G., Hartl, P., Besenfelder, U., Wolf, E., Zinovieva, N., and Pfaller, R. (1994). Expression of synthetic cDNA sequences encoding human insulin-like growth factor-1 (IGF-1) in the mammary gland of transgenic rabbits. *Gene* **149,** 351–355.

Brinster, R. L., Chen, H. Y., Trumbauer, M. E. *et al.* (1985). Factors affecting the efficiency of introducing foreign DNA into mice by microinjection eggs. *Proc. Natl. Acad. Sci. U.S.A.* **82,** 4438–4442.

Bylund, G. (1995). "Dairy Processing Handbook." Tetra Pak Processing Systems AB, Lund.

Campbell, K. H. S., McWhir, J., Ritchie, W. A., and Wilmut, I. (1996). Sheep cloned by nuclear transfer from a cultured cell line. *Nature (London)* **380,** 64–66.

Campbell, S. M., Rosen, J. M., Hennighausen, L. G. *et al.* (1984). Comparison of the whey acidic protein genes of rat and mouse. *Nucleic Acids Res.* **12,** 8685–8697.

Carver, A. (1996). Transgenics on trial. *Scrip Mag.* November, pp. 51–53.

Clark, A. J., Simons, P., Wilmut, I., and Lathe, R. (1987). Pharmaceuticals from transgenic livestock. *Trends Biotechnol.* **3,** 20–24.

Clark, A. J., Bessos, H., Bishop, J. O., Brown, P., Harris, S., Lathe, R., McClenaghan, M., Prowse, C., Simons, J. P., Whitelaw, C. B. A., and Wilmut, I. (1989). Expression of anti-haemophilic factor IX in the milk of transgenic sheep. *Bio/Technology* **7,** 487–492.

Cole, E. S., Higgins, E., Bernasconi, R., Garone, L., and Edmunds, T. (1994). Glycosylation patterns of human proteins expressed in transgenic goat milk. *J. Cell. Biochem. Suppl.* **18D,** 265.

Committee on Proprietary Medicinal Products (CPMP) (1995). "Use of Transgenic Animals in the Manufacture of Biological Medicinal Products for Human Use." Ad hoc Working Party on Biotechnology/Pharmacy, Directorate-General III/3612/93 Final.

Curling, E. M., Hayter, P. M., and Baines, A. J. (1990). Recombinant interferon-gamma. Differences in the glycosylation and proteolytic processing lead to heterogeneity in batch culture. *Biochem. J.* **272,** 333–337.

Degener, A., Belew, M., and Velander, W. H. (1996). High selectivity purification of recombinant proteins from milk using expanded bed chromatography. *Abstr. Recovery Biol. Prod.,* Tucson, AZ, *8th, 1996.*

Denman, J., Hayes, M., O'Day, C., Edmunds, T., Bartlett, C., Hirani, S., Ebert, K. M., Gordon, K., and McPherson, J. M. (1991). Transgenic expression of a variant of human tissue-type plasminogen activator in goat milk: Purification and characterization of the recombinant enzyme. *Bio/Technology* **9,** 839–843.

Devinoy, E., Thépot, D., Stinnakre, M.-G., Fontaine, M.-L., Grabowski, H., Puissant, C., Pavirani, A., and Houdebine, L.-M. (1994). High level production of human growth hormone in the milk of transgenic mice: The upstream region of the whey acidic protein (WAP) gene targets transgene expression to the mammary gland. *Transgenic Res.* **3,** 79–86.

Drohan, W. N., Zhang, D., Paleyanda, R. K., Chang, R., Wroble, M., Velander, W. H., and Lubon, H. (1994). Inefficient processing of human protein C in the mouse mammary gland. *Transgenic Res.* **3,** 355–364.

Ebert, K. M., and Schindler, J. E. S. (1993). Transgenic farm animals: Progress report. *Theriogenology* **39,** 121–135.

Ebert, K. M., Selgrath, J. P., Di Tullio, P., Denman, J., Smith, T. E., Memon, M. A., Schindler, J. E., Monastersky, G. M., Vitale, J. A., and Gordon, K. (1991). Transgenic production a variant of human tissue-type plasminogen activator in goat milk: Generation of transgenic goats and analysis of expression. *Bio/Technology* **9,** 835–838.

Ebert, K., Di Tullio, P., Barry, C. A., Schindler, J. E., Ayres, S. L., Smith, T. E., Pellerin, L. J., Meade, H. M., Denman, J., and Roberts, B. (1994). Induction of human tissue plasminogen activator in the mammary gland of transgenic goats. *Bio/Technology* **12,** 699–702.

Echelard, Y. (1996). Recombinant protein production in transgenic animals. *Curr. Opin. Biotechnol.* **7,** 536–540.

Edmunds, T., Higgins, E., Bernasconi, R., Garone, L., and Cole, E. S. (1994). Tissue specific and species differences in the glycosylation pattern of Antithrombin III. *J. Cell. Biochem., Suppl.* **18D,** 265.

Food and Drug Administration (FDA) (1995). "Points to Consider in the Manufacture and Testing of Therapeutic Products for Human Use Derived from Transgenic Animals." FDA, CBER.

Gordon, J. W., and Ruddle, F. H. (1981). Integration and stable germ-line transmission of genes injected into mouse pronuclei. *Science* **214,** 1244–1246.

Gordon, J. W., Scangos, G. A., Plotkin, D. J., Barbosa, J. A., and Ruddle, F. H. (1980). Genetic transformation of mouse embryos by microinjection of purified DNA. *Proc. Natl. Acad. Sci. U.S.A.* **77,** 7380–7384.

Gordon, K., Lee, E., Vitale, J. A., Smith, A. E., Westphal, H., and Hennighausen, L. (1987). Production of human tissue plasminogen activator in transgenic mouse milk. *Bio/Technology* **5,** 1183–1187.

Gutierrez, A., Meade, H. M., Ditullio, P., Pollock, D., Harvey, M., Jimenez-Flores, R., Anderson, G. B., Murray, J. D., and Medrano, J. F. (1996). Expression of a bovine k-CN cDNA in the mammary gland of transgenic mice utilizing a genomic milk protein gene as an expression cassette. *Transgenic Res.* **5,** 271–279.

Halter, R., Carnwath, J., Espanion, G., Herrmann, D., Lemme, E., Niemann, H., and Paul, D. (1993). Strategies to express Factor VIII gene constructs in the ovine mammary gland. *Theriogenology* **39,** 137–149.

Hammer, R. E., Pursel, V. G., Rexroad, C. E., Jr., Wall, R. J., Bolt, D. J., Ebert, K. M., Palmiter, R. D., and Brinster, R. L. (1985). Production of transgenic rabbits, sheep and pigs by microinjection. *Nature (London)* **315,** 680–683.

Hansson, L., Edlund, M., Edlund, A., Johansson, T., Marklund, S. L., Fromm, S., Stromqvist, M., and Tornell, J. (1994). Expression and characterization of biologically active human extracellular superoxide dismutase in milk of transgenic mice. *J. Biol. Chem.* **269,** 5358–5363.

Harris, S., Ali, S., Anderson, S., Archibald, A. L., and Clark, A. J. (1988). Complete nucleotide sequence of the ovine b-lactoglobulin gene. *Nucleic Acids Res.* **19,** 10379–10380.

Henninghausen, L. (1990). The mammary gland as a bioreactor: Production of foreign proteins in milk. *Protein Express. Purif.* **1,** 1–6.

Henninghausen, L., Ruiz, L., and Wall, R. (1990). Transgenic animals—production of foreign proteins in milk. *Curr. Opin. Biotechnol.* **1,** 74–78.

Hogan, B., Costantini, F., and Lacy, E. (1986). "Manipulating the Mouse Embryo." Cold Spring Harbor Press, Cold Spring Harbor, NY.

Houdebine, L.-M. (1994). Production of pharmaceutical proteins from transgenic animals. *J. Biotechnol.* **34,** 269–287.

Houdebine, L.-M. (1995). The production of pharmaceutical proteins from the milk of transgenic animals. *Reprod. Nutr. Dev.* **35,** 609–617.

Iannaccone, P. M., Taborn, G. U., Garton, R. L., Caplice, M. D., and Brenin, D. R. (1994). Pluripotent embryonic stem cells from the rat are capable of producing chimeras. *Dev. Biol.* **163,** 288–292.

James, D. C., Freedman, R. B., Hoare, M., Ogonah, O. W., Rooney, B. R., Larionov, O. A., Dobrovolsky, V. N., Lagutin, O. V., and Jenkins, N. (1995). N-glycosylation of recombinant human interferon-g produced in different animal expression systems. *Bio/Technology* **13,** 592–596.

Jänne, J., Hyttinen, J.-M., Peura, T., Tolvanen, M., Ahlonen, L., and Halmekytö, M. (1992). Transgenic animals as bioproducers of therapeutic proteins. *Ann. Med.* **24,** 273–280.

Jänne, J., Hyttinen, J.-M., Peura, T., Tolvanen, M., Ahlonen, L., Sinervirta, R., and Halmekytö, M. (1994). Transgenic bioreactors. *Int. J. Biochem.* **26,** 859–870.

Jenkins, N., Parekh, R. B., James, D. C. (1996). Getting the glycosylation right: implications for the biotechnology industry. *Nat. Biotechnol.* **14,** 975–981.

Koczan, D., Hobom, G., and Seyfert, H. M. (1991). Genomic organization of the bovine aS1-casein gene. *Nucleic Acids Res.* **19,** 5591–5596.

Krimpenfort, P., Rademakers, A., Eyestone, W., van der Schans, A., van den Broek, S., Kooiman, P., Kootwijk, E., Platenburg, G., Pieper, F., Strijker, R., and de Boer, H. (1991). Generation of transgenic dairy cattle using "in vitro" embryo production. *Bio/Technology* **9,** 844–847.

Lee, S. H., and de Boer, H. A. (1994). Production of biomedical proteins in the milk of transgenic dairy cows: State of the art. *J. Controlled Release* **29,** 213–221.

Limonta, J. M., Castro, F. O., Martinez, R., Puentes, P., Ramos, B., Aguilar, A., Lleonart, R. L., and de la Fuente, J. (1995). Transgenic rabbits as bioreactors for the production of human growth hormone. *J. Biotechnol.* **40,** 49–58.

Logan, J. S. (1993). Transgenic animals: beyond 'funny milk.' *Curr. Opin. Biotechnol.* **4,** 591–595.

Lonnerdal, B., and Iyer, S. (1995). Lactoferrin molecular structure and biological function. *Annu. Rev. Nutr.* **15,** 93–110.

Maga, E. A., and Murray, J. D. (1995). Mammary gland expression of transgenes and the potential for altering the properties of milk. *Bio/Technology* **13,** 1452–1457.

Maga, E. A., Anderson, G. B., Huang, M. C., and Murray, J. D. (1994). Expression of human lysozyme mRNA in the mammary gland of transgenic mice. *Transgenic Res.* **3,** 36–42.

McEvoy, T. G., and Sreenan, J. M. (1990). The efficiency of production, centrifugation, microinjection and transfer of one- and two-cell bovine ova in a gene transfer program. *Theriogenology* **33,** 819–828.

Meade, H., Gates, L., Lacy, E. *et al.* (1990). Bovine aS1-casein gene sequences direct high level expression of active urokinase in mouse milk. *Bio/Technology* **8,** 443–446.

Morcöl, T., Akers, R. M., Johnson, J. L., Williams, B. L., Gwazdauskas, F. C., Knight, J. W., Lubon, H., Paleyanda, R. K., Drohan, W. N., and Velander, W. H. (1994). The porcine mammary gland as a bioreactor for complex proteins. *Ann. N.Y. Acad. Sci.* **721,** 218–233.

Ninomiya, T., Hirabayashi, M., Sagara, J., and Yuki, A. (1994). Functions of milk protein gene 5′ flanking region on human growth hormone gene. *Mol. Reprod. Dev.* **37,** 276–283.

Nuijens, J. H., Geerts, M. E. J., van Berkel, P. H. C., Hartevelt, P. P., de Boer, H. A., van Veen, H. A., and Pieper, F. R. (1995). Characterization of recombinant human lactoferrin expressed in the milk of transgenic mice. *Proc. Int. Conf. Struct. Funct. Lactoferrin, 2nd.*

Palmiter, R. D., and Brinster, R. L. (1986). Germ-line transformation of mice. *Annu. Rev. Genet.* **20,** 465–499.

Persuy, M. A., Stinnakre, M. G., Printz, C., Mahe, M. F., and Mercier, J. C. (1992). High expression of the caprine b-casein gene in transgenic mice. *Eur. J. Biochem.* **205,** 887–893.

Persuy, M. A., Legrain, S., Printz, C., Stinnakre, M. G., Lepourry, L., Brignon, G., and Mercier, J. C. (1995). High-level, stage- and mammary-tissue specific expression of a caprine k-casein-encoding minigene driven by a b-casein promoter in transgenic mice. *Gene* **165,** 291–296.

Pinkert, C. A. (1994). "Transgenic Animal Technology: A Laboratory Handbook." Academic Press, San Diego, CA.

Platenburg, G. J., Kootwijk, E. P. A., Kooiman, P. M., Woloshuk, S. L., Nuijens, J. H., Krimpenfort, P. J. A., Pieper, F. R., deBoer, H. A., and Strijker, R. (1994). Expression of human lactoferrin in milk of transgenic mice. *Transgenic Res.* **3,** 99–108.

Prieto, P. A., Mukerji, P., Kelder, B., Erney, R., Gonzalez, D., Yun, J. S., Smith, D. F., Moremen, K. W., Nardelli, C., Pierce, M., Li, Y., Chen, X., Wagner, T. E., Cummings, R. D., and Kopchick, J. J. (1995). Remodeling of mouse milk glycoconjugates by transgenic expression of a human glycosyl transferase. *J. Biol. Chem.* **270,** 29515–29519.

Prunkard, D., Cottingham, I., Garner, I., Bruce, S., Balrymple, M., Lasser, G., Bishop, P., and Foster, D. C. (1996). High-level expression of recombinant human fibrinogen in the milk of transgenic animals. *Nat. Biotechnol.* **14,** 867–871.

Pursel, V., Hammer, R. E., Bolt, D., Palmiter, R. D., and Brinster, R. L. (1990). Integration, expression and germ-line transmission of growth-related genes in pigs. *J. Reprod. Fertil., Suppl.* **41,** 77–87.

Reddy, V. B., Vitale, J. A., Montoya-Zavala, M., and Robl, J. M. (1991). Expression of human growth hormone in the milk of transgenic mice. *Anim. Biotechnol.* **2,** 15–29.

Rexroad, C. R., Jr., and Wall, R. J. (1987). Development of one-cell fertilized sheep ova following microinjection into pronuclei. *Theriogenology* **27,** 611–619.

Rexroad, C. R., Jr., Powell, A. M., Behringer, R. R. *et al.* (1990). Insertion, expression and physiology of growth regulating genes in ruminants. *J. Reprod. Fertil., Suppl.* **41,** 119–124.

Riego, E., Limonta, J., Aguilar, A., Perez, A., de Armas, R., Solano, R., Ramos, B., Castro, F. O., and de la Fuente, J. (1993). Production of transgenic mice and rabbits that carry and express the human tissue plasminogen activator cDNA under the control of a bovine alpha S1 casein promoter. *Theriogenology* **39,** 1173–1185.

Roberts, B., Di Tullio, P., Vitale, J. *et al.* (1992). Cloning of the goat beta-casein gene and expression in transgenic mice. *Gene* **121,** 255–262.

Roschlau, K., Rommel, P., Andreewa, L., Zackel, M., Roschlau, D., Zackel, B., Schwerin, M., Hühn, R., and Gazarjan, K. G. (1989). Gene transfer experiments in cattle. *J. Reprod. Fertil., Suppl.* **38,** 153–160.

Rudolph, N. S. (1995). Regulatory issues relating to protein production in transgenic animal milk. *Genet. Eng. News* **15,** 16–18.

Selgrath, J. P., Memon, M. A., Smith, T. E., and Ebert, K. M. (1990). Collection and transfer of microinjectable embryos from dairy goats. *Theriogenology* **34,** 1195–1205.

Shani, M., Barash, I., Nathan, M., Ricca, G., Searfross, G. H., Dekel, I., Faerman, A., Givol, D., and Hurwitz, D. R. (1992). Expression of human serum albumin in the milk of transgenic mice. *Transgenic Res.* **1,** 195–208.

Simons, J. P., Wilmut, I., Clark, A. J., Bishop, J. O., and Lathe, R. (1988). Gene transfer into sheep. *Bio/Technology* **6,** 179–183.

Stice, S. L., Strelchenko, N. S., Keefer, C. L., and Matthews, L. (1996). Pluripotent bovine embryonic cell lines direct embryonic development following nuclear transfer. *Biol. Reprod.* **54,** 100–110.

Stinnakre, M. G., Vilotte, J. L., Soulier, S., l'Haridon, R., Charlier, M., Gaye, P., and Mercier, J. C. (1991). The bovine a-lactalbumin promoter directs expression of ovine trophoblast interferon in the mammary gland of transgenic mice. *FEBS Lett.* **284,** 19–22.

Subramanian, A., Paleyanda, R. K., Lubon, H., Williams, B. L., Gwazdauskas, F. C., Knight, J. W., Drohan, W. N., and Velander, W. H. (1996). Rate limitations in posttranslational processing by the mammary gland of transgenic animals. *Ann. N.Y. Acad. Sci.* **782,** 87–96.

Thépot, D., Devinoy, E., Fontaine, M. L., Stinnakre, M.-G., Massoud, M., Kann, G., and Houdebine, L.-M. (1995). Rabbit whey acidic protein gene upstream region controls high level expression of bovine growth hormone in the mammary gland of transgenic mice. *Mol. Reprod. Dev.* **42,** 261–267.

U.S. Department of Health and Human Services Public Health Service. (1993). "Pasteurized Milk Ordinance." FDA, Washington, D.C.

Uusi-Oukari, M., Hyttinen, J-M., Korhonen, V-P., Västi, A., Alhonen, L., Jänne, O. A., and Jänne, J. (1997). Bovine a-s1 casein gene sequences direct high level expression of human granulocyte-macrophage colony-stimulating factor in the milk of transgenic mice. *Transgenic Res.* **6,** 74–84.

Velander, W. H., Johnson, J. L., Page, R. L., Russell, C. G., Subramanian, A., Wilkins, T. D., Gwazdauskas, F. C., Pittius, C., and Drohan, W. (1992). High-level expression of a heterologous protein in the milk of transgenic swine using the cDNA encoding human Protein C. *Proc. Natl. Acad. Sci. U.S.A.* **89,** 12003–12007.

Vilotte, J. L., Soulier, S., Stinnakre, M. G., Massoud, M., and Mercier, J. C. (1989). Efficient tissue-specific expression of bovine a-lactalbumin in transgenic mice. *Eur. J. Biochem.* **186,** 43–48.

Wall, R. J., Hawk, H. W., and Nel, N. (1992). Making transgenic livestock: Genetic engineering on a large scale. *J. Cell. Biochem.* **49,** 113–120.

Wei, Y., Yarus, S., Greenberg, N. M., Whitsett, J., and Rosen, J. M. (1995). Production of human surfactant protein C in milk of transgenic mice. *Transgenic Res.* **4,** 232–241.

Wheeler, M. B. (1994). Development and validation of swine embryonic stem cells: A review. *Reprod. Fertil. Dev.* **6,** 563–568.

Whitelaw, C. B. A., Archibald, A. L., Harris, S. *et al.* (1991). Targeting expression to the mammary gland: Intronic sequences can enhance the efficiency of gene expression in transgenic mice. *Transgenic Res.* **1,** 3–13.

Whitelaw, C. B. A., Harris, S., McClenaghan, M. *et al.* (1992). Position-independent expression of the ovine b-lactoglobulin gene in transgenic mice. *Biochem. J.* **286,** 31–39.

Wilmut, I., Archibald, A. L., McClenaghan, M., Simons, J. P., Whitelaw, C. B. A., and Clark, A. J. (1991). Production of pharmaceutical proteins in milk. *Experientia* **47,** 905–912.

World Health Organization (1996). "Emerging and Other Communicable Diseases." Report of a WHO Consultation on Public Health Issues Related to Human and Animal Transmissible Spongiform Encephalopathies. WHO, Geneva. http://www.who.ch

Wright, G., Carver, A., Cottom, D., Reeves, D., Scott, A., Simons, P., Wilmut, I., Garner, I., and Coleman, A. (1991). High level expression of active human alpha-1-antitrypsin in the milk of transgenic sheep. *Bio/Technology* **9,** 830–834.

Yarus, S., Greenberg, N. M., Wei, Y., Whitsett, J. A., Weaver, T. E., and Rosen, J. M. (1997). Secretion of unprocessed surfactant protein B in milk of transgenic mice. *Transgenic Res.* **6,** 51–57.

Young, M. W., Okita, W. B., Brown, M., and Curling, J. M. (1997). Production of biopharmaceutical proteins in the milk of transgenic dairy animals. *BioPharmacology* (in press).

Yull, F., Harold, G., Wallace, R., Cowper, A., Percy, J., Cottingham, I., and Clark, A. J. (1995). Fixing human factor IX (fIX): Correction of a cryptic RNA splice enables the production of biologically active fIX in the mammary gland of transgenic mice. *Proc. Natl. Acad. Sci. U.S.A.* **92,** 10899–10903.

Ziomek, C. A. (1996). Minimization of viral contamination in human pharmaceuticals produced in the milk of transgenic goats. *Dev. Biol. Stand.* **88,** 263–266.

15

RECOMBINANT PROTEIN EXPRESSION IN PLANTS

Andreas E. Voloudakis, Yanhai Yin, and Roger N. Beachy
Department of Cell Biology, The Scripps Research Institute, La Jolla, California 92037

Gene Expression Systems: Using Nature for the Art of Expression

General Introduction

Advances in cellular and molecular biology provide the opportunity to engineer plants to produce a wide variety of products. There are a variety of applications of recombinant proteins in transgenic plants: (1) change endogenous metabolic pathways to obtain new products, e.g., engineering rapeseed (*Brassica napus*) to produce an elevated content of lauric acid; (2) confer novel traits to the modified plants such as improved nutrition and pathogen resistance; and (3) use plants as "factories" to produce industrial enzymes or products for pharmaceutical applications such as antibodies, antigens, and other useful proteins and even plastics. Basic procedures to produce recombinant proteins in plants include molecular cloning of genes of interest; construction of expression vectors with the appropriate promoter and other regulatory sequences; introduction of genes into plant cells and regeneration of transgenic plants (or cell lines); and detection, characterization, and purification of the final products. The most widely used transformation methods are discussed in the next section followed by description of the promoters that are commonly used to express recombinant proteins in plants. Examples are given to illustrate the applications of recombinant protein expression in agricultural biotechnology and for obtaining novel products from plants. Specific targeting and the accumulation of the recombinant proteins are also mentioned. Finally, methods and applications of plant virus-mediated transient expression systems are reviewed, followed by a discussion.

Transformation Methods

Several methods for gene transfer, and stable integration into the plant genome, have been developed since the early 1980s. These transformation techniques include *Agrobacterium tumefaciens*-mediated gene transfer, transfer of purified DNA via microparticle bombardment, electroporation of protoplasts and whole cells, chemical treatments of protoplasts, and microinjection or use of silicon fibers to facilitate penetration and transfer of DNA into the cell. Although these systems have been employed successfully to produce transgenic plants, the first two methods are the most widely used and will be described in greater detail.

Plant transformation currently involves the use of antibiotic-resistant genes for the selection of transformants; which may be of general concern in some quarters. However, a positive selection technique (Joersbo and Okkels, 1996) has been developed that will reduce the concerns of potential horizontal gene transfer of antibiotic markers (used for selection) to other organisms, including bacteria.

To successfully apply transformation techniques in plants it is essential to use appropriate DNA sequence elements to control transgene expression. A number of examples of transgene silencing (manifested by low steady-state levels of specific mRNAs) have been reported since 1993. The mechanisms of gene silencing are not fully understood, but it was shown that it can occur before or after transcription (Depicker and Van Montagu, 1997; Meyer, 1995). Gene silencing occurs more frequently when multiple gene sequences are integrated and is less frequent following the integration of single genes. Therefore, it is preferable to use transformation systems that result in the integration of a single or very low numbers of the target DNA sequences in the plant genome. In addition, multiple integration events give rise to gene segregation patterns that are more complex and thus less desirable from the perspective of crop breeding. Several factors should be considered before attempting a transformation process in order to minimize the possibility of inducing gene silencing (Koziel et al., 1996a). However, a strategy has been described for the consistent activation of gene silencing in trangenic plants (Angel and Baulcombe, 1997).

Agrobacterium tumefaciens-Mediated Transformation

The ability of *A. tumefaciens* to transfer part of its tumor-inducing (Ti) plasmid DNA into the host genome was exploited to develop the first transformation vehicle (Barton *et al.*, 1983; Herrera-Estrella, 1983); it has been widely used for the production of transgenic plants. *A. tumefaciens,* which causes the crown gall disease in a variety of dicoteledonous plants, has been utilized successfully for the stable transformation of most of the dicotyledonous plants. The *A. tumefaciens* strains that have been used most commonly in the authors' laboratory include GV3001, ABI, EH101, and LBA4404 (commercially available from Life Technologies Inc.).

The first transgenic plant engineered by *A. tumefaciens* was tobacco (*Nicotiana tabacum*) (Barton *et al.*, 1983); the basic protocol that is now common in many laboratories was described by Horsch *et al.* (1985). Since that report many different types of plants have been transformed, including *Arabidopsis*, rapeseed, alfalfa, soybean, potato, white clover, tomato, cotton, celery, carrot, mung bean, canola, and chickpea, to name but a few. Although monocotyledonous plants are not infected by *A. tumefaciens* in the wild, it has been used successfully for the transformation of several cereals (Christou, 1996). Agrobacterium-mediated transformation has bee achieved for rice (Hiei *et al.*, 1994; Toki, 1997) and maize (Ishida *et al.*, 1996).

A number of different transformation vectors have been designed for use with *A. tumefaciens*. These consist of two types: cointegrating vectors (Rogers *et al.*, 1987) and binary vectors (An, 1987). Cointegrating vectors are designed to recombine in *Agrobacterium* cells by homologous recombination with common sequences in the helper plasmid to produce a megaplasmid from which the T-DNA area is integrated during the transformation process into the nuclear plant DNA. In contrast, binary vectors do not require recombination with a helper plasmid. A suitable vector should be selected carefully as incompatibility with the *Agrobacterium* strain can restrict replication of the vector and reduce frequency of transformation. Certain vectors are available commercially (e.g., pBI vector series, Clontech, Inc.).

In general, the application of Agrobacterium-mediated transformation results in low copy integration events (1–3 copies) compared with biolistic methods (1–10 copies). Although it seems that there is a greater control of DNA incorporation via *Agrobacterium* transformation, controlling the site of incorporation of foreign DNA in plants, although highly desirable, is not yet possible. It is more probable, however, that site-specific integration of foreign genes will be accomplished with the use of *Agrobacterium* transformation rather than with biolistic methods.

The size of DNA transferred into the plant genome can vary, as integration of the left border of the target DNA is not always precise (Bakkeren *et al.*, 1989). In addition, the roles of the several virulence genes of the Ti plasmid that are important for DNA transfer are not yet determined [e.g., virE2 required to preserve the integrity of the T-DNA (Rossi *et al.*, 1996), virD2 binds to 5′ of nicked T-DNA (Jasper *et al.*, 1994)]. There is likely an upper limit on the maximum size of DNA that can be transferred into the plant genome. This depends in part on the ability of the vector-plasmid to retain large pieces of DNA and, in part, to the transfer of DNA from *Agrobacterium* to the plant cell. For the purpose of transferring single genes or few genes by *Agrobacterium* transformation, size consideration is not a significant issue. However, one of the challenges facing scientists is to transfer large pieces of DNA (BIBAC libraries; Hamilton *et al.*, 1996) into plants for the purpose of screening genomic libraries (shotgun transformation).

Particle Bombardment

An alternative method that is often used to introduce foreign DNA into plant genomes is the direct introduction of DNA molecules via particle bombardment. This method was developed primarily for the transformation of monocotyledonous plants as, initially, the use of *A. tumefaciens* for stable monocot transformation was not possible. Transgenic plants have been produced from a number of different plant species (maize, rice, wheat, barley, soybean, cotton, cassava) via particle bombardment (for review, see Christou, 1996). Considerations to be made when using particle bombardment techniques are (1) the choice of suitable target tissue to be bombarded, (2) the choice of a selectable marker, (3) the type and size of particles that will be used, and (4) the preparation and concentration of DNA that is used to coat the microparticles. The types of explant tissues that have been used for bombardment vary. However, the tissue of choice, in many situations, seems to be embryogenic callus and suspension cultures; regeneration of transformants from these tissues has been achieved for several plant species. In the authors' laboratory, efficient transformation systems via particle bombardment have been established for rice (Li *et al.*, 1993; Zhang *et al.*, 1996) and cassava (Schöpke *et al.*, 1996; Taylor *et al.*, 1996) and are used routinely to produce transgenic plants for fundamental research (Yin and Beachy, 1995; Song *et al.*, 1995; Yin *et al.*, 1997). Improvements in the method include the use of embryogenic suspension cultured cells or calli as target tissues for transformation, optimization of DNA concentration and plasmid ratios, bombardment conditions, and selection and plant regeneration procedures (Schöpke *et al.*, 1997). With such improvements, 10–20 transgenic lines of rice can be obtained from the bombardment of 100 embryogenic calli. The method has certain disadvantages, including that the procedure results in the integration of multiple copies of DNA and the incorporation of fragments instead of intact genes. In addition, the cost of the acquisition and maintenance of the particle gun (e.g., PDS 1000/He system from Bio-Rad, Inc.) may inhibit its use by small laboratories, particularly those in developing countries. However, inexpensive particle bombardment devices have been produced (Takeuchi *et al.*, 1992).

In some cases, it is desirable to deliver foreign DNA into the genomes of chloroplasts (Boynton *et al.*, 1988) or mitochondria (Johnston *et al.*, 1988) rather than into the nuclear DNA. In such cases, biolistic methods are the method of choice. The integration of a gene(s) in the chloroplast genome often results in higher amounts in expression of the desired protein because the chloroplast contains multiple genome copies. McBride *et al.* (1994) developed an elegant method to overexpress genes in chloroplasts that involved the T7 promoter. These genes were expressed at high levels with the aid of the T7 RNA polymerase that was expressed as a nuclear gene and targeted to chloroplasts using an appropriate signal peptide in the protein (McBride *et al.*, 1994).

The size of DNA to be integrated via particle bombardment is of great importance. If the piece of DNA is very large (>15 kb) the technique may be limited due to the breakage of big DNA molecules. However, this method is highly successful for the simultaneous integration of multiple genes from different plasmids that are cobombarded (Huet *et al.*, 1996).

Promoters Used for Recombinant Protein Accumulation in Plants

An important consideration before starting an experiment to express foreign genes in plants is the choice of promoter; constitutive, tissue-specific, or inducible promoters may be used, depending on the application.

Constitutive Promoters

The most widely used constitutive promoter is the 35S promoter of cauliflower mosaic virus (CaMV). This confers a high level of gene expression in most cell types in transgenic plants, especially in dicots (reviewed in Benfey and Chua, 1990). Analogous promoters from related plant viruses [including figwort mosaic virus (Maiti *et al.*, 1997) and cassava vein mosaic virus (CsVMV) (Verdaguer *et al.*, 1996)] are also used. Combinatorial interactions of tissue-specific subdomains confer the constitutive expression of the CaMV 35S (Benfey, 1990) and CsVMV promoters (Verdaguer *et al.*, 1996). For monocotyledonous crops, such as rice, maize and wheat, the maize ubiquitin promoter (Christensen *et al.*, 1992) or the rice actin–1 promoter (Zhang *et al.*, 1991) are widely used to drive the expression of the inserted genes.

Tissue-Specific Promoters

For Specific applications, tissue-specific promoters are available and can be used instead of constitutive promoters. For example, a seed-specific promoter should be employed to change fatty acid synthesis pathways in oil seeds (see later). Cell-specific gene expression in plants and a description of the different tissue-specific plant promoters are reviewed in detail by Edwards and Coruzzi (1990). Examples of several tissue-specific promoters are listed.

Embryo Many seed storage proteins are expressed exclusively in embryos, such as the soybean β-conglycinin gene (Chen *et al.*, 1989), legumin genes from common bean (*Phaseolus vulgaris*)

(Shirsat *et al.*, 1989), β-phaseolin gene (Bustos *et al.*, 1989), napin and cruciferin genes from rapeseed (DeLisle and Crouch, 1989). Many of these promoters were shown to retain tissue specificity when expressed in heterologous plant species.

Endosperm Some seed storage proteins accumulate solely in endosperm tissue, such as maize zein genes (Scherntaner *et al.*, 1988), wheat glutenin genes (Colot *et al.*, 1987), and barley hordein genes (Marris *et al.*, 1988).

Fruit The ethylene-responsive E8 gene from tomato is expressed in a fruit-specific manner during the ripening process (Deikman and Fischer, 1988).

Tuber The glycoprotein patatin accounts for up to 40% of the soluble protein of potato tubers, and the class I patatin promoter is 100- to 1000-fold more active in tubers than in other vegetative organs (Rocha-Sosa *et al.*, 1989; Wenzler *et al.*, 1989).

Leaf (Light Inducible) The ribulose–1,5-bisphosphate carboxylase small subunit (rbcS) gene promoter (Fluhr *et al.*, 1986) and the chlorophyll *a/b* binding protein (Cab) gene promoter (Fluhr *et al.*, 1986; Simpson *et al.*, 1985) confer leaf-specific and light-dependent expression of genes.

Inducible Promoters

In some cases it is preferable to use inducible promoters. The tetracycline-inducible promoter is the most advanced system used to regulate the expression of transgenes in plants (Gatz, 1996; Gatz *et al.*, 1996). In this system, the bacterial Tet repressor (tetR) is expressed constitutively in transgenic plants. The tet operator, to which the repressor binds, is introduced into the CaMV 35S promoter; systematic analysis of the effects of repressor–operator complexes in different positions within the CaMV 35S promoter led to the design of a tightly repressible CaMV 35S promoter that contains one tet operator directly upstream of the TATA box and two tet operators downstream of the TATA box. After treatment of transgenic plant tissues that contain the repressor and the target gene with low, nontoxic levels of tetracycline, the promoter activity can increase up to 500-fold (Gatz *et al.*, 1996). This system has been used successfully to control the expression of the *Agrobacterium rolb* gene (Roder *et al.*, 1994) and a dominant-negative mutant of the TGA family of transcription factors (Rieping *et al.*, 1994). One drawback of the system is that the constitutively expressed repressor may not be tolerated or may inhibit the growth of transgenic plants.

Another inducible system is based on the induction of an α-amylase promoter by carbohydrate starvation (Chan *et al.*, 1994). It was shown that sugars suppress the expression of α-amylase genes in cultured cells (Yu *et al.*, 1991; 1992) as well as in germinating seeds (Karrer and Rodriguez, 1992). The promoter region of an α-amylase gene from barley was fused with the β-glucuronidase (GUS) reporter gene and the fusion gene was introduced into rice, tobacco, and potato cell lines (Chan *et al.*, 1994). In transformed cells, GUS expression is suppressed in the presence of sucrose and is induced by its absence. Furthermore, a signal peptide was fused to the GUS gene, leading to the secretion of GUS into the medium, simplifying purification of the target protein. A yield of 370 mg of GUS/g of cells was achieved with this system (Chan *et al.*, 1994).

In yet another system, the mammalian glucocorticoid receptor (GR), which activates transcription in the presence of steroids such as dexamethasone, was used. The ligand-binding domain of GR was fused with the maize transcription factor R; the fusion factor (R-GR) is steroid inducible and can complement the *ttg* mutant phenotype in transgenic *Arabidopsis* in a steroid-dependent manner (Lloyd *et al.*, 1994). This system was used to determine the function of an *Arabidopsis* transcription activator (Aoyama *et al.*, 1995). However, the expression of the GR fusion transcription factor, whose activity is tightly controlled by steroid through the GR ligand-binding domain, is not always achieved in each transgenic event (Aoyama *et al.*, 1995).

Finally, the regulation of the yeast metallothionine gene was exploited to develop a copper-inducible expression system in plants. In the presence of copper ions, in subtoxic levels, the transcriptional activator ACE1 changes conformation, binds to, and activates the expression of the desirable gene located downstream of a chimeric promoter (Mett *et al.*, 1993).

Expression of Recombinant Proteins in Plants and Agricultural Biotechnology

The genetic improvement of plants for desirable traits is possible due to the advances in plant transformation and molecular biology techniques. A recent review article presents the case for commercialization of transgenic plants (Kridl and Shewmaker, 1996).

Resistance against Pathogens and Pests

Plant breeders and molecular biologists have spent considerable effort to map and isolate genes for resistance against pathogens and pests in order to use these in conventional breeding projects, thereby increasing crop yields and decreasing the amounts of agricultural chemicals used. Natural genes have been cloned for resistance against viruses [N gene, against the tobacco mosaic virus (TMV) (Whitham *et al.*, 1994)], fungi [Cf–2, gene resistance against *Cladosporium fulvum* (Dixon *et al.*, 1996); L6 gene, for resistance against flax rust (Lawrence *et al.*, 1995)], bacteria [RPS2 gene, for resistance against *Pseudomonas syringae* carrying the avirulence gene avrRpt2 (Mindrinos *et al.*, 1994); Xa21 gene, for resistance against *Xanthomonas campestris* pv. oryzae (Song *et al.*, 1995), Pto gene, for resistance against *Pseudomonas syringae* pv. tomato (Martin *et al.*, 1993)], and nematodes [resistance against the sugarbeet cyst nematode (Cai *et al.*, 1997)]. The majority of the genes mentioned provide resistance to single pathogens. The isolation of naturally occurring genes opens up the opportunity to transfer them to different varieties or species of plants that are infected by the same pathogen. For example, the N gene of tobacco has been introduced successfully into tomato and shown to confer protection against TMV infection (Whitham *et al.*, 1996). The use of natural resistance genes is highly desirable as they are considered to be environmentally "safer" than the use of chemical control agents.

Other strategies have been investigated to improve the resistance of plants against pathogens via stable transformation. Pathogen-derived resistance (PDR) has been developed to confer resistance against plant viruses (Wilson, 1993; Fitchen, 1993). Antiviral proteins have also been tested (Lodge *et al.*, 1993). For the control of fungal pathogens, genes encoding antifungal proteins [chitinase/

glucanase/RIP (Jach *et al.*, 1995)], small cysteine-rich proteins (Terras *et al.*, 1995), and enzymes involved in phytoalexin biosynthesis (Hain *et al.*, 1993) have been employed. To control bacterial pathogens, lysozyme (Duering, 1996) and the H_2O_2 generating enzyme glucose oxidase (Wu *et al.*, 1995) were shown to be partially effective.

For insect control, the majority of research has concentrated on the use of *Bacillus thuringiensis* (Bt) δ-endotoxins (Altman *et al.*, 1996). Bt toxin formulations have been used, over the years, as insecticides against a variety of insects (field sprays). When expressed as transgenes, these proteins are insecticidal; different δ-endotoxins can be used to control insect species in the families of Lepidoptera, Diptera, and Coleoptera. A review of insecticidal spectra of different δ-endotoxins was presented by Höfte and Whiteley (1989). Because Bt δ-endotoxins have a high A/T base content, some gene modification may be required to optimize codon usage in the gene. Optimization generally results in an increase of insecticidal protein that is produced in plants (Estruch *et al.*, 1997; Perlak *et al.*, 1991). Several gene constructs (full length and truncated forms of the toxins) have been introduced into plants and proven to be useful for the control of insects. Crops that are targets for Bt engineering include cotton (Altman *et al.*, 1996), maize (Bolin *et al.*, 1996), (Koziel *et al.*, 1996b), potato (Perlak *et al.*, 1993), tomato (Fischhoff *et al.*, 1987), and tobacco (Vaeck *et al.*, 1987). However, the use of Bt δ-endotoxins in the near future will expand to other plant species, including tree species (Moffat, 1996b).

Field trials have been performed with a number of crops; release of new varieties of cotton containing the Bt protein to reduce the damage by cotton bollworm took place in 1996. Unfortunately, the first release of the new Bt cotton varieties coincided with unusual environmental conditions and rendered the crop protection less effective than anticipated in certain geographical regions (Fox, 1996). Nevertheless, the product performed well in a number of areas in the United States. Although there have not been signs of resistance to Bt cotton in the field by the bollworm, it is well accepted that the selective pressure that is applied when using a single resistance agent (in this case, the transgenic plant) is higher than when several resistant traits are employed (Altman *et al.*, 1996). To avoid constitutive expression of Bt toxins, it is possible to introduce the gene downstream of an inducible promoter (e.g., pathogenesis-related protein 1a promoter, PR–1a) (Williams *et al.*, 1992). An integrative pest management program (combining the use of transgenic plants with the use of reduced amounts of pesticides) (Schell, 1997), the availability of refugia, and developing pyramiding strategies (using multiple host-plant resistant traits) (Altman *et al.*, 1996) are considered effective ways to prolong the use of Bt crops.

Other insecticidal proteins that have been identified include the polyphenol oxidases, proteinase inhibitors, α-amylase inhibitors, chitinases, and lectins (Estruch *et al.*, 1997). However, the effectiveness of these proteins has not been as promising as the Bt toxins. Nevertheless, development of the second generation of insecticidal proteins is underway, including the vegetative insecticidal proteins (VIPs) and cholesterol oxidases (Estruch *et al.*, 1997; Koziel *et al.*, 1996b).

Bioremediation

Both microorganisms and plants have been employed to alleviate man-made pollutants from the environment. Microorganisms are used primarily in situations where the degradation of certain pollutants (primarily petroleum hydrocarbons) is necessary. In other circumstances, plants may be preferred for phytoremedation processes where absorption of heavy metals and radionucleotides are the contaminants (Cunningham *et al.*, 1995; Cunningham and Ow, 1996). The cost of extraction of heavy metals from soil and standing water can be extremely high, estimated to have an average cost of $1,000,000 per acre (Raskin *et al.*, 1997). There is a need, therefore, to find more cost-effective methods to decontaminate soils. Several plant species that have the ability to grow in soils that have elevated concentrations of certain heavy metals have been identified by looking in the wild for plants that hyperaccumulate pollutants. In other laboratories, genetic transformation is being used to develop hyperaccumulator plants. Such plants should have rapid growth rates, high biomass, selectivity for the elements to be absorbed (especially Pb, Cd, Cr, As, Ni, Zn, Cu), and large root structure and density. Genetic engineering will aid in the development of varieties with increased levels of absorption of the heavy metals.

A mercuric ion reductase (of bacterial origin) has been incorporated into *Arabidopsis* and found to increase the tolerance to mercury in transgenic plants. In addition, these plants volatilized mercuric ions (Rugh *et al.*, 1996). Phytoremediation is in its infancy and the metabolic capabilities of the plants need to be studied further and perhaps modified in order to select plants that accumulate metals in high amounts.

Improvement of Tree Species

Until now, most of the expression of foreign proteins in plants has been carried out in herbaceous plants. However, the next generation of agricultural biotechnology will include the improvement of tree species (Moffat, 1996a). Substantial effort is currently being put into developing trees that grow faster, produce more and better pulp wood, produce wood with superior fuel capabilities, and are resistant to pests, diseases, and adverse environmental conditions (Moffat, 1996b).

Other Biotechnological Applications

In some cases, specific consumer and/or industrial demands could be satisfied by the application of biotechnology. For example, genetic transformation was used to develop naturally colored cotton fibers to reduce or replace the dying process (Francisco, 1996).

To control the ripening, and increase the shelf-life, of tomato fruits, polygalacturonase (PG) activity was reduced via an antisense technology to produce the FLAVR SAVR tomato (Kramer and Redenbaugh, 1994). In an alternative approach, the antisense strategy was employed to reduce the aminocyclopropane carboxylase (ACC) oxidase, ACC synthase, or ACC deaminase (Grierson and Fray, 1994); in each case, the ripening of tomato fruits was retarded or blocked.

The S-RNase expression in *Nicotiana* was suppressed using antisense technology. As a result, two species that exhibit self-incompatibility were able to sexually cross with each other (Murfett *et al.*, 1995).

Sometimes the induction of male sterility in plants is highly desirable in breeding programs. The production of male sterile plants was accomplished by utilizing a chimeric ribonuclease gene (Mariani *et al.*, 1990). With the use of a chimeric ribonuclease-inhibitor gene, fertility to male sterile plants was restored (Mariani *et al.*, 1992). Furthermore, it was possible to produce female sterile plants using a stigma-specific cytotoxic gene (Goldman *et al.*, 1994).

Significant progress has been made in developing genes to confer tolerance to herbicides in crop plants. Herbicide-resistant crops are desirable as herbicide applications would eliminate weeds but will have no effect on crop plants. For example, the *bar* gene has been introduced into several plant species and provided resistance to the herbicide bialaphos (Gallo-Meagher and Irvine, 1996), and genes for resistance to glyphosate (Round-up) have been developed (Mannerlof *et al.*, 1990). In general the source for these genes is soil bacteria. Crop plants, such as soybeans and corn, that are resistant to "environmentally friendly" herbicides have been commercialized. The use of such genes has raised concern in certain sectors as gene transfer to a weedy relative species (if one exists in the region of application) may confer resistance to a particular herbicide. It was shown that interspecific crosses resulted in the transfer of a transgene for resistance to bialaphos from oilseed rape to its weedy relative (Mikkelsen *et al.*, 1996). Although this was not a surprising result, it is an important reminder of the potential consequences that could follow if adequate care is not taken.

Recombinant Protein Expression in Plants to Obtain New Products

Transgenic plants are being used to develop novel products, such as modified oils, fibers, starches, and other carbohydrates, proteins with high contents of certain amino acids, enzymes, and other proteins of medical importance as well as antibodies, antigens, and even plastics.

Oils

About 20 billion pounds of vegetable oil are used in the United States each year for food (72%) and industrial (28%) applications (Kridl and Shewmaker, 1996). Plant storage oils consist of triglycerols, a glycerol molecule with three fatty acids esterified to the backbone. The chain length, the degree of desaturation of the fatty acids, and the order of the molecules on the triglyceride determine the functional properties of the oil. Most plant oils contain 16 and 18 carbon molecules, either saturated (16:0, palmitic; 18:0, stearic)

Table 1
Expression Yields of Foreign Proteins in Transgenic Plants

Product	Vector	Expression level	Reference
Tobacco mosaic virus coat protein in tobacco	pTM319	0.001–0.05% (w/w)	Clark *et al.* (1995)
γ and κ immunoglobulins in tobacco	pMON530	1.3% of total leaf protein	Hiatt *et al.* (1989)
SIgA-G in tobacco		200–500 μg/g leaf	Ma *et al.* (1995)
T7 RNA polymerase in tobacco	pCGN4026	0.01–2.25 units/mg total protein; 0.01 is 40% above background	McBride *et al.* (1994)
PHB in *Arabidopsis thaliana*		10 mg/g fresh weight	Poirier *et al.* (1995)
LT-B enterotoxin in tobacco		Without KDEL signal, 2–5 μg/g fresh weight With KDEL signal, 6–14 μg/g fresh weight	Haq *et al.* (1995)
Escherichia coli ADPGPP in potato		Starch varied from 7.33 to 12.74% fresh weight, vs untransformed (3.7–9.16 %)	Stark *et al.* (1996)
cDNA encoding phytoene desaturase	TMV based vector	2% of soluble protein	Kumagai *et al.* (1995)
cry1Ab in cotton		Hundreds of ng/mg protein	Estruch *et al.* (1997)
Brazil nut methionine-rich seed protein in tobacco		Increase of 30% in methionine content	Altenbach *et al.* (1989)
cDNA encoding glucocerebrosidase in tobacco		10% of total soluble protein	Cramer *et al.* (1996)

or with one (18:1, oleic), two (18:2, linoleic), or three (18:3, linolenic) double bonds (Kridl and Shewmaker, 1996). The biosynthesis of fatty acids involves three kinds of enzymes; ketoacyl synthases (Kas) which elongate the growing fatty acyl chains (generally up to 18 carbons in length); desaturases; and thioesterases. The activities and specificities of these three enzymes determine the chain length and saturation levels of fatty acids (Kridl and Shewmaker, 1996). Genes encoding most of the enzymes in these pathways have been cloned (reviewed in Knauf, 1993; Kridl and Shewmaker,

1996; Murphy, 1996; Ohlrogge and Brown, 1995; Töpfer *et al.*, 1995), providing the basis for engineering plants to produce specific fatty acids. For example, a ketoacyl-CoA synthase gene from jojoba has been cloned and expressed as a transgene in canola and was found to restore the biosynthesis of long chain fatty acids that are of industrial interest (Lassner *et al.*, 1996). Thioesterases have also been used to modify chain lengths as well as levels of saturation. Expression of a *Cuphea hookeriana* thioesterase gene (Ch FatB2) in canola seeds resulted in accumulation of high levels of 8:0 and 10:0 fatty acids that are easily digested by humans and other animals and thus have potential applications in the food industry (Dehesh *et al.*, 1996). Desaturases have been decreased by antisense and gene cosuppression strategies to produce high oleic and low linolenic canola and soybean (Kinney, 1995). These alternations produced oils with reduced polyunsaturates, which have important applications in the food industry (Kridl and Shewmaker, 1996).

Fiber

Cotton is the premier source of fiber for the textile industry and is a biological composite of cellulose, small quantities of hemicellulose, pectins, and proteins that provide excellent wearability and aesthetics. However, further improvements in strength, length, chemical reactivities for dye binding, water absorption, thermal properties, and wrinkle and shrinkage resistance are desirable for textile and other industrial applications (John and Keller, 1996). Details of cellulose biosynthesis have been reviewed (Brown *et al.*, 1996; Delmer and Amor, 1995). The molecular cloning of the cotton *cel*A gene that encodes the catalytic subunit of cellulose synthase provides the opportunity to engineer the pathways of cellulose biosynthesis in higher plants (Pear *et al.*, 1996).

A natural thermoplastic polyester compound, poly-$\text{D}^{(-)}$-3hydroxybutyrate (PHB), is synthesized in bacteria (*Alcaligenes eutrophus*) by enzymes encoded by three genes, *pha*A, *pha*B, and *pha*C. Superior insulating characteristics and thermal properties in cotton fibers were achieved by the expression of *pha*B and *pha*C genes. This expression resulted in the accumulation of PHB in transgenic fibers (John and Keller, 1996).

Biodegradable Plastics

Another use of the PHB is production of plastics in plants. The latter was accomplished with the use of bacterial genes from the bacterium *A. eutrophus* (Nawrath *et al.*, 1994). In this case the polymer accumulated in *A. thaliana* plastids. Characteristics of these plastics are similar to those developed by other methods. In addition, they have the potential to be degraded by soil microorganisms that contain specific hydrolases.

Starch

In plants, three enzymes are involved in starch biosynthesis: ADPglucose pyrophosphorylase (ADPGPP), starch synthase, and one or more branching enzymes. The controlling and rate-limiting enzyme in this pathway is ADPGPP. A highly active *Escherichia coli* ADPGPP enzyme that is insensitive to cellular control was expressed in transgenic potato tubers from the tuber-specific patatin promoter, and a 20–30% increase in starch and total dry matter was observed (Stark *et al.*, 1992). Expression of the same gene in tomato and canola also resulted in increased starch content (Stark *et al.*, 1996).

Heterologous Proteins

There is a great interest in expressing recombinant proteins in plants, either as an approach to improve food quality (Sun *et al.*, 1996) or as bioproduction systems for proteins (Cramer *et al.*, 1996). A methionine-rich 2S protein (18% methionine and 8% cysteine) in the Brazil nut (BN2S) gene was cloned (Altenbach *et al.*, 1987), expressed in transgenic tobacco plants by the seed-specific phaseolin promoter, and enhanced the methionine content of transgenic tobacco seeds by up to 30% (Altenbach *et al.*, 1989). A similar enhancement of the methionine content was later observed in rapeseeds (Altenbach *et al.*, 1992). Similar approaches can be used to increase the lysine content by lysine-rich proteins or sweetness by using the sweet protein mabinlin (Sun *et al.*, 1996).

Compared to microorganisms, plants have several advantages as organisms in which to overproduce proteins of medical importance

(Cramer *et al.*, 1996), including large-scale biomass production at low cost, efficient and low-cost procedures to produce and maintain transgenic plants, similarity of posttranslational protein modification in plants and animals, and increased medical safety of the product (plants do not serve as hosts for human or animal infectious agents) (Table 2). For example, Cramer and colleagues (1996) developed a system to produce active human glucocerebrosidase in transgenic tobacco plants; this enzyme, normally isolated from human placenta, is used to treat Gaucher disease with an annual cost of \$70,000 to \$300,000. Thus the search for an alternative source of this enzyme is demanding. The human cDNA for glucocerebrosidase was tagged at the 3′ end with sequences for the FLAG epitope to facilitate detection and purification of the transgene product (Sorge *et al.*, 1985). The gene fusion was cloned downstream of a proprietary inducible plant promoter (MeGA promoter, CropTech, Inc.) and was introduced into transgenic plants via *Agrobacterium*-mediated transformation. Following gene activation, glucocerebrosidase accumulated to 10% of the total soluble protein in transgenic tobacco leaves. More importantly, the plant-produced enzyme was found to be enzymatically active (Cramer *et al.*, 1996).

Kobayashi and colleagues (1996) expressed the human epidermal factor (hEGF) in fruit trees. This is the first example of the production of a bioactive peptide, nonfused with other proteins in fruit trees. This suggests the possibility of the production of other peptides or proteins in such a way that the useful components are produced in fruits (Kobayashi *et al.*, 1996).

Table 2
Advantages and Disadvantages of Transgenic Plant Expression

Advantages	Disadvantages
Integration of single or multiple gene copies, usually at a single locus	Control of copy number not well established when particle bombardment is used
Stable transformation and genetic recombination via crossing	Site-specific integration not yet available
Easy storage procedure of transgenic progeny (seeds)	Depending on the plant species, lengthy transformation procedure
Processing of foreign genes equivalent to mammalian cells	Expression level is not controllable and can vary significantly
Protein production on a large scale	
Very competitive cost of production	

Antibody/Antigen Production in Plants

Significant progress has been made since 1990 to produce antibodies in plants (Ma and Hein, 1995, 1996). A murine monoclonal antibody κ chain, a hybrid immunoglobin A-G heavy chain, a murine joining chain, and a rabbit secretory component were produced in different transgenic tobacco plants; sexual crosses between these plants resulted in plants that produce all four proteins simultaneously. A surprisingly high molecular weight secretory immunoglobin was assembled from these four chains and bound the antigen; in this case, the native streptococcal antigen I/II cell surface adhesion molecule (Ma *et al.*, 1995).

Another application is the expression of vaccines in transgenic plants. For example, the capsid protein of Norwalk virus, a calicivirus that causes epidemic acute gastroenteritis in humans, was expressed in tobacco and tomato plants using the CaMV 35S promoter and patatin promoter, respectively (Mason *et al.*, 1996). The capsid protein was assembled into virus-like particle. Furthermore, these authors found that giving extracts of tobacco leaves to mice by gavage and feeding of mice with potato tubers that express the viral protein caused the mice to develop antibodies against the target materials. These results indicate the potential applications of plants for production and delivery of edible vaccines.

Protein Targeting and Accumulation

The use of targeting sequences to localize foreign proteins to different compartments of the plant cell has emerged as an important aspect of the attempts to accumulate foreign proteins, especially in cases where the accumulated proteins are susceptible to degradation. It is known that signal peptides are needed to target and retain proteins in specific organelles. It has been shown that different peptides are required to target proteins to chloroplasts (Nawrath *et al.*, 1994), mitochondria (Silva Filho *et al.*, 1996), vacuoles (Matsuoka and Nakamura, 1991), and endoplasmic reticulum (Schouten *et al.*, 1996; Denecke *et al.*, 1992; Wandelt *et al.*, 1992). Signal peptides to direct proteins to seed protein bodies have been proposed based on sequence comparisons of several seed storage proteins (Roy and Mandal, 1996).

Virus-Mediated Expression Systems

An alternative method to express proteins in plants relies on transient expression from virus-based vectors (Scholthof *et al.*, 1996). Proteins have been expressed as single proteins or as fusions to a viral protein [e.g., fusions with capsid proteins (Beachy, 1996; Fitchen *et al.*, 1995)]. Viruses replicate with high efficiency and, in some cases, can infect the entire host plant, creating the potential to express a foreign protein in high levels. Advantages of virus-mediated expression, compared to transgenic expression, are the relative ease in manipulation of the viral genome, avoidance of time-consuming procedures to develop transgenic plants, and the potential to produce very high concentrations of proteins. However, in some cases, there have been problems of instability of the gene that is introduced in the viral genome. However, in most cases, the genome of the viral vector can be modified to accommodate the expression of the foreign gene. In some cases, viral genes can be deleted and replaced by a gene and the virus can be propagated on transgenic host plants that produce the missing protein (e.g., Padgett *et al.*, 1996; Kaplan *et al.*, 1995). A variety of vectors and possible modes of gene expression have been described by Scholthof *et al.* (1996). Tobamovirus, potexvirus, and potyvirus genomes have been used most commonly in this type of expression. A general concern of the expression of proteins in biological systems is the ability to purify the expressed proteins. A fusion of a 38-kDa protein was made to the PVX coat protein using an elegant translation system; the fusion did not inhibit the formation of virion particles and the particles were decorated with the foreign protein (Santa Cruz *et al.*, 1996). Such an approach has a great advantage in the protein purification process as particles can be purified readily. The foreign protein can then be subsequently cleaved by proteolysis from the external surface of the virion.

Another approach that has been developed is the use of proteases. A polyprotein containing the protease recognition sites can be coexpressed with an appropriate protease to facilitate *in vivo* cleavage and release of mature protein. In the authors' laboratory, they analyzed the potential to express multiple proteins in plants via transgenes and from a TMV-based vector using an expression cassette that contains multiple cleavage sites that are attacked by the protease of tobacco etch virus [(Marcos and Beachy, 1994, 1997; Ceriani *et al.*, 1997). This approach has the advantage that multiple proteins can be expressed at the same level.

It is possible to alter the biosynthetic pathway of a cell process by employing RNA viral vectors. Inhibition of the carotenoid biosynthesis was shown by expressing a cDNA encoding phytoene synthase and a partial cDNA for phytoene desaturase from a subgenomic promoter of a tobamovirus viral vector (Kumagai *et al.*, 1995).

Other vehicles that can be anticipated in the near future to express proteins in plants might include self-replicating DNA molecules that can be introduced to chloroplasts or mitochondria.

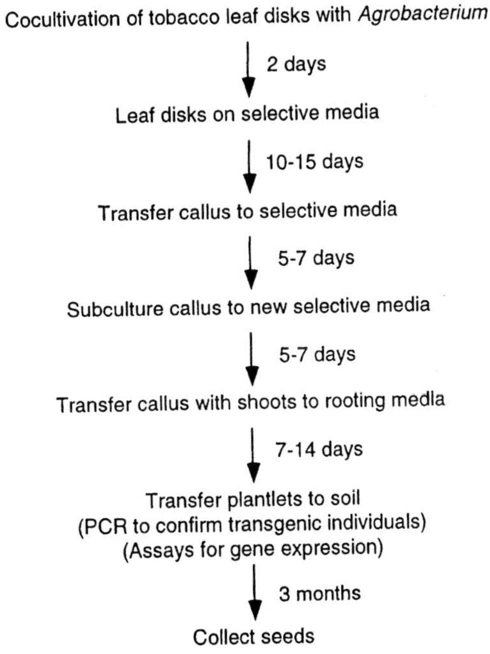

Figure 1 Flow diagram to develop transgenic plants of *Nicotiana tabacum*.

Summary/Discussion

It is well accepted that agricultural biotechnology has advanced since the early 1990s and that it will continue to grow in the future. This chapter described some of the methods and the tools and gave various examples in which transgenic technology has been applied. The use of plants as "small factories" for the expression of foreign proteins is an attractive concept, and the adaptation of tools for stable transformation of foreign genes and transient viral expression vectors can be readily adapted to a variety of systems. A flow diagram for the time frame of developing transgenic plants is depicted in Figs. 1–3 for comparison.

It should be noted that a relative minor segment of the public has expressed some concern about the use of the first generation of

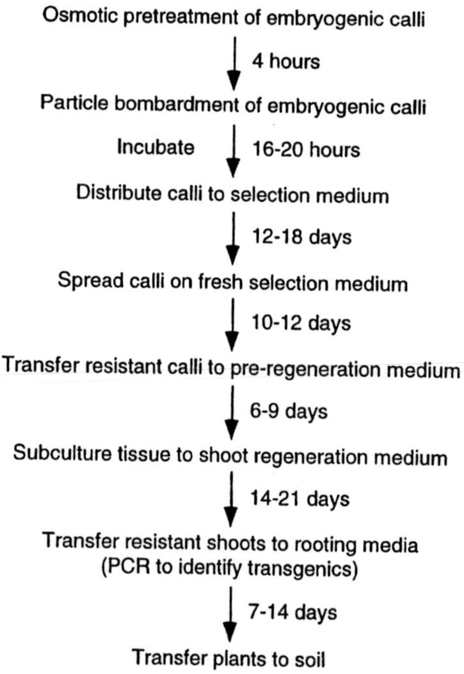

Figure 2 Flow diagram to develop transgenic rice (*Oryza sativa*) plants.

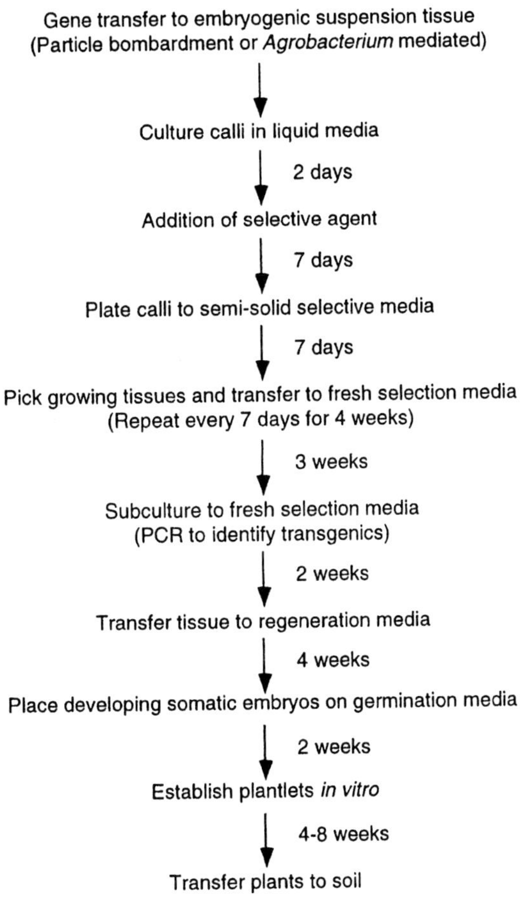

Figure 3 Flow diagram to develop transgenic cassava (*Manihot esculenta*) plants.

transgenic plants. For example, the use of antibiotic resistance genes, used as marker genes, and the potential impact of the genes are often raised as an issue of perceived risk (Wadman, 1996); however, it is widely accepted that this technology will improve the life of humankind. Substantial additional fundamental research of the technology is needed and will require continuous funding to accumulate sufficient evidence to convince the public of the use and safety of the products that are developed.

Improving the nutritional value of crops for human or animal use and the use of plants for pharmaceutical purposes are attractive targets for agricultural biotechnology that will be exploited in the future. Of course, a variety of economic factors will determine whether the production of the proteins in plants and transgenic technologies will compete with other production methods.

Acknowledgments

We thank Drs. Christoph Reichel and Hal Padgett for providing helpful comments and for reviewing this manuscript.

References

Altenbach, S. B., Pearson, K. W., Leung, F. W., and Sun, S. S. M. (1987). Cloning and sequence analysis of a cDNA encoding a Brazil nut protein exceptionally rich in methionine. *Plant Mol. Biol.* **8,** 239–250.

Altenbach, S. B., Pearson, K. W., Meeker, G., Staraci, L. C., and Sun, S. S. M. (1989). Enhancement of the methionine content of seed proteins by the expression of a chimeric gene encoding a methionine-rich protein in transgenic plants. *Plant Mol. Biol.* **13,** 513–522.

Altenbach, S. B., Kuo, C. C., Staraci, L. C., Pearson, K. W., Wainwright, C., Georgescu, A., and Townsend, J. (1992). Accumulation of a Brazil nut albumin in seeds of transgenic canola results in enhanced levels of seed protein methionine. *Plant Mol. Biol.* **18,** 235–245.

Altman, D. W., Benedict, J. H., and Sachs, E. S. (1996). Transgenic plants for the development of durable insect resistance. *Ann. N.Y. Acad. Sci.* **792,** 106–114.

An, G. (1987). Binary Ti vectors for plant transformation and promoter analysis. *In* "Methods in Enzymology" (R. Wu and L. Grossman, eds.), Vol. 153, pp. 292–305. Academic Press, Orlando FL.

Angel, S. M., and Baulcombe, D. C. (1997). Consistent gene silencing in transgenic plants expressing a replicating potato virus X RNA. *EMBO J.* **16,** 3675–3684.

Aoyama, T., Dong, C.-H., Wu, Y., Carabell, M., Sessa, G., Ruberti, I., Morelli, G., and Chua, N.-H. (1995). Ectopic expression of the Arabidopsis transcriptional activator Athb-1 alters leaf cell fate in tobacco. *Plant Cell* **7,** 1773–1785.

Bakkeren, G., Z., K.-N., N., G., and B., H. (1989). Recovery of Agrobacterium tumefaciens T-DNA molecules from whole plants early after transfer. *Cell (Cambridge, Mass.)* **57,** 847–857.

Barton, K. A., Binns, A. N., Matzke, A. J., and Chilton, M. D. (1983). Regenaration of intact tobacco plants containing full length copies of genetically engineered T-DNA, and transmission of T-DNA to R1 progeny. *Cell (Cambridge, Mass.)* **32,** 1033–1043.

Beachy, R. N. (1996). Use of plant viruses for delivery of vaccine epitopes. *Ann. N.Y. Acad. Sci.* **792,** 43–49.

Benfey, P. N. and Chua, N.-H. (1990). The cauliflower mosaic virus 35s promoter: Combinatorial regulation of transcription in plants. *Science* **250,** 959–966.

Bolin, P. C., Hutchison, W. D., and Davis, D. W. (1996). Resistant hybrids and *Bacillus thuringiensis* for managment of European corn borer (Lepidoptera: Pyralidae) in sweet corn. *J. Econ. Entomol.* **89,** 82–91.

Boynton, J. E., Gillham, N. W., Harris, E. H., Hosler, J. P., Johnson, A. M., Jones, A. R., Randolph-Anderson, B. L., Robertson, D., Klein, T. M., Shark, K., and Sanford, J. C. (1988). Chloroplast transformation in *Chlamydomonas* with high velocity microprojectiles. *Science* **240,** 1534–1538.

Brown, R. M., Saxena, I. M., and Kudlicka, K. (1996). Cellular biosynthesis in higher plants. *Trends Plant Sci.* **1,** 149–155.

Bustos, M. M., Guiltinan, M. J., Jordano, J., Begum, D., Kalkan, F. A., and Hall, T. C. (1989). Regulation of beta-glucuronidase expression in transgenic tobacco plants by an A/T rich cis-acting sequence found upstream of a French bean beta-phaseolin gene. *Plant Cell* **1,** 839–853.

Cai, D., Kleine, M., Kifle, S., Harloff, H.-J., Sandal, N. N., Marcker, K. A., Klein-Lankhorst, R. M., Salentijn, E. M. J., Lange, W., Stiekema, W. J., Wyss, U., Grundler, F. M. W., and Jung, C. (1997). Positional cloning of a gene for nematode resistance in sugarbeet. *Science* **275,** 832–834.

Ceriani, F. M., Marcos, J. F., Hopp, H. E., and Beachy, R. N. (1997). Simultaneous accumulation of multiple viral coat proteins from a TEV-NIa based expression vector. *Plant Mol. Biol.* **36,** 239–248.

Chan, M.-T., Chao, Y.-C., and Yu, S.-M. (1994). Novel gene expression system for plant cells based on induction of α-Amylase promoter by carbohydrate starvation. *J. Biol. Chem.* **269,** 17635–17641.

Chen, Z.-L., Naito, S., Nakamara, I., and Beachy, R. N. (1989). Regulated expression of genes encoding soybean beta conglycinins in transgenic plants. *Dev. Genet.* **10,** 112–122.

Christensen, A. H., Sharrock, R. A., and Quail, P. H. (1992). Maize polyubiquitin genes: structure, thermal perturbation of expression and transcript splicing, and promoter activity following transfer to protoplasts by electroporation. *Plant Mol. Biol.* **18,** 675–689.

Christou, P. (1996). Transformation technology. *Trends Plant Sci.* **1,** 423–431.

Clark, W. G., Fitchen, J. H., and Beachy, R. N. (1995). Studies of coat protein-mediated resistance to TMV. I. The PM2 assembly defective mutant confers resistance to TMV. *Virology* **208,** 485–491.

Colot, V., Robert, L. S., Kavanagh, T. A., Beven, M. W., and Thompson, R. D. (1987). Localization of sequences in wheat endosperm protein genes which confer tissue-specific expression in tobacco. *EMBO J.* **6,** 3559–3564.

Cramer, C. L., Weissenborn, D. L., Oishi, K. K., Grabau, E. A., Bennett, S., Poncc, E., Grabowski, G., and Radin, D. (1996). Bioproduction of human enzymes in transgenic tobacco. *Ann. N.Y. Acad. Sci.* **792,** 62–71.

Cunningham, S. D., and Ow, D. W. (1996). Promises and prospects of phytoremediation. *Plant Physiol.* **110,** 715–719.

Cunningham, S. D., Berti, W. R., and Huang, J. W. (1995). Phytoremediation of contaminated soils. *Trends Biotechnol.* **13,** 393–397.

Dehesh, K., Jones, A., Knutzon, D. S., and Voelker, T. A. (1996). Production of high levels of 8:0 and 10:0 fatty acids in transgenic canola by overexpression of Ch FatB2, a thioesterase cDNA from *Cuphea hookeriana*. *Plant J.* **9,** 167–172.

Deikman, J., and Fischer, R. L. (1988). Interaction of a DNA binding factor with the 5′-flanking region of an ethylene-responsive fruit ripening gene from tomato. *EMBO J.* **7,** 3315–3320.

DeLisle, A. J., and Crouch, M. (1989). Seed storage protein transcription and mRNA levels in *Brassica napus* during development and in response to exogenous abscisic acid. *Plant Physiol.* **91,** 617–623.

Delmer, D. P., and Amor, Y. (1995). Cellulose biosyhthesis. *Plant Cell* **7,** 987–1000.

Denecke, J., De Ryke, R., and Botterman, J. B. (1992). Plant and mammalian sorting signals for protein retention in the endoplasmic reticulum contain a conserved epitope. *EMBO J.* **11,** 2345–2355.

Depicker, A., and Van Montagu, M. (1997). Post-transcriptional gene silencing in plants. *Curr. Opin. Cell Biol.* **9,** 373–382.

Dixon, M. S., Jones, D. A., Keddie, J. S., Thomas, C. M., Harrison, K., and Jones, J. D. G. (1996). The tomato Cf-2 disease resistance locus comprises two functional genes encoding leucine-rich repeat proteins. *Cell (Cambridge, Mass.)* **84**(3), 451–459.

Duering, K. (1996). Genetic engineering for resistance to bacteria in transgenic plants by introduction of foreign genes. *Mol. Breed.* **2**(4), 297–305.

Edwards, J. W., and Coruzzi, G. M. (1990). Cell-specific gene expression in plants. *Annu. Rev. Genet.* **24,** 275–303.

Estruch, J. J., Carozzi, N. B., Desai, N., Duck, N. B., Warren, G. W., and Koziel, M. G. (1997). Transgenic plants: An emerging approach to pest control. *Nat. Biotechnol.* **15,** 137–141.

Fischhoff, D. A., Bowdish, K. S., Perlak, F. J., Marrone, P. G., McCormick, S. M., Niedermaeyer, J. G., Dean, D. A., Kusano-Kretzmer, K., Mayer, E. J., Rochester, D. E., Rogers, S. G., and Fraley, R. T. (1987). Insect tolerant transgenic tomato plants. *Bio/Technology* **5,** 807–813.

Fitchen, J., Beachy, R. N., and Hein, M. B. (1995). Plant virus expressing hybrid coat protein with added murine epitope elicits autoantibody response. *Vaccine* **13,** 1051–1057.

Fitchen, J. H. (1993). Genetically engineered protection against viruses in transgenic plants. *Annu. Rev. Microbiol.* **47,** 739–763.

Fluhr, R., Kuhlemeier, C., Nagy, F., and Chua, N.-H. (1986). Organ-specific and light-induced expression of plant genes. *Science* **232,** 1106–1112.

Fox, J. L. (1996). Bt cotton infestations renew resistance concerns. *Nat. Biotechnol.* **14,** 1070.

Francisco, M. (1996). Calgene moves into the black, brown, blue and red. *Nat. Biotechnol.* **14,** 1072.

Gallo-Meagher, M., and Irvine, J. E. (1996). Herbicide resistant transgenic sugarcane plants containing the *bar* gene. *Crop Sci.* **36,** 1367–1374.

Gatz, C. (1996). Chemically inducible promoters in transgenic plants. *Curr. Opin. Biotechnol.* **7,** 168–172.

Gatz, C., Frohberg, C., and Wendenburg, R. (1996). Strigent repression and homogeneous de-repression by tetracycline of a modified CaMV 35S promoter in intact transgenic tobacco plants. *Plant J.* **2,** 397–404.

Goldman, M. H. S., Goldberg, R. B., and Mariani, C. (1994). Female sterile tobacco plants are produced by stigma-specific cell ablation. *EMBO J.* **13,** 2976–2984.

Grierson, D., and Fray, R. (1994). Control of ripening in transgenic tomatoes. *Euphytica* **79,** 251–263.

Hain, R., Reif, H. J., Krause, E., Langebartels, R., Kindl, H., Vornam, B., Wiese, W., Schmelzer, E., Schreier, P. H., and *et al.* (1993). Disease resistance results from foreign phytoalexin expression in a novel plant. *Nature (London)* **361,** 153–156.

Hamilton, C. M., Frary, A., Lewis, C., and Tanksley, S. D. (1996). Stable transfer of intact high molecular weight DNA into plant chromosomes. *Proc. Natl. Acad. Sci. U.S.A.* **93**(18), 9975–9979.

Haq, T. A., Mason, H. S., Clements, J. D., and Arntzen, C. J. (1995). Oral Immunization with recombinant bacterial antigen produced in transgenic plants. *Science* **268,** 714–716.

Herrera-Estrella, L. (1983). Expression of chimeric genes transferred into plant cells using Ti-plasmid-derived vector. *Nature (London)* **303,** 209–213.

Hiatt, A., Cafferkey, R., and Bowdish, K. (1989). Production of antibodies in transgenic plants. *Nature (London)* **342,** 76–78.

Hiei, Y., Ohta, S., Komari, T., and Kumashiro, T. (1994). Efficient transformation of rice (*Oryza sativa* L.) mediated by *Agrobacterium* and sequence analysis of the boundaries of the T-DNA. *Plant J.* **6,** 271–282.

Höfte, H., and Whiteley, H. R. (1989). Insecticidal crystal proteins of *Bacillus thuringiensis*. *Microbiol. Rev.* **53,** 242–255.

Horsch, R. B., Fry, J., Hoffman, N. L., Wallroth, M., Eichholtz, D., Rogers, S. G., and Fraley, R. T. (1985). A simple and general method for transferring genes into plants. *Science* **227,** 1229–1231.

Huet, H., Sivamani, E., Ong, C. A., Chen, L., de Kochko, A., Beachy, R. N., and Fauquet, C. M. (1996). Genetically engineered protection against the rice tungro spherical virus in transgenic rice. *Abstr., Int. Cong. Virol.*, Jerusalem, Israel, *10th 1996*, p. 184.

Ishida, Y., Saito, H., Ohta, S., Hiei, Y., Komari, T., and Kumashiro, T. (1996). High efficiency transformation of maize (*Zea mays* L.) mediated by *Agrobacterium tumefaciens*. *Nat. Biotechnol.* **14,** 745–750.

Jach, G., Gornhardt, B., Mundy, J., Logemann, J., Pinsdorf, E., Leah, R., Schell, J., and Maas, C. (1995). Enhanced quantitative resistance against fungal disease by combinatorial expression of different barley antifungal proteins in transgenic tobacco. *Plant J.* **8**(1), 97–109.

Jasper, F., Koncz, C., Schell, J., and Steinbiss, H.-H. (1994). Agrobacterium T-strand production in vitro: Sequence-specific cleavage and 5′ protection of single-stranded DNA templates by purified VirD2 protein. *Proc. Natl. Acad. Sci. U.S.A.* **91,** 694–698.

Joersbo, M., and Okkels, F. T. (1996). A novel principle for selection of transgenic plant cells: Positive selection. *Plant Cell Rep.* **16,** 219–221.

John, M. E., and Keller, G. (1996). Metabolic pathway engineering in cotton: Biosynthesis of polyhydroxybutyrate in fibre cells. *Proc. Natl. Acad. Sci. U.S.A.* **93,** 12768–12773.

Johnston, S. A., Anziano, P. Q., Shark, K., Sanford, J. C., and Butow, R. A. (1988). Mitochondrial transformation in yeast by bombardment with microprojectiles. *Science* **240,** 1538–1541.

Kaplan, I. B., Shintaku, M. H., Li, Q., Zhang, L., Marsh, L. E., and Palukaitis, P. (1995). Complementation of virus movement in transgenic tobacco expressing the cucumber mosaic virus 3a gene. *Virology* **209,** 188–199.

Karrer, E. E., and Rodriguez, R. L. (1992). Metabolic regulation of rice alpha-amylase and sucrose synthase genes in planta. *Plant J.* **2,** 517–523.

Kinney, A. J. (1995). Genetic modification of the storage lipids of plants. *Curr. Opin. Biotechnol.* **6,** 144–151.

Knauf, V. C. (1993). Progress in the cloning of genes for plant storage lipid biosynthesis. *In* "Genetic Engineering" (J. K. Setlow, ed.), pp. 149–164. Plenum, New York.

Kobayashi, S., Nakamura, Y., Kaneyoshi, J., Higo, H., and Higo, K. I. (1996). Transformation of kiwifruit (*Actinidia chinensis*) and trifoliate orange (*Poncirus trifoliata*) with a synthetic gene encoding the human epidermal growth factor (hEGF). *J. Jpn. Soc. Hortic. Sci.* **64,** 763–769.

Koziel, M. G., Carozzi, N. B., and Desai, N. (1996a). Optimizing expression of transgenes with an emphasis on post-transcriptional events. *Plant Mol. Biol.* **32,** 393–405.

Koziel, M. G., Carozzi, N. B., Desai, N., Warren, G. W., Dawson, J., Dunder, E., Launis, K., and Evola, S. V. (1996b). Transgenic maize for the control of european corn borer and other maize insect pests. *Ann. N.Y. Acad. Sci.* **792,** 164–171.

Kramer, M. G., and Redenbaugh, K. (1994). Commercialization of a tomato with an antisense polyglacturonase gene: The FLAVR SAVR™ tomato story. *Euphytica* **79,** 293–297.

Kridl, J. C., and Shewmaker, C. K. (1996). Food for thought: Improvement of food quality and composition through genetic engineering. *Ann. N.Y. Acad. Sci.* **792,** 1–12.

Kumagai, M. H., Donson, J., Della-Cioppa, G., Harvey, D., Hanley, K., and Grill, L. K. (1995). Cytoplasmic inhibition of carotenoid biosynthesis with virus-derived RNA. *Proc. Natl. Acad. Sci. U.S.A.* **92,** 1679–1683.

Lassner, M. W., Lardizabal, K., and Metz, J. G. (1996). A jojoba b-ketoacyl-CoA synthase cDNA complements the canola fatty acid elongation mutation in transgenic plants. *Plant Cell* **8,** 281–292.

Lawrence, G. J., Finnegan, E. J., Ayliffe, M. A., and Ellis, J. G. (1995). The L6 gene for flax rust resistance is related to the Arabidopsis bacterial resistance gene RPS2 and the tobacco viral resistance gene N. *Plant Cell* **7,** 1195–1206.

Li, L., Qu, R., De Kochko, A., Fauquet, C., and Beachy, R. N. (1993). An improved rice transformation system using the biolistic methods. *Plant Cell Rep.* **12,** 250–253.

Lloyd, A. M., Schena, M., Walbot, V., and Davis, R. W. (1994). Epidermal cell fate determination in Arabidopsis: Patterns defined by a steroid-inducible regulator. *Science* **266,** 436–439.

Lodge, J. K., Kaniewski, W. K., and Tumer, N. E. (1993). Broad-spectrum virus resistance in transgenic plants expressing Pokeweed antiviral protein. *Proc. Natl. Acad. Sci. U.S.A.* **15,** 7089–7093.

Ma, J. K-C., and Hein, M. B. (1995). Immunotherapeutic potential of antibodies produced in plants. *Trends Biotechnol.* **13,** 522–527.

Ma, J. K.-C., and Hein, M. (1996). Antibody production and engineering in plants. *Ann. N.Y. Acad. Sci.* **792,** 72–81.

Ma, J. K-C., Hiatt, A., Hein, M. B., Vine, N., Wang, F., Stabila, P., VanDolleweerd, C., Mostov, K., and Lehner, T. (1995). Generation and assembly of secretory antibodies in plants. *Science* **268,** 716–719.

Maiti, I. B., Gowda, S., Kiernan, J., Ghosh, S. K., and Shepherd, R. J. (1997). Promoter-leader deletion analysis and plant expression vectors with the figwort mosaic virus (FMV) full length transcript (FLt) promoter containing single or double enhancer domains. *Transgenic Res.* **6,** 143–156.

Mannerlof, M., Tuvesson, S., Steen, P., and Tenning, P. (1990). Transgenic sugar beet tolerant to glyphosate. *Euphytica* **94,** 83–91.

Marcos, J. F., and Beachy, R. N. (1994). In vitro characterization of a cassette to accumulate multiple proteins through synthesis of a self-processing polypeptide. *Plant Mol. Biol.* **24,** 495–503.

Marcos, J. F., and Beachy, R. N. (1997). Transgenic accumulation of two plant virus coat proteins on a single self-processing polypeptide. *J. Gen. Virol.* **78,** 1771–1778.

Mariani, C., de Beuckeleer, M., Truettner, J., Leemans, J., and Goldberg, R. B. (1990). Induction of male sterility in plants by a chimaeric ribonuclease gene. *Nature (London)* **347,** 737–741.

Mariani, C., Gossele, V., de Bauckeleer, M., de Block, M., Goldberg, R. B., de Greef, W., and Leemans, J. (1992). A chimaeric ribonuclease-inhibitor gene restores fertility to male sterile plants. *Nature (London)* **357,** 384–387.

Marris, C., Gallois, P., Copley, J., and Kreis, M. (1988). The 5′ flanking region of a barley B hordein gene controls tissue- and developmental-specific CAT expression in tobacco plants. *Plant Mol. Biol.* **10,** 359–366.

Martin, G. B., Brommonschenkel, S. H., Chunwongse, J., Frary, A., Ganal, M. W., Spivey, R., Wu, T., Earle, E. D., and Tanksley, S. D. (1993). Map-based cloning of a protein kinase gene conferring disease resistance in tomato. *Science* **262,** 1432–1436.

Mason, H. S., Ball, J. M., Shi, J.-J., Jiang, X., Estes, M., and Arntzen, C. J. (1996). Expression of Norwalk virus capsid protein in transgenic tobacco and potato and its oral immunogenicity in mice. *Proc. Natl. Acad. Sci. U.S.A.* **93,** 5335–5340.

Matsuoka, K., and Nakamura, K. (1991). Propeptide of a precursor to a plant vacuolar protein required for vacuolar targeting. *Proc. Natl. Acad. Sci. U.S.A.* **88,** 834–838.

McBride, K. E., Schaaf, D. J., Daley, M., and Stalker, D. M. (1994). Controlled expression of plastid transgenes in plants on a nuclear DNA-encoded and plastid-targeted T7 RNA polymerase. *Proc. Natl. Acad. Sci. U.S.A.* **91,** 7301–7305.

Mett, V. L., Lochhead, L. P., and Reynolds, P. H. S. (1993). Copper-controllable gene expression system for whole plants. *Proc. Natl. Acad. Sci. U.S.A.* **90,** 4567–4571.

Meyer, P. (1995). Understanding and controlling transgene expression. *Trends Biotechnol.* **13,** 332–337.

Mikkelsen, T. R., Andersen, B., and Jorgensen, R. B. (1996). The risk of crop transgene spread. *Nature (London)* **380,** 31.

Mindrinos, M., Katagiri, F., Yu, G. L., and Ausubel, F. M. (1994). The *A. thaliana* disease resistance gene RPS2 encodes a protein containing a nucleotide-binding site and leucine-rich repeats. *Cell (Cambridge, Mass.)* **78**(6), 1089–1099.

Moffat, A. S. (1996a). Moving forest trees into the modern genetics era. *Science* **271,** 760–761.

Moffat, A. S. (1996b). Genetic engineering turns to trees. *Science* **271,** 761.

Murfett, J., Bourque, J. E., and McClure, B. A. (1995). Antisense suppression of S-RNase expression in *Nicotiana* using RNA polymerase II- and III- transcribed gene constructs. *Plant Mol. Biol.* **29,** 201–212.

Murphy, D. J. (1996). Engineering oil production in rapeseed and other oil crops. *Trends Biotechnol.* **14,** 206–213.

Nawrath, C., Poirier, Y., and Somerville, C. (1994). Targeting of the polyhydroxybutyrate biosynthetic pathway to the plastids of *Arabidopsis thaliana* results in high levels of polymer accumulation. *Proc. Natl. Acad. Sci. U.S.A.* **91,** 12760–12764.

Ohlrogge, J., and Brown, J. (1995). Lipid biosynthesis. *Plant Cell* **7,** 957–970.

Padgett, H. S., Epel, B. L., Kahn, T. W., Heinlein, M., Watanabe, Y., and Beachy, R.

N. (1996). Distribution of tobamovirus movement protein in infected cells and implications for cell-to-cell spread of infection. *Plant J.* **10,** 1079–1088.

Pear, J. R., Kawagoe, Y., Schreckengost, W. E., Delmer, D. P., and Stalker, D. M. (1996). Higher plants contain homologues of the bacteria *cel*A genes encoding the catalytic subunit of cellulose synthase. *Proc. Natl. Acad. Sci. U.S.A.* **93,** 12637–12642.

Perlak, F. J., Fuchs, R. L., Dean, D. A., McPherson, S. L., and Fischhoff, D. A. (1991). Modification of the coding sequence enhances plant expression of insect control protein genes. *Proc. Natl. Acad. Sci. U.S.A.* **88,** 3324–3328.

Perlak, F. J., Stone, T. B., Muskopf, Y. M., Petersen, L. J., Parker, G. B., McPherson, S. A., Wyman, J., Love, S., Reed, G., Biever, D., and Fischhoff, D. A. (1993). Genetically improved potatoes: Protection from damage by Colorado potato beetles. *Plant Mol. Biol.* **22,** 313–321.

Poirier, Y., Nawrath, C., and Somerville, C. (1995). Production of polyhydroxyalkanoates, a family of biodegradable plastics and elastomers, in bacteria and plants. *Bio/Technology* **13,** 142–150.

Raskin, I., Smith, R. D., and Salt, D.E. (1997). Phytoremediation of metals: Using plants to remove pollutants from the environment. *Curr. Opin. Biotechnol.* **8,** 221–226.

Rieping, M., Fritz, M., Prat, S., and Gatz, C. (1994). A dominant negative mutant of PG13 suppresses transcription from a cauliflower mosaic virus 35S truncated promoter in transgenic tobacco plants. *Plant Cell* **6,** 1087–1098.

Rocha-Sosa, M., Sonnewald, U., Frommer, W., Stratmann, M., Schell, J., and Willmitzer, L. (1989). Both developmental and metabolic signals activate the promoter of a class 1 patatin gene. *EMBO J.* **8,** 23–29.

Roder, F. T., Schmulling, T., and Gatz, C. (1994). Efficiency of the tetracycline dependent gene expression system-complete suppression and efficient induction of the RolB phenotype in transgenic plants. *Mol. Gen. Genet.* **243,** 32–38.

Rogers, S. G., Klee, H. J., Horsch, R. B., and Fraley, R. T. (1987). Improved vectors for plant transformation: Expression cassette vectors and new selectable markers. *In* "Methods in Enzymology" (R. Wu and L. Grossman, eds.), Vol. 153, pp. 253–277. Academic Press, Orlando, FL.

Rossi, L., Hohn, B., and Tinland, B. (1996). Integration of complete transferred DNA units is dependent on the activity of virulence E2 protein of *Agrobacterium tumefaciens. Proc. Natl. Acad. Sci. U.S.A.* **93,** 126–130.

Roy, B., and Mandal, R. K. (1996). Sequence analysis of signal peptides of seed storage proteins. *J. Plant Biochem. Biotechnol.* **5,** 113–116.

Rugh, C. L., Wilde, H. D., Stack, N. M., Thompson, D. M., Summers, A. O., and Meagher, R. B. (1996). Mercuric ion reduction and resistance in transgenic *Arabidopsis thaliana* plants expressing a modified bacterial *merA* gene. *Proc. Natl. Acad. Sci. U.S.A.* **93,** 3182–3187.

Santa Cruz, S., Chapman, S., Roberts, A. G., Roberts, I. M., Prior, D. A. M., and Oparka, K. J. (1996). Assembly and movement of a plant virus carrying a green fluorescent protein overcoat. *Proc. Natl. Acad. Sci. U.S.A.* **93,** 6286–6290.

Schell, J. (1997). Cotton carrying the recombinant insect poison Bt toxin: No case to doubt the benefits of plant biotechnology. *Curr. Opin. Biotechnol.* **8,** 235–236.

Scherntaner, J. P., Matzke, M. A., and Matzke, A. J. M. (1988). Endosperm-specific activity of a zein gene promoter in transgenic tobacco plants. *EMBO J.* **7,** 1249–1256.

Scholthof, H. B., Scholthof, K. B. G., and Jackson, A. O. (1996). Plant virus gene vectors for transient expression of foreign proteins in plants. *Annu. Rev. Phytopathol.* **34,** 299–323.

Schöpke, C., Taylor, N., Carcamo, R., Konan, N. K., Marmey, P., Henshaw, G. G., Beachy, R. N., and Fauquet, C. (1996). Regeneration of transgenic cassava plants (*Manihot esculenta* Crantz) from microbombarded embryogenic suspension cultures. *Nat. Biotechnol.* **14,** 731–735.

Schöpke, C., Taylor, N. J., Carcamo, R., Beachy, R. N., and Fauquet, C. (1997). Optimization of parameters for particle bombardment of embryogenic suspension cultures of cassava (*Manihot esculenta* Crantz) using computer image analysis. *Plant Cell Rep.* **16,** 526–530.

Schouten, A., Roosien, J., van Engelen, F. A., de Jong, G. A. M., Borst-Vrenssen, A. W. M., Zilverentant, J. F., Bosch, D., Stiekema, W. J., Gommers, F. J., Schots, A., and Bakker, J. (1996). The C-terminal KDEL sequence increases the expression level of a single-chain antibody designed to be targeted to both the cytosol and the secretory pathway in transgenic tobacco. *Plant Mol. Biol.* **30,** 781–793.

Shirsat, A., Wilford, N., R., C., and Boulter, D. (1989). Sequences responsible for the tissue specific promoter activity of a pea legumin gene in tobacco. *Mol. Gen. Genet.* **215,** 326–331.

Silva Filho, M. D. C., Chaumont, F., Leterme, S., and Boutry, M. (1996). Mitochondrial and chloroplast targeting sequences in tandem modify protein import specificity in plant organelles. *Plant Mol. Biol.* **30,** 769–780.

Simpson, J., Timko, M. R., Cashmore, A. R., Schell, J., Van Montague, M., and Herrera-Estrella, L. (1985). Light-inducible and tissue-specific expression of a chimeric gene under control of the 5′ flanking sequence of a pea chlorophyll a/b-binding protein gene. *EMBO J.* **4,** 2723–2729.

Song, W. Y., Wang, G. L., Chen, L. L., Kim, H. S., Pi, L. Y., Holsten, T., Gardner, J., Wang, B., Zhai, W. X., Zhu, L. H., Fauquet, C., and Ronald, P. (1995). A receptor kinase-like protein encoded by the rice disease resistance gene, Xa21. *Science* **270,** 1804–1806.

Sorge, J., West, C., Westwood, B., and Beutler, E. (1985). Molecular cloning and nucleotide sequence of the human glucocerebrosidase gene. *Proc. Natl. Acad. Sci. U.S.A.* **82,** 7289–7293.

Stark, D. M., Timmerman, K. P., Barry, G. F., Preiss, J., and Kishore, G. M. (1992). Regulation of the amount of starch in plant tissues by ADP-glucose pyrophosphorylase. *Science* **258,** 287–292.

Stark, D. M., Barry, G. F., and Kishore, G. M. (1996). Improvement of food quality traits through enhancement of starch biosynthesis. *Ann. N.Y. Acad. Sci.* **792,** 26–36.

Sun, S. S. M., Zuo, W., Tu, H. M., and Xiong, L. (1996). Plant proteins: Engineering for improved quality. *Ann. N.Y. Acad. Sci.* **792,** 37–42.

Takeuchi, Y., Dotson, M., and Keen, N. T. (1992). Plant transformation a simple particle bombardment device based on flowing helium. *Plant Mol. Biol.* **18,** 835–839.

Taylor, N. J., Edwards, M., Kiernan, R. J., Davey, C. D. M., Blakesley, D., and Henshaw, G. G. (1996). Development of friable embryogenic callus and embryogenic suspension culture systems in cassava (*Manihot esculenta* Crantz). *Nat. Biotechnol.* **14,** 726–730.

Terras, F. R. G., Eggermont, K., Kovaleva, V., Raikhel, N. V., Osborn, R. W., Kester, A., Rees, S. B., Torrekens, S., Van Leuven, F., Vanderleyden, J., Cammue, B. P.

A., and Broekaert, W. F. (1995). Small cysteine-rich antifungal proteins from radish: Their role in host defense. *Plant Cell* **7**, 573–588.

Toki, S. (1997). Rapid and efficient *Agrobacterium*-mediated transformation in rice. *Plant Mol. Biol. Rep.* **15**, 16–21.

Töpfer, R., Martini, N., and Schell, J. (1995). Modification of plant lipid synthesis. *Science* **268**, 681–686.

Vaeck, M., Reynaerts, A., Hofte, H., Jansens, S., De Beuckeleer, M., Dean, C., Zabeau, M., and Leemans, J. (1987). Transgenic plants protected from insect attack. *Nature (London)* **327**, 33–37.

Verdaguer, B., De Kochko, A., Beachy, R. N., and Fauquet, C. (1996). Isolation and expression in transgenic tobacco and rice plants, of the cassava vein mosaic virus (CVMV) promoter. *Plant Mol. Biol.* **31**, 1129–1139.

Wadman, M. (1996). Genetic resistance spreads to consumers. *Nature (London)* **383**, 564.

Wandelt, C. I., Khan, M. R. I., Craig, S., Schroeder, H. E., Spencer, D., and Higgins, T. J. V. (1992). Vicilin with carboxy-terminal KDEL is retained in the endoplasmic reticulum and accumulates to high levels in the leaves of transgenic plants. *Plant J.* **2**, 181–192.

Wenzler, H., Mignery, G., Fisher, L., and Park, W. (1989). Sucrose regulated expression of a chimeric potato tuber gene in leaves of transgenic tobacco plants. *Plant Mol. Biol.* **13**, 347–354.

Whitham, S., Dinesh-Kumar, S. P., Choi, D., Hehl, R., Corr, C., and Baker, B. (1994). The product of the tobacco mosaic virus resistance gene N: Similarity to toll and the interleukin-1 receptor. *Cell (Cambridge, Mass.)* **78**, 1101–1115.

Whitham, S., McCormic, S., and Baker, B. (1996). The N gene of tobacco confers resistance to tobacco mosaic virus in transgenic tomato. *Proc. Natl. Acad. Sci. U.S.A.* **93**, 8776–8781.

Williams, S., Friedrich, L., Dincher, S., Carozzi, N., Kessmann, H., Ward, E., and Ryals, J. (1992). Chemical regulation of *Bacillus thuringiensis* δ-endotoxin expression in transgenic plants. *Bio/Technology* **10**, 540–543.

Wilson, T. M. (1993). Strategies to protect crop plants against viruses: Pathogen-derived resistance blossoms. *Proc. Natl. Acad. Sci. U.S.A.* **90**, 3134–3141.

Wu, G., Shortt, B. J., Lawrence, E. B., Levine, E. B., Fitzsimmons, K. C., and Shah, D. M. (1995). Disease resistance conferred by expression of a gene encoding H-2O-2-generating glucose oxidase in transgenic potato plants. *Plant Cell* **7**, 1357–1368.

Yin, Y., and Beachy, R. N. (1995). The regulatory regions of the rice tungro bacilliform virus promoter and interacting nuclear factors in rice (*Oryza sativa* L.). *Plant J.* **7**, 969–980.

Yin, Y., Zhu, Q., Dai, S., Lamb, C., and Beachy, R. N. (1997). Rf2a, a bZIP transcriptional activator of the phloem-specific rice tungro bacilliform virus promoter, functions in vascular development. *EMBO J.* **16**, 5249–5259.

Yu, S.-M., Kuo, Y.-H., Sheu, Y.-J., and Liu, L.-F. (1991). Metabolic depression of a-amylase gene expression in suspension-cultured cells of rice. *J. Biol. Chem.* **266**, 21131–21137.

Yu, S. M., Tzou, W. S., Lo, W. S., Kuo, Y. H., Lee, H. T., and Wu, R. (1992). Regulation of alpha amylase-encoding gene expression in germinating seeds and cultured cells of rice. *Gene* **122**, 247–253.

Zhang, S., Chen, L., Qu, R., Marmey, P., Beachy, R. N., and Fauquet, C. (1996). Regeneration of fertile transgenic indica (group 1) rice plants following microprojectile transformation of embryogenic suspension culture cells. *Plant Cell Rep.* **15,** 465–469.

Zhang, W., McElroy, D., and Wu, R. (1991). Analysis of rice Act1 5′ region activity in transgenic rice plants. *Plant Cell* **3,** 1155–1165

INDEX

Printed in the United States
60067LVS00002B/271